Key Technologies on New Energy Vehicles

Key Technologies on New Energy Vehicles publishes the latest developments in new energy vehicles - quickly, informally and with high quality. The intent is to cover all the main branches of new energy vehicles, both theoretical and applied, including but not limited:

- Control Technology of Driving System
- Hybrid Electric Vehicle Coupling Technology
- Cross Disciplinary design optimization technology
- Single and Group Battery Technology
- Energy Management Technology
- Lightweight Technology
- New Energy Materials and Device
- Internet of Things (IoT)
- Cloud Computing
- 3D Printing
- Virtual Reality Technologies

Within the scopes of the series are monographs, professional books or graduate textbooks, edited volumes, and reference works purposely devoted to support education in related areas at the graduate and post-graduate levels.

Keqiang Li · Mingyuan Bian · Yugong Luo ·
Jianqiang Wang

The Intelligent Environment Friendly Vehicle

Concept, Architecture and Implementation

机械工业出版社
CHINA MACHINE PRESS

Springer

Keqiang Li
School of Vehicle and Mobility
Tsinghua University
Beijing, China

Mingyuan Bian
School of Vehicle and Mobility
Tsinghua University
Beijing, China

Yugong Luo
School of Vehicle and Mobility
Tsinghua University
Beijing, China

Jianqiang Wang
School of Vehicle and Mobility
Tsinghua University
Beijing, China

ISSN 2662-2920 ISSN 2662-2939 (electronic)
Key Technologies on New Energy Vehicles
ISBN 978-981-19-4850-3 ISBN 978-981-19-4851-0 (eBook)
https://doi.org/10.1007/978-981-19-4851-0

Jointly published with China Machine Press Co., Ltd.
The print edition is not for sale in China (Mainland). Customers from China (Mainland) please order the print book from: China Machine Press Co., Ltd.

Translation from the Chinese language edition: "智能环境友好型车辆: 概念、技术架构与工程实现" by Keqiang Li et al., © China Machine Press 2020. Published by China Machine Press. All Rights Reserved.

This Springer imprint is published by the registered company Springer Nature Singapore Pte Ltd.
The registered company address is: 152 Beach Road, #21-01/04 Gateway East, Singapore 189721, Singapore

Preface

As "machines that change the world," automotives have greatly improved the traffic efficiency and the convenience of human travel since they emerged on scene. At the same time, they have also caused various social problems such as traffic safety, energy consumption, emission pollution and urban traffic congestion. With the development of intelligent and electric vehicle technologies, especially the emerging automotive electronics, information and communication, big data and cloud computing technologies, "energy and power revolution", "intelligent revolution" and the "Internet revolution" are propelling a disruptive transformation in the automotive technology and industry. The intelligent and connected vehicle technology in the context of intelligent transportation and the clean-energy vehicle technology represented by the electric drive are the two major technical roadmaps to solve the increasingly serious social problems such as traffic accidents, road congestion, environmental pollution, energy shortage and so on. However, the current researches on intelligent vehicles and clean-energy vehicles mainly focus on the outstanding problems in their respective fields. Despite the overlap between each other, the fields of power and intelligence lack deep fusion. From the perspective of multi-system integration/optimization and system reconstruction, the study of new vehicle configurations with optimal comprehensive performances including safety, comfort, energy saving and environmental protection as well as relevant common fundamental theories will present major opportunities and challenges in the future automotive development.

As a carrier of automotive power revolution and intelligent integration, the intelligent environment-friendly vehicle (i-EFV) is a new-concept apparatus of transport featuring new architecture forms, technical systems and application characteristics, rather than a traditional intelligent vehicle based on an electric vehicle. It is an urgent need to systematically explore and study the theoretical foundation, key technologies and engineering applications of this new vehicle and business form in the context of the "four development trends" of the automotive industry (electrification, intelligence, sharing and connectivity).

To satisfy the overall performance needs of modern vehicles, such as safety, ride comfort, energy saving and environmental protection, this book studies the fundamental concepts, system architectures and key technologies of the i-EFV from the perspective of interdisciplinary fusion and multi-system integration. By focusing on the system architecture and three key technologies of the i-EFV: structure sharing, information fusion and control collaboration, the book not only probes into the multi-physical-process coupling mechanism of the complex electromechanical system of the i-EFV and the fundamental theoretical methods for the modeling of the generalized mechanical system dynamics, but also systematically introduces the engineering realization and application of the three key technologies with cases of the adaptive cruise control (ACC) system of intelligent hybrid vehicles, the travel planning of connected electric vehicles, eco-run and charging/battery-swapping scheduling to achieve the safety, economy and comfort objectives.

Keqiang Li, the first author of this book, began to work on the dynamic design and controlling technologies of vehicle systems during his academic visit and work in Japan in the 1990s. After he returned to China in 2000, he established a research group on intelligent vehicles in the Department of Automotive Engineering of Tsinghua University to systematically and consecutively study the energy-saving control of electric vehicles, advanced driver assistant control and intelligent vehicle technologies around the topics of vehicle safety and energy efficiency, and to systematically explore the integration of vehicle electrification and intelligence. In 2005, he put forward the concept of intelligent environment-friendly vehicle (i-EFV) for the first time, which based on three systems (clean-energy power, electronic control chassis and intelligent information exchange) and three key technologies (structure sharing, information fusion and collaborative control) that aim to comprehensively achieve four major functions (safety, ride comfort, energy efficiency and environmental protection). In the following dozen years, the authors continued to deepen, enrich and improve the theoretical system of this new-concept vehicle. By presenting the theoretical foundation, key technologies and engineering applications of i-EFV, this book systematically reviews and summarizes the research achievements of the authors' team in the field of intelligent vehicles over the past 20 years.

The intelligent Environment-friendly Vehicle—Concept, Architecture and Implementation is divided into nine chapters to systematically discuss the concept, system architecture, fundamental theories, key technology systems, typical engineering application and realization of i-EFV from the perspective of multi-disciplinary integration. During the compilation of this book, graduates including Chen Tao, Liu Li, Zou Guangcai, Wang Xiao, Zheng Yang, Huang Bin, Zhang Shuwei, Cao Kun, Zhu Tao, Xie Laiqing, Bian Yougang, Qin Xiaohui and Xu Biao and postdoctoral researchers including Kong Weiwei and Huang Chen of the intelligent vehicle research group provided valuable materials for relevant chapters. Dr. Xie Laiqing has spent a lot of time and efforts on the compilation of the manuscript materials. I would like to extend my heartfelt thanks to them for their hard work.

As an interdisciplinary, deeply intertwined and integrated system, i-EFV will keep evolving in terms of architecture, theoretical foundation and key technologies. Although this book is far from providing the complete picture of all scientific, technical and engineering issues of i-EFV, we still hope that the book can serve as an introduction to this new-concept vehicle by providing the industry with a sample case for reference. Since there might be some mistakes and flaws in this book due to the author's limited knowledge, any kind opinion or criticisms from industry peers and readers are welcome.

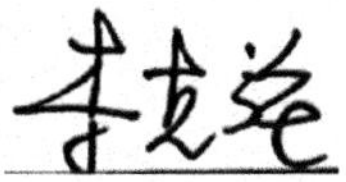

Beijing, China
June 2020

Contents

Chapter 1
Introduction to i-EFV

1.1 Introduction

Safety, energy efficiency and environmental protection are the eternal themes in the history of the automotive industry. A booming automotive industry has brought great convenience to human transport over the years, but an increasing number of vehicles has also caused frequent traffic accidents, excessive energy consumption, serious environmental pollution and other social issues. It has become the consensus of global research institutions and automotive enterprises to solve these problems by integrating new technologies from multiple fields.

Intelligent vehicles (IVs) based on the intelligent transportation system (ITS) can help prevent accidents, improve vehicle active safety and driving comfort by recognizing vehicle and environment state and performing intelligent and active safety control, so as to ensure the safety and comfort required by modern vehicles, which has become one of the important directions of research. Another hot subject studied by worldwide scholars to solve energy and environmental problems is clean-energy vehicles (CEVs) propelled by natural gas, hydrogen, electric energy and other unconventional automotive energy source.

IVs usually adopt advanced on-board sensors, controllers, actuators and other devices and they also integrate modern communication and network technologies to realize intelligent information exchange and sharing between vehicles and people, other vehicles, infrastructures and clouds. With the ability of complex environment perception, intelligent decision-making and coordinated control supported by big data, cloud computing and artificial intelligence (AI), IVs ensure safe, efficient, comfortable and energy-saving driving experiences with partially or completely automated driving functions. This will eventually replace human operation, and gradually transform vehicle from a simple means of transportation to an intelligent mobile space and application terminal.

Current researches on ITS-based intelligent vehicle technologies generally focus on conventional internal combustion engine vehicles (ICEVs), which can hardly

K. Li et al., *The Intelligent Environment Friendly Vehicle*,
Key Technologies on New Energy Vehicles,
https://doi.org/10.1007/978-981-19-4851-0_1

solve fuel economy and emission problems in a fundamental way. It will become an important direction of IV technology in the future to integrate new powertrains with new vehicle structures to make breakthroughs in energy efficiency and environment performances. Current studies of the active safety system mainly concentrate on vehicle stability control, longitudinal/lateral assistant driving based on intelligent active safety technology and pedestrian protection. How to take longitudinal and lateral driving safety and pedestrian safety into consideration to ensure vehicle safety, vehicle-infrastructure and vehicle–pedestrian safety will be another difficult problem that the IV technology needs to solve.

CEVs use unconventional clean fuels as power sources and integrate advanced power control technologies and driving technologies with new structures. CEVs include hybrid electric vehicles (HEVs), battery electric vehicles (BEVs), fuel cell electric vehicles (FCEVs), hydrogen internal combustion engine vehicles (HICEVs) and other new energy vehicles (NEVs), such as high-efficiency energy accumulator and dimethyl ether vehicles. BEVs, especially, have become the current research focus around the world [1] for several advantages such as high popularity as energy source (electricity), zero emission and low running noise. Electric drive HEVs reduce vehicle fuel consumption and improve system efficiency through idle stop, regenerative braking and optimization of engine displacement and working range, enabled by the coupling and collaborative optimization of the electric energy provided by the battery and the mechanical energy generated from engine combustion. As the transition from traditional ICEVs to BEVs, HEVs have gradually entered the industrial phase.

Researches on electric vehicles (EVs) mainly focus on vehicle energy optimization and management, low cost and high-energy–density battery, motor and driving technology and other aspects related to vehicle design and control. One of the key features of EVs is their closer connection with the environment, compared with conventional internal combustion vehicles. For HEVs, the optimal distribution of electric energy and mechanical energy is closely related to the traffic environment, and the optimal energy distribution strategy may vary dramatically under different driving conditions [2]. Limited by battery charging, battery energy density and other issues, the large-scale application of BEVs depends on EV charging navigation and the integrated energy management system design of power grids [3]. Therefore, how to conduct unified scheduling and optimal distribution of EV energy flow among multiple systems based on environment information and further improve EV operating economy is an important direction of future EV research. Small energy density and large mass of the on-board battery cause the prominent conflict between energy efficiency, environmental protection and vehicle lightweighting. In addition, problems such as EV driving safety, service safety of high-voltage battery pack and collision safety also need to be solved effectively. How to improve EV safety, active safety in particular, is an urgent problem to be solved before its large-scale application.

Current researches on ITS-based intelligent vehicles and CEVs are mainly focused on individual prominent problems, but it will be a future challenge to design vehicles with the best safety, comfort, energy efficiency and environment performances from the perspective of multi-system integration and optimization while exploring relevant general basic theories.

Therefore, the automotive industry in the future should focus on the technologies related to safety, energy efficiency and environmental protection, and the design of a multi-level and efficient way of transportation. With the progress of automotive technologies and the rapid development of intelligent transport systems, the automotive industry is witnessing a redefined revolution of energy, intelligence, and the Internet of things. The development of green (electrification), intelligence, connectivity and sharing technologies will bring revolutionary changes to transportation, which is also the main development direction of IVs in the future. As an internationally recognized future development direction and focus, intelligent and connected CEVs can provide better solutions of convenient transportation while ensuring safety, energy efficiency and environmental protection.

In terms of product structure and form of future vehicles, NEVs, especially electric drive vehicles, are the ideal technology platform for IVs. First of all, energy conservation and environmental protection have become the basic requirements for future transportation due to issues including fuel consumption and environmental pollution, so CEVs and electric drive vehicles have become the direction of industrial development. Therefore, the IV technology is bound to be closely combined with NEVs, especially electric drive vehicles, and intelligent NEVs are the inevitable trend of future development. IVs and NEVs are the important breakthroughs in the technological progress and upgrading of China's automotive industry. Future IVs are not only transportation tools, but also mobile energy storage carriers. Intelligent NEV technology equipped with V2G/V2V energy exchange is the main development trend in the future.

Secondly, Internet of Vehicles (IoV) is an important tool to empower intelligent vehicles. With the help of high-speed communication, data analysis, cloud computing platform, intelligent infrastructure and other new-generation revolutionary information technologies, intelligent vehicles will serve as a key node of information exchange in the Internet of things (IoT) of smart cities, a mobile data processing terminal, personal business space and information service platform in the whole life cycle of the vehicle. With the development and industrial applications of IoV which includes automotives as key information exchange nodes, enterprises in various industries such as automotives, information, communication, transportation, security, finance and logistics are fully coordinated as service content providers and technical supports for the automotive aftermarket and people's daily life. In terms of function and form, future IVs are not only mobile energy storage carriers, but also mobile data processing terminals, personal business spaces and information service platforms. Intelligent and connected NEVs are important breakthroughs in the transformation and upgrading of China's automotive industry in terms of both size and strength.

Intelligent connected vehicles (ICVs) and electric vehicles (EVs) in the context of intelligent transportation are the two main technical roadmaps to address increasingly severe environmental pollution, energy shortage, road congestion, traffic accidents and other social issues. With the continuous maturity of intelligent and electric vehicle technologies, especially the progress and application of emerging automotive electronics technologies, communication technologies and big data platforms, new advanced vehicle platforms that integrate intelligent and electric features and related technologies have received extensive attention and become a research focus. ITS technology enables comprehensive identification of driving environment, intelligent active safety control, and intelligent and efficient optimal management of traffic, which improve driving safety and comfort, and NEV's energy distribution strategy can be further optimized to improve vehicle fuel economy based on identified driving environment information [4]. Research institutions at home and abroad have begun preliminary technical exploration on information sharing and function integration between EVs and ITS technologies in recent years.

Throughout domestic and overseas researches, applying ITS to solve driving safety, comfort, energy efficiency and environment issues during BEVs' driving process and improve the overall vehicle performance has become a consensus among scholars, and some progress has been made in overall system structure design, longitudinal/lateral assistant driving and traffic system scheduling. How to solve issues in the security of system components, the integrated 'human-vehicle-infrastructure' safety, vehicle fuel economy, battery charging and battery life optimization, and collaborative optimization of driving comfort, entertainment, traffic and power systems still depends on the further researches on overall system architecture design, key theories and technology systems.

Modern automotives are not only important mechanical and electrical products, but also carriers of advanced technologies. They involve basic sciences and generic technologies in many fields and their future development is restricted by many social requirements such as safety, energy conservation and environmental protection. At present, breakthroughs are being made in IVs and NEVs. However, how to organically combine them to form a new type of intelligent vehicle structure is a problem worth exploring.

1.2 Concept and Connotation of i-EFV

I-EFVs (intelligent environment-friendly vehicle) are next-generation advanced vehicles that integrate clean energy power, electronically-controlled chassis and intelligent information exchange. With structure sharing, information fusion and control collaboration, i-EFVs can ensure safety, comfort, energy efficiency and environmental protection [5–7]. A vehicle based on this new concept is also a typical complex electromechanical system that integrates mechanical, hydraulic, electronic, electric power and communication subsystems.

Key terms involved include i-EFV, intelligent information exchange, clean-energy power, electronically-controlled chassis, structure sharing, information fusion, control collaboration, safety, ride comfort, energy efficiency and environmental protection, and next-generation advanced vehicles.

Intelligent environment friendly: a combination of ITS and electric-drive system's characteristics that is able to optimize vehicle energy, ensure security and undertake the collaborative optimization of the transportation system and the electric system to deliver safety, comfort, energy efficiency and environmental protection, and ensure the harmonious operation of the transportation system and the energy system.

Intelligent environment-friendly vehicle: "intelligent" refers to the integration of conventional intelligent vehicle technologies and functions with focus on the application of ITS technology to solve issues including driving safety and entertainment, and to improve vehicle comfort and safety, especially the active safety. "Environment-friendly" refers to the integration of clean-energy vehicle technologies with electric drive platforms to achieve energy efficiency and environmental protection. "Intelligent environment friendly" means to make use of the characteristics and advantages of ITS and electric drive system in structure, information and control to improve the vehicle's comprehensive performance, achieve harmony and optimization among vehicle, road traffic environment and power supply system, and solve problems including energy shortage, environment deterioration, traffic congestion, frequent accidents brought by large-scale vehicle application and related power supply problems.

Intelligent information exchange: ITS is used for the complete perception of driving environment and vehicle-vehicle, vehicle-infrastructure information exchange to realize vehicle-centered environmental information perception, road-centered traffic system monitor and management, and comprehensive scheduling and management of related systems.

Clean-energy power: an electric drive system is used as the platform to realize clean and efficient driving through the collaborative optimization of multiple energy systems.

Electronically-controlled chassis: fast, accurate and coordinated control of braking, steering and suspension systems with control-by-wire technologies.

Intelligent information exchange, clean energy power and electronically-controlled chassis constitute the i-EFV's hardware platform, which is the basis for optimal vehicle driving. Intelligent information exchange first perceives vehicle and environment information before information exchange, which is the premise for intelligent and optimal control. Designed on the basis of control-by-wire technology, clean energy power and electrically-controlled chassis form coupled driving, braking, steering and suspension systems to support safe, efficient and comfortable driving.

In an era where information technologies rapidly develop, the IoV concept is first applied in automotives. With the growing maturity of EV technology and people's urgent need for intelligent EV systems, from the perspective of long-timescale planning, obtaining road traffic environment and terrain information of the road network ahead can help EVs optimize travel strategy and path, reduce driving time and energy consumption. During EVs' driving process, the power output and vehicle speed can be optimized while ensuring safety and energy efficiency by obtaining current moment information of vehicles ahead and around, such as position, velocity, acceleration and pedal signal, modeling and forecasting surrounding vehicles' movement behavior, and combining it with slope change information of roads ahead. Therefore, thanks to the effective use of information from various ITSs, there is still space for further improvement of EV travel strategy optimization and driving control. Since

BEVs have many inherent limitations, how to properly utilize the ITS information to improve vehicle performance and meet drivers' needs is an important research topic and frontier research direction.

From the perspective of inter-vehicle network, an important feature of Internet of electric vehicles is the information exchange between vehicles, so each vehicle can be regarded as an individual node. For each object in the Internet of electric vehicles, it is far from enough to just obtain vehicles' identity, and the wireless sensor network can obtain the dynamic characteristics of vehicles, including paths and locations. According to the definition of Internet of electric vehicles, its architecture is shown in Fig. 1.1 [8].

The overall structure of the Internet of electric vehicles can be divided into the "terminal layer", the "network layer" and the "cloud layer".

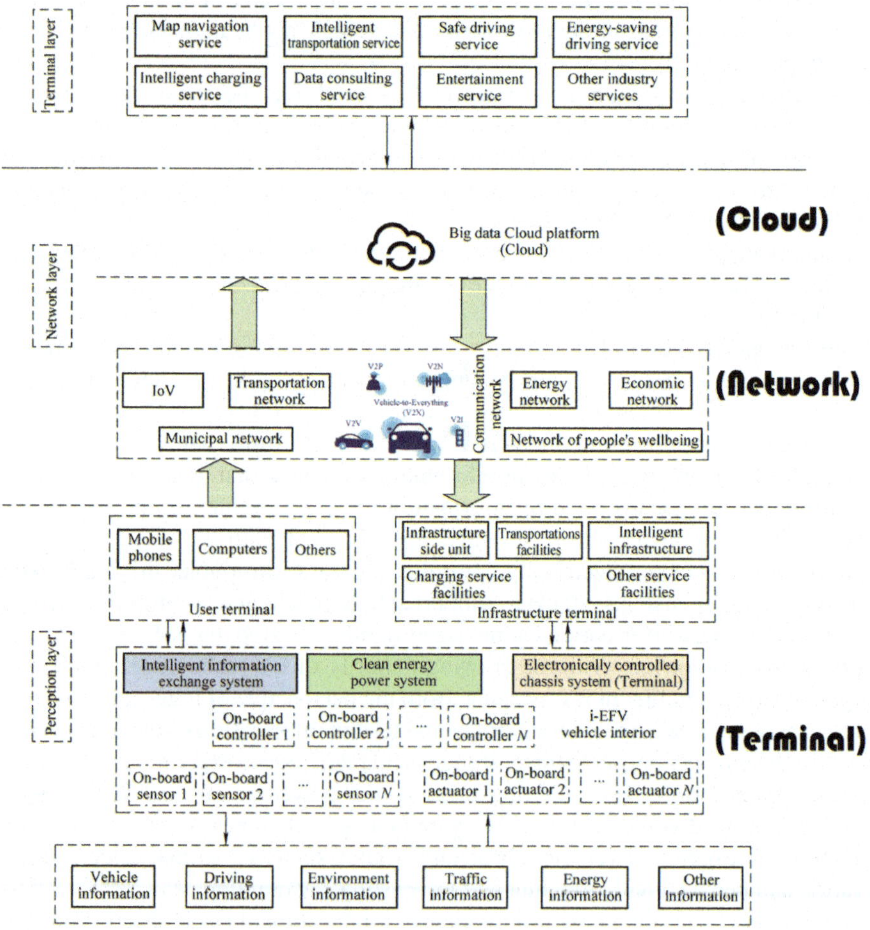

Fig. 1.1 Architecture of the internet of electric vehicles

The terminal system, or the first layer, is the hardware platform of the Internet of electric vehicles, which consists of a variety of intelligent sensors and controllers. Sensors connected to the EV's central control system are responsible for acquiring vehicle state and environment information, and converting them into electrical signals that can be recognized and processed by the control center. The terminal layer is the basis of in-vehicle communication, inter-vehicle communication and on-board mobile Internet communication.

Serving as the second layer, the network system is the communication basis of the Internet of electric vehicles. Cellular communication technology and short-range wireless communication technologies such as Wi-Fi and bluetooth are to used to enable the information exchange and data communication between vehicles (V2V), vehicles and infrastructures (V2I), vehicles and the networks (V2N), and vehicles and pedestrians (V2P), which connect the vehicle's ad-hoc network and a variety of heterogeneous networks to ensure the real-time and effective detection and control of the Internet of electric vehicles.

The third layer, or the cloud system, refers to the management system of the Internet of electric vehicles based on a cloud architecture, which provides services based on users' needs, and makes use of information exchange between the cloud system and vehicles to enable convenient travel.

IoV is an extension of the Internet of Things (IoT) technology in the transportation field, and based on its definition, the Internet of electric vehicles is divided into the perception layer, the network layer and the application layer, as shown in Fig. 1.1.

The perception layer at the bottom is the foundation of the entire IoV system. It consists of sensors, controllers, control chips and other hardware deployed inside the vehicle, and the terminal equipment in the driving environment responsible for the perception and transmission of the acquired information about humans, vehicles and the environment. The perception layer receives and executes relevant instructions to control the vehicle.

The network layer in the middle is responsible for data transmission and provides a unified interface for the perception layer. Due to the complexity and diversity of devices in the perception layer, the network layer includes all kinds of communication network technologies, such as cellular mobile communication network, wireless local area network, Bluetooth/RF and other short-range wireless communication technologies, as well as the Internet.

The application layer on the top provides various types of services to calculate, process, monitor and manage the acquired data. The role of the application layer is to provide users with visual service of the Internet of electric vehicles and reduce operation complexity.

With advantages of intelligent and connected vehicles and NEVs, and based on ITS and electric drive systems, the i-EFV collaboratively optimizes multiple complex electromechanical systems and multi-performance goals to achieve safety, ride comfort, energy efficiency and environmental protection. Considering its system structure and design objectives, the i-EFV is recognized as the integrated carrier and engineering realization of future development directions including intelligence, connectivity and electrification, as shown in Fig. 1.2.

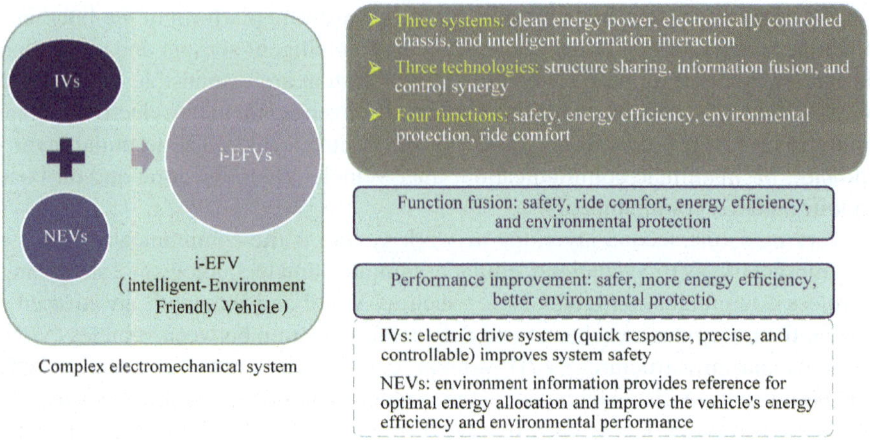

Fig. 1.2 Concept of the i-EFV

Structure sharing, information fusion and control collaboration are three key technologies that support the i-EFV. The i-EFV system involves complex "mechanical, electronic, hydraulic, sensing and communication" multi-system coupling, so it is necessary to use the structure sharing theory to guide structure optimization, integrate and simplify its complex strong coupling vehicle system. Since the information acquired by on-board sensors and wireless communication devices is redundant, information fusion is put in place based on the theory of multi-source information fusion and state estimation to support the accurate identification and modeling of vehicle state and driving conditions. Control collaboration embodies the concept of multi-system coordination and multi-objective collaborative optimization, which coordinates multiple energy systems in energy output and controls braking, steering and suspension systems, and therefore enables optimization of multiple objectives such as safety, environmental protection and energy efficiency.

Structural sharing: emphasizing and applying the structure sharing optimization design theory for complex "mechanical, electronic, hydraulic, sensing and communication" systems such as intelligent information exchange, clean energy power, and electronically-controlled chassis, to integrate and optimize vehicle systems with complex coupling structures.

Information fusion: multi-source information fusion and multi-scale estimation theories are applied for redundant on-board sensing information and wireless communication information, to support the identification and prediction of vehicle state and traffic environment.

Control collaboration: multi-system dynamic coordination, multi-objective collaborative optimization and other control theories are applied to coordinate and control multi-energy driving, braking, steering and suspension systems, and optimize the safety, economy and comfort of "human-vehicle-infrastructure" integration during driving.

Safety, comfort, energy efficiency and environmental protection are the four major functions of the i-EFV, and also the composite demands of modern automotive development. Safety not only refers to the operation stability of the ego vehicle, but also the safety of surrounding vehicles, road boundaries and pedestrians. These safety issues often occur at the same time, for example, vehicle stability issues and collision between the ego vehicle and neighboring vehicles, road boundaries and pedestrians may occur simultaneously. Energy efficiency and environmental protection are closely related to vehicle powertrain control, and also related to power generation efficiency, integrated scheduling and management for electric drive platform. From the perspective of vehicle control, energy efficiency and environmental protection are also related to vehicle safety to some extent. For example, for the i-EFV adaptive cruise control, different longitudinal car-following safety performance requirements have certain impact on the vehicle's economy. Excessive emphasis on car-following safety causes frequent vehicle acceleration and deceleration, which results in unnecessary energy loss and goes against system energy optimization and vehicle economy, which further compromises the vehicle's energy efficiency and environmental performance. On the other hand, excessive emphasis on economy affects car-following safety to some extent. Comfort has a wide range of connotations, inducing pleasure of driving, convenience and subjective feelings of the driver during acceleration, deceleration, yaw, rolling and other movements. Take the i-EFV adaptive cruise control as an example, comfort is mainly reflected in the car-following effect and speed changes in acceleration and deceleration during car-following; excessive car-following acceleration may compromise comfort while improving following safety. Therefore, safety, comfort, energy efficiency and environmental protection may contradict with each other depending on the specific situation, and therefore the collaborative optimization by application scenario is needed during the integrated control of i-EFV.

Next-generation advanced vehicles: vehicles with the purpose to solve safety, comfort, energy efficiency and environmental protection problems faced by vehicles, the traffic and the power systems represented by the new-concept i-EFV. The concept focuses on the design of the vehicle structure and its integrated control, and the vehicle is a part of the coordination and optimal management of the traffic and the power systems in the form of node. The concept and basic theory involved will provide a strong technical support for the vehicle design in the future.

1.3 System Characteristics of i-EFV

As a type of new-concept vehicle based on the collaborative optimization of vehicle, the traffic and the power systems, i-EFV delivers safety, comfort, energy efficiency and environmental protection by integrating ITS-based intelligent vehicle technology and clean-energy vehicle technology based on the electric drive system, and its system structure characteristics include:

(1) Intelligent system

The intelligent system mainly covers two aspects: (1) accurate identification and prediction of vehicle state and driving environment enabled by the on-board sensing system, short-range and remote wireless communication system; (2) intelligent and optimal control of systems, such as local or global optimal control of the driving, braking, steering and suspension systems, which delivers vehicle stability and safety for the traffic environment, pedestrians and roads, optimizes energy efficiency, environmental protection and driving comfort, and ensures a coordinated management of vehicles, the traffic system and the power system by exchanging information with and performing integrated scheduling of these systems.

(2) Control-by-wire system

The driving, braking, steering and suspension systems of i-EFV are controlled through electrical connection, which communicate and cooperate with each other by bus. The control-by-wire technology allows fast system response and easiness to extend vehicle control functions and lower vehicle costs.

(3) Structured system

The increasingly complex functions of modern vehicles require cooperative control of multiple components to achieve vehicle control. For i-EFV that is characterized by distributed systems, the functions of components are taken as the criteria to design and execute the system structure, and vehicle functions are realized through the cooperative control of multiple execution systems, therefore i-EFV is characterized by modular and flexible structures, which is convenient for new vehicle structure integration and design of new control functions.

Intelligent emulation of human behavior and response, perception and prediction of environment information and vehicle state are the basis for integrated optimization of vehicles, the traffic and the power systems. Control-by-wire technology enables system components to be controlled by electrical connection, which allows the intelligent control system to command components through buses or electronic chips. The structured i-EFV has specific function for each component, so it's easy for the intelligent control system to directly command components and perform specific functions. Control-by-wire technology and structured i-EFV effectively support the goal of intelligent system and ensures optimal system performance.

Compared with ITS-based conventional intelligent vehicles (IVs), i-EFV takes demands of modern vehicles for energy efficiency, environmental protection, safety and comfort into account to optimize system performance. Control-by-wire and structured system design adopted by i-EFV facilitate convenient driving. Take ACC as an example, the delivery of IV ACC requires change in the structure and control mode of the accelerator pedal and the brake pedal, as the conventional driving and braking systems controlled by the accelerator/brake pedal are slow in response [1], while the control-by-wire engine and electrical and hydraulic braking systems adopted by the i-EFV can ensure ACC function and fast system response. Conventional IVs' structures are designed based on system functions and control the safety of longitudinal and lateral movement separately, while it is easier for i-EFV to collaboratively control driving, braking, steering and suspension with a structured control-by-wire design, which further ensures the vehicle's overall safety performance, including integrated stability, vehicle and pedestrian collision avoidance and lateral lane/road safety. From the perspective of system composition, since i-EFV combines the characteristics of electric drive systems, such as fast response and precise control, its system can have both accelerated response speed of driving and braking and improved vehicle safety, especially when an emergent brake occurs.

Compared with CEVs based on electric drive systems, i-EFV not only ensures energy efficiency and environment protection, but also integrates ICV technology to achieve overall performance of safety, comfort, energy efficiency and environment protection. Furthermore, thanks to the close connection between electric drive systems and environment systems, i-EFV can optimize management from the perspective of vehicle, traffic and power systems. Effective identification and interaction of traffic and environment information can further improve vehicle economy and ensure the safe and efficient operation of the power system and other support systems.

(4) Systematic operation

In a traditional sense, automotive is just an independent operation unit and information silo in the road traffic system. With the development of IoV technology, the integration of NEV and IoV technologies has become an important direction for future automotive development.

By adding information processing and transmission on top of NEV's original functions in its operation system, i-EFV can serve as a network unit in the road traffic system, a node in the information exchange network of the vehicle-infrastructure system, as well as a mobile storage unit in the energy network. Various electrical and mechanical devices within the i-EFV are connected by networks, while the vehicle itself depends on the application of on-board computing and control platforms, communication modules and sensor networks to support data transmission among users, EVs and the environment. This enables intelligent management and remote control of the vehicle by users and enterprises to provide users with more convenient and safer way to travel.

1.4 New Architecture of i-EFV

The overall system structure is designed from the perspective of general management and collaborative optimization of the vehicle cluster, the traffic and the power system (as shown in Fig. 1.3), in which i-EFV, the ITS center and the power grid scheduling center are three independent subsystems. These subsystems interact and

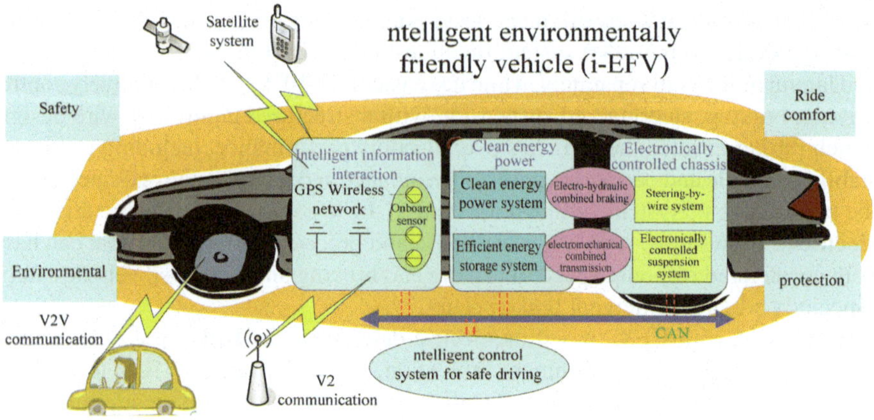

Fig. 1.3 System structure of i-EFV

share information through wireless communication, optical fiber and other communication methods to achieve collaboration and optimal control of each system. Each subsystem has different tasks: the power grid scheduling center optimizes power generation and scheduling through information exchange with the ITS center; the ITS center monitors and guides vehicles in the traffic system based on vehicle-infrastructure communication and feedback from the grid. As the execution node of large-scale transportation, i-EFV optimizes the control of operating components and ensures optimal safety, comfort, energy efficiency and environmental protection based on V2V (Vehicle to Vehicle) communication, V2I (Vehicle to Infrastructure) communication, remote wireless communication and other interactive information, as well as driving environment information acquired by the on-board sensor system.

Based on the collaborative optimization of vehicles, the traffic and the power systems, ITS-based intelligent vehicle technology and CEV technology with electric drive system, i-EFV aims to achieve safety, energy efficiency, environmental protection and comfort, and meanwhile ensure optimal overall performance of large-scale vehicles, the traffic and the power systems. These targets are supported by three key systems: intelligent information exchange, clean energy power and electronically-controlled chassis.

Researches on the i-EFV dynamics are relevant to different objects including IVs and NEVs, and researches on its basic theory and key generic technologies mainly involve the dynamics mechanism analysis, system modeling and dynamics control of the complex electromechanical system.

As one of the effective means to solve increasingly serious environmental and energy problems, NEV has been intensively studied, and ITS, an important approach to solve traffic safety and traffic congestion problems with intelligent vehicle technology, has also received great attention from various research institutions and automotive manufacturers. However, NEV and ITS researches are independent, and therefore impossible to achieve ideal multi-objective system functions through structure sharing, information sharing and collaborative control. In recent years, foreign research institutions have started to explore the technical possibility of combining the NEV and ITS systems, but only entered initial stage due to a lack of research on new concept vehicle systems that integrate NEV and ITS. The concept of i-EFV proposed in this book ensures the integrated performance of energy efficiency, environmental protection and ride comfort under the premise of driving safety, and meanwhile provides a brand new vehicle architecture and technical system by integrating the respective advantages of NEV and IV. Figure 1.4 shows the structural comparison between the i-EFV and existing intelligent and connected NEVs overseas.

1.4.1 Intelligent Information Exchange System

The intelligent information exchange system supports the identification of vehicle state and traffic environment information, information exchange between i-EFV and the environment system, and the comprehensive processing and display of multi-source information. As the structure diagram in Fig. 1.5 shows, the system includes a vehicle-vehicle (V2V) communication module, a vehicle–road (V2R) communication module, a remote wireless communication module, an on-board sensor system and a multi-source signal processing module. All the modules and sensors in the on-board sensor system interact with each other by the bus.

Vehicle-vehicle communication mainly transfers information about vehicle movement and working components through the on-board V2V communication module, including vehicle velocity, acceleration, accelerator pedal, brake pedal, geographical location, component status and fault messages.

Vehicle–road communication involves the on-board V2I communication module and wireless communication of roadside base stations. Vehicle-side mainly sends information about vehicle movement and working components, including vehicle speed, route, component status, fault messages, safety state (relative distance with the preceding vehicle, relative velocity and driving stability state) and geographical location. Roadside base stations mainly send information about traffic flow and road, including information about traffic flow, traffic light, intersection ahead, special road state (traffic accident ahead, road construction, speed limit and other prompt information) and road information (estimated road friction coefficient, road curvature and road grade).

The remote wireless communication module connects the vehicle with mobile phone and the Internet. The interactive information involved includes Internet-based vehicle system update, service query, mobile phone remote monitoring and remote control.

The on-board sensor system is equipped with GPS, radar, machine vision, vehicle movement sensors and vehicle component sensors to collect information about vehicle and its driving environment. GPS collects vehicle geographic location and speed information. Radar is used to detect the vehicle's relative velocity and distance

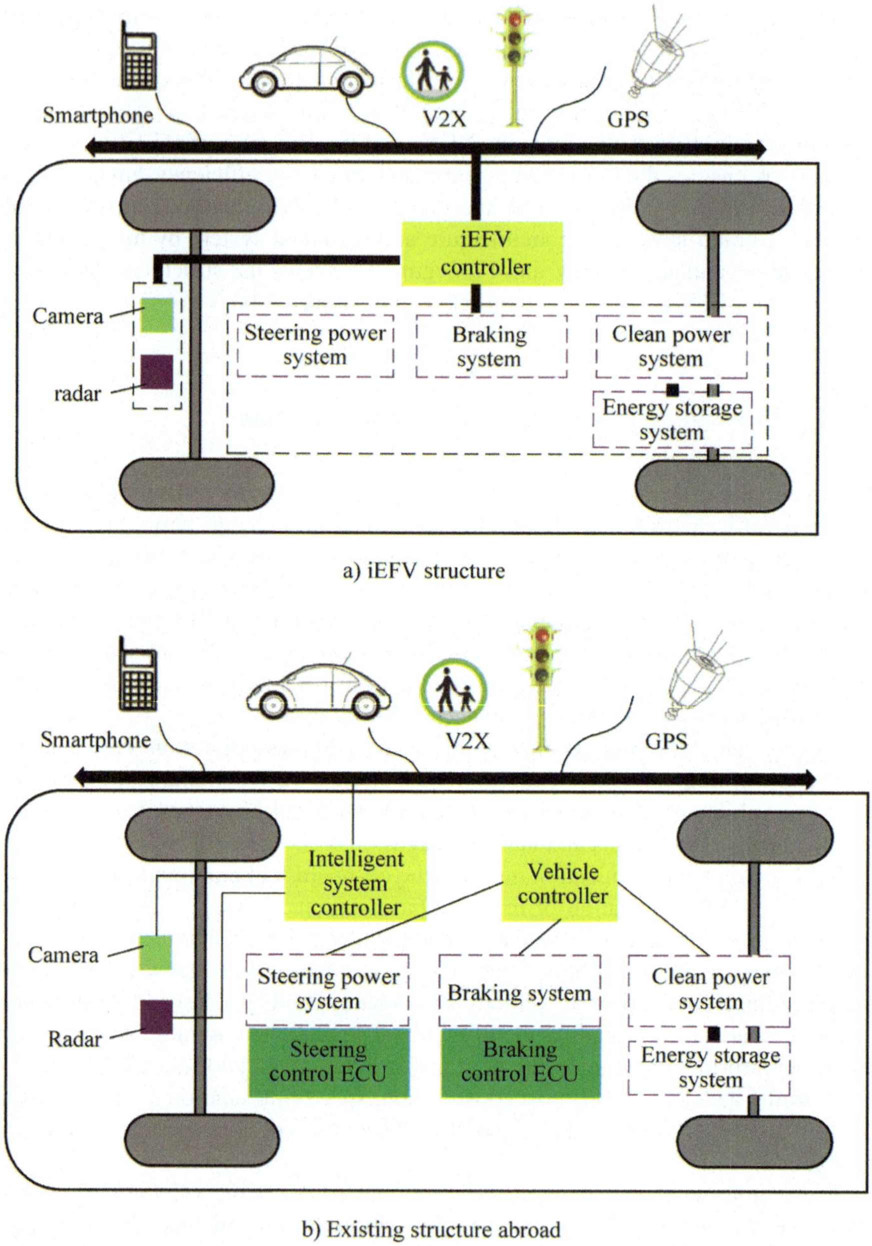

Fig. 1.4 Comparison between the structure of i-EFV and existing NEV abroad

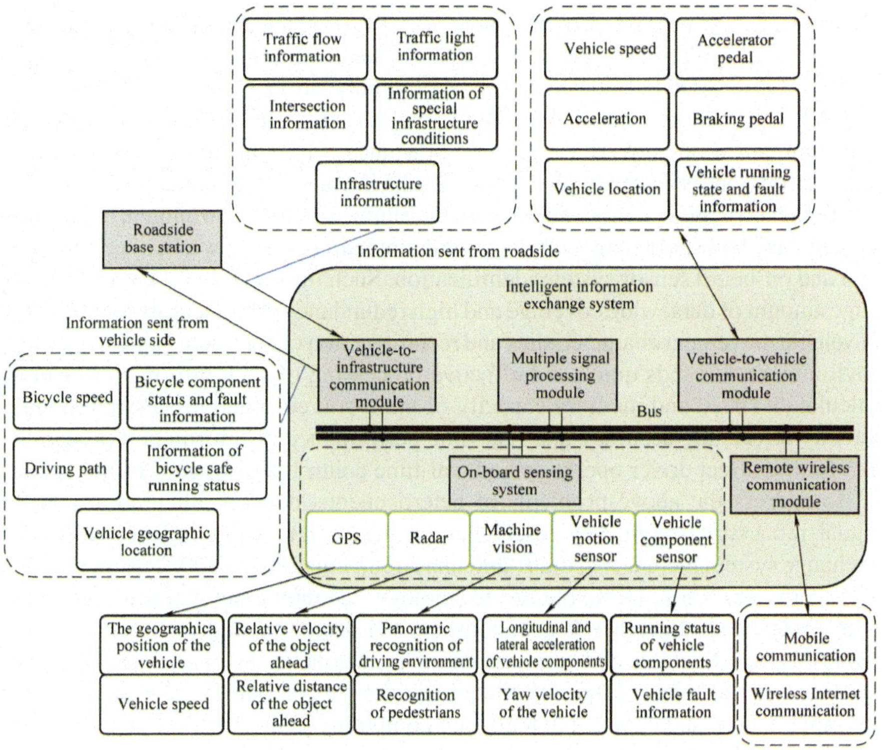

Fig. 1.5 Structure diagram of the intelligent information exchange system

to objects ahead. Machine vision identifies pedestrian and driving environment (such as lane line, road curvature, adjacent vehicles, road signs and other vehicle and traffic environment information), and extracts information by system function. Vehicle movement sensors collect vehicle movement information such as longitudinal/lateral acceleration and yaw velocity. Vehicle component sensors collect state information and fault messages of powertrain, driving system, braking system, suspension system, steering system and relevant accessories.

The multi-source signal processing module is responsible for signal monitoring, management and comprehensive processing based on V2V communication, V2R communication, remote wireless communication and on-board sensor system. This module provides precise information infrastructure for integrated vehicle control and driver operation, with information fusion as the basic signal processing theory.

The structural composition and transmission characteristics of the intelligent information exchange system reveal that it is characterized by multi-network heterogeneity and huge data processing capacity.

On the one hand, the network system integrates bus network, short-range wireless communication network and remote communication network. Depending on different application scenarios, the bus network may integrate CAN (Controller Area

Network), LIN (Local Interconnect Network), FlexRay and MOST (Media Oriented System Transport). V2V/V2R short-range wireless communication network mainly includes DSRC (Dedicated Short Range Communications) based on IEEE802.11p and WLAN (Wireless Local Area Networks) based on IEEE802.11a/b/g/n/p. The telecommunication network enables long-distance communication between vehicle and information service platform based on 4G/5G. On the other hand, i-EFV acquires comprehensive information about vehicle state, driving environment and traffic system based on short-range wireless communication, remote wireless communication and on-board sensor system identification. Such information is characterized by huge amount of data, wide coverage and high redundancy. The accurate identification of vehicle movement characteristics and reconstruction of complete vehicle operating environment demands quick and effective processing of such information, and the calculation speed and memory capacity of multi-source signal processing modules directly affects the accuracy and real-time performance of information processing, and further affect driver operation and real-time control performance of the vehicle.

To address the above problems of heterogeneous network integration and fast signal processor design, key researches on the design of intelligent information exchange system include: (1) multi-antenna circuit integration (V2V communication, V2R communication, GPS, 4G/5G) technology; (2) intelligent gateway technology with on-board multi-network collaboration; (3) design of multi-network information exchange standards and communication protocols for diversified needs; (4) new hardware and software system architecture with high computing speed and low cost; (5) modularized and expansible hardware computing platform.

1.4.2 Clean Energy Powertrain

The clean energy powertrain converts heat and electric energy into mechanical energy to drive the i-EFV. Its system structure includes a clean power driving system and an efficient energy storage system. This section presents the powertrain's structure from the perspective of integrated system performance optimization, based on the current development of various clean energy powertrains and their respective advantages. As shown in Fig. 1.6, the efficient energy storage system includes an external-charging enabled high-voltage battery pack and a super capacitor, while the clean power driving structure can be serial or parallel. In the parallel structure shown in Fig. 1.6a, the clean power driving system includes an engine and a motor, which are dynamically coupled by mechanical connection for power output. In the serial structure shown in Fig. 1.6b, the clean power driving system includes an on-board power generation unit and a motor, in which the clean power driving system is directly electrically coupled with the battery, and the vehicle's power output comes from the motor.

Figure 1.6 shows that the super capacitor is connected to the high-voltage battery pack through the DC/DC inverter. Considering the energy density and power density of the vehicle, it is an optimal combination option to integrate a high-voltage battery

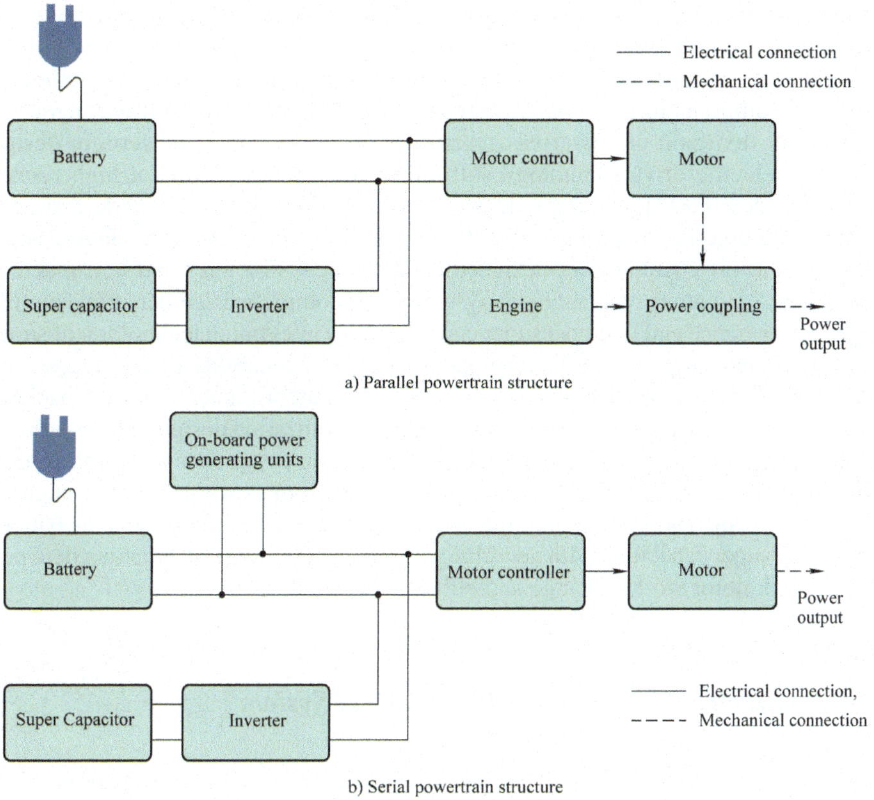

Fig. 1.6 Schematic diagram of clean energy powertrain structure

pack and a super capacitor. The large specific capacity provided by the high-voltage battery pack can supply the vehicle with sufficient driving energy, but the downside is small instantaneous charge–discharge power and poor performance at low temperatures. However, this can be offset by the super capacitor' advantages of high instantaneous charge–discharge power and good low-temperature working performance. Coordinated control of the battery pack and the super capacitor can lower the instantaneous charge–discharge current of the battery pack and improve its cycle life. In addition, the super capacitor's advantage of instantaneous high-power charging can further improve the recovery rate of braking energy and optimize vehicle economy.

In the parallel power structure, the clean power driving system includes a motor, a controller and an engine, in which the engine is a clean powertrain system designed in accordance with energy efficiency and environmental protection requirements, such as an efficient diesel engine, an improved gasoline engine, an alternative fuel engine and a hydrogen fuel engine. In the serial power structure, the on-board power generation unit is coupled with the energy storage system to generates electric energy through the internal combustion engine/generator unit or the fuel cell engine and other powertrains. Various types of improved engines and on-board generator units can solve current problems of high battery cost and insufficient driving range. Considering the current cost and technology maturity, it is feasible to adopt an optimally matched traditional internal combustion engine or generator unit as drive engine.

Current domestic and overseas researches on clean energy powertrain design based on electric drive technology still fail to solve the problems of high energy storage system cost, low energy density, low power density and imperfect system matching. Large-scale application still requires key researches on: (1) high energy density and low cost batteries designed based on new structures and new reaction principles; (2) battery management systems with comprehensive performance; (3) high-performance and low-cost super capacitors; (4) integration technology of super capacitors and high-voltage battery packs with high reliability and low cost; (5) stable, effective and applicable collaborative control algorithms for super capacitors and high-voltage battery packs; (6) combination optimization design of high-voltage battery packs and super capacitors, which delivers the comprehensive optimization of system cost, as well as power and energy density of powertrain; (7) design of new drive forms and system architecture for multi-energy systems, such as battery packs and super capacitors with new characteristics; (8) optimized matching between engine and motor working range and efficiency characteristics.

1.4.3 Electronically-Controlled Chassis System

The electronically-controlled chassis system delivers driving safety and ride comfort on top of vehicle driving objectives by controlling the vehicle's longitudinal/lateral/vertical movement. The electrically-controlled chassis mainly includes a electro-hydraulic braking system, a steer-by-wire system and an electronic suspension system. The electro-hydraulic braking system combines a motor and a hydraulic brake to deliver the active adjustment and rapid response of braking pressure through coordinated control of each system, and serves as the shared braking pressure regulating structure to meet the requirements of driving safety, energy efficiency and environmental protection. As an active steering system, the steer-by-wire system actively adjusts the steering angle according to the driver's demand, vehicle state and environment state, which is a shared steering torque regulating structure to meet driving safety and comfort requirements. As an active suspension system, the electronic suspension adjusts vehicle rolling and pitching to change the vehicle's moving posture and characteristics, and improve driving stability and comfort. The chassis system is an important execution system to complete driving task and satisfy comfort and safety requirements. The traditional steer-by-wire, braking and electronic suspension systems are quite mature so far, but the new power platform of the entire vehicle and the characteristics of its electrical system enabled by the introduction of the electric drive system have brought changes and opportunities to chassis system design: (1) it's easier to deliver 24 V or 42 V vehicle power supply; (2) high-power dynamic source; (3) functional redundancy in the conventional hydraulic braking, steering and electric drive systems (such as motor regenerative braking, direct yaw moment steering by motor). How to optimize chassis system design based on the vehicle' new electrical characteristics and new powertrain distribution will become the focus of structure redesign of the new generation of vehicle chassis system.

1.5 Typical System Application of i-EFV

As a new concept vehicle that integrates ITS and clean energy system, i-EFV utilizes vehicle structure optimization design, vehicle and environmental information fusion and identification, and multi-objective and multi-system cooperative control technology to achieve optimal comprehensive performance. Its research has a wide range of connotations and involves mechanical design, data processing, optimal control and the wide application of different types of vehicles such as i-HEVs (intelligent Hybrid Electric Vehicles) and i-BEVs (intelligent Battery Electric Vehicles) in different fields.

1.5.1 i-HEV Based on the Exchange of Driving Environment Information

Hybrid electric vehicles that have gradually entered the industrialization stage can solve current energy and environmental problems to a certain extent by improving vehicle fuel economy and reducing emission. However, the absence of effective identification of traffic information and the energy management and distribution solely based on current vehicle state make it difficult to optimize fuel economy under all operating conditions. The intelligent active safety control technology of the hybrid electric system that integrates active safety and driving comfort also needs further research. In light of these problems of hybrid electric vehicles, it is proposed to design i-HEVs that combine the advantages and characteristics of hybrid electric system and ITS to improve the overall vehicle performance in safety, comfort, energy efficiency and environmental protection.

Compared with conventional intelligent vehicles, the i-HEV has both a traditional internal combustion engine and a motor battery pack, which are interrelated but work independently. The engine and the motor control the torque directly through the CAN bus, so it is easier for an i-HEV to realize driver assistant control without separate design of underlying controllers and actuators such as throttle and brake pedal [9]. Compared with conventional intelligent vehicles and traditional hybrid electric vehicles, i-HEV is safer and more cost-effective: firstly, i-HEV integrates the hybrid power system's characteristics including fast response and accurate control, which can further enhance the response speed of driving and braking, as well as vehicle safety, especially safety under emergency braking; secondly, i-HEV improves fuel economy by optimizing the powertrain's energy distribution strategy based on identified traffic environment information while ensuring the vehicle's active safety control.

Compared with conventional vehicles, the advantages of i-HEV are mainly reflected in the fuel economy and power performance in longitudinal movements, thanks to the multi-energy powertrain integrating the coupled engine and the motor battery. Therefore, key technologies involved in longitudinal movement control should be the focus of future study. The overall structure of the designed i-HEV system mainly includes intelligent information exchange, clean energy power and electronically-controlled chassis, as shown in Fig. 1.7. The intelligent information exchange system identifies the vehicle's movement state, inter-vehicle movement state and information about traffic, road and environment, and the system includes radars, cameras, GPS, GIS and DSRC vehicle–road communication modules. As the vehicle's power supply, the hybrid power system includes an engine, a drive motor PM (Permanent Motor), a generator BSG (Belt Starter Generator) and an AMT with 5 gearboxes. Since the engine is connected to the AMT through the clutch, the driving motor is located on the back-end of the transmission and the generator is connected to the engine by a belt, the system can work in different modes such as engine drive, engine drive simultaneously with power generation, engine/driving motor co-drive, and driving-motor drive. In the longitudinal movement control, the electronically-controlled chassis mainly includes an EVB (Electronic Vacuum Booster) hydraulic braking system and a motor regenerative braking system, which are distributed according to the total required braking torque.

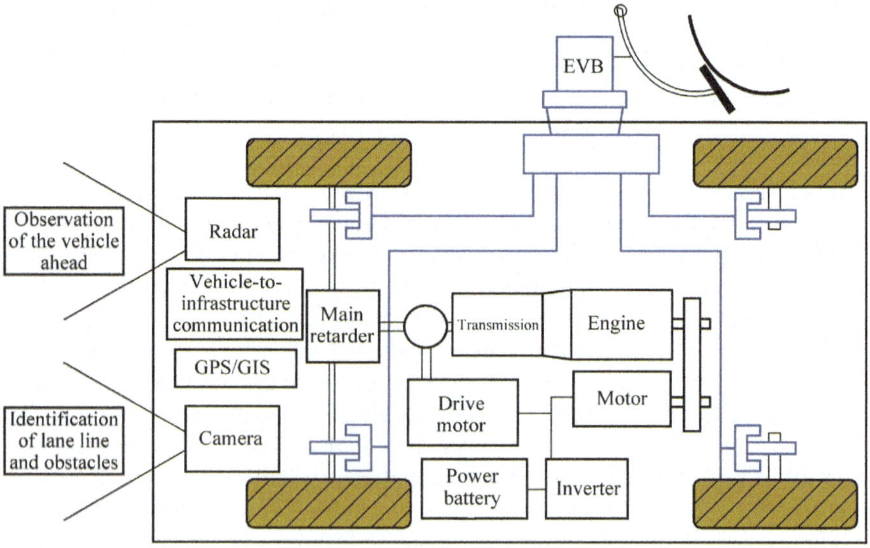

Fig. 1.7 System structure of an i-HEV

Since the real-time control of the driving and braking systems is required by different driving modes, the i-HEV execution and control system architecture with dual CAN structure is proposed, as shown in Fig. 1.8.

Based on the communication structure of the dual CAN network, the i-HEV system is divided into sub-CAN networks for the power system and the sensor system. The sub-CAN network for the power system includes the engine, the motor hydraulic braking, the AMT and relevant power system nodes. The sub-CAN network for the sensor system includes radars, cameras, GPS/GIS, DSRC vehicle–road communication, such as yaw velocity, lateral acceleration, longitudinal acceleration, steering wheel angle and components' sensor nodes.

Components of the i-HEV system are built in a structure of distributed and layered execution and centralized control, and vehicle functions are realized through coordination of three layers. The upper layer is a coordinated multi-system vehicle controller, which formulates optimal control command according to driver operation, current vehicle state and driving environment. The middle layer is distributed

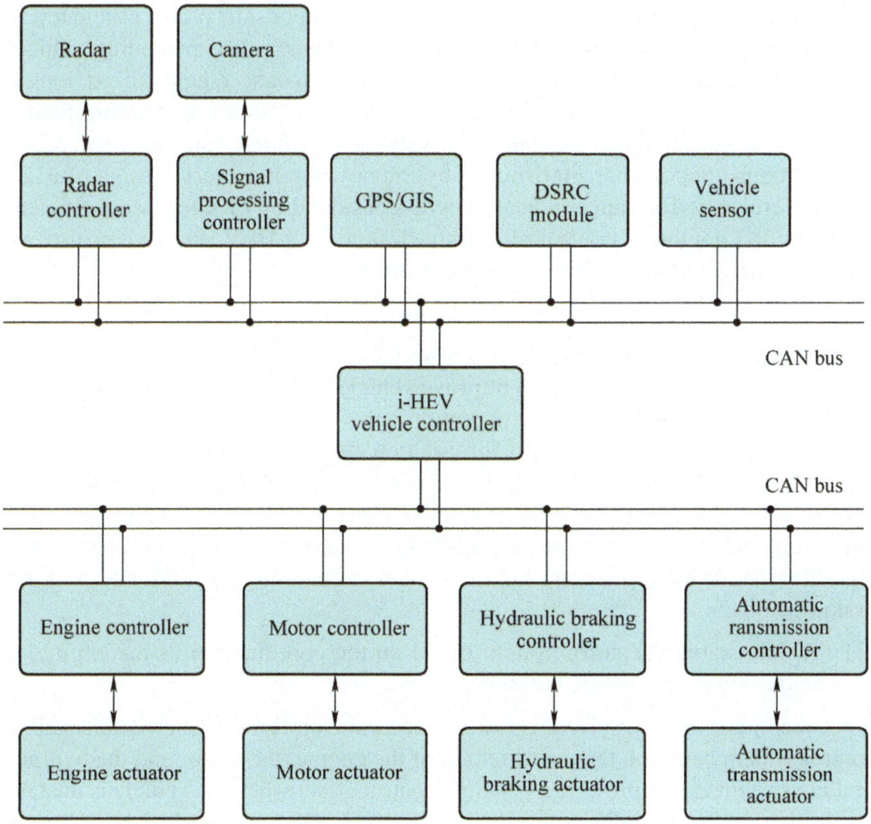

Fig. 1.8 Topological structure of the i-HEV "integrated-distributed" dual CAN system

execution system controllers that receive control instructions from the CAN bus and drive the action of the execution system with the help of execution units for the engine, the motor, the hydraulic braking and the AMT system in the lower layer. Execution units of each system use the unified interface definition and carry out corresponding operations according to instructions from middle layer controllers.

For the purpose of integrated safety, cost efficiency and comfort performance, the i-HEV system is divided into the following three sub-systems by function: the adaptive cruise control (ACC) system that coordinates safety, cost efficiency and comfort; the vehicle control system based on real-time traffic information; the vehicle stability control system based on the coordination of the engine, the motor and the hydraulic braking system.

(1) i-HEV adaptive cruise control (ACC) system that coordinates multiple objectives of safety, cost efficiency and comfort

The composite power system provides i-HEV with enough cost optimization potentials. Combined with the collaborative optimization of vehicle driving/braking power and multi-source power allocation in cruise control process, i-HEV adaptive cruise control can achieve better overall performance in system safety, cost efficiency and comfort. Key researches include algorithm of recognizing the preceding vehicle's movement based on radar, camera and multiple sensors; a generalized longitudinal dynamics modeling for i-HEV adaptive cruise considering coupling features of system nonlinear dynamics and inter-vehicle dynamics; estimation of equivalent fuel consumption characteristics of hybrid powertrain battery; design of i-HEV adaptive cruise performance indicators with consideration of safety, cost efficiency and comfort; nonlinear control algorithm design for i-HEV adaptive cruise with multi-objective and multi-energy system coordination.

(2) i-HEV vehicle control system based on real-time traffic information

The hybrid power system of i-HEV optimizes the energy allocation based on real-time traffic environment information to improve vehicle fuel economy. Key research will focus on the recognition of traffic operation conditions and the preceding vehicle's movement feature based on multi-information fusion of radar, camera, GPS/GIS and on-board sensor system; the optimal energy allocation algorithm for the i-HEV hybrid system based on inter-vehicle relative movement information and characteristics of global traffic conditions; the multi-system dynamic and coordinated control algorithm for switching among hybrid system engine, motor, AMT and hydraulic braking mode.

(3) Vehicle stability control system based on the coordination of the engine, the motor and the hydraulic braking system

Since the motor driving system is fast and accurate in response, the vehicle stability control system based on the coordination of the engine, the motor and the hydraulic braking system can improve the system's control effect when the vehicle is unstable, and further optimize driving stability. Key research will focus on the i-HEV acceleration slip regulation (ASR), the anti-lock brake system (ABS) and the control system of electronic stability program (ESP).

1.5.2 i-BEV Based on Vehicle-Infrastructure-Network Interaction

A BEV is directly driven by the electric energy stored in the battery without any fuel consumption and emission. With increasingly mature clean power generation technology, it has become the development trend of the next generation of vehicles. Thanks to the fast response speed, high precision and low noise characteristics of BEV enabled by the motor drive, an ADW vehicle with independent electric drives has independently controllable multi-wheel driving/braking forces, which provide the vehicle with outstanding acceleration, steering stability and ride comfort. At present, BEV development is encountering the following challenges: low energy density makes the battery a big part of the vehicle mass, so it is urgent to develop lightweight BEVs and improve vehicle safety; limited battery storage capacity requires close monitoring and planning of the battery's state of charge (SOC), driving path and charging path during driving process to optimize cost efficiency and sustainability.

If the BEV movement is abstracted as flowing energy points, then the large-scale energy flow in the transport system and energy supply flow in the power system interact and influence each other, forming an inseparable energy supply entirety. Compared with conventional internal combustion engine vehicles, the most outstanding characteristic of BEV is that the vehicle movement is not only about the control of individual vehicles, the resultant large-scale electric energy flow also has a serious impact on the distribution, operation safety and efficiency of the corresponding power grid.

To sum up, it is necessary to ensure the integrated management and optimal control of the entire driving process of BEV from the perspective of vehicle cluster, ITS and the power grid system to achieve the safe, economical, comfortable and sustainable large-scale application. Therefore, the concept of intelligent battery electric vehicles (i-BEVs) based on vehicle-infrastructure-network interaction is proposed, which coordinates energy scheduling, optimizes EV intelligent driving control and realizes comfortable, safe, economic and sustainable operation of large-scale EV cluster based on the temporal-spatial distribution and information exchange of energy flow in the power grid and the transportation network. As shown in the definition of i-BEV, it not only involves the control of the vehicle, but also requires an intelligent control from the perspective of power management coordination between traffic system and the power grid system, which is a new demand and challenge brought by large-scale application of BEVs.

To comprehensively explore and study key technologies involved in i-BEVs, the overall structure of i-BEV is designed based on the four-wheel independent electric drive structure as shown in Fig. 1.9. The intelligent information exchange system is composed of the on-board environment sensing system including radars and cameras, the traffic-road sensing system GPS/GIS and the DSRC-based vehicle-infrastructure communication system. The vehicle-infrastructure communication system allows the cluster of running i-BEVs, the ITS center and the power grid scheduling center to constitute an interactive, collaborative and integrated system. The clean energy

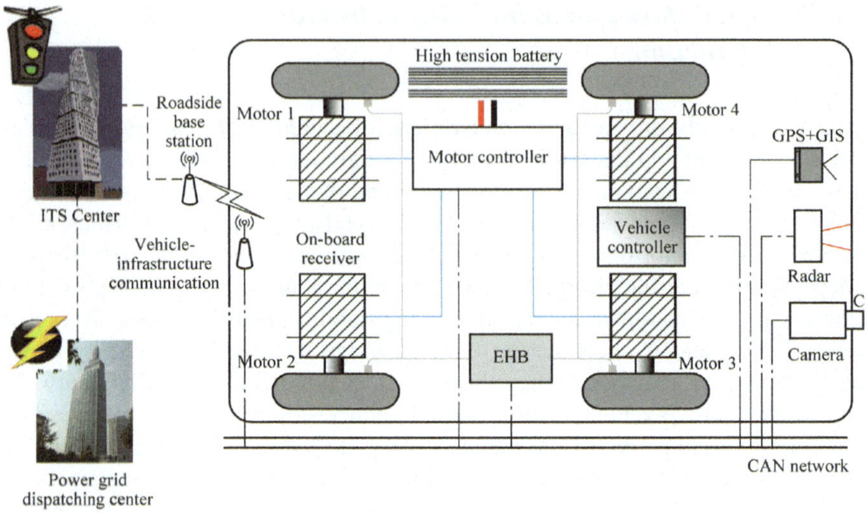

Fig. 1.9 Structure of the i-BEV system

power system consists of a high-voltage battery pack and a distributed electric drive system composed of independent driving motors for four wheels. The electronically-controlled chassis system consists of the electro-hydraulic brake (EHB) system and the motor system used for braking. The intelligent information exchange system, the motor controller, the high-voltage battery pack, the hydraulic braking system EHB and the vehicle controller all communicate through the CAN bus.

As shown in Fig. 1.9, the i-BEV system has the following characteristics: a shared motor driving/braking system structure, fast and accurate response, and independent control of four wheels; equipped with comprehensive environment identification sensors and communication systems, which can acquire real-time information about traffic and local driving environment. From the perspective of system control, the integrated control of i-BEV is enabled by the following three systems: the driver assistant system based on driving environment identification; the active safety control system based on distributed electric drive chassis; the integrated management and control system based on the large-scale EV cluster, the ITS traffic network and the smart grid.

(1) Driver assistant system based on driving environment identification

The driver assistant system of the i-BEV includes longitudinal and lateral driving assistance.

1. Longitudinal driver assistant system

The coordinated low-speed start-stop control, high-speed cruise control and adaptive cruise control systems for EVs are designed based on the response characteristics of the distributed electric drive chassis system and the power distribution characteristics of the driving systems, with consideration of longitudinal tracking performance, system economy and driving comfort. Key researches include the expected acceleration determination method based on the longitudinal tracking performance, system economy and driving comfort; the velocity tracking dynamic control based on distributed electric drive characteristics; the multi-mode switching principle and method for low-speed start-stop, high-speed cruise and adaptive cruise.

2. Lane-keeping driver assistant system

This system identifies the position of the vehicle and the lane line based on multi-source information fusion in the intelligent information exchange system, and warns the driver if vehicle deviates from the target lane. If the danger persists, the system will coordinate the control of front-wheel steer-by-wire system and the distributed driving system to actively correct the driver's operation and keep the vehicle moving in the lane. Key researches include the identification method of vehicle and lane line position based on multi-source information fusion; lane departure warning based on EV dynamic characteristics and active intervention mode switching features; lane-keeping control coordinated by distributed electric drive system based on front-wheel steer-by-wire and direct yaw moment control.

(2) Active safety control system based on the distributed electric chassis

With the integrated ITS and intelligent active safety control, vehicle dynamics stability control (VDSC) and EV high-pressure safety control, the i-BEV's active safety control system features high-pressure safety active control, longitudinal emergency collision avoidance and active control that integrates expected safety state of "human-vehicle-infrastructure".

1. Active control of the safety of high-voltage (HV) battery based on "human-vehicle- infrastructure" state information

Based on the relative movement information of vehicle-vehicle and vehicle-infrastructure, as well as the HV battery pack information obtained by the intelligent information exchange system, the active safety control system identifies and predicts the vehicle's HV safety and collision risks, controls the HV battery pack and the power components of the vehicle to realize in-vehicle HV safety control in anticipated vehicle collisions and faults of the HV battery. Key researches include HV battery insulation, electric leakage and other safety state monitoring methods; safety state identification and prediction upon vehicle collision based on the information of relative movement between vehicle–pedestrian, vehicle-vehicle and vehicle-infrastructure; active safety control based on coordination of battery pack and distributed electric drive system under anticipated collision safety state and HV safety state.

2. Longitudinal collision avoidance system based on the electric chassis of distributed driving

This system identifies collision dangers, warns the driver and controls the driving and braking systems to prevent collision between the vehicle and the obstacle, as well as reverse collision, based on the relative movement between vehicle and objects in front and rear of the vehicle. Key researches cover the identification of longitudinal driving behavior characteristics, including the longitudinal distance and safety margin of longitudinal relative speed; early warning and vehicle intervention measures under different safety levels and driver acceptance; the cooperative control algorithm based on stable motor and braking system in case of emergent braking.

3. Integrated safety control of vehicle chassis based on expected safety state

Based on the movement state of the preceding vehicle, the curvature of roads ahead, road friction coefficient, velocity, acceleration, yaw velocity and other vehicle state and environment information, this system predicts future vehicle trajectory and expected safety state of vehicle movement, and controls the vehicle's driving/braking/steering systems in dangerous situations, and therefore ensures the longitudinal and lateral driving safety over a period of time. Key research focuses on the vehicle state and driving environment recognition algorithm based on the fusion of multi-sensor information from GPS, radars, machine vision and the on-board sensing system; the vehicle's expected trajectory and safety state monitoring method based on vehicle state and driver operation; optimal calculation of vehicle moving target based on the expected vehicle safety state; the integrated cooperative control algorithm of multi-system chassis based on distributed electric drive chassis, steer-by-wire and hydraulic braking system.

(3) Integrated management and control system based on large-scale EV clusters, the ITS traffic network and the smart grid

Charging demand brought by large-scale EV application will have a great impact on the operating load, safety and efficiency of the power grid. Therefore, large-scale EV clusters, the ITS traffic network and the smart grid should be regarded as an organic whole for optimal management and real-time control based on the overall system optimization theory. From the perspective of vehicle dispatch and dynamics control, it mainly includes the design of an integrated communication platform and an information exchange system for intelligent interaction among vehicles, the road network and the power grid, and the design of EV charging management system based on information exchange among vehicles, the road network and the power grid.

1. Design of the integrated communication platform and information exchange system for intelligent interaction among vehicles, the road network and the power grid

This requires establishing a standard, reliable and flexible communication platform for vehicles, the road network and the power grid to enable their real-time, fast and reliable information exchange. Key research focuses on the integrated communication among EVs, ITS, the power grid and charging stations, as well as integrated communication interface and protocol standards for each component; multi-source information fusion method for EVs, the power grid and ITS with information redundancy and transmission safety taken into consideration; on-board intelligent information management system based on human–computer interaction.

2. EV charging management system based on vehicle- road network-grid information exchange

This system allows the ITS center to remotely monitor and manage the driving state of a large number of EVs and guide them for charging through information exchange among vehicles, the road network and the power grid. Key researches include EV safety state remote monitoring and early warning technology; remaining mileage estimation based on real-time traffic; energy flow and charging demand estimation based on large-scale EV operation; EV charging guidance considering the sustainable operation of vehicles, the power grid safety and traffic efficiency.

References

1. Wang B (2004) Development of clean energy vehicles in China. In: International conference on alternative fuel vehicles, pp 90–98
2. Beck R, Bollig A, Abel D (2007) Comparison of two real-time predictive strategies for the optimal energy management of a hybrid electric vehicle. Oil Gas Sci Technol Rev IFP 62(4):635–643
3. Li Z, Sun H, Guo Q et al (2012) GPF-based method for evaluating EVs' free charging impacts in distribution system. In: 2012 IEEE power and energy society general meeting, San Diego
4. Manzie C, Watson H, Halgamuge S (2007) Fuel economy improvements for Urban driving: hybrid versus intelligent vehicles. Transp Res Part C 15(1):1–16
5. Li K, Chen T, Luo Y et al (2009) Structure of intelligent environmental-friendly vehicle: China, 200810223099. 2009-02-04
6. Li K, Chen T, Luo Y et al (2010) Intelligent environmental-friendly vehicle–concept, architecture and engineering implementation. Automot Eng 32(9):743–748
7. Li K, Zhang S, Luo Y et al (2013) The concept of entelligent environmental-friendly vehicle and its latest development. J Automot Safety Energy Conserv 4(2):109–120
8. Li K, Bian M (2019) Editor-in-chief. Electric vehicle engineering handbook (6): intelligent network. China Machine Press, Beijing, p 12
9. Beiker SA, Vachenauer RC (2009) The impact of hybrid-electric powertrains on chassis systems and vehicle dynamics. SAE, 2009-01-0442

Chapter 2
Key Challenges and Key Technology Systems of i-EFV

As a mobile transport platform and embodiment of complex electromechanical system, the i-EFV integrates complicated coupling structure and actuating components of sensing, communication and mechanical-electronic-electrical-hydraulic systems. Meanwhile, it also represents the interdisciplinary technology integration of intelligence, network and electric vehicle. Through collaboration of multi-intelligent agents, such as the road network, the information network, the transportation network and the energy network, the i-EFV operation involves multi-physical-process coupling of multi-level, multi-objective and multi-system optimal configuration. The exploration of the mechanism of multi-physical-process coupling of complex electromechanical systems and the multi-system cooperative control in "human-vehicle-infrastructure" environment is the foundation to achieve the operation performance and targets of i-EFVs.

2.1 Pain Points of i-EFV

The operation process of each system in the i-EFV's driving process can be divided into three levels: execution and sensing, signal processing, and integrated vehicle control. Execution and sensing are performed by the i-EFV's power system, chassis system, sensing system and communication system. The power and chassis systems empower vehicle movement through energy flow between components; depending on information flow among inter-vehicle short-range communication, long-distance communication and on-board sensors, sensing and communication systems obtain and exchange vehicle and environment information. Vehicle signal processing carries out accurate vehicle and environment identification and prediction through comprehensive processing of multi-source information acquired by sensors and communication systems. Based on the identified and predicted information, the integrated

K. Li et al., *The Intelligent Environment Friendly Vehicle*,
Key Technologies on New Energy Vehicles,
https://doi.org/10.1007/978-981-19-4851-0_2

vehicle control can ensure safe, comfortable, energy-saving and eco-friendly driving performance.

Complex system structure, diverse identified information and control system coupling are difficult problems in the design and application of i-EFVs for their control-by-wire, structured and intelligent characteristics [1].

(1) Complex system structure

On the one hand, the i-EFV has to ensure safety, ride comfort, energy efficiency and environmental protection with consideration of people, vehicles and infrastructures. The product integrates complex vehicle sensors, the communication system and the mechanical–electrical-hydraulic coupling system, whose function and structure may be coupling and overlapping, resulting in waste of resources and increase of system costs, and therefore the structure optimization theory shall be applied in system design. On the other hand, the i-EFV has characteristics such as "wire-controlled system" and "distributed system components", which result in frequent information exchange among multiple components and large amount of data exchange, so reliable design theories and methods should be adopted to ensure safe collaboration of all components.

(2) Diverse identified information

As an integrated complex electromechanical system confronted with time-varying traffic environment, three main problems in the vehicle state and environment information identification need to be solved to realize the multi-performance objective control of the i-EFV. Firstly, massive information acquired through vehicle-vehicle communication, vehicle-infrastructure communication, remote wireless communication, and on-board sensing system is overlapping and redundant, thus requiring integrated analysis and processing to form a unified description of the vehicle and the environment. Secondly, the vehicle's driving environment is complex and changeable, and the information obtained by the existing sensor system is seriously disturbed, so it is necessary to integrate redundant information from multi-source sensors to obtain an accurate target information. Thirdly, sensors cannot directly obtain accurate driver-vehicle-infrastructure traffic environment information, so multi-source information needs to be integrated for synthetic analysis. To effectively use multi-source and redundant data information and acquire the accurate identification and prediction of vehicle state and traffic environment, it is necessary to adopt the systematic signal processing method.

(3) Control system coupling

On the one hand, the i-EFV aims at safety, comfort, energy efficiency and environmental protection. On the other hand, it also has a variety of driving modes, including longitudinal/lateral assisted driving and "human vehicle infrastructure" integrated safety control with consideration of vehicle and driving environment. At the same time, the i-EFV integrates complex mechanical-electronic-hydraulic coupling systems. In view of above multi-objective, multi-mode and multi-system cooperative control requirements, it is necessary to establish a complete control system architecture and theoretical system.

2.2 Key Scientific Issues Involved in i-EFV

As a complex electromechanical system, the i-EFV has a highly complicated physical structure, high level of integration and highly variable dynamic coupling relationship. One of the core issues in the study of the i-EFVs is how to carry out dynamic characteristics and mechanism analysis on the generalized mechanical dynamic system consist of drivers, vehicles and infrastructures, as well as the coupled system consist of new clean energy power system and electronic-controlled chassis system in the complex traffic environment, to find and establish a unified and flexible generalized mechanical dynamics system model that is capable of accurately express interdisciplinary coupling mechanism and system's global features.

In addition, current control principle and decision-making of vehicle dynamics mainly focus on individual researches on each important property, such as vehicle safety, fuel economy, emission behavior control and ride comfort. Although such researches can address individual dynamics problems better, it's difficult to support comprehensive driving performances in the presence of complex system coupling, such as multi-objective ride safety, fuel economy and comfort that take driver, vehicle and road traffic safety into account at the same time. How to take i-EFV driver-vehicle-infrastructure as an organic whole to consider system dynamics characteristics in the presence of multi-system coupling and realize cooperative control during driving process is also one of the most challenging questions in the research on the dynamic performance matching of complex electromechanical systems.

2.2.1 Coupling Mechanism of Multi-physical-Process of the i-EFV's Complex Electromechanical System

The i-EFV acquires and exchanges vehicle and environment information through sensing and communication systems, based on inter-vehicle short-range communication, remote communication, and information flow among on-board sensors. The data acquired by sensing and communication systems are used to identify and predict surrounding vehicle and environment after comprehensive processing of multi-source information. The identified and predicted information is also the basis of the vehicle's integrated control and decision-making, and is used to further control the vehicle to move towards the expected destination through the energy flow between power and chassis system components while ensuring safety, comfort, energy efficiency and environmental protection. In addition to the same complex electromechanical system properties as those in traditional new energy vehicle platforms, as an integrated embodiment of intelligence and network, the i-EFV is also a mobile node of the ubiquitous sensor and information exchange network, a mobile intelligent unit of the road traffic network, and a mobile storage unit of the energy network. Therefore, from the perspective of transportation, the i-EFV's operation process is a "human-vehicle-infrastructure" multi-physical-process coupling in the complex electromechanical system. Its features such as safety, comfort, energy efficiency and environmental protection are essential to ensure the optimal flow of information, energy and material in operating environment and its own systems. As shown in Fig. 2.1, the "human-vehicle-infrastructure" generalized mechanical dynamics system composed of drivers, i-EFVs and roads forms an organic whole through inter-coupling, interaction and feedback action. Based on its interaction mechanism, the generalized mechanical dynamic system of "human-vehicle-infrastructure" coupling can be decoupled into an information and energy coupling dynamic system in which "human-vehicle", "human-road" and "vehicle–road" interacts with each other. In terms of interaction medium, there is a real-time two-way energy and information exchange between the driver and the i-EFV, while the information exchange from roads to the driver and the energy exchange from the vehicle to roads are one-way, and there are both energy and information exchanges between the i-EFV and roads in real time.

From the perspective of vehicle operation control, the vehicle's longitudinal/lateral/vertical movement and rotation in the three directions of its coordinates (rolling, pitching, yawing movement) are inter-connected and influenced by each other, and its final movement is the result of the comprehensive action of complex component systems with mechanical-electronic-electrical-hydraulic coupling such as the driving, braking, steering and suspension systems, and it's also a multi-physical-process of the coupling of the information and energy flows in the optimally configured mechanical and electrical systems of the i-EFV, as shown in Fig. 2.2.

Coupling among "human-vehicle-infrastructure" system elements mainly includes human–machine (assisted driving system) coupling, human-road coupling, vehicle-infrastructure coupling, electromechanical coupling and dynamic coupling

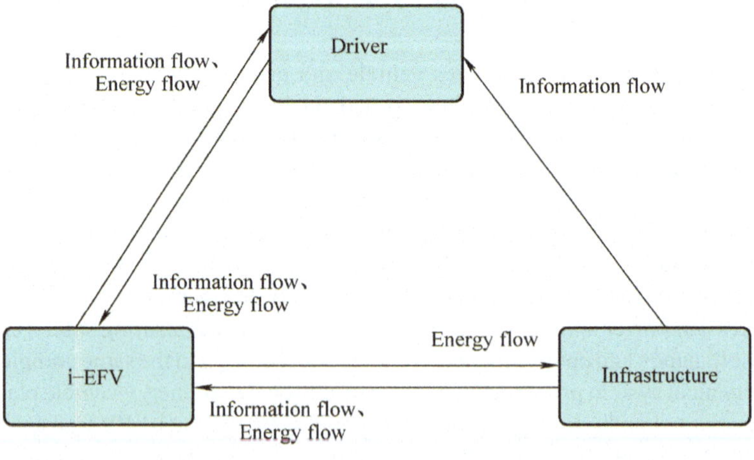

Fig. 2.1 Generalized "human-vehicle-infrastructure" coupling system with multiple physical processes

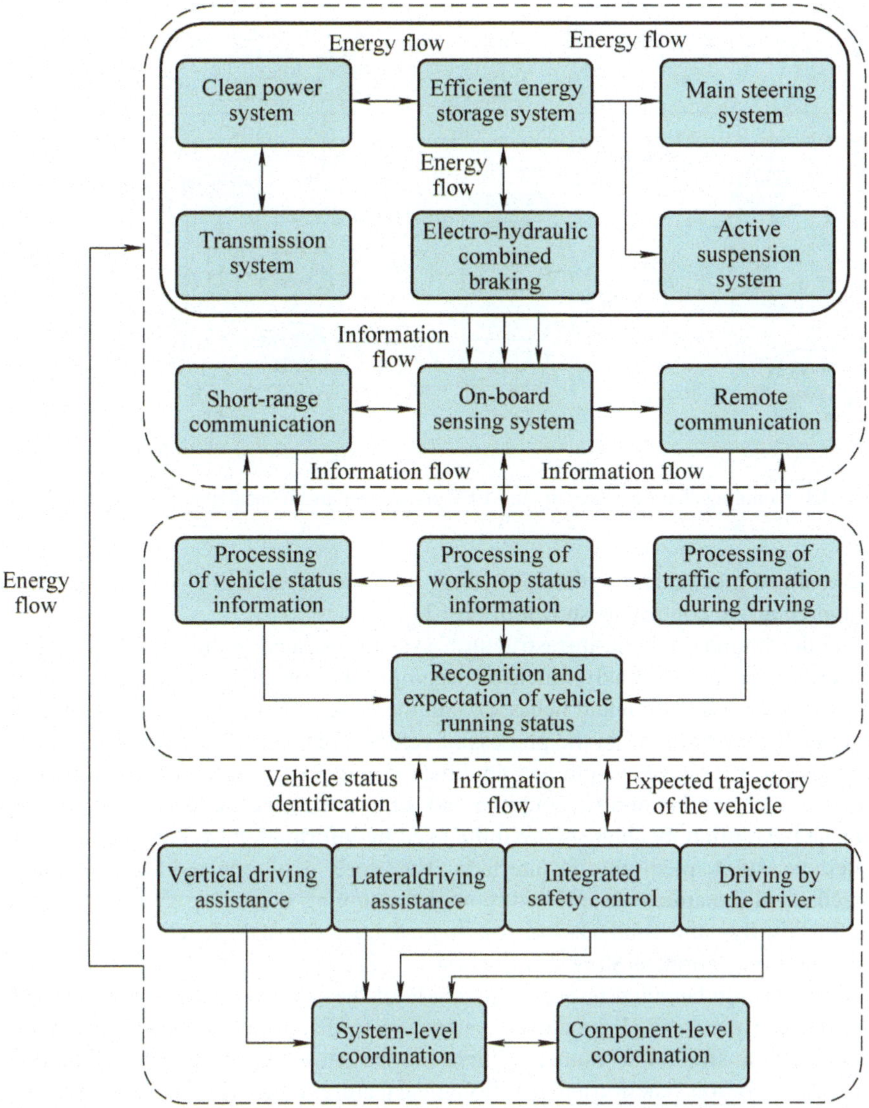

Fig. 2.2 Coupling structure of i-EFV generalized mechanical dynamics systems with multiple electromechanical systems and multiple physical processes

in the i-EFV system. As an integrated product of electromechanical system, the i-EFV is designed with the characteristics of multi-physical-process coupling, such as driver assistance, hybrid drive with multiple power sources, mechanical transmission, and electro-hydraulic combined braking. System efficiency and efficiency of different components are also inter-coupled in the "human-vehicle-infrastructure"

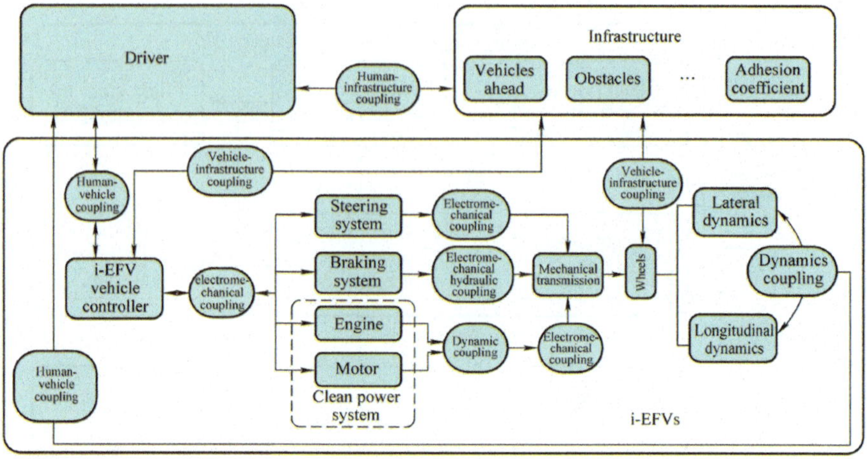

Fig. 2.3 Coupling characteristics of the i-EFV generalized mechanical dynamics system

generalized mechanical dynamics system involved in the i-EFV movement and its comprehensive control, as shown in Fig. 2.3.

In the "human-vehicle-infrastructure" system operated in the i-EFV, the vehicle, the driver and the road environment have complex nonlinear dynamics characteristics in terms of composition and behavior, and there are nonlinear constraints and inter-couplings among them. In the process of vehicle control and movement, the driver-vehicle-infrastructure system has nonlinear dynamics transmission characteristics, such as energy transmission coupling and information exchange, and the operation law and evolution mechanism are influenced by information exchange relationship, interactive unit, medium and interface. Therefore, it is reflected as a generalized mechanical dynamics system with complex coupling of vehicle system dynamics, traffic flow dynamics and network topological dynamics at the level of vehicle, traffic and communication, as shown in Fig. 2.4.

The key fundamental issues in the study of the complex coupling of the i-EFV generalized mechanical dynamics system are to discuss the coupling mechanism of energy transfer in the "human-vehicle-infrastructure" system; analyze changes of driver's operation, reaction and psychological features under varying vehicle and road dynamic states and parameters; understand the influence of driver's characteristic changes on vehicle dynamic state; reveal mutual constraint characteristics among relative vehicle distance, relative speed, acceleration and external environmental factors, as well as the nonlinear part of the "human-vehicle-infrastructure" system caused by the dynamics coupling relationship between vehicles; explore the nonlinear dynamics evolution mechanism of bifurcation and abrupt changes in the dynamics system with "human-vehicle-infrastructure" coupling, under the condition of driver response lag and change of uncertain vehicle and road parameters.

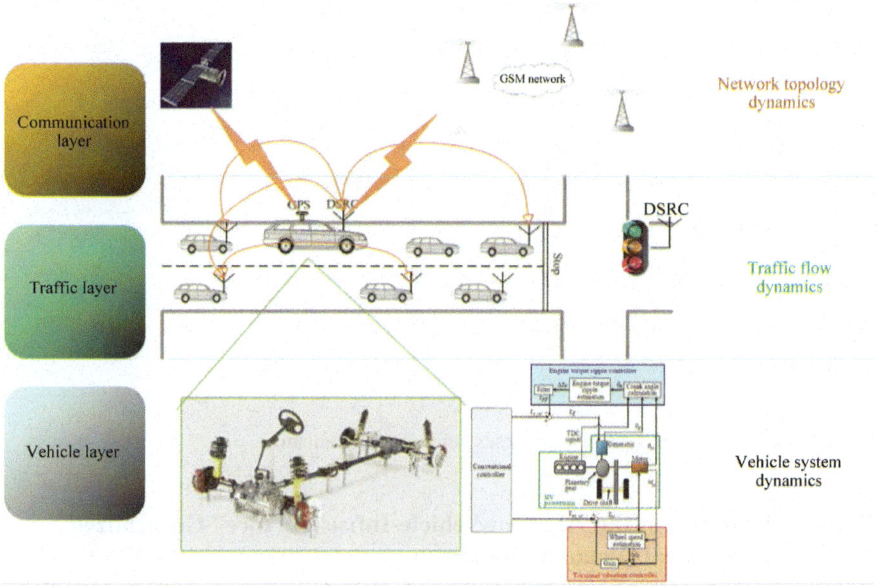

Fig. 2.4 Coupling of the "human-vehicle-infrastructure" dynamics system

2.2.2 *Modeling and Collaborative Control of the "Human-Vehicle-Infrastructure" Generalized Mechanical Dynamics System*

Most existing researches on the mechanism and characteristics of conventional vehicles' dynamics behavior are independent researches with focus on a specific aspect or a single system. Though such researches can provide in-depth understanding in dynamics behavior mechanism and characteristics of traditional vehicles and parts, the dynamics behavior mechanism and characteristics of "human-vehicle-infrastructure" complex multi-system under multi-physical-process and multi-system coupling (as shown in Fig. 2.5), and the resultant dynamics problems are still unclear. Therefore, for the i-EFV with complex structure, integration and coupling, it is one of the core problems for the researchers of the next generation of vehicle dynamics to study multidisciplinary unified modeling theory for complex electromechanical systems and multi-system multi-objective cooperative control in complicated traffic environment, with the aim to clearly express the system's cross-energy-domain coupling mechanism and describe the global properties of the system.

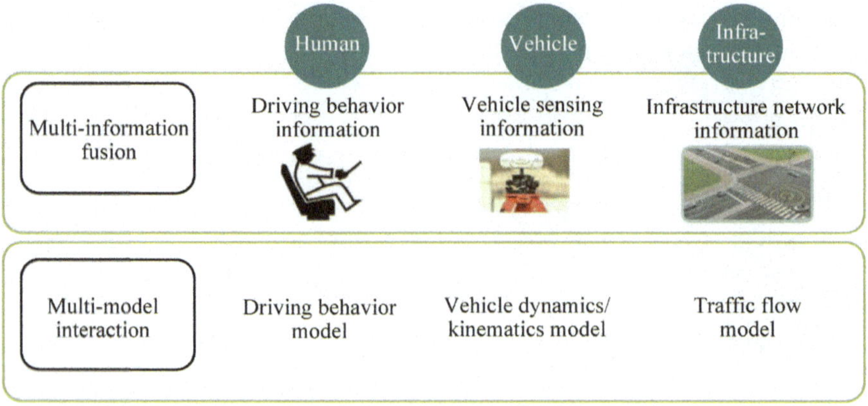

Fig. 2.5 i-EFV complex system's dynamics behavior mechanism

2.2.2.1 Modeling of the "Human-Vehicle-Infrastructure" Generalized Mechanical Dynamics System

Mechanism and modeling technology research of existing traditional vehicles' dynamics behavior mainly concentrated on the strength, stiffness, reliability and durability experiment of power, chassis and body system, or CAD/CAE analysis, finite element modeling and dynamic response characteristics analysis of dynamics subsystems, such as vehicle transmission, steering, suspension, braking and wheel, or vehicle dynamics property research. Such independent researches on a specific aspect or a single system provide in-depth understanding in dynamics behavior mechanism and characteristics of traditional vehicles and parts, but the dynamics behavior mechanism and characteristics of "human-vehicle-infrastructure" complex multi-system under multi-physical-processes and multi-system coupling, and the resultant dynamics problems are still unclear. Therefore, for the i-EFV with complex structure, integration and coupling, it is one of the core problems for i-EFV vehicle dynamics research to analyze the dynamic characteristics and mechanism of the "mechanical-electronic-hydraulic" coupling system composed of the "human-vehicle-infrastructure" generalized mechanical dynamics system, the new energy power system, and the electronic-controlled chassis system; to study unified modeling theories and methods for the complex electromechanical system, to find and establish unified and flexible models for the complex electromechanical system, with the aim to clearly express the system's coupling mechanism and describe the global properties of the system.

Building behavior model for the "human-vehicle-infrastructure" generalized mechanical dynamics system, first of all, requires an in-depth analysis on the driver, i-EFV, road characteristics and interaction mechanism. Therefore, the modeling study mainly focuses on the following two aspects.

(1) Analysis on movement and interaction mechanism of the "human-vehicle-infrastructure" complex system

When driving in complex road traffic environment, the i-EFV may be running multiple processes, such as assisted driving, integrated safety control, electric drive, hybrid drive and braking energy recovery. The driver may adopt different driving operations and even stress reaction in the time-varying traffic environment and various vehicle working modes. Meanwhile, the driver system, the vehicle system and the road system are coupling and restraining each other, so studying the movement mechanism of this kind of vehicle with complex system coupling is the foundation of realizing the integrated vehicle control.

A comprehensive analysis on vehicle movement characteristics related to i-EFV's safety, comfort, energy efficiency and environmental protection should focus on: (1) multi-body dynamics characteristics and comprehensive efficiency characteristics of the multi-source energy system under different driving cycles; (2) the i-EFV's longitudinal/lateral dynamic behavior pattern under driver-vehicle interaction and influence in the process of multi-degree-of-freedom vehicle movement; (3) the driver' longitudinal/steering behavior characteristics under different traffic environments and vehicle states; (4) the driver's behavior response characteristics caused by changes in vehicle and road traffic environment parameters and the driver's stress response characteristics under abruptly changed operation conditions; (5) nonlinear longitudinal/lateral kinematic coupling characteristics of the i-EFV and among vehicles as well as the mechanism of information and energy transfer of the driver-vehicle-infrastructure dynamics closed-loop system in the presence of the coupling of driver operations and changes of road condition.

(2) Modeling of the "human-vehicle-infrastructure" generalized mechanical dynamics behavior characteristics

Since i-EFV complex system dynamics control not only touches upon vehicle movement dynamics modeling, but also needs to consider the driver's driving habits, behavior characteristics, as well as road and traffic environment, therefore the integrated modeling of the i-EFV system involves the closed-loop generalized mechanical system consisting of the driver, the vehicle and the road traffic environment.

On the basis of dynamics coupling mechanism study, the research on the unified modeling of the "human-vehicle-infrastructure" complex coupling system should focus on: (1) exploration of the integrated system framework and joint modeling methods of different types of "human-vehicle-infrastructure" object systems; (2) driver behavior characteristics modeling based on experience and experimental data; (3) space equation description of complex dynamics behavior state considering time-delay, time-varying, coupling, multi-input, multi-output and strong nonlinear characteristics of the vehicle dynamics system; (4) mathematical modeling of the information flow and energy flow transfer and conversion of the "human-vehicle-infrastructure" generalized mechanical closed-loop coupling system; (5) global mathematical description of the dynamics system model of the "human-vehicle-infrastructure" generalized mechanical closed-loop coupling considering micro-road and macro-traffic stratification and uncertain relationship.

2.2.2.2 Comprehensive Dynamics Control of Multi-objective and Multi-system Coordination

Studies of regular control principle and decision-making of vehicle dynamics mainly focus on separate studies of each key performance, such as vehicle safety, fuel economy and emission behavior control, and operation noise and vibration control related to ride comfort. Although such researches can solve individual dynamics problems, it is quite difficult to collaboratively solve the dynamics synthesis problem in the presence of complex system coupling of the i-EFV, such as the collaboration of multiple objectives including safety, fuel economy and ride comfort with consideration of the driver, the vehicle and the road traffic, and the collaboration of multiple systems including the driving, braking, suspension and steering systems.

Aiming at the comprehensive control of safety, ride comfort, energy efficiency and environmental protection objectives, the i-EFV has a variety of driving modes, including longitudinal/lateral assisted driving and the integrated "human-vehicle-infrastructure" safety control with the vehicle and driving environment taken into consideration; it also integrates a complex "mechanical-electronic-hydraulic" coupling system.

Considering the i-EFV's characteristics such as controlled object coupling, various control functions and complex vehicle objects, how to treat "human-vehicle-infrastructure" as an organic whole, take complex system dynamics characteristics under multi-movement process and multi-system coupling into consideration, and realize the optimal dynamic control of multi-object and multi-system collaboration is another important subject in the dynamics control study of the i-EFV integrating complex electromechanical systems. This cooperative control requirement demands a complete i-EFV control system architecture and theoretical system.

There are three key focuses in the study of the two following scientific problems: the i-EFV multi-physical-process system's coupling mechanism and its dynamic behavior generation mechanism and modeling, as well as the theories and methods for the complex electromechanical system's multi-objective and multi-system cooperative control.

1. Longitudinal movement mechanism of the i-EFV "human-vehicle-infrastructure" electromechanical system with multi-physical-process and multi-system coupling

When driving in a complex road traffic environment, the i-EFV may be running multiple physical processes, such as assisted driving, electric drive, internal-combustion drive, hybrid drive and braking energy recovery. Since there are nonlinear coupling and restriction relationship among multiple driver-vehicle-infrastructure systems, studying vehicle movement mechanism based on complex system's multi-physical-process and multi-system coupling is the foundation of the comprehensive vehicle performance control. With focus on driving safety, fuel economy, comfort, and other important performance, this research explores longitudinal movement's dynamics behavior pattern under driver-vehicle interaction and inter-influence in the process of multi-degree-of-freedom, as well as driver's behavior response characteristics caused by changes in vehicle and road traffic environment parameters, on the basis of multi-body dynamics characteristics under different driving cycles, driver's operation behavior, dynamics characteristics of human-vehicle and vehicle-infrastructure coupling.

2. Movement state recognition and multi-domain unified modeling theory and method of the i-EFV "human-vehicle-infrastructure" generalized mechanical dynamics system.

Accurate recognition of the i-EFV "human-vehicle-infrastructure" generalized mechanical dynamics system state depends on the research of the identification method of driving characteristics, vehicle dynamics state, inter-vehicle dynamics state and road state, as well as the dynamics characteristics of steady and dynamic car-following behavior, approaching leading vehicle behavior and abnormal behavior. It also needs to study the model parameter adaption method of multi-system-fusion-based vehicle velocity estimation, vehicle yaw velocity redundancy estimation and state estimation, as well as the identification method of road friction coefficient, road curvature and other parameters with good instantaneity and high accuracy.

Since the i-EFV's complex system dynamics control not only involves the modeling of vehicle dynamics, but also needs to consider the driver's driving habits and behavior characteristics, as well as road conditions and traffic environment. Therefore for the closed-loop generalized mechanical system consisting of the driver, vehicles and the road traffic environment, the research should be focused on the "human-vehicle-infrastructure" complex coupling system's multidisciplinary unified modeling, besides the study of its dynamics coupling mechanism and dynamics characteristics identification. The research also needs to explore system integration, framework construction and joint modeling methods for different "human-vehicle-infrastructure" object systems; modeling of drivers' behavior characteristics based on empirical and experimental data; space equation description of complex dynamics behavior state considering time-delay, time-varying, coupling, multi-input/output and strong nonlinear characteristics of the vehicle dynamics system.

3. Multi-objective cooperative control method for the vehicle's longitudinal movement considering "human-vehicle-infrastructure" coupling characteristics.

The control of the i-EFV's movement needs to meet multiple mutually constraint and inter-coupling targets, such as safety, energy efficiency, environmental protection, and ride comfort. Therefore, the i-EFV control system must be able to realize real-time scheduling and coordination optimization while taking driver behavior characteristics, real-time vehicle state and road environment changes into consideration. Such a goal requires the study to be focused on the multi-system, multi-physical-process and multi-objective cooperative control structure with complex "human-vehicle-infrastructure" coupling characteristics that combines aggregation and dispersion, and explore the mathematical quantification of driving safeguard, acceleration performance, fuel economy and driver experience targets, and collaborative design of multi-performance optimal objectives and multi-performance objective collaborative control optimization method.

2.3 Key Technical Systems [2, 3] of i-EFV

The i-EFV involves the multi-sensor system, the hybrid electric drive system, the electrical-controlled braking system, the composite transmission system, the active steering system and the active suspension system, which support longitudinal/lateral driving assistance, integrated safety control, optimal energy distribution and other functions, and ultimately achieve safety, comfort, energy conservation and environmental protection. Compared with existing intelligent vehicles and new energy vehicles, the i-EFV is more complex and diverse in structure, function and objective. To address the difficulties of the i-EFV such as complex system structure, diverse identification information and control system coupling, we propose 3 key technologies: structure sharing, information fusion and control collaboration. Structure sharing addresses the hardware design problem of the i-EFV complex system's structural integration, optimization and functional security. Information fusion processes "human-vehicle-infrastructure" complex information recognition and prediction signals, and control collaboration coordinates multi-objectives of safety, comfort, energy conservation and environment protection, multi-driving mode of longitudinal and lateral driving assistance and integrated safety control, as well as integrated control of complex "mechanical-electronic-hydraulic" coupling's multi-system dynamics coordination, as shown in Fig. 2.6.

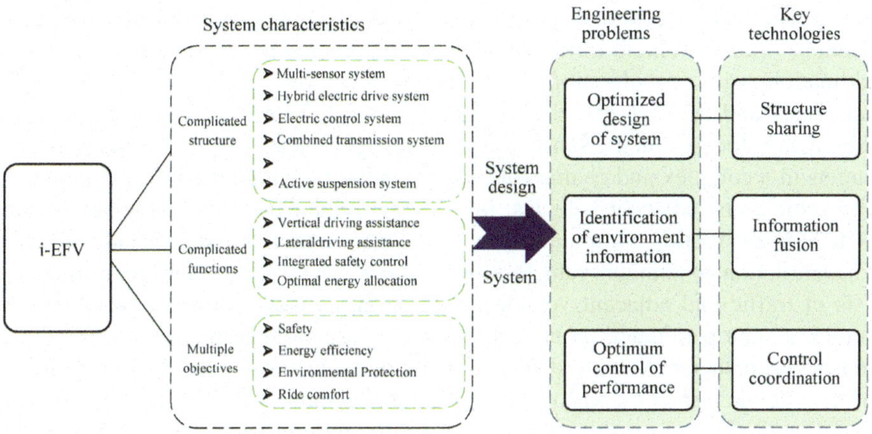

Fig. 2.6 Key technical systems of the i-EFV

2.3.1 *Structure Sharing*

As one of the important parts of sharing technology, structure sharing refers to the basic technology that delivers multiple system functions by using the same structure. By simplifying the redundant structure and optimizing the overall system scheme, structure sharing maximizes the system's functions and efficiency with minimum structures and costs. The i-EFV's complex execution and sensing systems require structural sharing and optimization, and key technologies involved include: (1) i-EFV execution units' structural sharing design for integration and optimization of "mechanical-electric-hydraulic" system structure; (2) composition and layout design optimization of the complex sensor system considering the redundancy of the i-EFV multi-sensor system; (3) sharing design of execution units and senor system in diverse driving tasks; (4) safety and reliability design of vehicle and its execution units under multi-system collaboration considering system structure and composition characteristics under structure sharing.

2.3.2 *Information Fusion*

Information fusion refers to spatial and temporal combinatorial optimization of complementary and redundant multi-source interactive information to produce a consistent description of the observed environment. A comprehensive and consistent identification of vehicle and environmental state provides the essential information basis for the i-EFV's control over safety and energy efficiency. To solve practical problems in the information extraction and fusion process, the i-EFV's information fusion technology is divided into three levels: target information identification, "human-vehicle-infrastructure" characteristics extraction and vehicle state estimation, with consideration of intelligent interactive information system characteristics and demand of vehicle optimal control. Target information identification mainly focuses on the sensor system's recognition accuracy over measurement targets in a complex and changeable environment, including the information of the ego vehicle and adjacent vehicles, information obtained from vehicle-vehicle and vehicle-infrastructure communication, traffic information of roads ahead and so on. The extraction of "human-vehicle-infrastructure" characteristics refers to the analysis of traffic and adjacent vehicle information, as well as the fusion of driving vehicle's state parameters after identifying target information based on the multi-sensor fusion theory. Vehicle state estimation refers to the assessment and prediction of the vehicle's safety, comfort, energy efficiency and environment performances based on the analysis results of collected information by using the i-EFV's system model, which is used as the target input of the control process to predict the future driving process.

2.3.3 Control Collaboration

Control collaboration aims to achieve the optimal i-EFV performance by integrating multiple objectives including safety, energy efficiency, environmental protection and ride comfort, and coordinating multiple systems including driving, braking, steering and suspension. It needs to solve two basic problems: the coupling mechanism of the i-EFV's complex electromechanical system and the modeling and analysis of the "human-vehicle-infrastructure" generalized mechanical dynamics behavior; the integrated dynamics control of the i-EFV's multi-objective and multi-system coordination.

The "human-vehicle-infrastructure" generalized mechanical dynamics behavior modeling is to address the dynamics behavior mechanism and characteristics of the i-EFV's complex electromechanical system in the context of multi-system coupling and the resultant dynamic problems, which are still unclear so far. With focus on the movement and interaction mechanism of the "human-vehicle-infrastructure" complex system, the research will include the modeling of the "human-vehicle-infrastructure" generalized mechanical dynamics system based on existing analysis results.

Structure sharing, information fusion and control collaboration constitute the key technology systems for the system design, the vehicle and environment information identification and the integrated optimal control of the i-EFV. These key technologies are not independent from each other; instead, they form a circular and optimized organic whole that is interrelated and mutually supporting. As shown in Fig. 2.7, structure sharing technology focuses on the optimum structural design of the execution system and sensing system, and the design process is based on the information

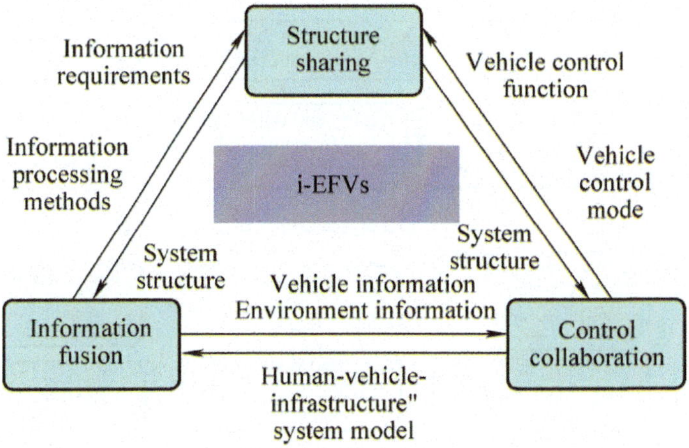

Fig. 2.7 Diagram of the i-EFV key technologies' coupling relationship

identification demands and information processing methods provided by information fusion, as well as vehicle control functions and control modes provided by control collaboration. Information fusion processes the information of the i-EFV's complex system and the environment, which requires an optimized system structure supported by structure sharing and "human-vehicle-infrastructure" system model and vehicle control characteristics supported by control collaboration. Control collaboration focuses on the comprehensively optimized control of the system, which needs system structure features supported by sharing structure and vehicle and environment information supported by information fusion.

To achieve the objectives of safety, energy efficiency, environmental protection and ride comfort, the i-EFV needs to have an improved structural design and optimized sensing, execution and other complex electromechanical systems, complex environment information identification during operation, and the optimal control of vehicle performance, which rely on the technical support of sharing structure, information fusion and control collaboration. While development of these key technologies depends on exploring the behavioral mechanism of the complex system dynamics of the i-EFV "human-vehicle-infrastructure" generalized mechanical dynamics system with multi-physical-process and multi-system coupling, the multi-system and multi-objective cooperative control theory for complex electromechanical systems, the design theory of vehicle sensing and execution system structure sharing, the multi-source sensing information fusion theory and other interdisciplinary fundamental scientific problems. The framework of basic theories and key technologies for the i-EFV is shown in Fig. 2.8.

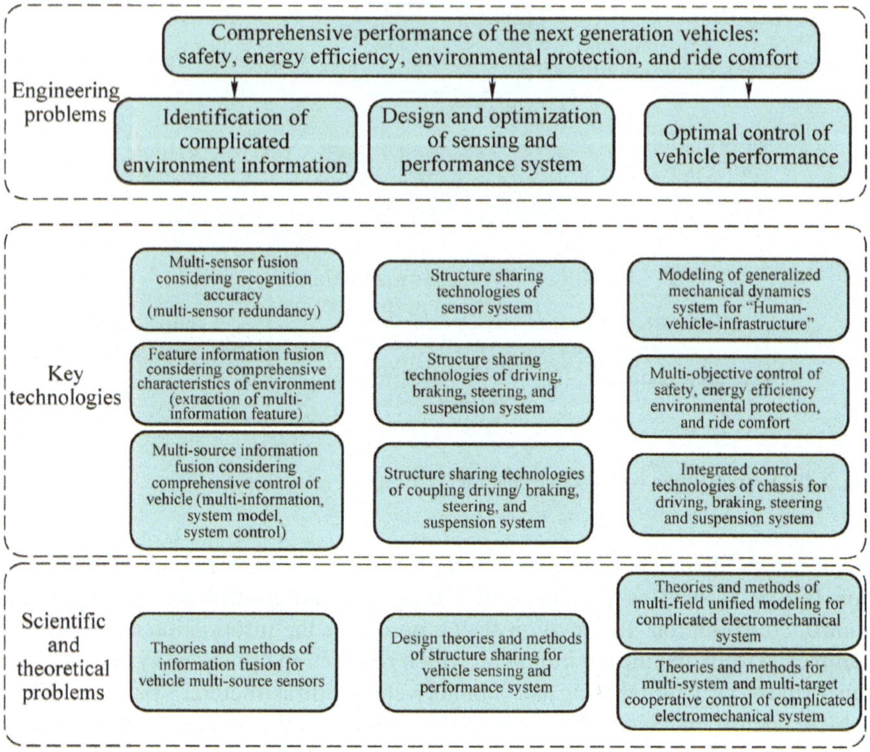

Fig. 2.8 Framework of basic theories and key technologies for the i-EFV

References

1. Chen T (2012) Intelligent environmental-friendly vehicle and its adaptive cruise control. Tsinghua University Press, Beijing
2. Li K, Chen T, Luo Y et al (2010) Intelligent environmental-friendly vehicle—concept, architecture and engineering implementation. Automot Eng 32(9):743–748
3. Li K, Zhang S, Luo Y et al (2013) The concept of intelligent environmental-friendly vehicle and its latest development. J Autom Saf Energy Conserv 4(2):109–120

Chapter 3
Structure Sharing Technology for i-EFV

As a typical complex electromechanical system [1, 2], i-EFV integrates mechanical, hydraulic, electronic, electric, sensing, communication, and control subsystems. Although the adoption of traditional functional superposition for its system integration solution occupies the advantage of simple design, such as an intelligent driving system can be directly superimposed on an EV, a continuous upgrade of intelligent driving system leads to more subsystems and complex system structure with inter-coupling and overlay of functions and structures. Especially with increasing type, model, and quantity of key components, integration by superposition not only causes structural redundancy, waste of resources and increasing system cost, but also makes it difficult to further improve overall system performance, due to inadequate use of key components. In order to optimize the i-EFV structure and improve resource utilization, it is proposed to develop optimal integration technology based on structure sharing, with the focus on cooperative optimization of multi-function integration and resource sharing of sensing, control and actuation to deliver optimized structure and function.

3.1 Overview of Structure Sharing Optimization and Integration Technology

3.1.1 Concept of Structure Sharing Integration

The concept of structure sharing was defined in early studies as "using the same physical structure to implement multiple system functions". As one of the four sharing technologies in the conceptual design of product development [3–5], it reflects the corresponding relationship between the number of structures and the number of functions in a system. The other three sharing technologies are function sharing,

K. Li et al., *The Intelligent Environment Friendly Vehicle*,
Key Technologies on New Energy Vehicles,
https://doi.org/10.1007/978-981-19-4851-0_3

function redundancy, and multi-mode integration. Function sharing refers to using several physical structures to achieve the same function. Function redundancy is to realize the same function by using several interchangeable structures. Multi-mode integration means that the same physical structure can realize interchangeable functions. Definitions and differences of these four types of technologies are shown in Fig. 3.1. The structure here refers to a physical structure composed of various internal elements, including single component and complex hardware devices or systems consisting of multiple components. Function refers to the ability of a system to act on the environment, and all physical structures need to perform their function in a specific environment. Conceptually, as a kind of product design idea, structure sharing can optimize system hardware composition, reduce the cost of system hardware and improve system resource utilization, and therefore it is more suitable for overall design and optimal integration of complex systems.

Starting from the original concept of structure sharing, it is not difficult to find that we have implemented the idea of structure sharing consciously or unconsciously in the design of vehicle systems. For example, the driving motor in EV can be used for vehicle driving, braking, and energy recovery. The engine in conventional internal combustion engine vehicles can be used for driving and auxiliary braking, and the heat dissipated from the engine can be used to heat the cab. In an intelligent hybrid electric vehicle, forward radar detects preceding vehicles, which can be used not only for the

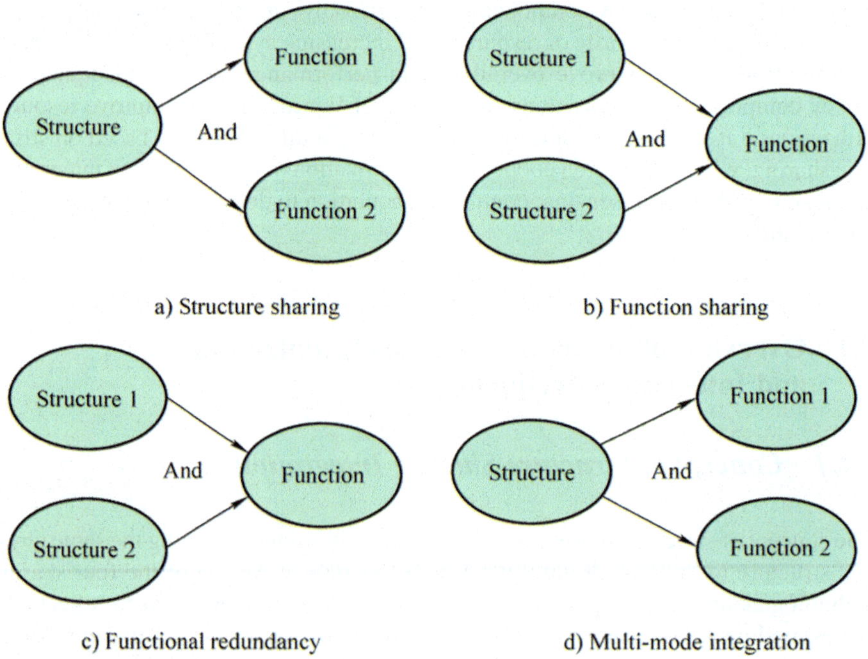

Fig. 3.1 Definition diagrams of four types of sharing technologies [2]

vehicle's active safety control but also for energy management and optimal energy-saving control. All the examples reflect that a single structure can deliver multiple system functions, which is actually an application of the structure sharing concept. The examples also show that no matter using motor energy flow in vehicle driving and braking or using information flow from forward radar in vehicle active safety and energy-saving control, the structure of hardware device remains unchanged, and multi-function is achieved by just making full use of hardware devices in different environments.

When introducing the concept of structure sharing to intelligent EV, we expand the original concept and summarize application cases to provide the definition of structure sharing integration of intelligent EV: an optimal integration process and method that makes full use of physical properties, including information flow, energy flow, and basic machine, of a hardware structure in an intelligent EV's integrated system, to deliver multiple functions or expand its function application in different subsystems, and therefore optimize system structure and improve overall performance. This defines the concept of structure sharing integration within the scope of intelligent EV system integration and avoids reconstruction of original chassis design. Defining structure sharing as an optimal integration method answers the question "what is structure sharing?" This definition keeps the part of "using the same structure to deliver multiple functions" in the original definition, and extends the implication of physical structure according to characteristics of system integration, which means it includes not only mechanical structure in a narrow sense, but also information flow and energy flow generated by basic machine (or hardware device), and based on which "different functions of the same structure in different subsystems" is also considered as structure sharing. Keywords involved in this definition mainly include intelligent EV, function integration, physical structure, information flow, energy flow, structure optimization, function expansion, and integration method.

Structure sharing is conceptually similar to function integration. Integration refers to the process in which some isolated things or elements change their original scattered state in some way and get together to form an organic whole. Function integration is the integration of several isolated function units to form a multi-functional body. In this sense, the common ground between structure sharing and function integration is that both of them can deliver multi-functionality, reduce system size and quality and improve resource utilization. The difference between the two is that structure sharing is to make full use of physical characteristics of existing hardware structures to realize multiple functions, while function integration is to combine multiple hardware structures organically through function merging and complementation. The former expands the original function without changing the main structure or affecting the original function, while the latter may change the original structure in the process of combination, and therefore decrease original functional features. In summary, the process of function integration contains the idea of structure sharing, and structure sharing can optimize function integration. Therefore, structure sharing integration is a method of system optimization and integration.

Structure sharing and structure optimization are similar in achieving their goals. Structure optimization design is to find the best design scheme according to some optimization objectives under given constraints. In the narrow sense, structure optimization generally refers to the optimization of the structure itself, and structure optimization of complex systems also considers the shape and combination of the structure. Structure sharing integration is to optimize and combine key components by merging repeated structures and reducing structure redundancy, without changing the structure of key components in the system. Therefore, structure sharing can be classified as a generalized structure optimization method. Structure sharing integration can be applied to structural optimization of complex systems, by simplifying and optimizing system composition to improve resource utilization and reduce system cost.

3.1.2 Connotation of Structure Sharing Integration

According to the concept of structure sharing integration, its application in intelligent EV mainly includes two aspects. The first is that different subsystems share the same physical structure or key components to reduce configuration redundancy and optimize structure; the second is to extract physical characteristics of some hardware structure in the system and reuse them in other subsystems, to improve overall system performance without adding extra hardware. Therefore, two methods for structure sharing integration in intelligent driving systems are proposed: combinatorial optimization method and function expansion (extension) method.

(1) Implementation of structure sharing integration based on combinatorial optimization

Combinatorial optimization, also known as combinatorial programming, is a kind of mathematical programming to find an optimal subset in all the subsets with certain characteristics in a given finite set according to a certain object. It uses mathematical methods to manage optimal arrangement, grouping, order, or screening of discrete events. Combinatorial optimization, in essence, is to select the best state from a finite number of discrete states and find the optimal solution in a finite set of feasible solutions. Combinatorial optimization methods have been widely used in the optimum structural design of complex systems and can also be applied to structure sharing integration of EV intelligent driving system.

EV intelligent driving system integrates complex control functions. In the conventional integration design of function superposition, each subsystem independently selects, arranges, and designs hardware based on its own functional needs. However, with the constant increase of system functions and complexity of system structure, structure redundancy of individual subsystems inevitably appears during independent design and superposition. Structure sharing integration based on combinatorial optimization can effectively integrate redundant resources, guarantee the overall

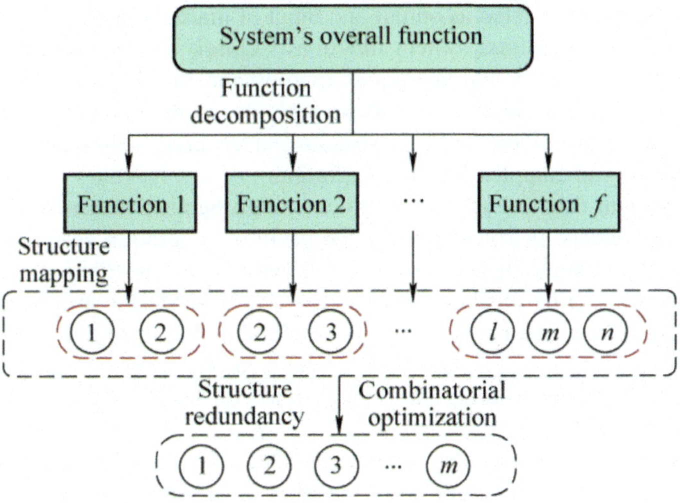

Fig. 3.2 Structure sharing integration based on combinatorial optimization

performance of the system and achieve the best cost performance, as shown in Fig. 3.2.

The basic idea of structure sharing integration based on combinatorial optimization is to start from the system's overall function, and then follow the technical route of "functional requirements definition → functional decomposition → structure mapping → structure combinatorial optimization → structure sharing". Under the premise that the needs for the overall function of the system are met, the overall function is decomposed top-down to individual function units. T hen the structural units corresponding to sub-function modules are combinatorially optimized and structures with functionally overlapping structures are merged to achieve sharing structure. In this process, a large number of structural units formed after function decomposition and structure mapping can be regarded as a finite set of feasible solutions, so the structure sharing problem can become the combinatorial optimization problem and mathematical methods can be used to obtain the optimal solution, which is the system composition after structure sharing integration.

This method can reduce structure redundancy, simplify structure composition and reduce system cost by combining structures with repeated functions, and therefore it is suitable for the optimum structural design of EV intelligent driving system.

(1) Implementation of structure sharing integration based on the function expansion

Function expansion, also known as function extension, refers to the reuse of a physical structure from the original system in other subsystems or new environments to show new functions or performances. Physical structure not only refers to mechanical hardware structure but also includes information flow and energy flow generated by the physical structure. The system function is the effect and capability that the system shows in a certain operating environment. It is the external connection of the system, which is manifested as output and input of material flow, information flow, and energy flow. Therefore, system function is closely related to elements of the system, system structure, and operating environment. The same physical structure may perform different functions in various systems or environments. Reusing these physical characteristics of hardware structure in different subsystems can realize structure sharing in complex system integration.

EV intelligent driving system integrates a large number of hardware structures and key components. In conventional superposition integration design, these physical structures belong to different subsystems and take on specific functions in their subsystems. Extracting physical characteristics of a hardware structure, including mechanical structure, energy flow, and information flow, and applying them to other subsystems may reveal new functions or applications, and thus improve system performance. Structure sharing integration based on function expansion is shown in Fig. 3.3.

The basic idea of structure sharing integration based on function expansion follows the technical route of "structure decomposition of original system → extracting physical characteristics from structure unit → function expansion in a new system or operating environment → structure sharing", which makes full use of physical characteristics of hardware structure in the original system without adding any physical structure and expands such physical properties to other subsystems to enhance overall system performance after integration.

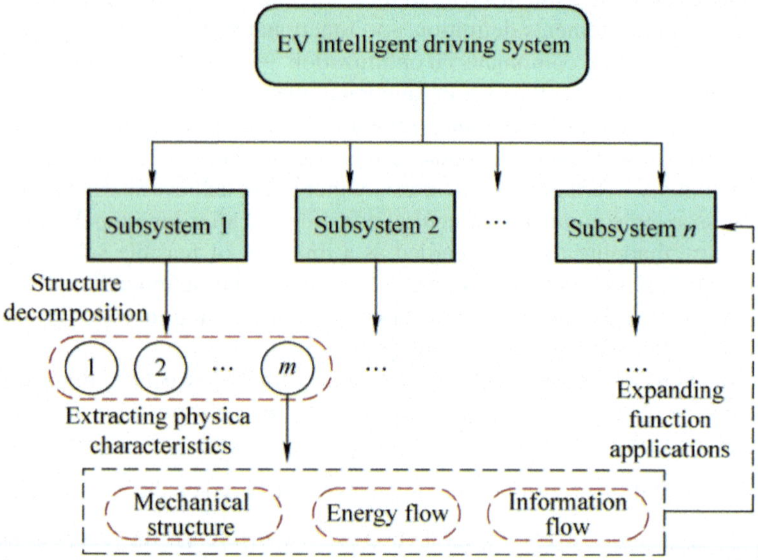

Fig. 3.3 Structure sharing integration based on the function expansion

This method improves overall system performance by expanding function application of physical structures, and therefore it is suitable for function integration and optimization of EV intelligent driving system.

3.2 Structure Sharing Technology Applied in the Vehicle System

As a basic technology that uses the same structure to realize multiple system functions, structure sharing is one of the important research contents of sharing technology [3]. By reducing resource utilization and system size and weight, structure sharing can effectively improve hardware efficiency and reduce the cost of mass production, especially for a system with complex structures, and therefore it has attracted growing interest at home and abroad [4].

The idea of structure sharing came from program design, which used structure sharing to simplify complex algorithm code to compress its memory space and improve operating efficiency. Later, structure sharing, as a method of structural optimization design, was gradually extended to the design of some products with simple structures, such as pressure sensors and acceleration sensors [6–8]. Up to now, there are few studies on basic theories and methods of using structure sharing technology in the optimal design of mechanical systems with complex structures. I-EFV with comprehensive function and complicated system integration demands requires research on structure sharing theory and its application technology, to simplify and optimize vehicle hardware system and achieve the most system functions and highest efficiency with the least structure and lowest cost. It is of great importance to drive the advancement of future automotive technology and provide a theoretical foundation for structure optimization design in other manufacturing industries.

From the perspective of system security, for i-EFV with complex structures, optimal design of different components or different accessories in the same component changes spatial distribution, operation mode, and control mode of each subsystem to a great extent, and resultant system security problems will become more prominent. Therefore, it is necessary to study further the functional safety of the i-EFV system under structure sharing theory.

Considering characteristics of i-EFV system structure and composition of control function, i-EFV structure sharing technology involves the following research contents:

(1) Structure sharing design of the multi-sensor system

The key focus is to develop an optimal design for the composition and layout of a complex sensor system to solve redundancy in the i-EFV multi-sensor system.

(2) Structure sharing design of vehicle control

The key focus is to develop a sharing design for actuation components and sensor systems in different driving tasks.

(3) Structure sharing design of actuation system

The key focus is to develop integration and optimal design for complex "mechanical-electric-hydraulic" system structures, as i-EFV has complex actuator structures.

(4) System functional safety design under structure sharing

The key focus is to develop safety and reliability design for the vehicle and its actuation components under multi-system cooperation, based on system structure and composition characteristics.

3.2.1 *Structure Sharing of a Multi-sensor System*

There is a lot of redundant information in multi-sensor and communication system, including vehicle-to-vehicle communication, vehicle-to-infrastructure communication, remote wireless communication and communication between on-board sensors, which provides the possibility to further reduce the number of sensors, optimize the combination of sensor systems, realize structure sharing of sensor systems and reduce their cost.

Structure sharing design of the multi-sensor system is very similar to that of actuation components in terms of process, as shown in Fig. 3.4: (1) comprehensive analysis of i-EFV's information identification requirements on vehicle components, vehicle movement and driving environment; (2) decoupling the above information into original sampling signals corresponding to vehicle-to-vehicle communication, vehicle-to-infrastructure communication, remote wireless communication and on-board sensor communication; (3) redundancy analysis about sensing and communication system with the help of information fusion technology, as well as reliability analysis and cost evaluation, for simplification and optimization of the sensing system.

Using structure sharing theory to optimize sensor type, number and location distribution can effectively improve the utilization rate of sensing information and reduce sensor system cost, but there are very few relevant studies at the moment. Based

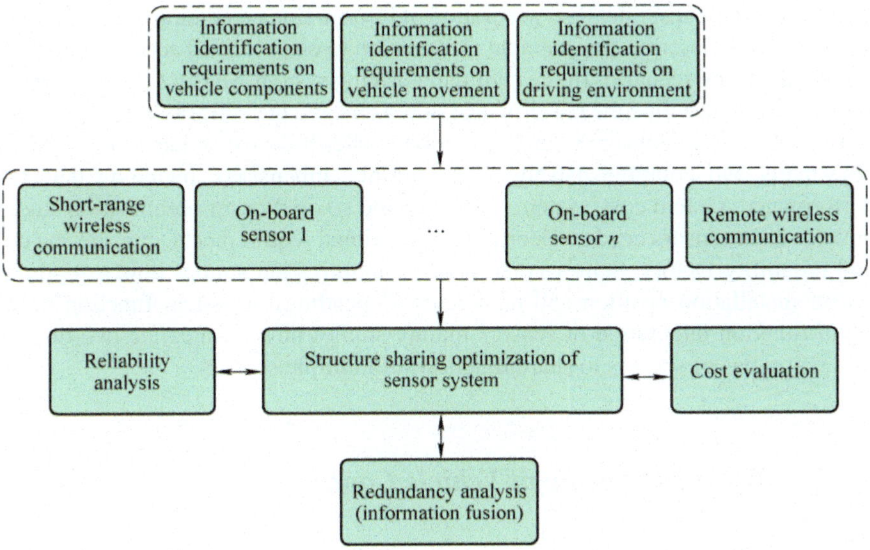

Fig. 3.4 Structure sharing design flow of the multi-sensor system

on functional requirements of the intelligent driving system and physical properties of multi-sensor, and also with consideration of cost and reliability of sensors, structure sharing optimization of multi-sensor system studies composition, type and arrangement of sensors in the intelligent driving system, to optimize sensor system composition, reduce the number of redundant sensors and decrease total cost of the sensor system.

Key research contents of structure sharing optimization of the multi-sensor system are shown in Fig. 3.5, including: (1) generation of information perception demand; (2) sensor system modeling and evaluation; (3) optimal sensor configuration; (4) multi-function sensor integration.

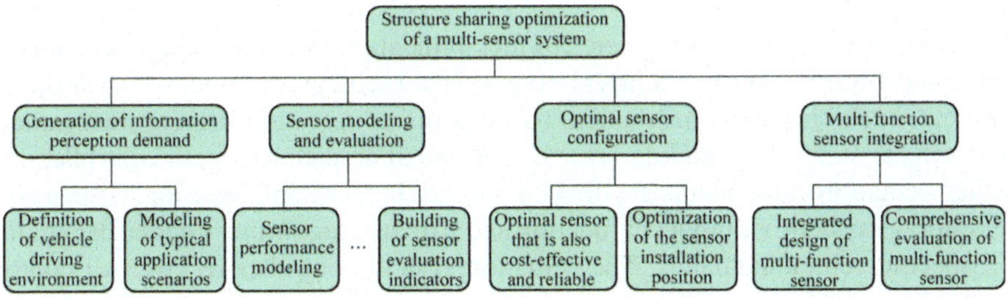

Fig. 3.5 Key research contents of structure sharing optimization of a multi-sensor system

As for technical application, generation of information perception demand mainly studies how to describe environment information to be perceived according to functions and driving scenarios of the intelligent driving system. Sensor system modeling studies how to describe the performance of sensors and the entire system after multi-sensor integration. Sensor system evaluation studies how to evaluate sensor systems objectively and comprehensively from multiple dimensions, including detection performance, system cost, system reliability and so on. Sensor optimization studies how to select sensors needed according to functional requirements, considering cost and reliability factors. Sensor configuration optimization studies how to optimize sensor installation position and parameters depending on system function needs. Multi-function integration of sensors mainly studies how to integrate two or more different types of sensors to share functions of multiple sensors.

3.2.2 Structure Sharing of Vehicle Control

Longitudinal/lateral assistant driving, real-time "human-vehicle-infrastructure" integrated safety control and a variety of other driving tasks require control of driving, braking, steering, suspension and other actuation systems. For example, an active steering system is required to be used in integrated safety control and lateral assistant driving. The electro-hydraulic composite braking system must fulfill braking requirements of vehicle stability control and adaptive cruise control. Since the i-EFV driving system has complex tasks and system structure, it is required to design an optimized component operation and control network considering system reliability to meet the demand of diversified driving modes on system component control. It is of great significance to simplify vehicle structure, reduce system control difficulty and extend system function.

Based on characteristics of the control-by-wire system, it is proposed to use the system structure of "integrated-distributed" actuation and control as a reference to the design of vehicle control structure sharing, as shown in Fig. 3.6. The designed structure can effectively enable sharing of actuation components and sensor systems in driving tasks.

Vehicle driving tasks are delivered by hierarchical control in the designed actuation and control system. The upper layer is a centralized vehicle controller which develops optimized vehicle control instructions based on driver operation, driving environment and vehicle states. The middle layer is distributed actuation system controllers, in which actuation components are divided into driving, braking, steering, suspension, and storage system. Component function and structure are further divided depending on specific vehicle demands. The actuation system controller receives all control instructions from the bus and controls the operation of components. The bottom layer is the actuation components of each system, which adopts a unified interface definition and carries out operations according to instructions from controllers in the middle layer.

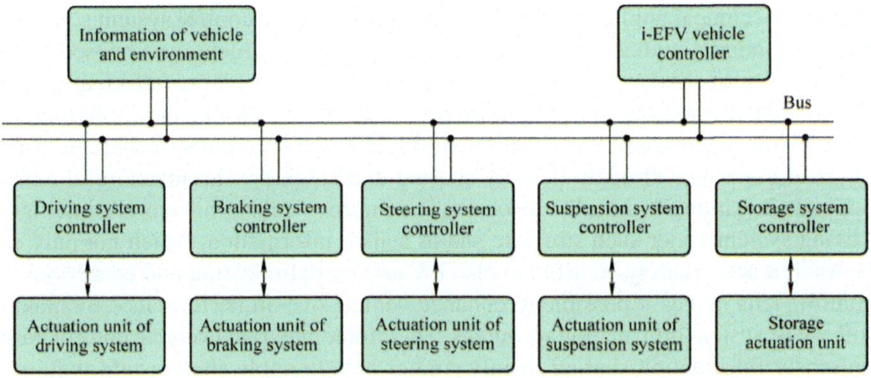

Fig. 3.6 Topological structure of "integrated-distributed" actuation and control system

At present, conventional integrated chassis control is a typical example [9] to realize control over the same actuation component in different driving tasks by applying centralized vehicle control and control-by-wire and distributed components control, as shown in Fig. 3.6. For i-EFV with complex sensor and control system structure, design of vehicle components in each layer and its control system with consideration of reliability and safety still needs further in-depth study.

Structure sharing optimization of vehicle control is proposed in response to complex driving tasks and integrated control requirements, which mainly studies optimal design of vehicle control network based on structure sharing, controller resource sharing under different functions and sharing of actuator and sensor systems, to meet system component control demands under diversified driving modes, and meanwhile balance system operation reliability and improve resource utilization. Structure sharing optimization of vehicle control mainly studies (as shown in Fig. 3.7): (1) optimal design of vehicle control system's network structure; (2) controller resource sharing and optimal allocation; (3) integrated vehicle control based on sensor information sharing.

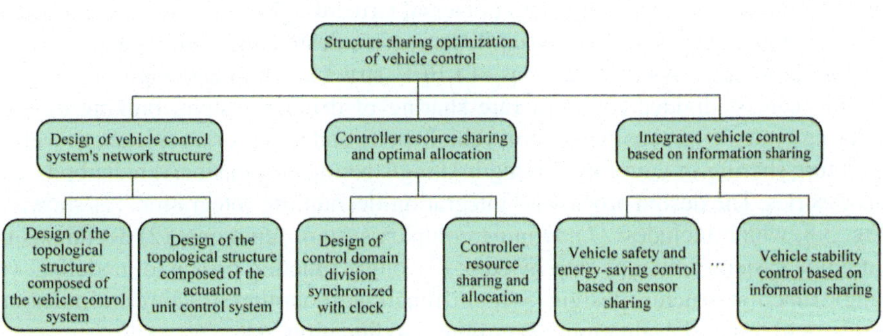

Fig. 3.7 Research contents of structure sharing optimization of vehicle control

At the technical application level, the design of vehicle control system's network structure mainly studies how to design the topological structure composed of the vehicle control system and actuation unit controllers to ensure effective control of distributed actuators in different driving tasks. Controller resource sharing and optimal allocation study how to allow multiple processing units to share the same physical controller through memory sharing and clock synchronization, to realize optimal allocation of controller resources and improve utilization rate. EV intelligent driving system under such structure shares sensor information, which not only can be used in active safety control but also EV energy optimization and energy-saving control. This makes it possible to enhance vehicle overall performance by making full use of system information resources. Therefore, integrated vehicle control based on sensor information sharing mainly studies how to apply sharing information to vehicle safety, energy conservation and stability control, without any extra hardware cost.

3.2.3 Structure Sharing of Actuation System

As i-EFV integrates complex control functions, the "mechanical-electronic-hydraulic" coupling structure has been widely used in the design of various vehicle systems. In conventional vehicles, each system is designed independently based on vehicle driving function, but system structure becomes increasingly complex along with the growing number of vehicle functions, and a large number of redundant structures appear in the design and superposition process of each system. Structure sharing of actuation components can effectively reduce vehicle design cost and manufacturing cost with optimal design of "mechanical-electronic-hydraulic" structures in the driving system, braking system, storage system, steering system, suspension system, vehicle body and accessories, and structure sharing of underlying actuation components.

The basic idea of structure sharing design of actuation system is to decompose system's corresponding functions and each function's corresponding components, analyze the universal principle and universal structure of each system, components and component substructures for unified structure design and optimization.

Based on basic system functions of i-EFV, structure sharing design of actuation system can be divided into structure sharing of driving system, braking system, steering system, suspension system, vehicle body and accessories system, as well as structure sharing of coupling driving/braking/steering/suspension/vehicle body and accessories. The design process is "integration-decoupling-integration", as shown in Fig. 3.8, which includes: (1) defining complex system functions; (2) decomposing complex system functions step by step; (3) single function principle modeling; (4) single function structure design; (5) multi-functional coupling structure optimization; (6) system test and evaluation; (7) circling modification.

Structure sharing of actuation systems is of great significance to reduce system cost and promote vehicle commercialization. Figure 3.8 shows the design process

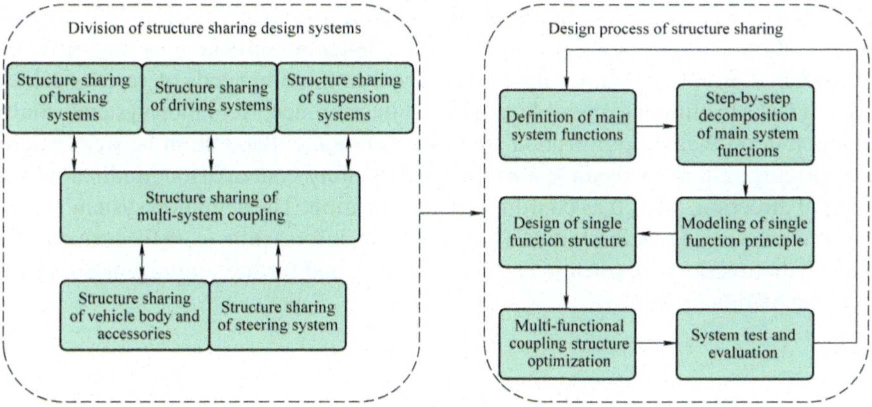

Fig. 3.8 Design process of actuation system structure sharing

and basic ideas, but currently theoretical research of structure sharing optimization design for i-EFV complex "mechanical–electronic-hydraulic" coupling system is still in the initial stage. The basic idea of the actuation system's structure sharing design is to decompose the system's corresponding functions and each function's corresponding components, analyze each system's universal principle and universal structure, components and component substructures for unified structure design and optimization. Through "integration-decoupling-integration", structure sharing effectively designs and optimizes driving system, braking system, storage system, steering system, suspension system, vehicle body and accessories system and other "mechanical-electronic-hydraulic" structure integrated by EV. Structure sharing of underlying components of actuation function can effectively reduce vehicle manufacturing costs.

Research on structure sharing optimal design of actuation system includes (as shown in Fig. 3.9): (1) actuation system's operation instruction coordination; (2) actuation system's multi-function coordinated optimization; (3) actuation system's information flow sharing.

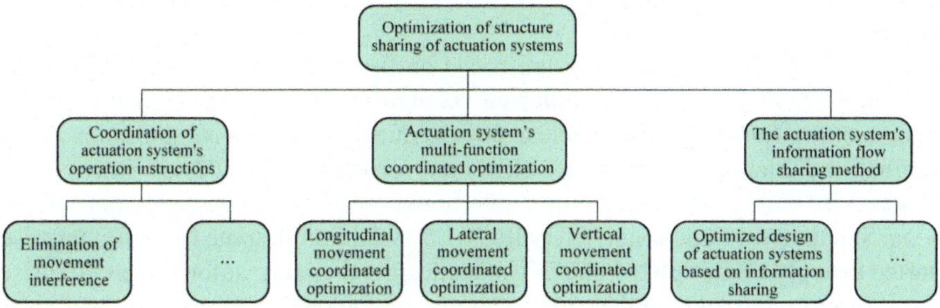

Fig. 3.9 Research on structure sharing optimal design of actuation system

At the technical application level, the actuation system's operation instruction coordination mainly studies how to coordinate operation instructions from different functions of an intelligent driving system according to the needs of various vehicle driving tasks to eliminate possible movement interference. Actuation system's multi-function coordinated optimization studies functional collaboration between intelligent driving system and vehicle's original active safety system in longitudinal/lateral/vertical direction, as well as coordinated optimization. The actuation system's information flow sharing method studies how to extract information from the actuator, and expand its functional application as added sensor, and further optimize the structure of the actuation system.

3.2.4 System Functional Safety Design Under Structure Sharing

The i-EFV system with structure sharing design in sensor, control and actuation system is characterized by a high degree of system integration, strong coupling, structured system and control-by-wire, which not only greatly increases structure utilization and degree of structure sharing, but also puts forward higher requests for function safety of each component in i-EFV, as any control error of a small part may result in multiple system security problems, and therefore affect vehicle safety. To reduce system safety problems caused by structure sharing design, it is required to develop functional safety (FS) of the i-EFV system under structure sharing design.

As a part of the overall system safety, functional safety ensures safety in case of vehicle or external environment errors, through system structure, sensor and control system hardware and software design based on the component failure and system safety loophole analysis, as well as components' sensor, diagnosis and handling of dangerous situations.

The basic design process of functional safety includes [10]: (1) decomposing system safety function and failure mode by applying FTA (Fault Tree Analysis) theory; (2) evaluating SIL (Safety Integrity Level); (3) system design based on functional safety and failure mode; (4) comprehensively evaluating SIL based on MTBF (Mean Time Between Failures) and SFF (Safe Failure Fraction); (5) functional safety testing.

As for step (3), i.e. system design based on functional safety and failure mode, there are established incomplete design standards overseas in aerospace, medical equipment, national defense, nuclear power, railway, automotive and other fields. International Standards Committee (ISO) put forward ISO26262, the international standard for functional safety in the automotive field, based on European International Standard for Functional Safety IEC 61,508. Although some foreign enterprises started to apply FS design to vehicle system design, domestic automobile enterprises have not mastered and applied this technology. In addition, FS analysis and corresponding system design under i-EFV complex system functions and structure sharing

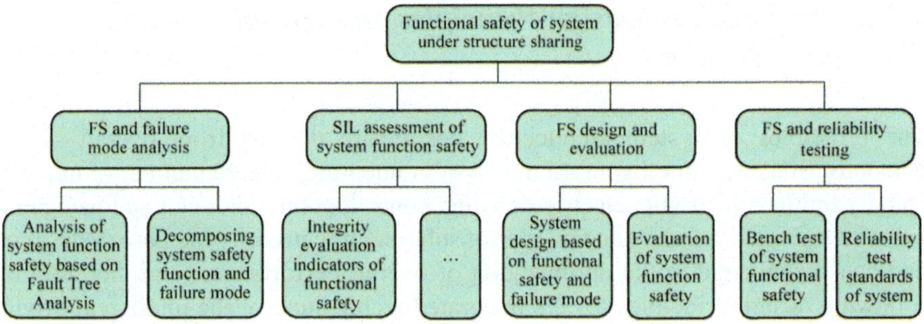

Fig. 3.10 Research on FS design under structure sharing

optimization still need further study. Research on FS design under structure sharing mainly includes (as shown in Fig. 3.10): (1) FS and failure mode analysis method; (2) SIL assessment method; (3) FS design and evaluation method; (4) FS and reliability testing method.

At the technical application level, FS and failure mode analysis research focuses on using FTA theory to decompose FS and failure mode; SIL assessment studies comprehensive evaluation of SIL based on MTBF and fault rate; FS design and evaluation studies methods to improve functional safety integrity and its effect evaluation; FS and reliability testing study mainly focuses on methods to test functional safety and reliability of the system after structure sharing integration, including bench test and real-life vehicle test standards.

3.3 Architecture Design for the Integrated Intelligent Driving System Based on Structure Sharing

The main task for architecture design of an intelligent driving system is to define the main elements (components) of the system and their functional requirements, and to plan logical relationships of the elements. This section confirms principles and overall composition of system architecture design according to intelligent driving system's optimization and integration requirements and implementation method of structure sharing integration; carries out functional partitioning and structural decomposition with the idea of "decomposition-integration-feedback"; uses the method of system analysis and deductive reasoning to design subsystem structure and plan the logical relationship of each element in subsystem; analyzes and summarizes features of the new architecture compared with the existing architecture.

3.3.1 Architecture Design Principles and Overall Composition

The principles of system architecture design are derived from overall system demands. Since EV intelligent driving system integrates a large number of mechanical, hydraulic, electronic, electric, sensing, control, communication and navigation subsystems, inter-coupling and overlap of subsystems can easily increase system cost and cause challenge to the optimization of overall performance, and therefore the following design principles for the integrated architecture of the intelligent driving system are proposed:

(1) The principle of optimal overall performance

Optimal integration of the system does not simply superpose subsystems, but optimizes overall system performance and achieves optimal system structure by optimizing and combining subsystem elements. The architecture of an intelligent driving system with shared structures is to improve overall system performance by making full use of system resources.

(2) The principle of minimum resource cost

Cost reduction is one of the objectives pursued by product design, and the premise of industrialization of the intelligent driving system. Making full use of or reusing existing system resources can improve resource utilization and save the cost of resources. An intelligent driving system architecture based on structure sharing integration aims to improve resource utilization and minimize resource cost with shared structure.

The idea of "decomposition-integration-feedback" adopted based on the above design principles will decompose vehicle intelligent driving system into subsystems by function which contain elements such as sensor, controller, actuator and other key components. Decomposed elements and components are re-integrated by function into the sensor system, the control system and the actuation system, corresponding to the sensor layer, the control layer and the actuation layer in system architecture. Communication and navigation devices are regarded as virtual sensors and therefore classified as sensor layers. The architecture of an EV intelligent driving system with a shared structure is shown in Fig. 3.11.

According to Fig. 3.11, the integrated architecture of an intelligent driving system with a shared structure includes sensor information sharing, controller resource sharing and actuator operation co-management. All sensor information flows are connected to vehicle control and actuation system through information sharing channels to facilitate sharing of system resources. The architecture design of the sensor layer, control layer and actuation layer will be discussed in the following sections.

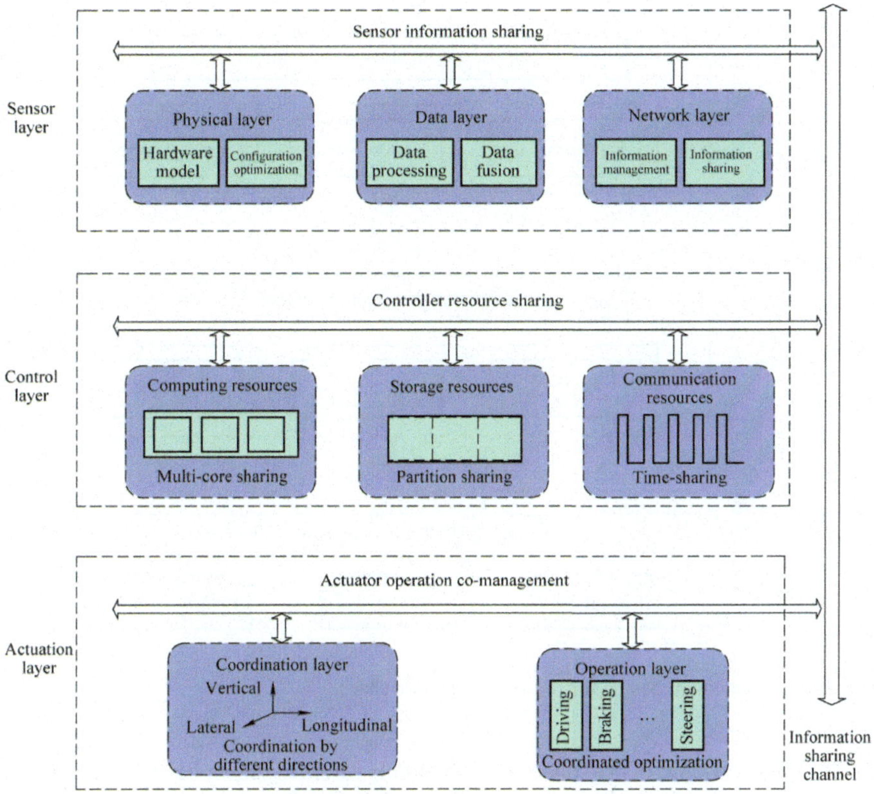

Fig. 3.11 Architecture of EV intelligent driving system with shared structure

3.3.2 Architecture Design for Sensor Information Sharing

In the superposition architecture of the existing intelligent driving system, sensors are selected and positioned according to functional requirements of their subsystems, without consideration of multi-sensor information sharing. With the continuous upgrading of intelligent driving function and increasing functional requirements, the number of subsystems and the type and number of sensors are also on the rise. Superposition architecture of sensors not only causes functional overlap and structure redundancy but also leads to low utilization of information resources.

Architecture design for sensor information-sharing aims to achieve maximum utilization of information resources by optimizing composition and configuration of the sensor system and sharing the information flow of sensors. For on-board sensor systems, sensor resources mainly include sensor hardware resources, data detected by sensors, and network resources formed after sharing of sensor information. Therefore, the architecture of the on-board sensor system based on information sharing is divided into the physical layer, data layer and network layer, as shown in Fig. 3.12.

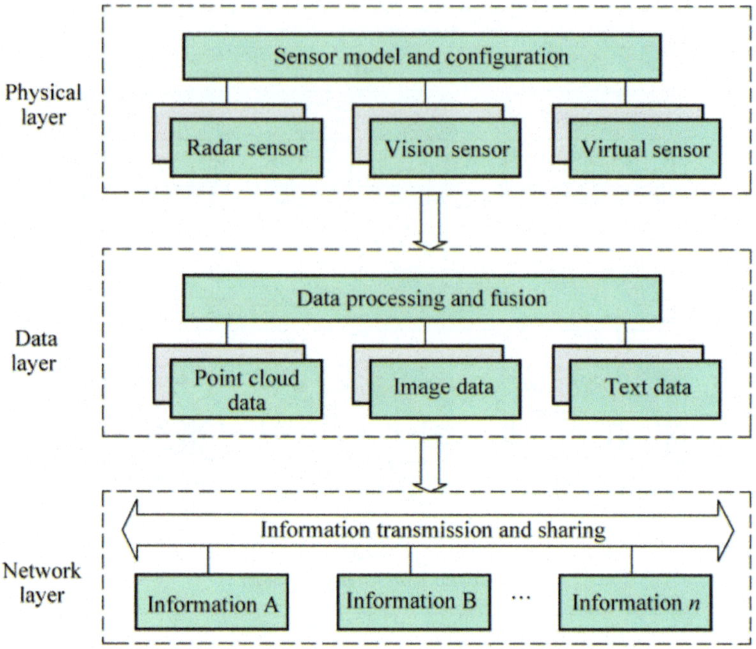

Fig. 3.12 Architecture design for sensor information sharing

The physical layer of the sensor system mainly refers to sensor hardware, including sensor models, hardware features, functions and installation location. The main functions of the physical layer are to optimize sensor configuration, reduce hardware cost and improve the utilization rate of hardware resources by optimizing the model and location of required sensors through physical modeling and comprehensive indicator evaluation according to the system's environment information perception demand.

The data layer of the sensor system mainly refers to data information output, including the type of detected information, data features and detected results after processing. Type and format of raw data output from various kinds of sensors are different, which make information sharing and data fusion quite difficult, and therefore it is necessary to code or standardize the structure of raw data or aggregate data value at the data layer, turning them into unified, structured or meaningful information.

The network layer of the sensor system, also known as the information transmission layer or information sharing layer, refers to the network of sensors connected by a shared transmission bus, which is responsible for unified management and cooperation scheduling of sensors and their information flow, and transmission of information to the vehicle control system. Its main functions include access to sensor hardware information, data detected by sensors and scheduling information.

The intelligent driving system's sensor workflow is shown in Fig. 3.13.

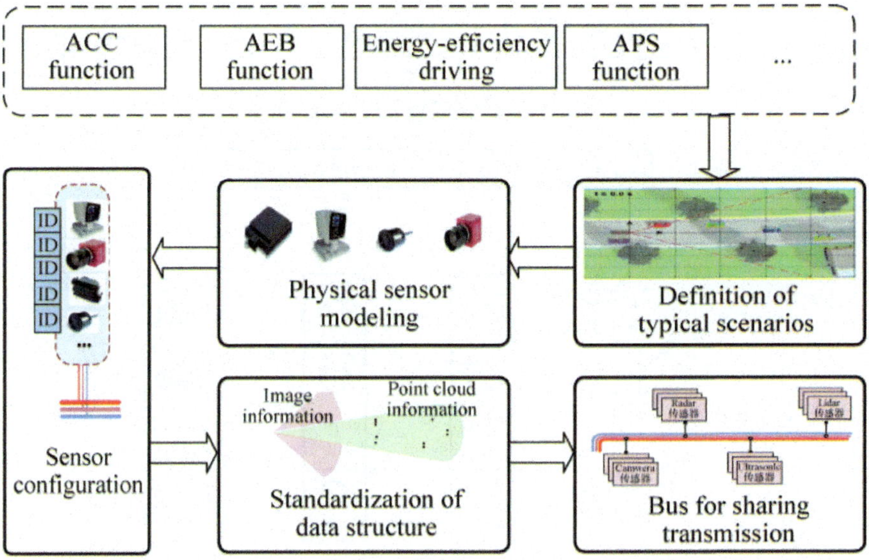

Fig. 3.13 Workflow of sensor information sharing architecture

3.3.3 Architecture Design for Controller Resource Sharing

Current superposition architecture consists of controllers with independent design and distributed arrangement in each subsystem. While in an intelligent driving system, subsystems are constantly switching to a working state instead of working full-time, depending on driving scenarios. For example, a park assist system or auto park system works in low-speed scenarios, while an adaptive cruise control system usually works in high-speed scenarios. Superposition architecture with independent controller design and configuration in subsystems not only increases system structure complexity but also underutilizes controller resources as some of them remain idle for a long time. Moreover, information transmission delay between controllers and failure of information sharing between functional modules will make it difficult to synchronize sensors with different frequencies and asynchronization in various subsystems. In order to make full use of controller resources, the controller resource sharing architecture of the intelligent driving system is designed as shown in Fig. 3.14.

Architecture design of controller resources sharing aims to share the same physical controller among multiple processing units and achieve a reasonable allocation of processing resources by memory sharing and clock synchronization. Structure sharing architecture usually selects one or a small number of controllers and divides control domains, according to the need of overall control function, making the domains relatively independent virtual processing units, and then achieves sharing of processing resources through memory sharing and clock synchronization. The architecture of controller resource sharing mainly includes multi-core sharing of computing resources, partition sharing of storage resources and time-sharing of communication resources and its workflow is shown in Fig. 3.15.

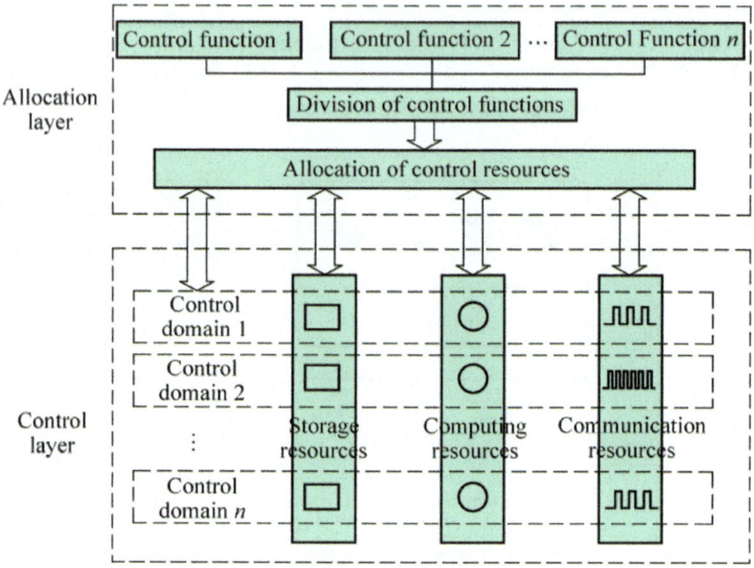

Fig. 3.14 Architecture of controller resources sharing

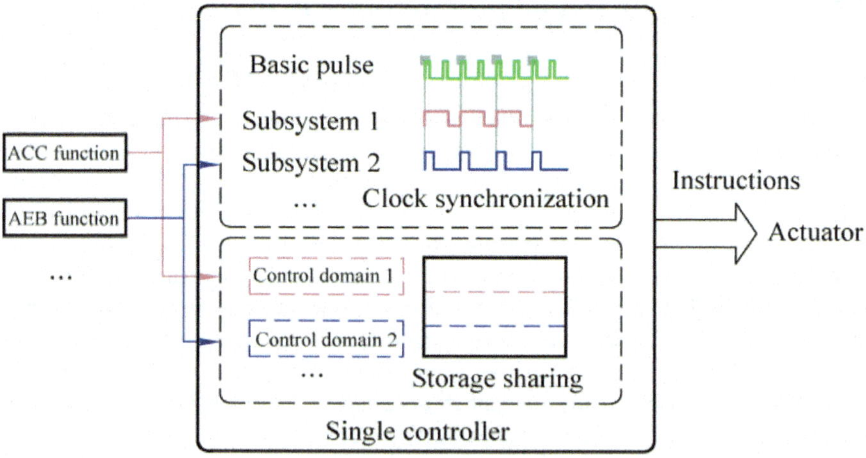

Fig. 3.15 Workflow of controller resource sharing architecture

3.3.4 Architecture Design for Actuator Operation Co-management

In the functional superposition architecture of the existing intelligent driving system, sensors and controllers of each subsystem are designed independently, and each controller sends control instructions to the actuator according to its own functional needs. When the intelligent driving function is relatively simple or control objects are in different directions, the actuators can complete longitudinal or lateral motion control. However, with increasing functions of intelligent driving system, action performance may interfere with each other when multiple control instructions are transmitted to the actuation system in superposition architecture, resulting in ineffective system function and performance. In order to eliminate action interference and coordinate functions of various subsystems, the architecture of actuation operation co-management is designed and shown in Fig. 3.16.

Architecture design of actuator operation co-management aims to manage control instructions sent to the actuation system, establish an instruction coordinator, identify and coordinate instruction sets, send operation instructions to actuators according to hierarchical priority, to eliminate possible interference actions and optimize overall system performance. Actuator operation co-management architecture makes it not only convenient for coordinated optimization of actuator operation, but also convenient to extract information flow in the actuation system and make full use of the information flow to play the part of the role of sensors, and therefore ensures sharing of actuation components and sensors and further simplifies system structure. The schematic diagram of actuator operation co-management architecture of an intelligent driving system is shown in Fig. 3.17.

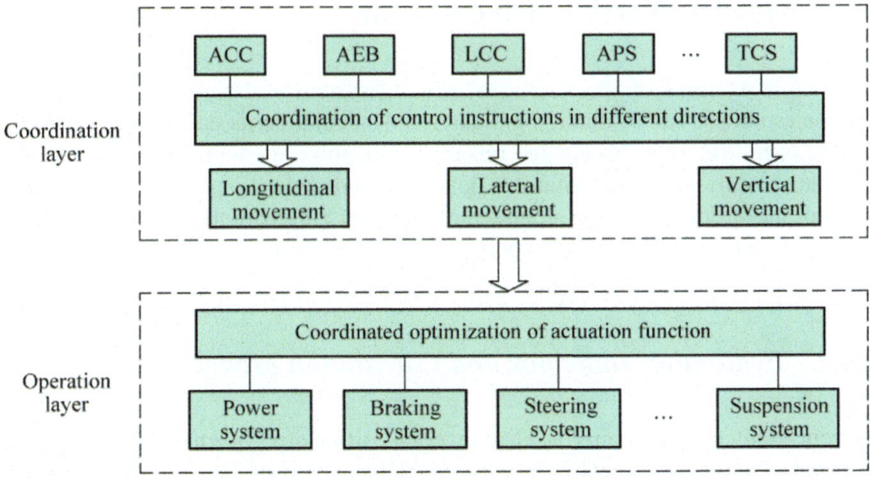

Fig. 3.16 Architecture of actuator operation co-management

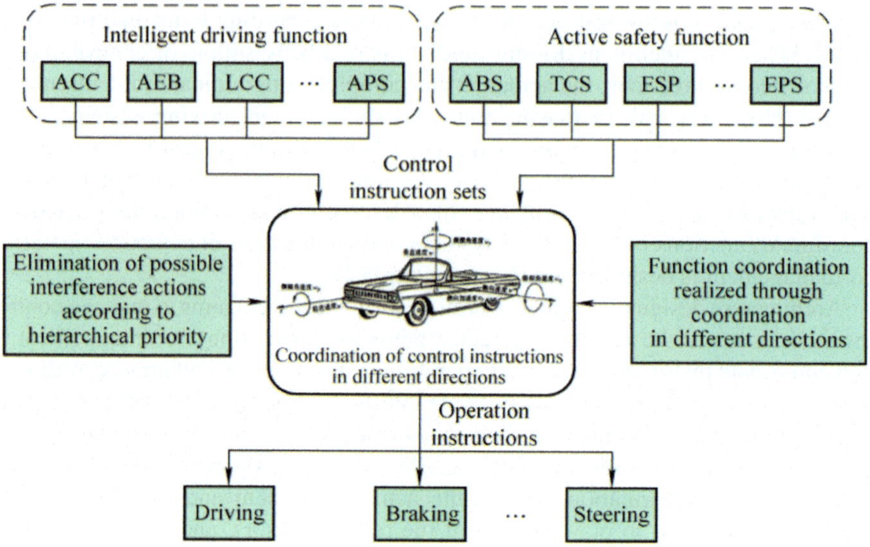

Fig. 3.17 Schematic diagram of actuator operation co-management architecture workflow

3.4 Analysis on Features of Integrated Structure Sharing of the Intelligent Driving System

To analyze features of the intelligent driving system with structure sharing, this section establishes an evaluation indicator and calculation model from the aspect of degree of information sharing and resource utilization rate with the combination of qualitative analysis and quantitative calculation, and analyzes and summarizes advantages of the structure sharing architecture by comparing with the existing intelligent driving system with functional superposition architecture.

3.4.1 Evaluation Indicator and Calculation Model

System architecture contains the element composition of a system and its logical relationships. For an intelligent driving system, element composition mainly refers to the selection and quantity of key components such as sensors, controllers, and actuators, and information interaction relationships among them can represent their logical relationships. Structure sharing architecture aims to improve overall performance and reduce resource cost. Overall system performance is composed of the performance of each subsystem, which is not only related to the configuration of key components, but also related to logical relationships among components. Resource cost mainly consists of the cost of system components. There are many factors affecting component cost, and a wide range of price changes and fluctuations, so it is difficult to accurately calculate the cost in architecture design stage, and for this reason, it is not suitable to use cost as an evaluation indicator of system architecture. But in general, the higher the system resource utilization, the lower the resource cost. Therefore, system information sharing degree and resource efficiency indicator are put forward to quantitatively evaluate the information interaction degree and resource utilization rate of the system.

(1) Degree of information sharing

The degree of information sharing reflects the degree (or sharing ability) of information interaction among all components in the system. To quantify the degree of information sharing, the indicator is defined as the number of information interaction per unit component in the system, namely:

$$\mathrm{DIS} = \frac{N_{\text{information}}}{N_{\text{component}}} \tag{3.1}$$

where DIS represents the degree of information sharing, $N_{\text{information}}$ represents the number of information interactions among components in the system, $N_{\text{component}}$ represents the number of components in the system. Components here mainly refer to sensors, controllers and actuators in the intelligent driving system.

Calculation method of the number of information interaction among components in the system is as follow: if there is no information flow transmission between two components in the system, the number of information interaction is 0; if there is information flow transmission, then the number of information interaction is 1; the number of information interaction increases by 1 for each information transmission between components in the system. In the case of the same component composition, the degree of information sharing reflects the performance potential of the system to some extent.

Without loss of generality, for any system with n components, if there is information interaction among all of them, then the number of information interaction is:

$$N_{\text{information}} = \mathrm{C}_n^2 = n \cdot (n-1)/2 \tag{3.2}$$

The degree of information sharing in the system at this moment is:

$$DIS = \mathrm{C}_n^2/n = (n-1)/2 \tag{3.3}$$

(2) Resource efficiency

Resource efficiency refers to the relative number of beneficial economic, social, ecological, and environmental effects produced per unit of resource. For an intelligent driving system, a unit of resource is defined as a single component, and the beneficial effect is defined as intelligent driving functions generated by components. Therefore, the resource efficiency indicator of an intelligent driving system is defined as the ratio of the number of intelligent driving functions to the number of components:

$$\mathrm{RE} = \frac{N_{\text{function}}}{N_{\text{component}}} \tag{3.4}$$

where *RE* represents the resource efficiency of the system, N_{function} represents the number of system functions, and $N_{\text{component}}$ represents the number of system components. System functions here refer to intelligent driving functions, and components refer to key components such as sensors, controllers and actuators in an intelligent driving system.

However, key components such as sensors, controllers and actuators play different roles in the intelligent driving system, and their resource costs vary greatly. When we calculate the number of components, the number of sensors, controllers and actuators should be calculated separately, instead of simply adding up their quantities, and then we calculate the integrated resource efficiency of the system by weighted sum:

$$\mathrm{RE} = \delta_1 \cdot \mathrm{RE_s} + \delta_2 \cdot \mathrm{RE_c} + \delta_3 \cdot \mathrm{RE_a} \tag{3.5}$$

where RE_{s}, RE_{c} and RE_{a} are the resource efficiency of sensors, controllers and actuators respectively which meet the conditions of $RE_{\mathrm{s}} = \frac{N_{\text{function}}}{N_{\mathrm{s}}}$, $RE_{\mathrm{c}} = \frac{N_{\text{function}}}{N_{\mathrm{c}}}$, $RE_{\mathrm{a}} = \frac{N_{\text{function}}}{N_{\mathrm{a}}}$, N_{s}, N_{c} and N_{a} are the number of sensors, controllers and actuators in the system respectively, $\delta_1, \delta_2, \delta_3$ are weight coefficients.

3.4.2 *Comparative Analysis with Functional Superposition Integration*

3.4.2.1 Calculation of Structure Sharing Architecture's Evaluation Indicator

In the structure sharing architecture of an intelligent driving system, the system is divided into the sensing layer, the control layer and the actuation layer. Through combinatorial optimization and function expansion, sharing of sensor information, controller resources and co-management of actuator operation are achieved, as shown in Fig. 3.18.

Figure 3.18 shows the architecture of an intelligent driving system integrated with n intelligent driving subsystems, and the numbers of sensors, controllers and actuators are N_{s1}, N_{c1}, N_{a1} respectively. The degree of information sharing and resource efficiency indicators of the system are calculated as below.

The degree of information sharing among all sensors, controllers and actuators in the system is:

$$\mathrm{DIS}_1 = \frac{C_{m_1}^2}{m_1} = \frac{m_1 \cdot (m_1 - 1)}{2} \cdot \frac{1}{m_1} = \frac{m_1 - 1}{2} \tag{3.6}$$

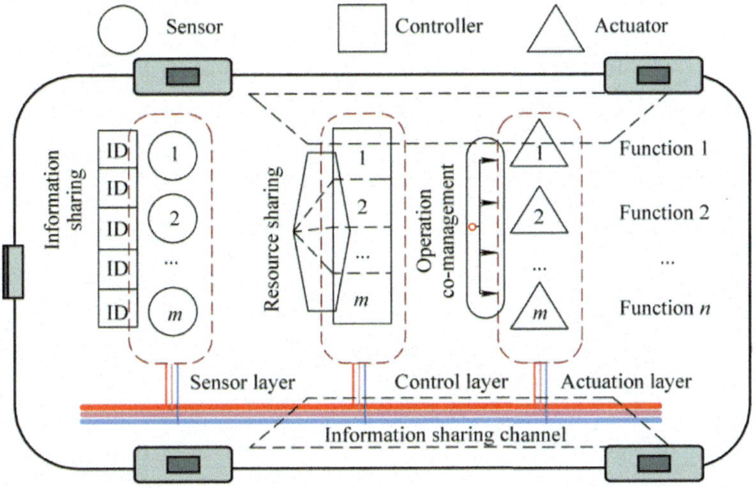

Fig. 3.18 Schematic diagram of structure sharing architecture

where $m_1 = N_{s1} + N_{c1} + N_{a1}$.

Resource efficiency indicators of sensors, controllers and actuators in the system are calculated respectively and weighted to sum. The resource efficiency indicator of system hardware is the weighted sum of resource efficiency of various components:

$$RE_1 = \delta_1 \cdot \frac{n}{N_{s1}} + \delta_2 \cdot \frac{n}{N_{c1}} + \delta_3 \cdot \frac{n}{N_{a1}} \tag{3.7}$$

3.4.2.2 Calculation of Functional Superposition Architecture's Evaluation Indicator

In the current vehicle intelligent driving system, sensors, controllers and actuators are designed and developed independently depending on system functions. In the case of system function integration, new systems are usually directly superimposed on top of the existing intelligent driving system to generate new functions. The system configuration scheme of "independent structure design and functional superposition" can be classified as functional superposition architecture, as shown in Fig. 3.19.

In Fig. 3.19, the system is composed of n intelligent driving subsystems in superposition architecture. The numbers of sensors, controllers and actuators are N_{s2}, N_{c2}, N_{a2} respectively, and the number of functions is n. The degree of information sharing and resource efficiency indicators of the system are calculated as below.

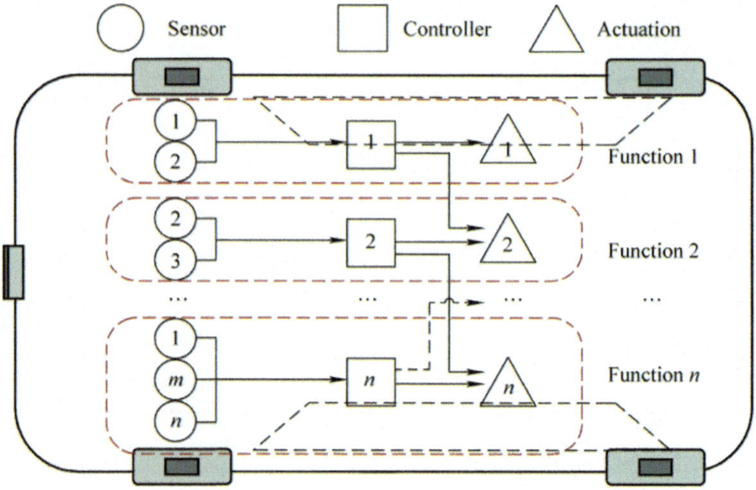

Fig. 3.19 Schematic diagram of functional superposition architecture

Information flow in the system exists between sensors and controllers, as well as controllers and actuators in subsystems. The number of information interactions between sensors and controllers is the same as the number of sensors, so the number of information interactions in the system can be expressed as:

$$N_{\text{information}} = N_{s2} + N_2 \tag{3.8}$$

where N_2 represents the number of information flow interactions between controllers and actuators, which is related to the number of controllers and the number of controlled actuators corresponding to a single controller.

Let $N_0 = N_{c2} + N_{a2} - N_2$, which represents the difference between the total number of controllers and actuators and the number of information flow interactions among them, reflecting the complexity of information interaction. The indicator of system information sharing degree can be expressed as

$$\text{DIS}_2 = \frac{N_{s2} + N_{c2} + N_{a2} - N_o}{N_{s2} + N_{c2} + N_{a2}} = 1 - \frac{N_0}{m_2} \tag{3.9}$$

where $m_2 = N_{s2} + N_{c2} + N_{a2}$.

Resource efficiency indicators of sensors, controllers and actuators in the system are calculated respectively and weighted to sum, and the resource efficiency indicator of system hardware is:

$$\text{RE}_2 = \delta_1 \cdot \frac{n}{N_{s2}} + \delta_2 \cdot \frac{n}{N_{c2}} + \delta_3 \cdot \frac{n}{N_{a2}} \tag{3.10}$$

3.4.2.3 Qualitative Comparison and Analysis

Characteristics of both architectures are compared based on the established calculation models of information sharing degree and resource efficiency indicator. Δ_D represents the difference of information sharing degree between the two architectures, which is expressed as follow:

$$\Delta_{\mathrm{D}} = \mathrm{DIS}_1 - \mathrm{DIS}_2 = \frac{m_1 - 1}{2} - \left(1 - \frac{N_{\mathrm{O}}}{m_2}\right) = \frac{m_1 - 3}{2} + \frac{N_{\mathrm{O}}}{m_2} \tag{3.11}$$

Components in an intelligent driving system include sensors, controllers and actuators. As shown in the above equation, $m_1 \geq 3$, so $\Delta_D \geq 0$, and the value of Δ_D increases as the number of components increases. Equation (3.11) shows that with an increasing number of intelligent driving functions and components, the advantages of structure sharing architecture in the degree of information sharing will be more obvious.

Δ_R represents the difference in resource efficiency between the two architectures:

$$\Delta_{\mathrm{R}} = \mathrm{RE}_1 - \mathrm{RE}_2 = \delta_1 \cdot \left(\frac{n}{N_{\mathrm{s1}}} - \frac{n}{N_{\mathrm{s2}}}\right) + \delta_2 \cdot \left(\frac{n}{N_{\mathrm{c1}}} - \frac{n}{N_{\mathrm{c2}}}\right) + \delta_3\left(\frac{n}{N_{\mathrm{a1}}} - \frac{n}{N_{\mathrm{a2}}}\right) \tag{3.12}$$

Since there are fewer key components such as sensors, controllers and actuators in the structure sharing architecture, which means $N_{\mathrm{s1}} \leq N_{\mathrm{s2}}, N_{\mathrm{c1}} \leq N_{\mathrm{c2}}, N_{\mathrm{a1}} \leq N_{\mathrm{a2}}$ in the above formula, so it is obvious that $\Delta_R \geq 0$.

Equation (3.12) shows that for the two types of intelligent driving system architectures with the same number of functions, the resource efficiency indicator of the structure sharing architecture is obviously higher than that of the functional superposition architecture.

According to the above comparative analysis and comparison of two architecture schematics in Figs. 3.18 and 3.19, it can be seen that:

The functional superposition architecture is relatively simpler and easier to implement, and therefore it is more practical for systems with simpler functions. However, with an increasing number of functions and subsystems in the intelligent driving system, system structure becomes more complex, and inter-coupled and overlapped system functions and structures may cause structure configuration redundancy, low resource utilization and high system cost. Furthermore, lack of information sharing among key components may result in inadequate utilization of resources and impede optimization and improvement of overall system performance.

Through combinatorial optimization and function expansion of key components, the structure sharing architecture of intelligent driving system solves the problems in functional superposition architecture at top-level design. Firstly, at the sensing layer, all sensor information flow in the system is processed and connected to the vehicle control system and the actuation system through an information-sharing channel, which makes information sharing more sufficient. Secondly, at the control layer, the subsystems improve hardware resource utilization by sharing a small number of controllers. Finally, at the actuation layer, operation co-management and coordination facilitate multi-function coordinated optimization.

3.4.2.4 Quantitative Comparison and Analysis

To quantitatively analyze characteristics of the structure sharing architecture, the information sharing degree and resource efficiency of both architectures are compared by taking functional integration of typical intelligent driving assistance systems (AVM, APS, AEB and ACC) as an example.

Key components' configuration and connection in an intelligent driving system with functional superposition architecture are shown in Fig. 3.20. The four subsystems are integrated by superposition. Calculation results of the number of sensors, controllers and relevant indicators of each subsystem are shown in Table 3.1.

Key components' configuration and connection in an intelligent driving system with structure sharing architecture are shown in Fig. 3.21. Integration and sharing

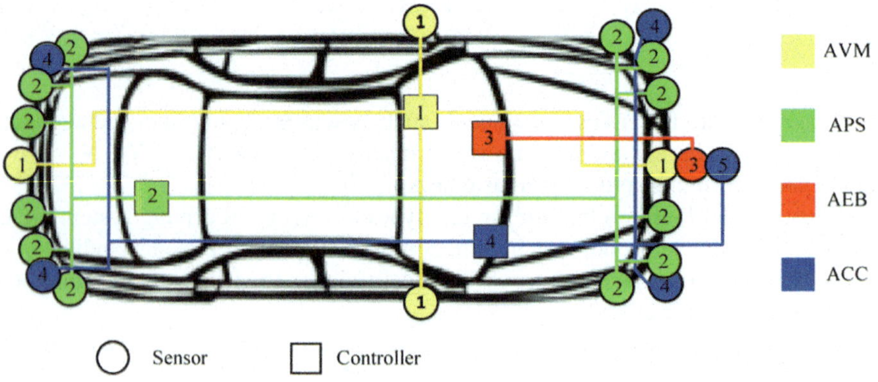

Fig. 3.20 Functional superposition architecture

Table 3.1 Calculation results of relevant indicators in functional superposition architecture

Function	Number of sensors	Number of controllers	Sensor resource efficiency	Controller resource efficiency	Information sharing degree
AVM	4	1	1/4	1	4/5
APS	12	1	1/12	1	12/13
AEB	1	1	1	1	1/2
ACC	5	1	1/5	1	5/6
Functional superposition	22	4	4/22	1	22/26

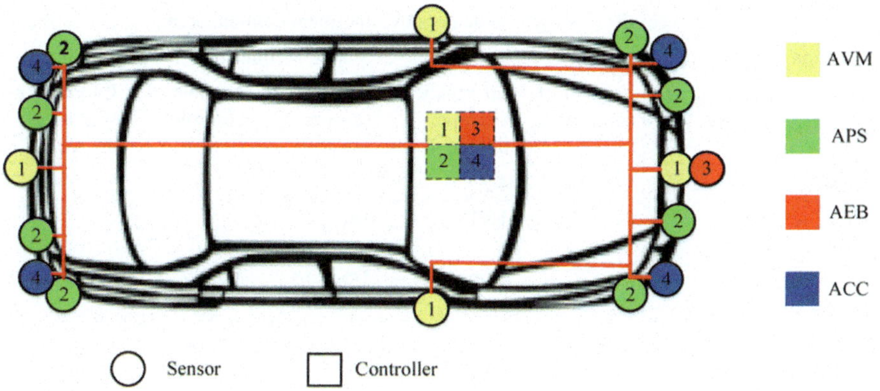

Fig. 3.21 Structure sharing architecture

reduce the number of sensors and controllers and ensure information sharing. Calculation results of information sharing degree and resource efficiency indicators are shown in Table 3.2.

Calculation results in Tables 3.1 and 3.2 show that the structure sharing architecture is evidently better than the functional superposition architecture in terms of information sharing degree and hardware resource efficiency, and the indicator of information sharing degree increases by over 10 times.

To further quantitatively analyze the relationship between increased ratios of information sharing degree and resource efficiency and the number of system functions under structure sharing architecture, relevant indicators of system with different functional combinations under both architectures are calculated respectively, and the results are shown in Table 3.3.

Compared with the functional superposition architecture, the calculation results of relevant indicators' ratio increase by different functional combinations under the structure sharing architecture are shown in Table 3.4 (when calculating overall resource efficiency, the calculation weights of sensors and controllers are both set as 1).

Comparative results in Table 3.4 show that the structure sharing architecture with multi-function combinations is significantly better than the functional superposition architecture in terms of resource efficiency and information sharing degree, and this advantage becomes more obvious with an increasing number of system functions. The resource efficiency of the structure sharing architecture increases by up to 256%,

Table 3.2 Calculation results of relevant indicators in structure sharing architecture

Function	Number of sensors	Number of controllers	Sensor resource efficiency	Controller resource efficiency	Information sharing degree
AVM + APS + AEB + ACC	19	1	4/19	4	10

Table 3.3 Calculation results of system indicators by functional combinations

Function	Superimposed architecture			Shared architecture		
	Number of sensors	Number of controllers	Information sharing degree	Number of sensors	Number of controllers	Information sharing degree
AVM	4	1	4/5	4	1	2
AVM + APS	16	2	16/18	14	1	7
AVM + APS + AEB	17	3	17/20	15	1	7.5
AVM + APS + AEB + ACC	22	4	22/26	19	1	10

Table 3.4 Relevant indicators' ratio increase by functional combinations under structure sharing architecture

Function	Resource efficiency improvement (%)			Information sharing degree improvement (times)
	Sensor	Controller	Overall	
AVM	0	0	0	1.50
AVM + APS	14.3	100	90	6.87
AVM + APS + AEB	13.3	200	172	7.82
AVM + APS + AEB + ACC	15.8	300	256	10.8

and the information sharing degree increases up to 10.8 times, compared with those of functional superposition architecture.

3.4.3 *Feature Summary of the Integrated Structure Sharing Architecture*

Based on the above comparison and analysis, the structure sharing architecture has the following characteristics and advantages compared with the existing intelligent driving system with integrated functional superposition architecture:

(1) Improved information sharing degree facilitates coordinated optimization of overall functions

An intelligent driving system with structure sharing architecture improves information sharing degree through component sharing and information sharing. Information sharing among sensors improves detection accuracy through multi-sensor data fusion. Multiple subsystems sharing the same controller make it easy to eliminate delay and realize synchronization. Sharing of actuation parts simplifies actuation operation and facilitates operation co-management and coordination. Furthermore, information flow in the actuator plays part of the senor's role. Information sharing of all components in the system provides an information basis for overall system performance enhancement and facilitates coordinated optimization of overall function.

(2) Improved resource utilization facilitates reduction of system resource cost

Through combinatorial optimization and function expansion of key components, the structure sharing architecture of an intelligent driving system solves the problems of structure configuration redundancy and low resource utilization in current functional superposition architecture at top-level design. With the structure sharing architecture, key components in the intelligent driving system, such as sensors and controllers, are integrated to optimize system structure composition and improve hardware resource efficiency. All sensor information flow in the system is processed and connected with the control and actuation system of the vehicle, which makes information sharing more sufficient and maximizes resource utilization, and therefore reduces resource cost of the intelligent driving system.

3.5 Optimal Configuration of Onboard Environment Perception Sensors Based on Structure Sharing

Onboard environment perception sensors are key components in an intelligent driving system, so proper model selection and arrangement of such sensors play an important role in the optimization and integration of the EV intelligent driving system with structure sharing, and also lay a hardware foundation to achieve high system performance with low cost. In the initial stage of intelligent driving development, simple intelligent driving functions and limited sensor models make it possible to obtain a good configuration scheme with a superposition configuration method based on qualitative analysis or engineering experience. However, with the increase of intelligent driving functions, as well as types and models of sensors, there is also a proliferation of sensor configuration options. It becomes difficult to obtain the optimal sensor configuration solution if solely depends on engineering experience and qualitative analysis. Therefore, an optimal sensor configuration method based on a quantitative solution of the optimization algorithm is needed to rapidly optimize sensor model selection and configuration scheme of the intelligent driving system.

Considering sensor modeling, multi-dimensional evaluation indicator, multi-objective optimization solution and other issues in sensor configuration, in this section we build a method for optimal configuration of environment perception sensors based on structure sharing and combinatorial optimization theory. The overall framework is shown in Fig. 3.22.

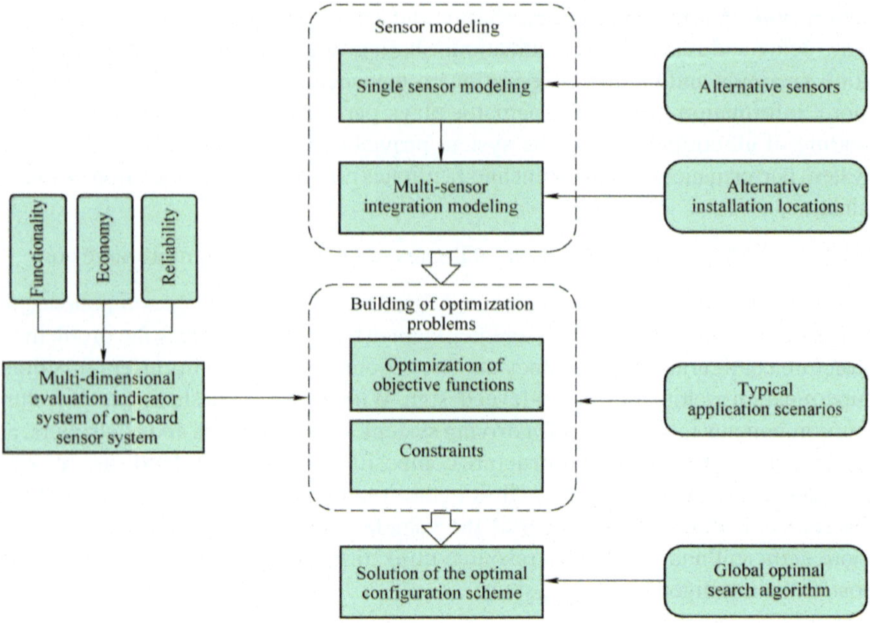

Fig. 3.22 Overall framework of intelligent driving system's sensor optimization and configuration

As shown in Fig. 3.22, the proposed method mainly includes optional sensor modeling, construction of a multi-dimensional evaluation indicator system, multi-objective combinatorial optimization building and global optimal search algorithm. Main steps are as follows: firstly, build a detection performance model and a reliability model of the onboard multi-sensor system, quantitatively describe the detection range, accuracy and reliability of multiple sensors integrated in the vehicle; secondly, build an evaluation indicator system of the onboard multi-sensor system from the perspective of functionality, economy and reliability to quantitatively evaluate sensor configuration scheme; thirdly, select typical application scenarios according to the functional positioning of the intelligent driving system to obtain unified description of environmental information perceived by the system, and then establish a multi-objective optimization problem based on objective functions and constraints designed according to the evaluation indicator system; finally, adopt a multi-objective global optimization algorithm to optimize alternative sensors and their installation locations, and achieve a computer solution of sensor configuration optimal scheme. The key steps of the proposed method are described in detail below.

(1) Modeling the onboard multi-sensor system

The sensor detection performance model is a mathematical description of the quantified detection function and performance of sensors, and the premise of using a computer to optimize sensors. As mentioned above, there are many types of frequently-used environment perception sensors with different sensing mechanisms.

How to ensure a universal description of their sensing function and performance is the key to sensor selection and configuration optimization. In this section, we build a single sensor detection performance model from the perspective of sensor performance based on ultrasonic radar, millimeter-wave radar and laser radar study (without consideration of the influence brought by different sensor detection mechanisms and detection algorithms) and also a detection performance model and a reliability model for integrated sensors considering different installation locations and integration methods.

(2) Establishing a multi-dimensional evaluation indicator system

To quantitatively evaluate the advantages and disadvantages of a sensor configuration scheme, it is necessary to establish an evaluation indicator system. From the perspective of actual scale application of the intelligent driving system, the cost of environment perception sensors accounts for a large proportion in that of the intelligent driving system, which is also the bottleneck restricting its scale application in low-end vehicles. Evaluation of sensor configuration scheme not only considers function and performance, but also cost and reliability of the system. Therefore, this section will construct an evaluation indicator system for the sensor configuration of intelligent driving system from multiple dimensions including functionality, economy and reliability, and with consideration of functional performance, purchase cost, service life and other factors.

(3) Establishing a multi-objective optimization problem

To transform the sensor optimization problem into a multi-objective combinatorial optimization problem is to establish the optimization objective function and constraints for sensor optimal configuration according to the environment information required by an intelligent driving system. Environment information includes coverage of the object area to be detected, traffic participants to be identified and relative motion information of the object. Coverage of the object area can be described by the coverage area of the front and rear sensing field in vehicle movement scene; traffic participants include vehicles, pedestrians, obstacles and traffic signs; relative motion information of measured object includes azimuth information, relative speed and relative distance of the object. Sensor system's demand for environment information perception comes from functional positioning of the intelligent driving system. However, it is difficult to quantify, verify and evaluate environment information demand just according to the functional positioning of an intelligent driving system designed by developers. Therefore, we propose in this section a description method of the environment information demand based on typical application scenarios, which generates area coverage of a sensing object according to function realization of an intelligent driving system in typical application scenarios. Based on the geometrical relationship between detection range and object area of a multi-sensor system, design detection coverage rate and other objective functions of a sensor configuration, and then select several performances or cost indicators from the multi-dimensional evaluation indicator system as constraints to establish a multi-objective optimization problem with constraints.

(4) Solving the multi-objective optimization problem

There are two methods to solve the multi-objective optimization problem: one is to transform it to a single objective problem through weighted sum, which not only requires normalization of each object value, but also needs to provide relative importance of objects as weight coefficient of each object; the other is to use multi-objective optimization algorithm to solve the multi-objective problem, i.e. Pareto front solutions, and the obtained results are a solution set composed of Pareto optimal solutions. The former method is relatively simpler and easier to get the only optimal solution, but the result is heavily dependent on weight coefficient, making it difficult to meet the needs of different users. Since different designers are concerned about different performances, the use of multi-objective respective optimization may select different optimization results in Pareto front solutions based on requirements. Therefore, this chapter adopts a multi-objective optimization method, i.e. using a multi-objective optimization algorithm to find Pareto front solutions that balance multiple optimization objectives, so that different users can choose solutions based on their needs, which enhances design flexibility.

3.5.1 Modeling the Onboard Multi-sensor System

The environment perception mechanism of radar sensors, including ultrasonic radar, millimeter-wave radar and laser radar is to irradiate objects with mechanical waves or electromagnetic waves and receive their echoes to obtain the distance (or relative velocity) between the object and the launching point, as well as azimuth information. Mathematical models of radar sensors can be divided into three types [11–16]: the first is micro-level modeling, i.e. a sensor model based on the finite element analysis, which calculates time difference by simulating transmission and reception of pulses, and accurately locates reflection points according to transmission and reflection features of simulated sound waves or electromagnetic waves; the second is system-level modeling, i.e. a sensor model based on the detection mechanism, including reflective distance calculation, signal attenuation model and effective object detection algorithm, which can classify and simulate error sources; the third is performance-level modeling, i.e. a sensor model based on the detection performance, which uses geometric parameters or probability to describe the detection range, accuracy and detectivity of sensor and is usually built based on experimental data. To unify the description of sensing function and performance of different sensors, this section will ignore the influence brought by sensing mechanism and detection algorithm of different sensors, and build a detection performance model and a reliability model for the multi-sensor system from the perspective of detection function performance.

3.5.1.1 Detection Performance Model of the Onboard Multi-sensor System

The purpose of sensor modeling is to describe the sensing function and detection performance of sensors. This section adopts a hierarchical modeling method and selects models according to different scenario requirements. The simplest is the true value layer, whose model only considers coverage range and directly outputs object distance and azimuth information within the detection range, so it's a geometric model. The second layer is the performance layer, and the model outputs object distance, azimuth information within the detection range, as well as detection error range and detectivity, so it's a performance parameter model. The most complex level is the energy model, which describes the reflected energy value of the sensor after detecting the object. It is a function related to transmitted energy, detection distance, antenna gain, the shape of the detection object, echo area, detection object reflectivity and weather attenuation. Different types of sensors have great differences in the energy model. This section will study sensor configuration based on good weather scenarios, so there will not be an exploration of the energy model. The geometric model and performance model can meet the requirements of model selection and use, without considering the influence of weather attenuation.

Due to the difference in detection mechanisms of different sensors, there is no universal description of actual detection performance. Therefore, the following simplifications and assumptions are made in the modeling process:

a. Regardless of the shape difference of actual detection coverage area of each sensor, it is assumed that the detection coverage area of each sensor in horizontal direction is approximately fan-shaped, and detection range is represented by effective detection distance and horizontal FOV (Field of View);
b. Detection accuracy and detectivity of sensors are only described at performance level, without considering the influence of sensor perception mechanism and different detection algorithms;
c. Modeling parameters are performance parameters in ideal weather conditions, without considering the influence of sensor energy attenuation in poor weather condition.

Construct a detection range model, a detection precision model and a detectivity model of onboard multi-sensors based on the above simplifications and assumptions, with consideration of different sensor installation positions.

(1) Detection range model

Installation position and direction of sensors affect the detection range of an integrated multi-sensor system. When sensors are installed at different positions and in different directions and the detection range of each sensor does not overlap in the object area to be measured, then the detection range of the integrated sensor system is the sum of the detection range of each sensor. When two sensors are installed in the same position to detect objects in the same direction, then the detection range is the union of the detection range of the two sensors. Similarly, it will be the same if two sensors are installed next to each other with overlapped detection range. Therefore, the detection range model of an integrated multi-sensor system needs to be discussed case by case.

a. Distributed sensor arrangement without overlapped detection coverage

When sensors are installed at different positions and in different directions without overlapped detection range, as shown in Fig. 3.23, then the detection range of the integrated multi-sensor system is the sum of the detection range of each sensor, and the detection accuracy and detectivity of the object are consistent with those of the single sensor at the same position. Thus, the detection range model with *m* (number) integrated sensors is established as follow:

$$S_{\text{total}} = \sum_{i=1}^{m} S_i = \sum_{i=1}^{m} \pi\theta_i \mathrm{r}_i^2/360 \tag{3.13}$$

where S_{total} is the detection coverage of the integrated sensor system; S_{i}, d_{i} and θ_{i} are the detection coverage, detection distance and detection angle of every single sensor.

b. Distributed sensor arrangement with overlapped detection coverage

When sensors are installed at different positions and in different directions with overlapped detection coverage between adjacent sensors, as shown in Fig. 3.24, then the detection range of the integrated sensor system is the union of the detection range of adjacent sensors:

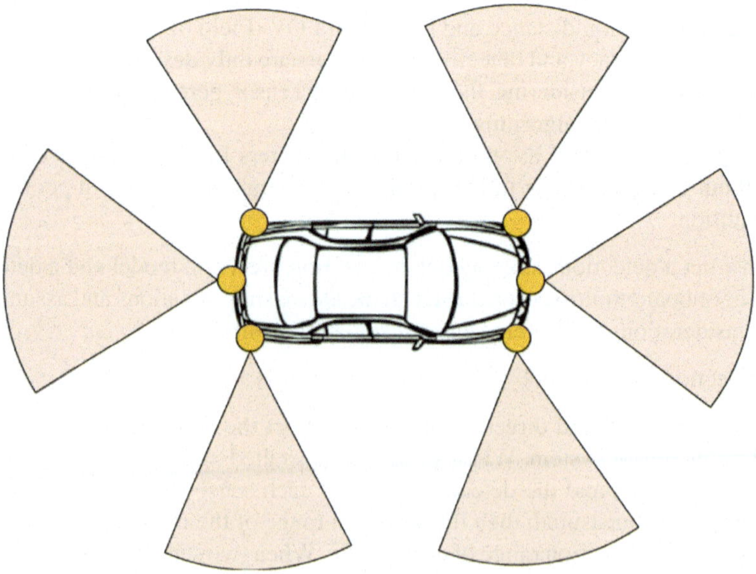

Fig. 3.23 Schematic diagram of each sensor installed in different positions

$$S_{\text{total}} = S_1 + S_2 - S_{1,2} \tag{3.14}$$

$$S_{1,2} = \begin{cases} 0, & if\ d_1 \sin\frac{\theta_1}{2} + d_2 \sin\left(\frac{\theta_2}{2} - \alpha\right) \le 1; \\ \pi \cdot \left(\frac{\theta_1+\theta_2}{2} - \alpha\right) \cdot r_{1,2}^2/360, & else \end{cases} \tag{3.15}$$

$$r_{1,2} = d_2 - \frac{l}{\sin\left(\frac{\theta_2}{2} - \alpha\right) + \cos\left(\frac{\theta_2}{2} - \alpha\right) \cdot tg\frac{\theta_1}{2}} \tag{3.16}$$

where S_{total} is the detection coverage of the integrated sensor system; $S_1, S_2, d_1, d_2, \theta_1$ and θ_2 are the detection coverage, detection distance and detection angle of every single sensor respectively; l and α are the relative distance and relative direction angle of two adjacent sensors respectively; $r_{1,2}$ is the radius of the approximately fan-shaped overlapping area.

c. Centralized sensor arrangement to detect objects in the same direction

When two sensors are installed at the same position to detect objects in the same direction, as shown in Fig. 3.25, the detection range of the integrated multi-sensor system is the union of the detection range of the two sensors:

$$\begin{cases} S_{\text{total}} = S_1 + S_2 - S_{1,2} \\ S_{1,2} = \pi \theta_{\min} d_{\min}^2/360 \\ \theta_{\min} = \min(\theta_1, \theta_2) \\ d_{\min} = \min(d_1, d_2) \end{cases} \tag{3.17}$$

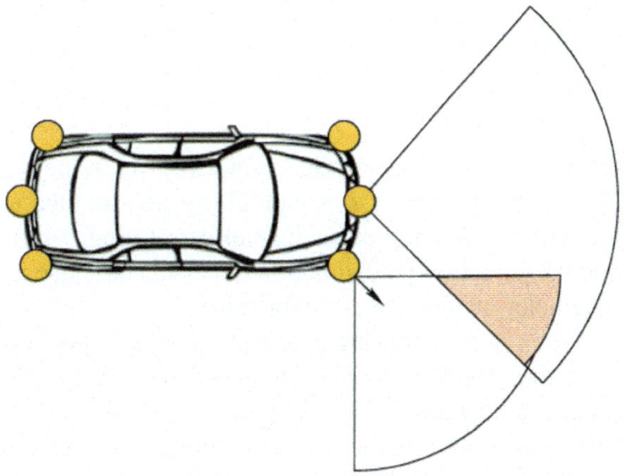

Fig. 3.24 Schematic diagram of overlapped detection range

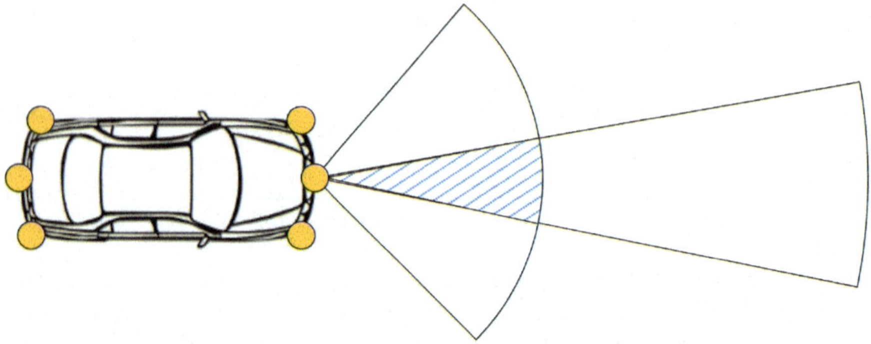

Fig. 3.25 Schematic diagram of two sensors installed at the same position

where S_{total} is the detection coverage of the integrated sensor system; S_1, S_2, d_1, d_2, θ_1 and θ_2 are the detection coverage, detection distance and detection angle of every single sensor respectively.

(2) Detection precision model

The sensor detection precision model describes the error of detection distance (or speed, direction). Errors are theoretically related to hardware's inherent error and detection algorithms, and they are not evenly distributed in the coverage area. Detection precision specified in the sensor's product manual is the inherent error, without considering the impact from detection distance and detection object. Regarding real-life vehicle experimental data, the detection error of the same detection object usually increases with the increase of detection distance within a certain limit, showing an approximately linear relationship, and therefore the detectivity model is established as follow:

$$\varepsilon = \pm(\varepsilon_u + kd_t) \tag{3.18}$$

where ε is the sensor's detection error, ε_u is the sensor's inherent error, d_u is the distance to the object, k is gain of the error related to detection distance, which can be obtained through experimental data fitting. But in practice, k can be negligible, and the detection precision of a radar sensor is directly expressed by range accuracy parameter, which is a constant value.

When two sensors are installed at the same position to detect an object in the same direction, data fusion of overlapped detection range should be considered. Detection precision and detectivity after fusion are both better than those of a single sensor. Two sensors installed next to each other with overlapped detection coverage will be in a similar situation.

For overlapped detection coverage, considering there will be less detection error after data fusion, the detection precision model is designed as follow:

$$\varepsilon_{1,2} = \pm\sqrt{\frac{\varepsilon_1^2\varepsilon_2^2}{\varepsilon_1^2\varepsilon_2^2}} \tag{3.19}$$

where $\varepsilon_{1,2}$ is the detection error of the integrated sensor system; ε_1 and ε_2 are the detection errors of two sensors installed at the same position respectively, and both of them use absolute value for easy calculation.

(3) Detectivity model

The sensor detectivity model describes the probability of a sensor detecting object, which is related to the reflected cross-sectional area and reflectivity of the object, and detection frequency, transmitted power and energy attenuation of the sensor. Relevant studies in Reference [17] show that detectivity of radar sensors falls rapidly in the transition region between its effective detection range and maximum detection range, showing a quartic function relationship, and therefore the sensor detectivity model for the same object is designed as follow:

$$P_s = \begin{cases} 1, & d \le r_0 \\ \dfrac{(r_m - d)^4}{(r_m - r_0)^4}, & r_0 < d \le r_m \\ 0, & d > r_m \end{cases} \tag{3.20}$$

where P_s, d, r_0, r_m represent detectivity, distance to the object, the effective detection range of a sensor and the maximum detection range of a sensor respectively. In practice, the sensor detectivity model is simplified to a binary probability model of (0, 1) without considering the transition region for easy calculation.

In overlapped detection coverage, the increase of detection probability after data fusion also needs to be considered, and therefore its detectivity model is designed as follow:

$$P_{s\_X} = P_{s1} + P_{s2} - P_{s1} \cdot P_{s2} \tag{3.21}$$

where P_{s_X} represents detectivity of the integrated sensor system; P_{s1} and P_{s2} represent detectivity of two sensors installed at the same position respectively.

3.5.1.2 Reliability Model of the Onboard Multi-sensor System

Reliability defines the capability of the system to perform specified functions under definite conditions and within a limited period, and reflects the average mean time to failure (MTTF) and failure probability in the process of using. Reliability is usually related to the reliability of system components and integrated installation technology. Constructing reliability model of the multi-sensor system is to calculate and evaluate system reliability quantitatively. Without consideration of differences in onboard sensors' integrated installation technology, sensor reliability models mainly include the degree of reliability model, failure density model, average failure rate model, MTTF model and mean time between failures (MTBF) model [18].

(1) Degree of reliability model

Sensor reliability degree defines the probability that a sensor performs specified functions under definite conditions and within a limited period. If limited working time is t_0, and time to failure is T, and if $T \geq t_0$, then the sensor is failure-free before t_0, and reliability function $R(t)$ is:

$$R(t) = P(T \geq t_0) \tag{3.22}$$

Assume there are N_0 sensors (considerable quantity), and the test starts from t_0 to t, during which period $r(t)$ sensors fail and $N_S(t)$ sensors work as normal, then the reliability degree of the sensor is:

$$R(t) = \frac{N_S(t)}{N_0} = \frac{N_0 - r(t)}{N_0} \tag{3.23}$$

(2) Failure density model

Failure density is also known as unreliability. On the contrary to the reliability, unreliability refers to the probability that a sensor fails to perform specified functions under definite conditions and within a limited period. The sum of reliability and unreliability is 1, and unreliability $F(t)$ is:

$$F(t) = P(T < t_0) \tag{3.24}$$

(3) Average failure rate model

Average failure rate refers to the number of failures of a sensor that has been working by t that occur after t in an average unit of time, which is represented by $\overline{\lambda}(t)$ and defined as:

$$\overline{\lambda}(t) = \frac{\Delta r(t)}{N_S(t) \cdot \Delta t} \tag{3.25}$$

(4) MTTF model

Mean time to failure (MTTF) is a basic reliability parameter for non-repairable products. In an intelligent driving system, sensors are purchased externally and therefore can be assumed as a non-repairable product. MTTF of sensors is defined as the ratio of the total number of life units to the total number of faulty products under specified conditions and within a specified time. If the failure time of N sensors is known to be $t_1, t_2, \ldots, t_N$, then

$$\text{MTTF} = \frac{1}{N}\sum_{i=1}^{N} t_i \tag{3.26}$$

If the failure density (unreliability) function $F(t)$ is known, then

$$\text{MTTF} = \int_{0}^{\infty} tF(t)\mathrm{d}t \tag{3.27}$$

When the sensor failure rate is λ, i.e.$F(t) = \lambda \cdot \exp(-\lambda \cdot t)$, then

$$\text{MTTF} = \int_{0}^{\infty} t, \lambda\exp(-\lambda t)\mathrm{dt} = \frac{1}{\lambda} \tag{3.28}$$

(5) MTBF model

Mean time between failures (MTBF) is the mathematical expectation of the time between two failures, and is mainly used to describe repairable products. The calculation formula is as follow:

$$\text{MTTF} = \frac{T(t)}{r(t)} \tag{3.29}$$

where $T(t) = \sum_{i=1}^{N} t_i$ is the total working time, $r(t)$ represents the total number of sensor failures within time t. For an intelligent driving system, MTBF is equal to MTTF, assuming sensors are non-repairable products.

After integration of a multi-sensor system composed of n sensors, system reliability depends on the integrated approach of n sensors, without consideration of mechanical reliability problems caused by integrated installation technology. When n sensors are integrated at different positions of the vehicle to detect objects in different directions, then the multi-sensor system is a reliability system in series. Assuming that MTTF of the number i sensor is $MTTF_i$ and its reliability is Ri (t), according to the definition, the $MTTF$ of a multi-sensor system is equal to the minimum $MTTF_i$ among all sensors, which is:

$$\text{MTTF} = \min(MTTF_1, MTTF_2, \ldots, MTTF_n) \tag{3.30}$$

The reliability degree model of a multi-sensor system is as follows:

$$\begin{aligned} R(t) &= P(MTTF \geq t) = P[\min(MTTF_1, MTTF_2, \cdots, MTTF_n) > t] \\ &= \prod_{i=1}^{n} P(MTTF_i > t) = \prod_{i=1}^{n} R_i(t) \end{aligned} \tag{3.31}$$

The reliability degree of a multi-sensor system is the product of the reliability of sensors connected in series.

Assume the failure rate of number i sensor is $\lambda_i(t)$, and take derivative of both sides of the above equation to obtain:

$$R'(t) = \prod_{i=1}^{n} R_i(t) \sum_{i=1}^{n} \frac{R_i'(t)}{R_i(t)} = -R(t) \sum_{i=1}^{n} \lambda_i(t) \quad (3.32)$$

Thus, the failure rate of the system is:

$$\lambda(t) = -\frac{R_i'(t)}{R_i(t)} = \sum_{i=1}^{n} \lambda_i(t) \quad (3.33)$$

It can be seen from above that the failure rate of a multi-sensor system is the sum of failure rates of all sensors connected in series.

When there are m sensors integrated at the same position in the vehicle to detect objects in the same direction, then m sensors can be considered as a reliability system in parallel. Similarly, the MTTF, reliability and failure rate of the parallel-connected multi-sensor system can be deduced as follow:

$$MTTF = \max(MTTF_1, MTTF_2, \ldots, MTTF_m) \quad (3.34)$$

$$R(t) = 1 - \prod_{i=1}^{n} [1 - R_i(t)] \quad (3.35)$$

$$\lambda(t) = \prod_{i=1}^{m} \lambda_i(t) \quad (3.36)$$

3.5.2 *Sensor configuration's Multi-dimensional Evaluation Indicator System*

The sensor selection and configuration for an EV intelligent driving system should consider not only the perceptive function realized by configuration scheme, but also configuration cost and system reliability. Therefore, it is necessary to construct an evaluation indicator system of sensor configuration from the dimension of functionality, economy and reliability, and design specific indicators under each dimension.

3.5.2.1 Functionality Indicators

System function is the ability of a system to act on the environment. As one of the most basic evaluation indicators, it reflects the physical attributes of a system to meet usage requirements. Radar and other onboard environmental perception sensors are mainly used to detect information about surrounding vehicles or obstacles. Their functions are mainly reflected in the detection performance of sensors, with evaluation indicators of the proportion of detection coverage, detection precision and detection frequency.

(a) Proportion of detection coverage

Detection coverage indicator describes a spatial scale that a sensor system can detect and cover. In a specific application scenario, sensor configuration needs to consider matching between the detection coverage and the object area to be measured. If the detection coverage of the onboard sensor system falls into the spatial scale of the object area, then it is effective coverage, otherwise it is invalid coverage. To quantify the effective coverage of sensor detection, the proportion of detection coverage (PDC) is defined as the proportion of area covered by sensor detection in the object area and expressed as follow:

$$\mathrm{PDC}(X) = S(X)/S_{\mathrm{R}} \tag{3.37}$$

where X represents the sensor set configured, which is a vector containing n elements, each element value represents the code of its selected sensor, and its position in the vector represents its installation position on the vehicle; $\mathrm{S}(X)$ represents the effective detection coverage under such sensor configuration,S_{R} is the object area to be detected. Similarly, the proportion of blind area (PBA) is:

$$\mathrm{PBA}(X) = 1 - S(X)/S_{\mathrm{R}} \tag{3.38}$$

(b) Detection precision

Detection precision indicator describes measurement error in the sensor detection results. From the perspective of perception mechanism, the detection precision of ultrasonic radar is low as it transmits mechanical waves, while that of laser radar is high as it radiates light waves. To facilitate quantitative comparison, we use range error-indicator to evaluate sensor detection precision, without consideration of different detection mechanisms of various radar sensors and gain of detection error with increasing detection distance.

(c) Detection frequency

As an important parameter to describe the detection performance of radar sensors, detectivity includes single-time detectivity and the number of detection per unit time. The former is the probability that a single-time continuous detection can detect the object, while the latter is the number of times that an object is detected per unit time. Single-time detectivity of radar is related to the signal-to-noise ratio and threshold value in the radar circuit, while the number of detection per unit time is related to single-time radar detectivity and detection frequency (detection cycle). Without consideration of radar detectivity's transitional region between the maximum detection range and the effective detection range, the number of times that an object is detected per unit time is only connected to detection frequency. Therefore, radar detection frequency is used to quantitatively evaluate sensor detectivity.

3.5.2.2 Economy Indicators

Economy indicators are the premise of selecting a low-cost sensor configuration scheme. The cost of sensor configuration includes the cost of sensors and onboard installation, so the evaluation indicator can be quantified by sensor price and installation cost.

(a) Sensor cost

For a configured sensor set X, if it is known that prices of onboard sensors configured at n positions are $Pr_{X(1)}$, $Pr_{X(2)}$, …, $Pr_{X(n)}$, then the cost of the multi-sensor system integrated on the vehicle is:

$$C_S(X) = \sum_{i=1}^{n} \text{Pr}_{X(i)} \tag{3.39}$$

where $X(i)$ represents the code of a sensor at position i, $i = 1,2,\ldots,n$.

(b) Installation cost

Sensor installation cost is connected not only to the selected sensor (size and weight), but also to the onboard installation position, as the installation cost of the same sensor at different positions is not the same. If it is known that the installation cost of sensor j at installation position i is L_{ij}, then for a configured sensor set X, installation cost of the integrated multi-sensor system is:

$$C_L = \sum_{i=1}^{n} L_{iX(i)} \tag{3.40}$$

(c) Sensor power (operation cost)

Considering energy consumption of the sensor system, sensor power can also be used as an indicator to evaluate the operation cost of a sensor system. If it is known that costs of sensors configured at n positions on the vehicle are $P_{CX(1)}$, $P_{CX(2)}$….$P_{CX(n)}$, then total power of the integrated multi-sensor system P_{CT} is:

$$P_{CT} = \sum_{i=1}^{n} P_{CX(i)} \tag{3.41}$$

3.5.2.3 Reliability Indicators

Reliability indicator reflects the MTTF of a sensor system and the probability of failure during use, which is related to the reliability of system components and integrated installation technology. According to the reliability model of the multi-sensor system established in the previous section, reliability evaluation indicators mainly include reliability, fault density, average failure rate, MTTF and MTBF. For an intelligent driving system, sensors are purchased non-repairable components, and therefore quantitative reliability evaluation indicators of the onboard multi-sensor system include reliability and MTTF, without considering mechanical reliability issues caused by integrated installation technology.

To sum up, a three-dimensional evaluation indicator system for onboard environmental perception sensors is shown in Fig. 3.26.

3.5.3 Establishing the Sensor Configuration Optimization Problem

The evaluation indicator system of sensor configuration is established to quantitatively evaluate the pros and cons of sensor configuration, which helps transform the problem of sensor optimization and configuration to a multi-objective combinatorial optimization problem with objective function and constraints based on the sensor system model and application scenarios.

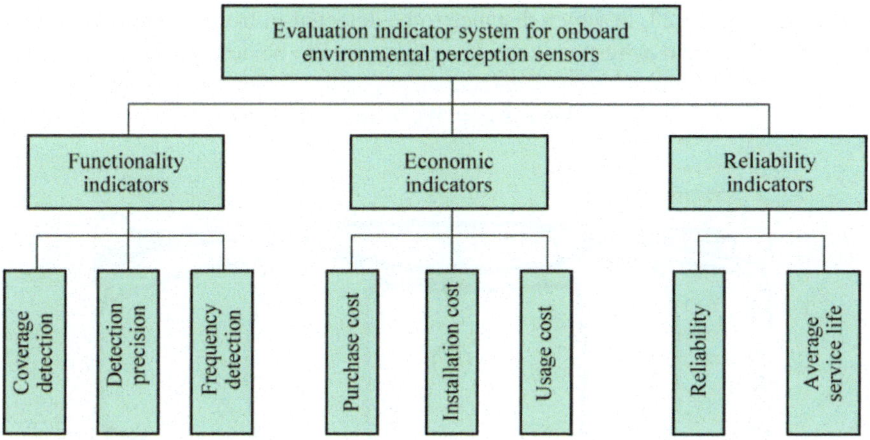

Fig. 3.26 Three-dimensional evaluation indicator system for onboard environmental perception sensors

3.5.3.1 Designing the Objective Function

The goal of sensor configuration optimization is to improve detection coverage and precision and reduce configuration cost as much as possible, under the condition that the configured onboard multi-sensor system satisfies basic detection function and reliability requirement, so it is necessary to design optimization objective function and constraints based on the established multi-dimensional evaluation indicator system and application scenarios.

To facilitate function representation of sensor selection and configuration scheme, sensor installation positions are divided into the front area, rear area, left area, right area, four corners and top of the vehicle, as shown in Fig. 3.27. Each area is divided into several portions by equidistance or equal angle as optional sensor installation position, and vehicle top is the optional installation position for a 360° multi-antenna laser radar. Number alternative installation positions and alternative sensors with 1 ~ n and 1 ~ m respectively, then sensor configuration scheme can be expressed as a 1 × n matrix, in which each element value represents the code of the sensor selected, and its position in the vector represents installation position on the vehicle. If the element value is 0, then it means that no sensor is arranged at this position.

Sensor detection coverage, range accuracy of front sensors and sensor costs in the established evaluation indicator system are selected as optimization objectives in this paper to design the objective function based on the object area to be detected. The object area is closely related to the application scenarios of an intelligent driving system, the object area to be detected in low-speed scenarios is within a safe distance around the vehicle, while in high-speed scenarios, it is related to the maximum vehicle speed. Considering typical application scenarios of an intelligent driving system, the object area to be detected by the onboard sensor system is simplified to a rectangle, as shown in Fig. 3.28, in which distance to be detected in front of and in the rear of the vehicle is d_f and d_r respectively. The object area to be detected is:

$$S_R = W_R \times L_R = W_R \times (d_f + d_r + L_V) \tag{3.42}$$

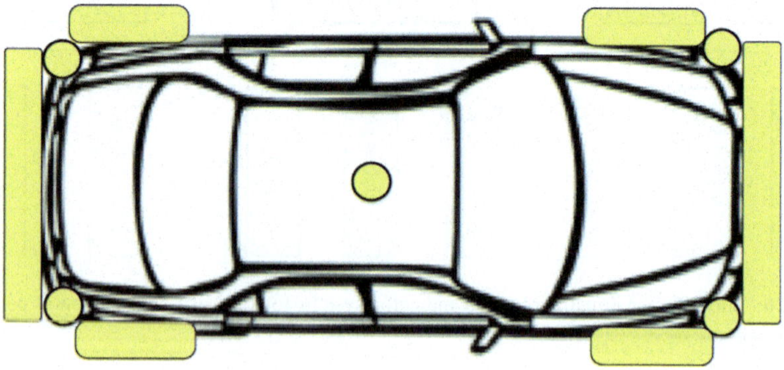

Fig. 3.27 Alternative sensor installation positions

where W_R is the width of the object area (total lane-width), L_R is the length of the object area and L_V is the vehicle length.

(1) Objective function of detection coverage

Deduce effective sensor detection coverage by divided area, according to the geometric relationship between the sensor detection coverage and the object area, and then get the objective function of the detection coverage.

a. Effective detection coverage of sensors at straight foreside/hind-side area

Effective detection coverage of the sensor at straight foreside of the vehicle is shown in Fig. 3.29, and according to the geometric relationship between the sensor detection coverage and the object area in the figure, it can be deduced that the effective sensor detection coverage S_1 is:

$$S_1 = \begin{cases} \pi \cdot \theta \cdot r^2/360, & if\ r \cdot \sin\frac{\theta}{2} < \frac{W_R}{2} \\ \left(d_f - \frac{W_R}{4} \cdot \mathrm{ctg}\frac{\theta}{2}\right) W_R, & else\,if\ \frac{W_R}{2} \cdot \mathrm{ctg}\frac{\theta}{2} < d_f \\ \frac{W_R}{2}\frac{W_R}{2}\mathrm{ctg}\frac{\theta}{2}, & others \end{cases} \tag{3.43}$$

where r, θ are the detection distance and horizontal view angle of the sensor respectively.

Effective coverage detected by a sensor at the straight hind-side of the vehicle is similar to that of Eq. (3.42), except that d_f in the equation is replaced by d_r, so there will be no separate discussion here.

b. Effective detection coverage of front/rear side sensors

Effective detection coverage of a sensor on the left front side of the vehicle is shown in Fig. 3.30, and according to the geometric relationship between the sensor detection coverage and the object area in the figure, it can be deduced that the sensor's effective detection coverage S_2 is:

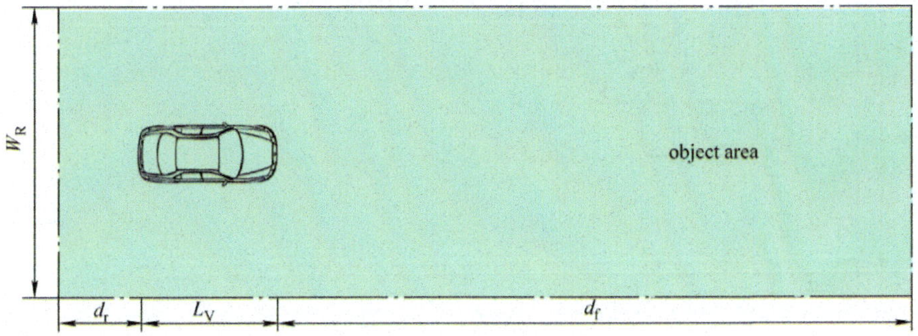

Fig. 3.28 Schematic diagram of vehicle and object area to be detected

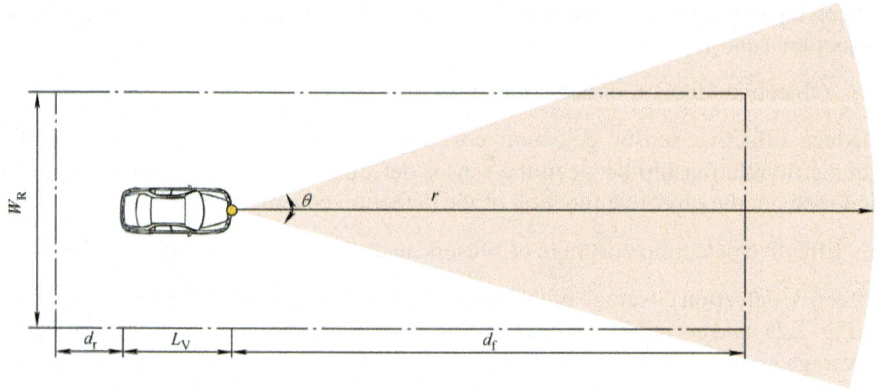

Fig. 3.29 Schematic diagram of effective coverage detected by a sensor at straight foreside of vehicle

$$S_2 = \begin{cases} \pi\theta r^2/360, & if\ r\sin\left(\frac{\theta}{2}+\alpha\right) < \frac{W_R}{2} \\ \frac{1}{2}d_L^2\left[\mathrm{ctg}\left(\alpha-\frac{\theta}{2}\right)-\alpha\right], & if\ r\sin\left(\alpha-\frac{\theta}{2}\right) \geq d_L \\ \pi r^2\left(\arcsin\frac{d_L}{r}-\alpha+\frac{\theta}{2}\right)/360+\frac{1}{2}d_L^2\left[\mathrm{ctg}\left(\arcsin\frac{d_L}{r}\right)-\mathrm{ctg}\left(\frac{\theta}{2}+\alpha\right)\right], & others \end{cases} \tag{3.44}$$

where $d_L = \frac{W_R - W_V}{2}$ is the distance between the sensor and broadside of the object area, W_V is the vehicle width, and α is the included angle between the installation direction of the sensor on the left front side and the longitudinal axis of the vehicle.

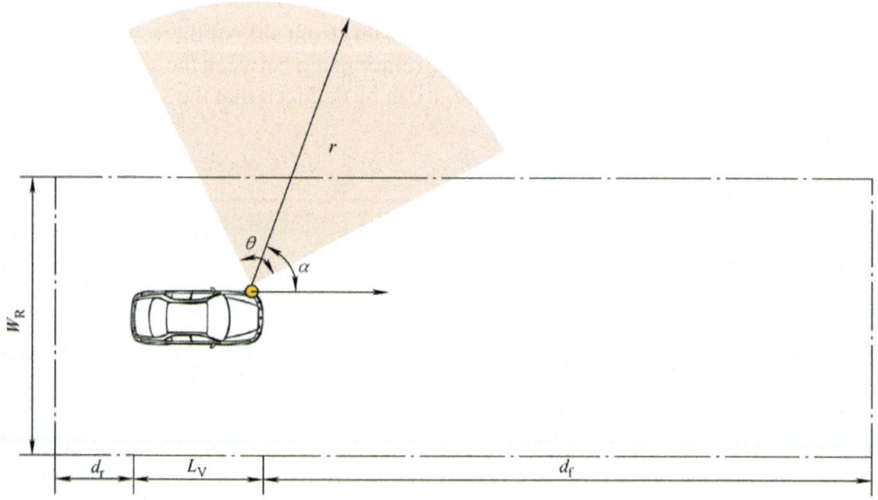

Fig. 3.30 Schematic diagram of effective coverage detected by a sensor at the left front position of the vehicle

Effective detection coverage of a sensor on the right front side of the vehicle is consistent with the above formula. Similarly, effective detection coverage of a sensor on the rear side of the vehicle (including left rear and right rear) can be obtained as follows:

$$S_3 = \begin{cases} \pi\theta r^2/360, \quad if \ r\sin\left(\frac{\theta}{2}+\beta\right) < \frac{W_R}{2} \\ \frac{1}{2}d_L^2\left[\mathrm{ctg}\left(\beta-\frac{\theta}{2}\right)-\beta\right], \quad if \ r\sin\left(\beta-\frac{\theta}{2}\right) \geq d_L \\ \pi r^2\left(\arcsin\frac{d_L}{r}-\beta+\frac{\theta}{2}\right)/360+\frac{1}{2}d_L^2\left[\mathrm{ctg}\left(\arcsin\frac{d_L}{r}\right)-\mathrm{ctg}\left(\beta-\frac{\theta}{2}\right)\right], \quad others \end{cases} \tag{3.45}$$

where β is the included angle between the installation direction of the rear-side sensor and the longitudinal axis of the vehicle.

c. Effective detection coverage of sensors on vehicle top

Considering installation requirements of a 360° multi-antenna laser radar, effective detection coverage of a sensor on vehicle top is shown in Fig. 3.31. According to the geometric relationship between the sensor detection coverage and the object area in the figure, it can be deduced that the effective sensor detection coverage S_7 is:

$$S_7 = \begin{cases} 2rW_R, \ if \ r < d_r \\ (r+L+d_r)W_R, \ if \ d_r < r < d_f \\ (d_f+L+d_r)W_R, \quad others \end{cases} \tag{3.46}$$

Based on the above formulas and with minimum PBA as the optimization object, the objective function of detection coverage is designed as follow:

$$J_1(X) = 1 - S(X)/S_R \tag{3.47}$$

where $S(X)$ is the union of effective detection coverage of sensors at different positions, which can be solved by relevant formulas in the multi-sensor system modeling section.

(2) Objective function of detection precision

Take detection precision of a straight-ahead object as one of the objective functions to improve straight-ahead ranging accuracy, and therefore ensure accurate longitudinal motion control (adaptive cruise control, emergency braking control). The objective function of detection precision is designed as follow:

$$J_2(X) = \varepsilon_{X(1)^2} \tag{3.48}$$

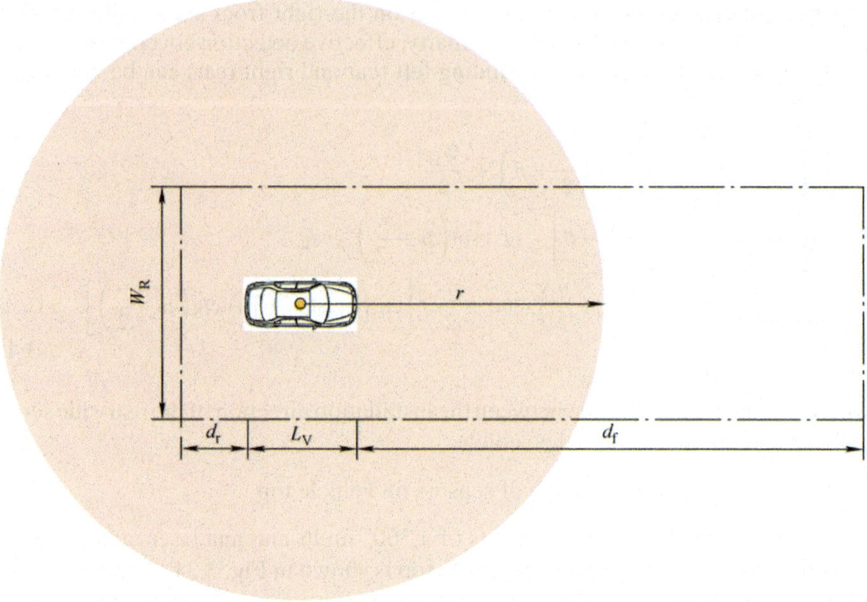

Fig. 3.31 Schematic diagram of effective coverage detected by a 360° sensor on vehicle top

where $\varepsilon_{X(1)}$ is the range error of the sensor at the right ahead position of the vehicle.

(3) Objective function of sensor cost

Total cost of an integrated onboard multi-sensor system includes configured sensor cost and installation cost. Installation cost is mainly related to sensor type, and for the convenience of unified quantification, the installation cost of sensors at different positions is not specified but it is included in sensor price, so the objective function of sensor cost is as follow:

$$J_3(X) = \sum_{i=1}^{n} \mathrm{Pr}_{X(i)} \tag{3.49}$$

3.5.3.2 Designing the Constraints

To transform the sensor optimal configuration problem into a multi-objective optimization problem, in addition to establish objective functions, it is necessary to consider performance, cost and reliability limit of the selected sensors and restricted conditions of sensor configuration, and then design the constraints of the multi-objective optimization problem accordingly. Constraints of the onboard sensor configuration mainly consider the shortest detection distance requirement in a certain direction (such as right ahead) and certain sensor's special requirements on alternative installation location. The limit value of one or several indicators in the evaluation indicator system can also be taken as constraints to fulfill basic detection performance or reliability.

First, priority for an intelligent driving system is the detection distance right in front of or behind the vehicle, so the constraints of effective sensor detection range at both positions are set up as follows:

$$d_{X(1)} \geq d_{\mathrm{f}} \tag{3.50}$$

$$d_{X(2)} \geq d_{\mathrm{r}} \tag{3.51}$$

where $X(1), X(2)$ represent codes of the sensors at straight foreside or hind-side of the vehicle (position number is 1 and 2 respectively), and $d_{X(1)}$, $d_{X(2)}$ represent detection distance of the sensors at such position, and d_{f}, d_{r} represent the distance that needs to be detected in front of and behind the vehicle.

Secondly, ultrasonic radar, millimeter-wave radar and laser radar sensors have different alternative installation positions on vehicles due to differences in overall dimensions, so the installation position constraints of each sensor need to be set as follow:

$$b \geq X(i) \geq a; \quad i \in 1, 2, \ldots, n \tag{3.52}$$

where i represents the code of sensor position, $X(i)$ represents the code of the sensor at this position, and a and b represent the number range of a certain type of sensor respectively.

Thirdly, MTTF of the sensor system is taken as a constraint condition to ensure the reliability of the multi-sensor system:

$$MTTF(X) \geq t_{\min} \tag{3.53}$$

where vector X represents the configured sensor set, and its element value represents the code of the selected sensor at this position; $\mathrm{t_{min}}$ represents the minimum MTTF allowed by the system.

In addition, it is also possible to set detection frequency, range accuracy and cost price of the selected single sensor as constraints to reduce the solution's search scope. For example, unit price of the selected sensor should not exceed the maximum value allowed, which is expressed as follow:

$$\mathrm{Pr}_{X(i)} \leq \mathrm{Pr}_{\max}; \quad i = 1, 2, \ldots, n \tag{3.54}$$

where Pr_i, $Pr_{\max}$ are the cost price at position i and the maximum price of a single sensor allowed by the system respectively.

Based on the above analysis, constraints of the sensor configuration optimization problem are established as follows:

$$\text{s.t.}\begin{cases} \left[d_{X(1)} \quad d_{X(2)}\right] \geq [d_{\mathrm{f}} \quad d_{\mathrm{r}}] \\ b \geq X(i) \geq a, \quad i \in 1, 2, \ldots, n \\ MTTF(X) \geq t_{\min} \\ \Pr_{X(i)} \leq \Pr_{\max}, \quad i = 1, 2, \ldots, n \end{cases} \tag{3.55}$$

3.5.4 Multi-objective Optimization Solution Algorithm

To solve the multi-objective optimization problem, it is relatively simpler to transform the multi-objective problem into a single objective problem by a weighted combination of object values, and the result obtained is the only solution. This method is simple and practical, but determination of the weight coefficients needs manual selection, and the obtained result is too dependent on the weight coefficients. Another commonly used method is to solve the Pareto front of a multi-objective problem by heuristic search algorithm, and the obtained result is a group of optimal solution set for different users to choose according to their needs. In this section, NSGA-II multi-objective optimization algorithm and GA-PSO combination algorithm are adopted respectively to solve the sensor configuration optimization problem.

3.5.4.1 Multi-objective Optimization Algorithm Based on NSGA-II

Pareto optimal solution set and Pareto front are defined as follows: for a multi-objective optimization problem as shown in Eq. (3.56), assume $\overline{x} \in X$, if there is no such condition of $x \in X$ to ensure $f(x) \leq f(\overline{x})$, then $\overline{x}$ is Pareto optimal solution (non-inferior solution) of the multi-objective optimization problem. The collection of all Pareto optimal solutions is called Pareto optimal solution set. Imaging of Pareto optimal solution set in object space is called Pareto front.

$$\begin{aligned} &\min f(x) = (f_1(x), f_2(x), \ldots, f_{\mathrm{m}}(x))^T \\ &\text{s.t.}\begin{cases} g_i(x) \geq 0, & i \in I \\ h_j(x) = 0, & j \in E \end{cases} \end{aligned} \tag{3.56}$$

The multi-objective genetic algorithm enables a non-dominated solution set constructed in each generation to gradually approach Pareto optimal boundary, by constructing and retaining individual non-dominated solution in the current evolutionary population during each generation of genetic operation. This method is more suitable for solving the multi-objective combinatorial optimization problem. Non-dominated sorting genetic algorithm (NSGA) is a commonly used multi-objective genetic algorithm, which can obtain Pareto front of the non-inferior solution set, but with the problem of high computation complexity when constructing Pareto optimal solution set, lack of optimal individual retention mechanism and difficulty in determining shared parameters. NSGA-II algorithm is the improved algorithm put forward by Deb et al. [19] against the above problems in the NSGA algorithm, which introduces elitist strategy, rapid non-dominated solution sorting and congestion comparison operator, and therefore improves the accuracy of optimization results, reduces NSGA complexity, ensures more uniform distribution of Pareto optimal solution and further improves computational efficiency and robustness of the algorithm.

Main solution steps for sensor optimal configuration: initial population configuration, genetic operation, sorting and grading of Pareto non-dominated solutions, congestion calculation, new individual selection, updating parent population and loop iteration.

(1) Configuration of the initial population

Number alternative sensors, read-in sensor parameters and randomly select sensors and installation positions as the initial population. The initial population is a set of vectors, each vector represents an initial sensor configuration scheme. It is better to select a variety of sensors and install them in different positions to ensure population diversity,

(2) Genetic operation

Genetic operation mainly includes selection, crossover and variation. The purpose of the genetic operation is to directly inherit the preferred sensor configuration to the next generation or generate a new sensor configuration scheme through paired crossover and mutation operation, and then inherit it to the next generation. Its selection operator, crossover operator and mutation operator are the same as those in general genetic algorithm. Genetic operation is based on individual fitness evaluation, and individual fitness function is the optimized objective function of sensor configuration.

(3) Pareto sorting and grading

After the genetic operation, the offspring population and parent population will go through Pareto sorting and grading together. The main steps of grading [20] are as follows:

a. Calculate n_p and S_p parameters of each individual p in the population, in which n_p represents the number of individuals that dominant individual p in the population, and S_p represents the individual set dominated by p in the population;
b. Search for the individual that makes $n_p = 0$ in the population and save the results to current set R_1;

c. For each individual i in current set R_1, the individual set dominated by i is S_i. Go through all individuals in S_i, execute $n_l = n_l - 1$, if $n_l = 0$, then save the individual l to set L;
d. Mark individuals in R_1 as the first non-dominant layer individuals, represented as $Rank = 1$, and then repeat the above steps with L as current set until the entire population is graded.

(4) Congestion calculation

Calculation of population congestion degree is an important part to ensure population diversity in the NSGA-II algorithm. For each objective function, sort population based on objective function first and then calculate individual congestion degree according to the following Eq. (3.57):

$$\begin{cases} D_1 = \infty \\ D_n = D_n + [f(n+1) - f(n-1)], \quad n = 2, 3, \ldots, N-1 \\ D_N = \infty \end{cases} \tag{3.57}$$

where D_n is the congestion degree of individual n, and its initial value is 0; $f(n)$ represents the objective function value of the number n individual.

(5) Individual selection and population renewal

After calculating Pareto sorting and grading and congestion degree, each individual in the population has two attributes: Pareto non-dominant grading $Rank_n$ and congestion degree D_n, which are the basis to distinguish the dominant and non-dominant relationship between any two individuals in the population. Select individuals based on the principle of small Pareto grading and small congestion degree, and then form a new parent population to renew the previous population.

(6) Loop iteration

Repeat the above step (2) to (5) and iterate until meet the maximum number of iterattions set or other termination conditions.

A basic flow chart of optimal sensor configuration solution based on the NSGA-II algorithm is shown in Fig. 3.32.

3.5.4.2 Multi-objective Weighted Combinatorial Optimization Algorithm Based on the GA-PSO Algorithm

PSO algorithm is a population-based random search evolutionary algorithm widely used in engineering, with advantages of swarm intelligence, inherent parallelism, simple iteration format and fast convergence speed. Similar to other evolutionary algorithms, the PSO algorithm and GA algorithm adopt the idea of "population" and "evolution" and operate based on the fitness values of individuals. However, there is no evolution operator in PSO, instead each individual is regarded as a particle without weight and volume and flying at a certain speed in search space, and the flying speed

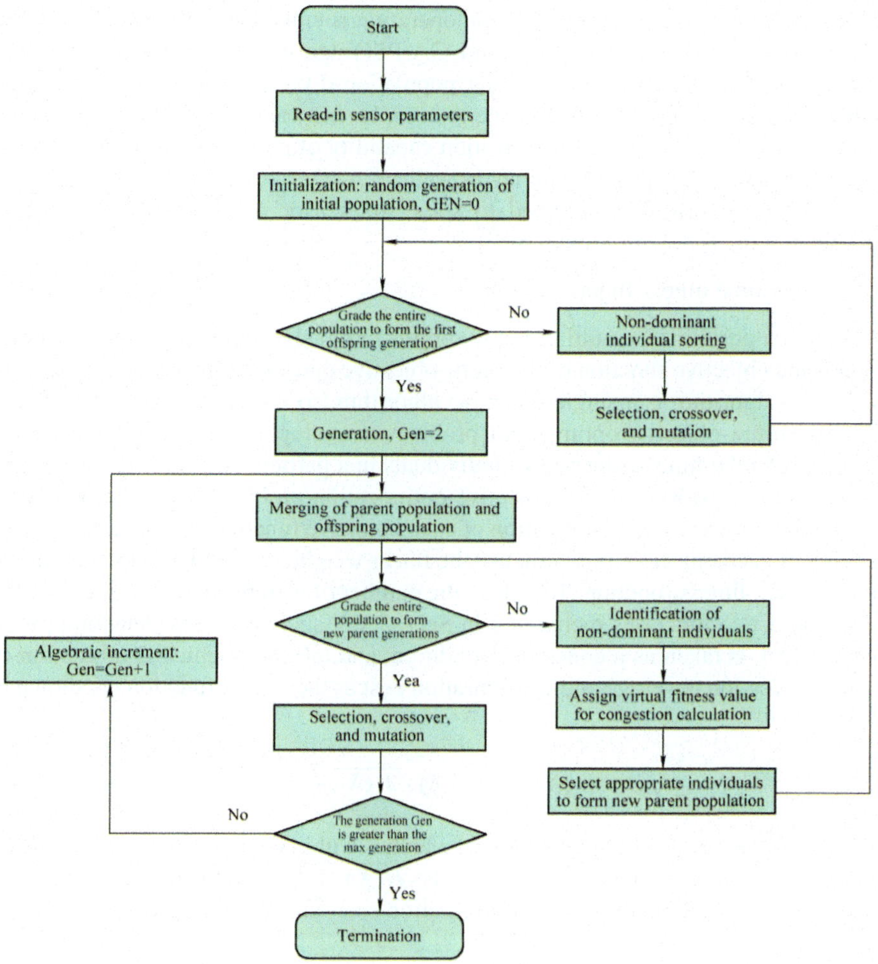

Fig. 3.32 Basic steps of NSGA-II algorithm

can be dynamically adjusted according to individual and group flying experience. Therefore, the PSO algorithm has disadvantages such as poor population diversity, small search range and premature local optimal value, especially when optimizing high-dimensional complex problems. GA algorithm uses selection, crossover and mutation operators to directly search information with the objective function, and therefore population diversity and global search ability are better. However, individuals in the GA algorithm have no memory but works in a probabilistic way, and no direction for genetic operation leads to slow convergence. The hybrid GA and PSO algorithm can overcome PSO algorithm's problem of easy fall into local optimum, and increase GA algorithm's convergence speed, thus improving algorithm performance. Their combination could be in serial, parallel, embedded and other ways.

In this section, GA algorithm's genetic operation is embedded into PSO algorithm, known as GA-PSO embedded algorithm. GA-PSO algorithm makes full use of advantages from both algorithms, including population diversity, random global search ability, memory and fast convergence to complement each other. Such a combination will enhance the global optimization capability of particle swarm optimization algorithm and convergence accuracy as well.

GA-PSO algorithm is adopted to solve the sensor preferential configuration problem in the following steps:

(1) Determine fitness function

NSGA-II algorithm is a multi-objective optimization algorithm, and its fitness function is the objective function of the multi-objective optimization problem. However, GA-PSO algorithm is a single objective algorithm, so it is necessary to first transform the multi-objective optimization problem into a single objective problem before determining the fitness function of individuals in a particle swarm. There are many methods to transform a multi-objective optimization problem. This section adopts the method of weighted combination of each objective function. After value normalization of each objective function, use the linear weighted sum method to determine individuals' fitness function. To reduce the number of weight coefficients, one of the three objective functions mentioned in Sect. 3.5.3, such as sensor detection precision $J_2(X)$, is taken as a constraint condition, and take a weighted combination of sensor detection coverage and configuration cost as the fitness function, as shown in Eq. (3.58):

$$fitness(X) = (1 - \delta) \cdot \overline{J_1(X)} + \delta \cdot \overline{J_3(X)} \tag{3.58}$$

where δ is the weight coefficient of cost, which is within the range of $0 \sim 1$ and can be selected by customers based on their needs; $\overline{J_1(X)}$, $\overline{J_3(X)}$ are the normalized sensor detection coverage function and configuration cost function respectively.

(2) Initial particle swarm generation

Number alternative sensors and initialize size of particle swarm, inertia weight, learning factors and number of iterations in the algorithm. Read-in sensor parameters, randomly select sensors and installation positions as the initial position of the particle swarm and set the initial velocity of the particle swarm.

(3) Calculate fitness value and select individual extremum and population extremum

Calculate the fitness value of each particle based on an individual fitness function, select optimal fitness value that can be obtained by the current particle as individual optimal value p_{best}, and select optimal fitness value that can be obtained by all particles up to now as population optimal value g_{best}.

(4) Individual particle's velocity and position update

Use Eq. (3.47) to implement particle evolution and update flying velocity and position of individual particles:

$$\begin{aligned} v_i^{t+1} &= wv_i^t + c_1 \times \text{rand} \times (p_{\text{best}} - x_i^t) + c_2 \times \text{rand} \times (g_{\text{best}} - x_i^t) \\ x_i^{t+1} &= x_i^t + v_i^{t+1} \end{aligned} \tag{3.59}$$

where v_i^t is the velocity of particle i in number t iteration, w is the inertia weight, x is the particle position, *rand* is a random number between 0 and 1, and c_1 and c_2 are learning factors.

(5) Genetic operation over the particle swarm

Genetic operation in GA algorithm mainly includes selection, crossover and mutation. Since evolutionary operation in GA-PSO algorithm already selects the next generation of particle swarm, so this step only carries out crossover and mutation operations, which means to generate new sensor configuration schemes through particle pairing, crossover and mutation operation, and therefore increases population diversity and expands search scope.

(6) Parameter cross-line test

Set search spacial scale of particle swarm according to constraints of the optimization problem, and check whether particle swarm is still within this search spacial scale after the update and genetic operation to ensure the optimal solution meets constraints of the optimization problem.

(7) Calculate fitness value and update individual extremum and population extremum

Calculate fitness value of all particles with individual fitness function, update optimal fitness value p_{best} available to current individual particle and optimal fitness value of particle swarm g_{best} up to now, and particle position corresponding to population extremum g_{best} is the optimal sensor configuration scheme of the current iteration.

(8) Loop iteration

Repeat step (4) to (7) and iterate until the set maximum number of iterations is reached or other termination conditions are met.

Based on the above steps, a basic flow chart of sensor optimal configuration solution based on GA-PSO algorithm is designed as shown in Fig. 3.33.

3.6 Structure Sharing Technology for i-EFV Driving/Braking System

As a future concept vehicle that integrates advanced technologies of intelligent vehicle and NEV, i-EFV also integrates its advantages in system composition. This team carries out system structure sharing technology research, including analysis on principle, function, application scope and dynamic characteristics of each subsystem actuator based on i-EFV system composition and its hierarchical control system, and

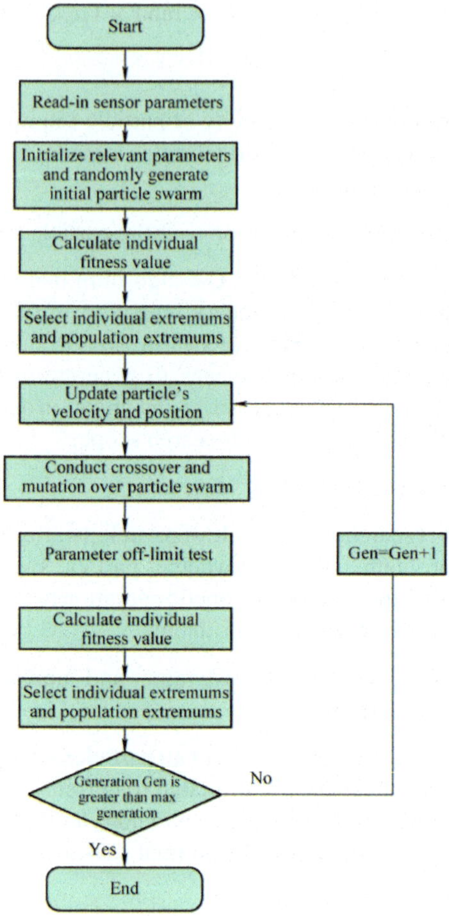

Fig. 3.33 Basic flow chart of GA-PSO algorithm

studies working state of driving and braking system. Thanks to these researches and system coordination control technology, integration and optimization of vehicle driving/braking systems are finally realized.

3.6.1 Structure Sharing Technology Based on Hybrid Driving System

I-EFV mainly includes intelligent hybrid electric vehicle (i-HEV) and intelligent battery electric vehicle (i-BEV). Since BEV only has electric driven system, this section mainly discusses structure sharing technology for the hybrid driving system.

In terms of driving system control mode, conventional internal combustion engine adopted by traditional intelligent vehicles generally uses the electronic throttle for control, so it is necessary to develop an under-layer controller based on electronic throttle control technology which converts expected driving torque to expected electronic throttle opening and passes it to the engine control unit to realize effective control of driving torque. Since the driving system in traditional hybrid electric vehicle (HEV) has two power sources and its dynamic coupling is torque coupling, its driving system control already adopts engine torque control and motor torque control, which directly passes the expected torque of engine and motor to their control units and realizes driving force control under hybrid electric driven state.

In terms of driving system control structure, conventional intelligent vehicles need an additional set of mechanical device for auxiliary control of electronic throttle to switch between manual driving and assistant driving; while the driving system in traditional hybrid electric vehicles adopts control-by-wire technology, which allows easy switch between various control states without any additional device.

That's why the i-HEV driving system adopts a dual power-source system composed of engine and motor, and the control system of the driving system adopts advanced distributed control system architecture, torque control mode and control-by-wire technology, as well as layered control system framework. This section studies and analyzes the principle, function, application scope and dynamic features of each subsystem, and designs powertrain working states under manual driving and intelligent assistant driving.

(1) Analysis of the powertrain's dynamic characteristics

There are two powertrains in i-HEV: engine and motor. The former's dynamic characteristics include slow response (over 100 ms) and low response accuracy, and the latter include fast response (less than 10 ms) and high response accuracy.

(a) Characteristics of engine dynamic process

Acceleration, deceleration, braking, cold start and heating of turbine are all dynamic working conditions of the engine. Taking gasoline engine as an example, engine intake will change immediately with the change of throttle opening in the process of increasing throttle opening for acceleration and decreasing throttle opening for deceleration, but the amount of fuel entering the cylinder cannot change immediately for two reasons: first, information delay causes a slower change of gasoline amount ejected by electronic fuel injector than the change of intake air amount in the engine; second is the impact of condensed oil film on the inlet pipe wall, as gasoline ejected from the injector is not entirely evaporated immediately and flows to cylinder together with air through inlet air pipe, but only a small amount of gasoline particles are in the airflow, and a lot of gasoline falls on the inlet pipe wall and forms an oil film that moves along the pipe wall towards the intake valve.

Therefore, gasoline sucked into the engine includes oil particles directly evaporated after ejected by the injector, oil vapor evaporated from oil film on the pipe wall and a small amount of oil film remaining at the intake valve. This makes the amount of fuel sucked into the cylinder does not equal to the amount of fuel being ejected from the injector in the same unit of time.

Figure 3.34 is the simulation result of engine torque response on the Matlab /Simulink platform, with an initial engine speed of 1600 r/min. There are three curves in the simulation result: solid line is the simulation result of complete control of transient air–fuel ratio, dotted line and dot-dash line are the simulation results of insufficient control of transient air–fuel ratio, as shown in Fig. 3.30e. In Fig. 3.30b, c, it can be seen that changes of throttle opening and intake manifold pressure in these three scenarios are basically consistent, which illustrates the same change rule for the amount of air coming into the cylinder, and change of transient air–fuel ratio is caused by fuel amount control. The dotted line represents the scenario of insufficient fuel amount, while the dot-dash line represents excessive fuel amount. The ideal situation is the transient air–fuel ratio under full control. As shown in the simulation result in Fig. 3.34a, the actual engine torque response time to the object torque step up stage is about 200 ms when the transient air–fuel ratio is under full control and convergence is fast. When the transient air–fuel ratio is in insufficient control, the excessive fuel amount may cause an overshoot of engine torque response and low fuel amount may cause slow engine torque response, both will affect dynamic response characteristics of engine torque. In addition, Fig. 3.34a also reveals that: (1) when object torque steps up, engine torque decreases first and then rises regardless of control mode; (2) when object torque steps down, engine torque response speeds under three control modes are slower than that of step-up stage. These torque response results will also affect power transmission stability of the power system.

(b) Characteristics of motor dynamic process

Permanent magnet asynchronous motor is used in intelligent hybrid electric vehicles. Relations among the stator current i, magnetic motive force ψ and voltage u are expressed as follows:

$$\begin{cases} i = i_d + ji_q \\ \psi = \psi_d + j\psi_q \\ u = u_d + ju_q \end{cases} \tag{3.60}$$

Stator flux-linkage equation and voltage equation of the permanent magnet synchronous motor in d-q coordinate system are as follows:

$$\begin{cases} \psi_d = L_d i_d + \psi_r \\ \psi_q = L_q i_q \end{cases} \tag{3.61}$$

$$\begin{cases} u_d = r_s i_d + p\psi_d - \omega\psi_q \\ u_q = r_s i_q + p\psi_q + \omega\psi_d \end{cases} \tag{3.62}$$

where ψ_r is the coupling flux linkage of rotor magnetic steel on the stator, L_d and L_q are the main inductance of d and q axes of the permanent magnet synchronous

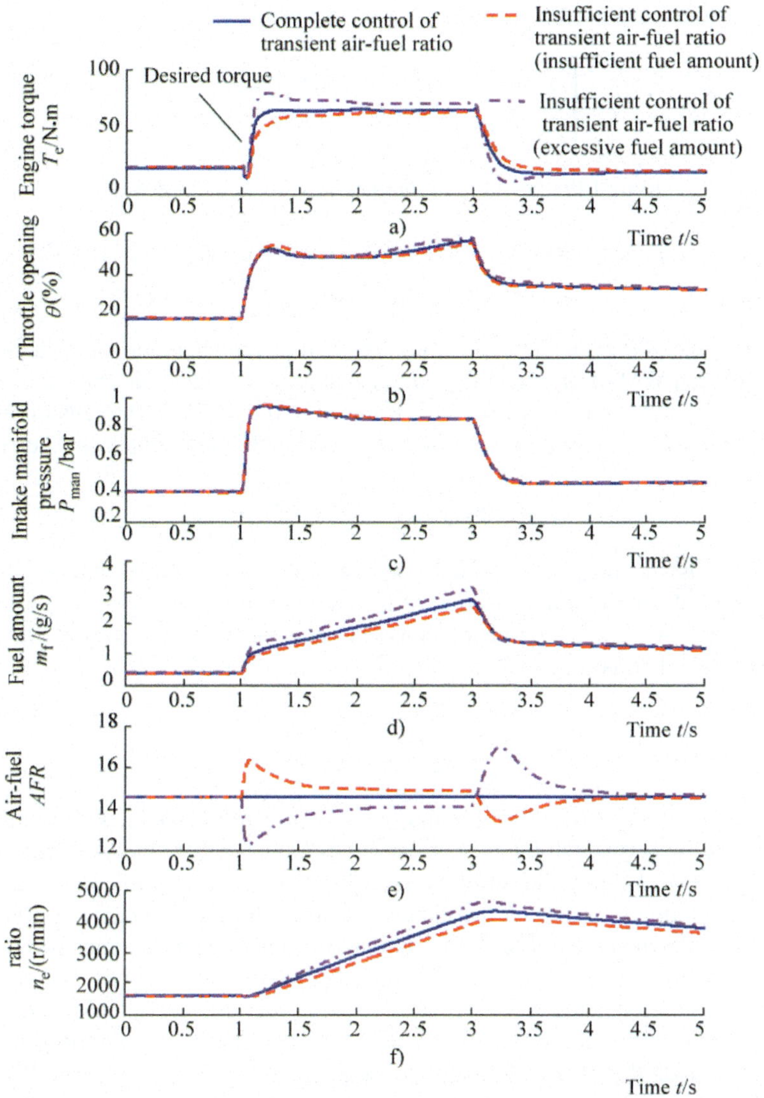

Fig. 3.34 Simulation results of engine torque dynamic response under different transient air–fuel ratio control

motor, i_d and i_q are the components of current vector of the stator in d and q axes, u_d and u_q are the components of voltage vector of the stator in d and q axes, ω is the rotor angular frequency, r_s is the internal resistance of stator and p is the differential operator.

By substituting Eq. (3.61) into Eq. (3.62), we can get:

$$\begin{cases} u_d = r_s i_d + L_d \frac{\mathrm{d}i_d}{\mathrm{d}t} - \omega L_q i_q \\ u_q = r_s i_q + L_q \frac{\mathrm{d}i_q}{\mathrm{d}t} + \omega L_d i_d + \omega \psi_r \end{cases} \tag{3.63}$$

Torque equation of the permanent magnet synchronous motor is:

$$\begin{aligned} T_m &= P \cdot \mathrm{Im}(\psi \cdot i^*) = P \cdot \mathrm{Im}[(i_d - j i_q)(\psi_d + j \psi_q)] \\ &= P \cdot [\psi_r i_q + (L_d - L_q) i_d i_q] \end{aligned} \tag{3.64}$$

where P is the pole pairs, and vector control of the permanent magnet synchronous motor adopts PWM inverter control, which makes the resultant current vector i of the stator's three-phase current i_A, i_B and i_C locate in q axis, $i_d = 0$ to ensure motor's output torque T_m is in proportion to the amplitude of the stator phase line current I_m:

$$T_{\mathrm{m}} = P\psi_{\mathrm{r}} i_q = P\psi_{\mathrm{r}}(1.5 \times \mathrm{Im}) = K_m \mathrm{I}_m \tag{3.65}$$

where K_m is the torque coefficient. As shown in the above, change of stator current and change in output torque of the permanent magnet synchronous motor is basically synchronous, and time constant between the stator current and PWM inverter-controlled voltage is:

$$\tau_{iu} = \frac{L_q}{r_{\mathrm{s}}} \tag{3.66}$$

The time constant is at millisecond level, so it can be considered that the time constant of the permanent magnet synchronous motor torque is also on millisecond scale. Figure 3.35 shows the simulation results on Matlab/Simulink platform. Simulation calculation is based on the change of object torque trajectory at given step up and step down stage in Fig. 3.35a, b, current in permanent magnet synchronous motor changes very fast, as shown in Fig. 3.35c, d, so torque of the permanent magnet synchronous motor can follow object torque curve very quickly, with a response time of only a few milliseconds and no obvious overshoot or delay like those in motor torque. Compared with gasoline engine torque's response time of several hundred milliseconds, the response time of a few milliseconds is very short, and therefore torque of the permanent magnet synchronous motor can be regarded as a transient variable.

(2) Designing the powertrain working state

The design of powertrain working state should facilitate driving system efficiency optimization as far as possible, which requires determination of vehicle working mode and optimal power allocation of the driving assembly under different modes.

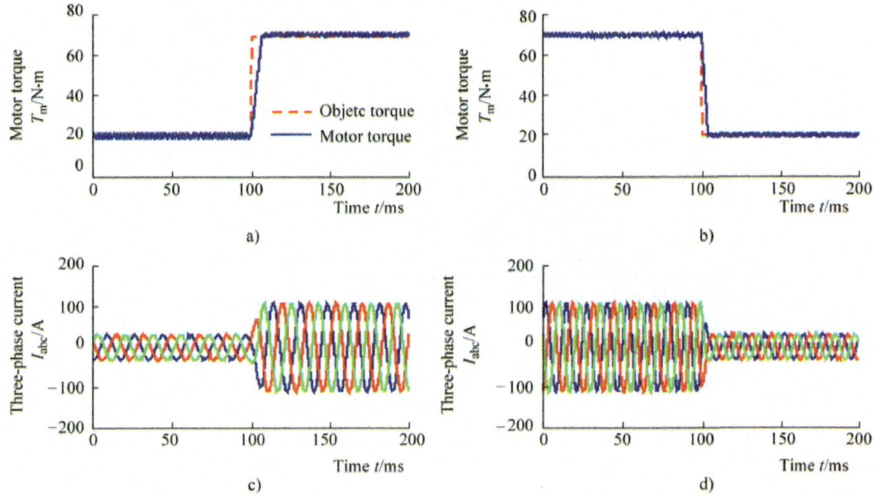

Fig. 3.35 Simulation results of motor torque dynamic response

Vehicle that just starts or runs at a low speed requires low torque, and the motor alone can provide the torque required for vehicle operation. Turning off the engine can avoid engine running in the non-economic zone of low speed and low torque.

In the condition of small acceleration or medium constant speed, the vehicle's demand for torque increases and then the engine starts. At the same time, the motor is in a power generation state, so that engine torque output is used to generate electricity while driving the vehicle. By adjusting the motor's power generation torque, two powertrains coordinate with each other and make sure that the engine works in the economic zone of high torque.

The vehicle's torque demand further increases in the condition of medium acceleration or high speed, and the vehicle driven by the engine alone can ensure that the engine runs in an efficient zone.

Significant acceleration or climbing further increases torque demand of the vehicle, and the motor is required to drive the vehicle together with the engine.

Figure 3.36 shows the schematic diagram of the driving assembly working state.

3.6.2 Structure Sharing Technology of Braking System Based on Motor Braking and Electronically-Controlled Hydraulic Braking

Since conventional intelligent vehicles adopt electronically-controlled hydraulic braking system based on ESP/EHB/EVB or additional electronically-controlled hydraulic braking system, it is necessary to develop a lower controller of braking

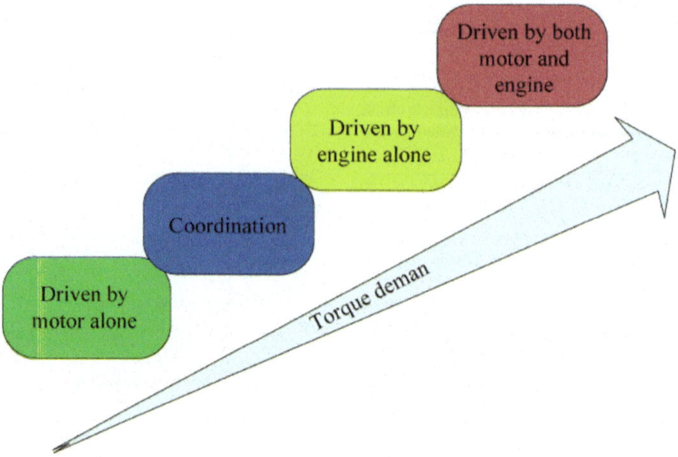

Fig. 3.36 Schematic diagram of driving assembly working state

control based on the inverse model of hydraulic system to effectively control braking torque, which converts expected braking torque to PWM control signals that control solenoid valve and then passes the signals to the hydraulic electronic control unit. While in conventional hybrid electric vehicles, due to the different complexity of braking systems, there are two types of hydraulic braking: controlled and uncontrolled. For the controlled mode of a hydraulic system, since motor braking and hydraulic braking are both in torque coupling mode, motor braking and hydraulic braking have adopted torque control in braking system control, which sends torque command directly to the motor control unit and hydraulic control unit to realize effective control of the braking system. For the uncontrolled mode of a hydraulic system, its motor still adopts torque control mode and its hydraulic system is completely operated by the driver.

Conventional intelligent vehicles generally adopt two types of braking system control structure to realize manual driving and assistant driving, either update the software of the existing ESP/EHB/EVB system, or add a set of single-channel/multi-channel hydraulic braking system on top of the original ABS/TCS/ESP hydraulic braking system. Conventional hybrid electric vehicles generally update the software of the existing EHB/EVB system in case of the controlled hydraulic braking system, or maintain the traditional hydraulic braking system in case of the uncontrolled hydraulic braking system.

Therefore, the i-HEV braking system adopts a dual braking system of motor and hydraulic braking based on EHB/EVB. Braking system control also adopts the advanced distributed control system architecture, torque control and control-by-wire technology, as well as layered control system framework. On this basis, this section studies and analyzes the principle, function, application scope and dynamic characteristics of each subsystem of the braking system, and designs the working state of braking assembly under manual driving state and intelligent assistant driving state.

(1) Dynamic characteristics analysis of two braking assemblies

(a) Dynamic characteristics of the electrically-controlled hydraulic braking system

The maximum braking force produced by a hydraulic braking system is much greater than that of a motor braking system, but it is difficult to accurately control its intensity due to the nonlinear influence of hydraulic system and mechanical friction. The hydraulic braking system has a certain response delay, which mainly comes from two aspects, one is the action lag of solenoid valve (about 10 ms); the other is the delay caused by the hydraulic pipeline between solenoid valve and cylinder. These two reasons cause some delay of 10–40 ms and first-order delay of 50–100 ms from sending object pressure instruction to real pressure establishment on wheel cylinder.

Table 3.5 compares time response characteristics of several different braking force regulating mechanisms, as shown in the Table 3.5, the response times of all the listed types of hydraulic braking systems are slower than that of the motor braking system.

(b) Dynamic characteristics of the motor braking system

A permanent magnet asynchronous motor is used as a generator in the braking process, which recycles kinetic energy of the vehicle to electric energy and stores it in the battery. Its torque equation is the same as the one in the driving process, and motor braking system can also respond to braking torque commands quickly and accurately.

However, electric braking has the following disadvantages as a braking system: (1) limited braking power and braking force; (2) great dynamic fluctuation of electric braking performance; (3) operational reliability is lower than that of a general mechanical braking system due to many limitations.

(2) Design of working state of two braking assemblies under normal braking state

Under a normal braking state, the sum of motor braking force and hydraulic braking force should be equal to the braking demand of the driver. During the braking process, due to the influence of the motor's constant power and battery capacity, the maximum braking force that the motor can provide varies with speed, battery *SOC* and other parameters, as shown in Fig. 3.37. Therefore, it is necessary to coordinate control of

Table 3.5 Response time comparison of different braking actuators

	τ_D	τ_m (ms)
Type I (motor brake)	100 μs	1
Type II (hydraulic brake)	5 ms	50
Type III (hydraulic brake)	10 ms	50
Type IV (hydraulic brake)	20 ms	100
Type V (hydraulic brake)	30 ms	100

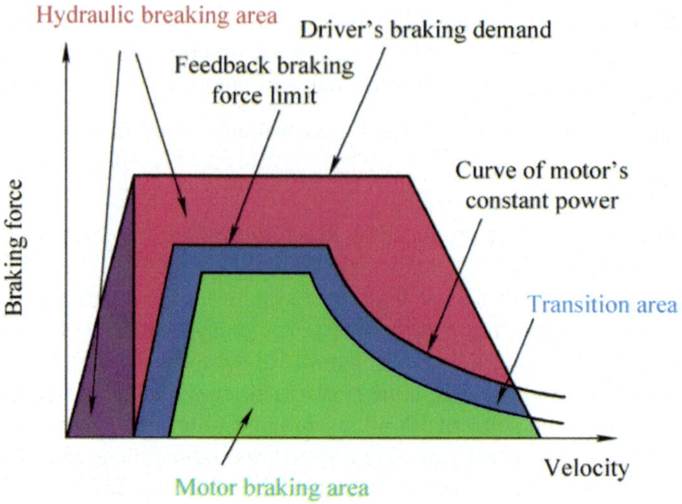

Fig. 3.37 Working states of the two braking assemblies under normal braking state

feedback braking force and friction braking force in the braking process and ensure the sum of the two is equal to the braking demand of the driver.

When the vehicle is running at low speed and the motor cannot provide enough braking torque, the hydraulic braking system works alone to provide the braking torque required by the vehicle.

When the vehicle increases its speed, if the required braking torque is small, it will be provided by motor braking system alone.

When the required braking torque increases, the vehicle enters the state of coordinated control of motor braking system and hydraulic braking system.

When the required braking torque continues to increase and exceeds maximum braking torque of motor braking system, then hydraulic braking system and motor braking system jointly provide braking torque to meet braking demand of the driver.

References

1. Li K, Chen T, Luo Y, Wang J (2012) Intelligent environment-friendly vehicles: concept and case studies. IEEE Trans Intell Transp Syst 13(1)
2. Li K, Chen T, Luo Y et al (2009) Structure of intelligent environmental friendly vehicle: China, 200810223099. 2009-02-04
3. Chakrabarti A (2001) Sharing in design: categories, importance and issues. In: Proceedings of international conference on engineering design, Glasgow
4. Chakrabarti A (2004) A new approach to structure sharing. J Comput Inf Sci Eng 4(1):11–19
5. Chakrabarti A (2007) A method for structure sharing to enhance resource effectiveness. J Eng Des 18(1):73–91
6. Ulrich K (1988) Computational and pre-parametric design. Artificial Intelligence Laboratory, MIT Cambridge

7. Sessler GM (1994) Silicon microphones. In: Proceedings of 27th Meeting Acoustical Society of America, MIT
8. Chakrabarti A, Johnson A, Kiriyama T (1997) An approach to automated synthesis of solution principles for micro-sensor designs. In: International conference on engineering design. Tampere
9. Mookhiamar O, Abe M (2006) How the four wheels should forces in an optimum cooperative chassis control. Control Eng Pract 14(3):295–304
10. Liu J (2008) Technology foundations to functional safety. China Machine Press, Beijing
11. Yu J (2016) Research on ultrasonic sensor array's perceptionon vehicle lateral object. Tsinghua University, Beijing
12. Liu Z, Deng W, Li Y, Jiang L et al (2016) Physical modeling method on ultrasonic sensors for virtual intelligent driving. SAE Technical Paper 2016-01-1901. https://doi.org/10.4271/2016-01-1901
13. Deng W, Dai J, Zhao Q, Litkouhi B, Moshchuk N, Nisonger R (2009) Modeling of range sensing and object detection for vehicle active safety. In: Proceedings of the 12th international IEEE conference on intelligent transportation systems, St. Louis, MO, USA, 3–7 Oct 2009
14. Deng W, Zeng S, Zhao Q, Dai J (2011) Modelling and simulation of sensor-guided autonomous driving. Int J Veh Design 56(1/2/3/4):341–326
15. Li Y, Wang Y, Deng W, Li X et al (2016) LiDAR sensor modeling for ADAS applications under a virtual driving environment. SAE Technical Paper 2016-01-1907. https://doi.org/10.4271/2016-01-1907
16. Wang C (2014) Study on optimal perception of sensor networks. Xidian University, Xi'an
17. Zhou G, Cai J (1986) Radar detection probability and SNR. Syst Eng Electron 12:45–60
18. Wang Q, Yang X, Yan X (2015) Introduction to integrated design of military vehicle-mounted electronic equipment. National Defence Industry Press, Beijing
19. Deb K, Pratap A, Agarwal S et al (2002) A fast and elitist multiobjective genetic algorithm: NSGA-II. IEEE Trans Evol Comput 6(2):182–197
20. Arroyo JEC, Armentano VA (2005) Genetic local search for multi-objective flowshop scheduling problems. Eur J Oper Res 167(3):717–738

Chapter 4
Information Fusion Technologies for i-EFV

Accurate identification of vehicle states and complex driving environment states is the necessary information base for i-EFV to carry out safe and energy-saving controls. For i-EFV that integrates complex electromechanical systems, faces time-varying traffic environment and implements multi-performance objective control, there are three main problems in the comprehensive identification of vehicle states and environment information: (1) the large amount of information data obtained based on V2V communication, V2I communication, remote wireless communication and onboard sensing system is overlapping and redundant, which needs integrated analysis and processing to form a unified description of vehicle and environment; (2) the vehicle driving environment is complex and changeable, and the information obtained from existing sensing systems is seriously disturbed, which makes it necessary to obtain accurate object information based on the characteristics of redundancy of multi-source sensors; (3) the accurate information on the characteristics of driver-vehicle-road traffic environment cannot be obtained directly through the sensors, which makes it necessary to fuse multi-source information and conduct comprehensive analysis. Therefore, in order to effectively use the multi-source and redundant data information to accurately identify and predict vehicle state and traffic environment, a systematic signal processing method is required.

The intelligent information exchange system realizes the information exchange of "human-vehicle-infrastructure" through onboard communication modules, so that i-EFV is no longer an information silo in the road traffic system. The perception of vehicle and driving environment state is not only based on the traditional onboard sensing system data, but also on the all-round multi-source sensing information data, which greatly expands the capability of environmental perception and raises the engineering problems of processing and applications of multi-source heterogeneous data fusion.

K. Li et al., *The Intelligent Environment Friendly Vehicle*,
Key Technologies on New Energy Vehicles,
https://doi.org/10.1007/978-981-19-4851-0_4

4.1 Multi-sensor Information Fusion Technology and Research State

Information fusion, also known as data fusion, sensor information fusion or multi-sensor information fusion, is the optimization of complementarity and redundancy in multiple sources of interactive information in space and time to produce a unified description of the observed environment. It is a process of correlating, associating and synthesizing data and information obtained from single and multiple information sources to obtain accurate location and identity estimates as well as a process of information processing to comprehensively and timely evaluate situations, threats and their significance.

Information fusion technology refers to the information processing technology that uses computer technology to automatically analyze and synthesize multi-source information detected from multiple sensors (of the same type or different types) according to certain rules and automatically generate the synthetic information that people expect. It includes a variety of intelligence information (such as data, photos and video images, etc.) obtained by multi-type, multi-source and multi-platform sensors for acquisition, transmission, aggregation, analysis, filtering, combination, correlation and synthesis of information, rapid intelligence processing and automatic graphic mapping. According to this definition, multi-sensor system provides the hardware foundation for information fusion, multi-source information is the object of information fusion processing and coordinated optimization and comprehensive processing is the core of information fusion.

Information fusion technology has the following advantages:

(1) Providing steady working performance

Each sensor in the system provides object information independently. When there is a failure of any sensor or when an object is not detected due to external interference, it does not affect the performance of other sensors.

(2) Improving spatial resolution

Compared with any single sensor, using multiple sensors enables the geometrical formation of a sensor aperture to obtain a higher resolution.

(3) Obtaining more accurate object information

The different information provided by multiple sensors reduces the set of assumptions about the objects or events. In addition, effective integration of the multiple times of (different time series of the same sensor) or multiple (using different sensors at the same time) independent measurements of the same object or event can enhance the confidence level and improve detection performance.

(4) Obtaining object information that single sensor cannot obtain

The complementarity among sensors can expand the spatial and temporal coverage, increase the dimensionality of the measurement space and reduce detection blindspots caused by electronic countermeasures (concealment, deception and camouflage) and meteorological and terrain-scattered interference. The inherent redundancy of multi-sensor systems will improve the reliability and fault tolerance of system operation.

The information fusion technology in i-EFV can synthesize multi-source information that provided by onboard sensing systems and wireless communication, optimally combine complementary and redundant information in space and time based on specific principles, and obtain an accurate description of the object and its changing characteristics [1].

In intelligent vehicle's environmental perception technology, sensors play an important role in collecting information about the surrounding environment. The common onboard sensors include Monocular camera, Binocular camera, Millimeter-wave Radar (Radar in short), Light Detection And Ranging (LiDAR in short), Ultrasonic Radar and Infrared sensor, etc. Different sensors have their own advantages, disadvantages and conditions of application.

Thanks to its cost-effective performance, the visual sensor based on the monocular camera is currently one of the mainstream on-board sensors for autonomous driving. The object identification technology based on visual sensors mainly includes traditional methods and deep learning-based methods. The former includes feature extraction methods such as Scale-invariant Feature Transform, Histogram of Oriented Gradient, Color Self-Similarity, and classification methods such as Support Vector Machine (SVM). The latter mainly uses models such as Convolution Neural Network, Sparse Autoencoder, and Recurrent Neural Networks to identify objects. Visual sensors boast strong target classification ability and lateral motion detection ability, but are poorly adapted to harsh environmental conditions such as rain and fog, bad lighting conditions and so on. Moreover, the monocular camera has low ranging accuracy and is not sensitive to the longitudinal motion of the object. Even Mobileye, one of the world-leading providers of machine visual solutions, clearly states in its product description that it cannot guarantee an effective operation of the product under poor lighting conditions.

Millimeter wave radar can make up for the inherent defects of visual sensors to a certain extent. It is mainly used in ADAS Adaptive Cruise Control, Autonomous Emergency Braking and other systems, and plays an important role in detecting dynamic objects in autonomous vehicles. However, the millimeter-wave radar has a narrow perception field of view, which makes it difficult to timely detect the vehicles in front in scenarios such as passing through a curve and the cut-in of adjacent vehicles, and it's insensitive to the lateral motion of the object.

Lidar plays an important role in autonomous driving schemes due to its wide field of view, high ranging accuracy, and strong object classification ability in object identification and tracking. However, the attenuation rate of laser decreases in bad weather such as rain, snow and fog, which will affect the perception distance of LiDAR, and result in a mediocre speed-measuring performance, since the measurement is based on the differentiation of the distance.

Single sensors have their unique advantages and disadvantages, they are not fully qualified for the perception of autonomous driving environment. At the same time, multi-sensor fusion technology is a mainstream method to ensure the complementary advantages of different sensors and improve the perception performance on the autonomous driving environment. Multi-sensor information fusion refers to theory and method of using computers to process multi-source perceptual information to enable a comprehensive utilization of such information [2]. In general, the information fusion method is adopted to rationally integrate and utilize the environmental perception data generated by each sensor, and then according to the perception model of each sensor (including recognition range, accuracy, etc.) and some optimization methods, all measurement data are described in a unified time and space, which can effectively solve the inherent shortcomings of single sensors' narrow fields of view and insufficient types of perception objects, to obtain more steady and accurate environmental perception results.

According to different levels of data abstraction, multi-sensor information fusion methods can be divided into three categories: data-level fusion, feature-level fusion and decision-level fusion [2, 3]. A diagram of structure of the three types of fusion methods is shown in Fig. 4.1.

The characteristics of these three fusion methods are:

(1) Data-level fusion, also known as low-level fusion. In data-level fusion, the raw measurement data from different sensors are directly utilized and are integrated into new measurement information for feature extraction and decision making for object identification. It has the highest rate of information utilization, largest amount of data processed and the longest processing time.
(2) Feature-level fusion, also known as middle-level fusion. Middle-level fusion first extracts the respective feature elements from the original data provided by different sensors and then combines them to obtain new feature elements for

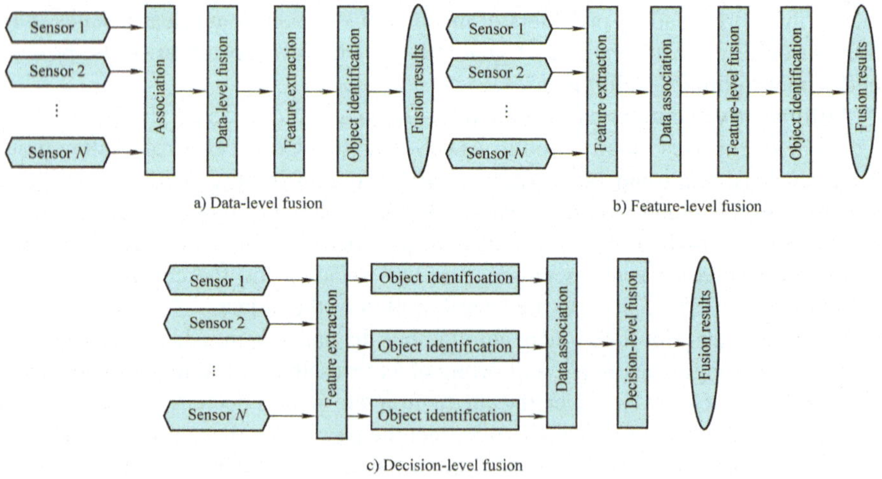

Fig. 4.1 Diagram of the structure of the three types of fusion methods [4]

subsequent decision making. This method can achieve some data compression without losing too much fusion accuracy.

(3) Decision-level fusion, also known as high-level fusion. In high-level fusion, each sensor first processes the raw data individually and makes decision, such as which category the road object fits into, and then combines these decisions to make an overall system decision. This processing method has the highest information loss rate but has low computational complexity and is conducive to the processing of heterogeneous sensor data.

The traditional estimation and statistical inference theories provide a good foundation for multi-sensor fusion technology, which mainly include Kalman Filter (KF), Extended Kalman Filter (EKF), Unscented Kalman Filter (UKF), Particle Filter (PF) based on random sampling, Markov Chain Monte Carlo (MCMC), Evidence theory based on statistical inference and Random set, etc. In addition, some new methods applied in the field of multi-sensor fusion have emerged and become an important force in the development of multi-sensor fusion technology, including artificial intelligence methods such as Artificial Neural Network (ANN) and Genetic Algorithm, and information theory methods such as maximum entropy methods. For example, the study [5] used ANN and SVM approaches based on information obtained from feature-level fusion from LiDAR and on-board cameras to detect passable road areas for autonomous vehicles for both structured and unstructured roads. In addition, the study [6] proposed an unsupervised learning (ULS)-based multi-sensor fusion method that uses observations from one reliable primary sensor to train the observation model of another auxiliary sensor and derived through theoretical inference the conclusion that the self-supervised learning-based sensor fusion outperforms any single one of the sensors when the prior knowledge of the object information is sufficient or the auxiliary sensor is sufficiently accurate.

In the environment perception technology of autonomous driving, there are quite a lot of mature research results at home and abroad, including vehicle self-positioning, static environment map construction and dynamic object detection and tracking etc.

In vehicle self-positioning, the solution combining Global Positioning System (GPS) and Inertial Measurement Unit (IMU) is the most mainstream and mature [7–10], in which the former obtains accurate latitude and longitude position information of the ego vehicle at a fixed frequency, the latter provides accurate angular and linear accelerations in all directions of space for vehicle positioning when the former signal is interfered or intercepted, and integrates them to obtain the quantity of position and attitude change. The methods to fuse these two information sources are mainly KF, EKF or Adaptive Kalman Filtering based on estimation theory and are data-level fusion.

In the static environment map construction, the autonomous vehicle estimates its own position and attitude and builds incremental maps of the surrounding static environment while moving in an unknown environment without relying on high-precision maps and GPS positioning, in which LiDAR or vision sensor-based simultaneous localization and mapping (SLAM) technology is generally used. The SLAM technology that integrates LiDAR and camera has attracted attention because it can obtain sufficient color information and accurate depth information simultaneously. To improve the closed-loop detection effects based on point cloud SLAM, Newman P. et al. used image information supplemented with closed-loop detection of map construction [11], and the combination of these two is decision-level fusion. Subsequently, Danelljan M. et al. proposed a SLAM method for camera and LiDAR fusion, based on the statistical properties of both camera and LiDAR to achieve inter-frame association of data, which is feature-level fusion [12]. In addition, Talos, a Massachusetts Institute of Technology (MIT)'s vehicle participating in the Defense Advanced Research Projects Agency (DARPA) Autonomoud Vehicle Challenge combines different sensing ranges of LiDAR to detect possible hazardous areas on unstructured roads using the data-level fusion approach [13, 14]. Similarly, Caroline, a vehicle participating in the Challenge from Technical University of Braunschweig, processes data from the stereo camera, monocular camera, and LiDAR to obtain a grip map of the vehicle's surroundings, where the stereo camera is used to determine the presence, type, and location of obstacles on the road ahead, the LiDAR is used to obtain the precise location of obstacles and the monocular camera assists the stereo camera and LiDAR in determining the type of obstacles on the road through color analysis [15]. Each of these three sensors outputs the probability of occupation of the grid map and their fusion is on the decision level.

In the aspect of dynamic object detection and tracking, most studies have used multi-sensor fusion approaches with feature-level fusion or decision-level fusion. Steux B. et al. used four onboard monocular cameras and one forward millimeter-wave radar for the research on object identification based on feature-level fusion to solve the problem of miss-detection of objects when a single sensor is used [16]. As for the decision-level fusion, multi-sensor fusion methods for dynamic object tracking are generally divided into track-level fusion and detection-level fusion [17, 18]. Track-level fusion means that the object trackers of each sensor send their respective tracking trajectories to the fusion center for integration and processing to improve the computational and communication efficiency of the fusion center; detection-level fusion means that each sensor sends its detected object measurements of each frame directly to the fusion center to generate the fused tracking trajectories, so as to improve the information utilization of the sensors.

At the application level, the Autopilot 8.0 system upgraded by Tesla has greatly increased the number of on-board cameras, ultrasonic radars and millimeter-wave radars compared to its past systems to achieve redundancy and fault tolerance for dynamic object detection. This reflects that the combination of multiple number and types of sensors will be the future trend of sensor layout on autonomous vehicles. Figure 4.2 shows the layout and detection range of Tesla's Autopilot system's sensors.

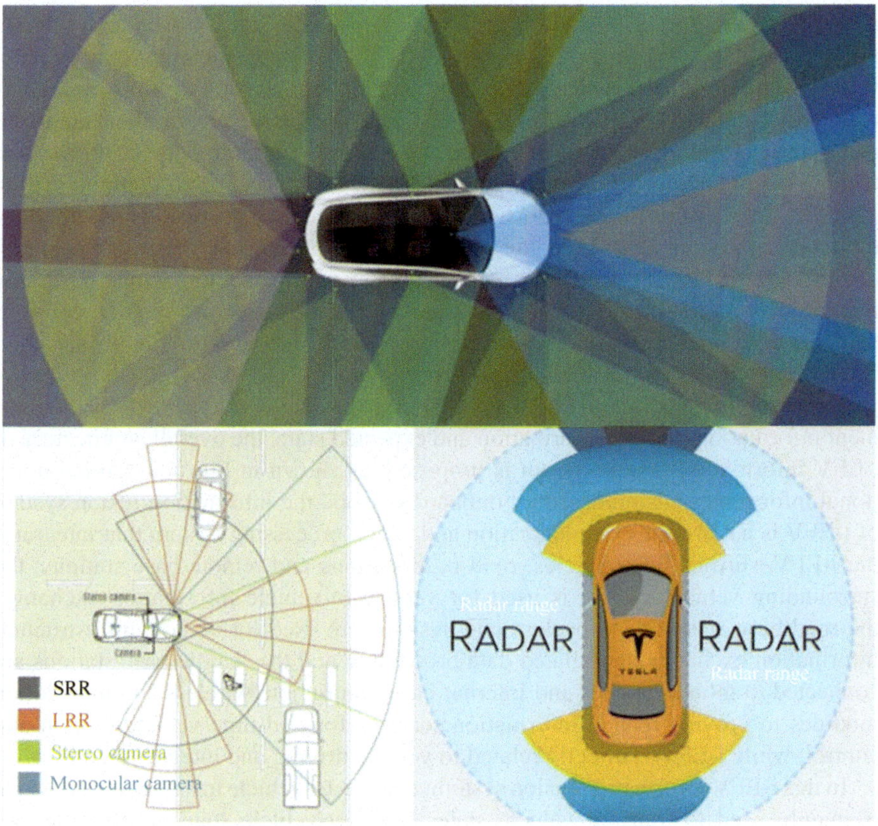

Fig. 4.2 Layout and detection range of sensors of Autopilot system of Tesla, USA

4.2 Information Fusion System Architecture of i-EFV

At present, multi-sensor information fusion technology has been partially studied and applied in ITS, such as preceding vehicle identification based on radar and camera fusion. However, the conventional intelligent vehicle's identification of environmental information and vehicle motion state is mainly targeted at single application function, such as adaptive cruise, lane keeping and vehicle stability control. This type of vehicle mainly relies on one or two sensors for fusion identification to obtain specific object information. As there is only one or two sources of sensing information source, its rate of comprehensive utilization of sensor information, especially the potential feature information is low and cannot realize the comprehensive identification of "human-vehicle-infrastructure" road information and feature information formed by the driver, the vehicle and the infrastructure. At the same time, the existing information fusion method is only used for specific control functions and cannot be applied in the environmental information identification support system integratedly controlled by multi-objective, multi-mode and multi-system. Therefore, for the i-EFV intelligent information exchange system that integrating vehicle-vehicle communication, vehicle–road communication, remote wireless communication and onboard sensing system, it is one of the keys for obtaining integrated and optimized performance of the vehicle to establish a multi-source information fusion system architecture which integrates multi-sensor and multi-communication systems. This system is enable to synthesize redundant information provided by multi-sensor systems and overcome interferences caused by environmental changes, obtain accurate "human-vehicle-infrastructure" state information, extract necessary feature information and anticipate "human-vehicle-infrastructure" safety, comfort, and energy efficiency and environmental protection state, and provide information and effective reference for the integrated optimization control of i-EFV.

Based on i-EFV's requirements for identifying vehicle state and traffic environment information, feature information and expected state, the overall architecture of i-EFV information fusion system is proposed, as shown in Fig. 4.3. Unlike traditional information fusion based on onboard sensors, the information fusion system of i-EFV is a real-time communication and signal processing system that integrates the i-EFV, surrounding vehicles, road base stations and remote base stations. the surrounding vehicle system is used for vehicle-to-vehicle information exchange, the road base stations and background systems are used for vehicle-infrastructure information exchange and related data processing and the remote base stations are connected to mobile phones and Internet based on remote wireless communication modules to provide related information services for real-time vehicle motion and control, while i-EFV is directly related to vehicle driving and integrated control.

In this i-EFV information fusion system, vehicle-to-vehicle information exchange is mainly used to transmit vehicle state and inter-vehicle state information, not including corresponding signal processing; road base stations and the ITS center in the background form a comprehensive system integrating vehicle-to-infrastructure information exchange and information processing responsible for receiving nearby vehicle motion information and fault state information and it's capable of acquiring current traffic flow and road state information through information fusion processing combined with GPS/GIS; remote base stations are connected to the Internet and can provide real-time query services based on user needs by uploading and downloading relevant information and at the same time apply cloud computing in the Internet processing background to realize large-scale data analysis and calculation, which further supports i-EFV multitasking; the i-EFV then applies layered information fusion methods based on vehicle-to-vehicle exchange, vehicle-to-infrastructure exchange, remote communication and on-board sensor information to identify and predict vehicle operation state and traffic environment information in real time.

Taking into account the composition of i-EFV intelligent information exchange system, the needs of optimal control of the vehicle and the main problems of i-EFV

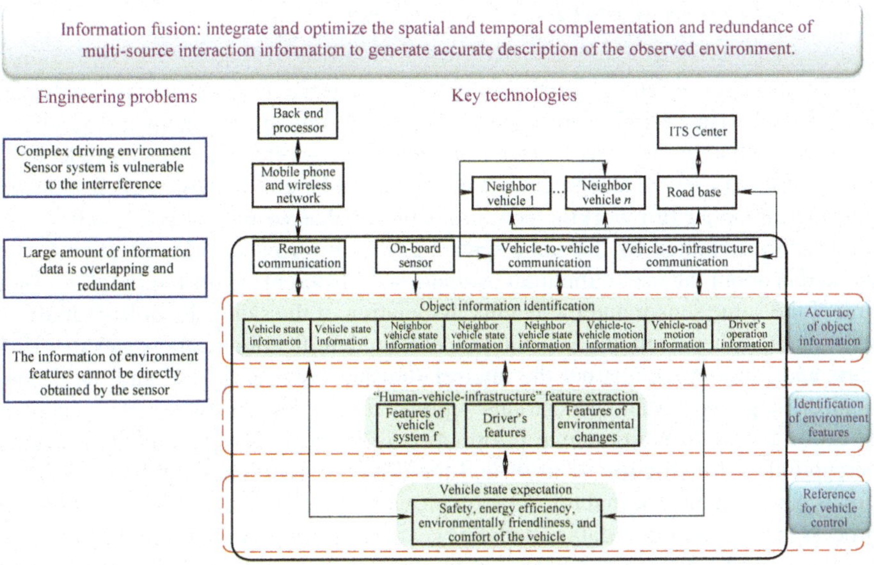

Fig. 4.3 Architecture of the i-EFV information fusion system

information identification, i-EFV information fusion is divided into the following three levels:

(1) Object information identification

It focuses on solving the problem of the object identification accuracy of the sensor system caused by complex time-varying environmental factors to obtain comprehensive "human-vehicle-infrastructure" multi-object information, including vehicle state information, surrounding vehicle state information, vehicle-to-vehicle motion information, vehicle-to-infrastructure motion information, pedestrian motion information, driver operation information, preceding traffic information and preceding road information etc.

(2) "Human-vehicle-infrastructure" feature extraction

Based on the multi-object information obtained at the former level, we use feature analysis and fusion to construct "human-vehicle-infrastructure" feature information, including vehicle system characteristic information, driver characteristic information and environment change characteristic information etc.

(3) Vehicle state expectation

Based on the "human-vehicle-infrastructure" multi-object information and "human-vehicle-infrastructure" feature information, the i-EFV system model and system control model are integrated to evaluate and predict the safety, comfort, energy-efficiency and environmental protection states of the vehicle as a reference for comprehensive vehicle control.

4.2.1 Object Information Identification

Each kind of sensor and communication equipment has its own specific applicable environment. The i-EFV's sensor system and communication system are vulnerable to complex changes in its driving environment, such as vehicle motion attitude, light, noise, buildings and weather etc. To achieve accurate identification of environmental information under full working conditions means that we cannot solely rely on the information obtained by the same sensor. Object information identification fusion method is about the use of the joint estimation of different spatio-temporal information of the same sensor and the feature association of the same object from multiple sensors and other methods, to achieve accurate identification of the states of i-EFV parts and motion, reconstruct the interrelationship between i-EFV and the driving road, driving vehicles and pedestrians within certain areas. The fusion of object identification based on multi-source information is the basis for the vehicle's accurate perception and prediction of itself and the driving environment and its identification is based on the original information provided by each sensor system, which can process large amounts of data and identify extensive information (including comprehensive identification of "human-vehicle-infrastructure" information). It is the core of i-EFV information fusion.

In order to realize the reconstruction of the vehicle and driving environment information, the object information identification based on multi-source information fusion mainly involves the three following key technologies: (1) "human-vehicle-infrastructure" environment element characterization and modeling; (2) object information identification based on specific environment elements; (3) vehicle and environment system modeling in geodetic coordinates.

The analysis and modeling on the characteristics of "human-vehicle-infrastructure" environment elements is mainly about analyzing the characteristics of the contour and motion of pedestrians, moving vehicles, fixed objects, road signs and other elements in the vehicle driving environment, establishing the database for classification of element characteristics and their models for sensor identification, which is the basis for reconstructing the information of vehicles and the environment. There is currently little research in this area and there is an urgent need for in-depth analysis and study of the properties of each element from a systematic perspective.

The object information identification based on specific environment elements uses a distributed data fusion structure as shown in Fig. 4.4. For the information obtained from sensor i ($i = 1, 2, \ldots$ N), signal pre-processing is first performed to filter and merge the data according to the observation time, sensor type and attributes to sort and merge the signals and compensate errors etc. Secondly, location-level fusion of data in time, space and spatio-temporal dimension based on information provided by each sensor, including data calibration, data interconnection, object tracking and state estimation etc., is performed to obtain accurate object information from a single sensor [4]. Finally, multi-sensor fusion is performed to combine and jointly estimate

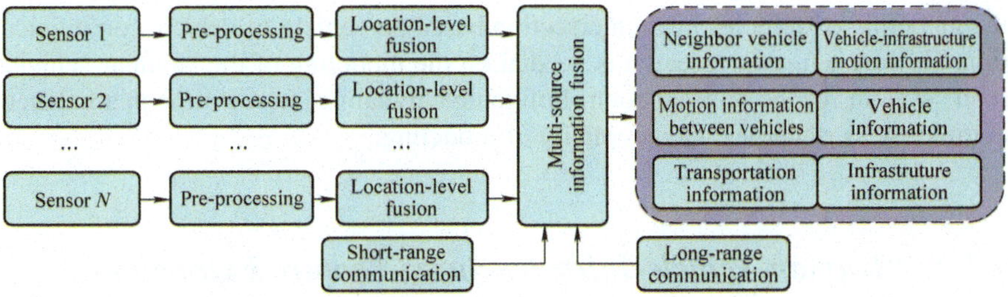

Fig. 4.4 Diagram of object information identification structure

data based on the attributes and applicability of multi-sensor information, vehicle-to-vehicle communication information and vehicle-to-infrastructure communication information to obtain object information adapted to complex traffic environments.

For the i-EFV's needs for vehicle and environmental information identification, the research on object information identification based on specific environment elements mainly contains the following contents: (1) identification of information about the relative motion between the vehicle and the surrounding vehicle, relative motion between the vehicle and pedestrians and relative motion between the vehicle and roads based on fusion of multi-source information from radar, camera, wireless communication and vehicle sensor system, including information on the relative distance between the ego vehicle and the preceding vehicle and relative speed of these two vehicles, relative distance between the ego vehicle and the adjacent vehicle and relative speed of these two vehicles, relative distance between the vehicle and the pedestrian and relative speed of the vehicle and the pedestrian, the lane position and lane marking position of the vehicle etc.; (2) vehicle motion attitude and road condition identification based on fusion of multi-source information from wireless communication, GPS, GIS and vehicle sensor system, including information on vehicle longitudinal/lateral acceleration, vehicle longitudinal/lateral speed, yaw rate, sideslip angle and road friction coefficient; (3) identification of working state of vehicle parts based on fusion of multi-source information from onboard sensor system, including information on the working state of engine, motor, battery and controller etc.

At present, some research has been carried out at home and abroad on object identification based on on-board multi-sensor information fusion. With the continuous maturity and application of communication technologies, further research on information fusion methods that make comprehensive use of vehicle-to-vehicle, vehicle-to-infrastructure and remote wireless communication is urgently needed.

After the identification information of each element of vehicle and the environment is acquired, it's required to model the vehicle and the environment system based on the geodetic coordinate system to obtain a unified mathematical description of the vehicle and its traffic environment. At present, there has been some research on vehicle system modeling based on some environment elements, such as research on the modeling of vehicles in an intersection based on vehicle-to-vehicle communication. However, further research is needed on the modeling of the "human-vehicle–road" system for the optimal control of vehicle dynamics in real time that balances completeness, accuracy and simplicity of modeling.

4.2.2 "Human-Vehicle-Infrastructure" Feature Extraction

Although the driving environment is complex and changeable, the change of fixed objects has a certain continuity and there is always an inherent connection between different object information. The feature information of "human-vehicle-infrastructure" system needs to be considered to realize the optimal integrated control of i-EFV. The basic structure of the multi-information feature extraction algorithm is shown in Fig. 4.5. Firstly, for the identified object information *i*, the feature vector is extracted; feature data is associated with each feature vector and the feature layer is fused to obtain the "human-vehicle-infrastructure" feature information based on the associated feature information.

The extraction of vehicle and environment characteristics such as traffic statistics, driver characteristics and some vehicle characteristic parameters based on partial sensor information has been studied, but the identification of environment change characteristics, driver longitudinal/lateral behavior characteristics and vehicle characteristic parameters based on complete "human-vehicle-infrastructure" multi-object information is to be further studied. The research on "human-vehicle-infrastructure" feature extraction mainly includes: (1) the statistical characteristics (speed and acceleration) of the preceding vehicle motion, pedestrian behavior characteristics, surrounding vehicle motion characteristics (lane changing, overtaking, speed and acceleration) and other environment change characteristics based on the fusion of multi-source features such as traffic information, surrounding vehicle information and relative motion between vehicles; (2) driver's longitudinal/lateral driving behavior information based on the fusion of traffic information, vehicle information, information on relative motion between vehicles, and surrounding vehicle information; and other multi-source features; (3) vehicle and road characteristic parameters

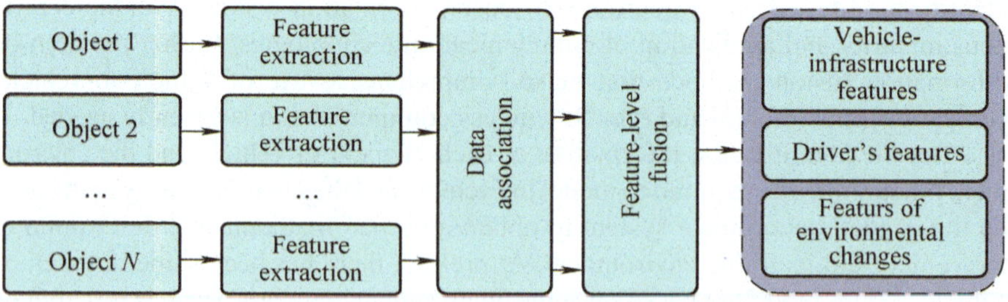

Fig. 4.5 Diagram of structure of "human-vehicle-infrastructure" feature extraction

such as vehicle weight, suspension stiffness, ground rolling resistance, road slope and road friction coefficient etc.based on the fusion of vehicle information and road information and other multi-source features.

4.2.3 Vehicle State Expectation

The vehicle state expectation is about the evaluation of the current and expected safety/energy efficiency/environmental protection/ride comfort states of the vehicle by integrating the object identification information and the "human-vehicle-infrastructure" characteristic information, as a reference for the vehicle's integrated coordinated control, as shown in Fig. 4.6. For the estimation of the expected vehicle state, it's required to consider all kinds of object information and "human-vehicle-infrastructure" characteristic parameters and combine the i-EFV system model and control strategy model, in order to estimate the expected vehicle state, including the basic vehicle motion, the working of components, the corresponding vehicle efficiency and driver perception to acquire expected vehicle safety, comfort, energy efficiency and environmental protection states by considering and comprehensively evaluating the system environmental constraints and driver characteristics etc.

At present, the system state evaluation and expectation methods based on multi-source information are mainly used in the military field. The design of such methods mainly relies on expert systems and practical experiences. In the automotive field, there are few studies that integrate comprehensive "human-vehicle-infrastructure" information to evaluate and anticipate the comprehensive operating states of vehicles in terms of safety, comfort, energy efficiency and environmental protection. It is

Fig. 4.6 Diagram of vehicle state expectation structure

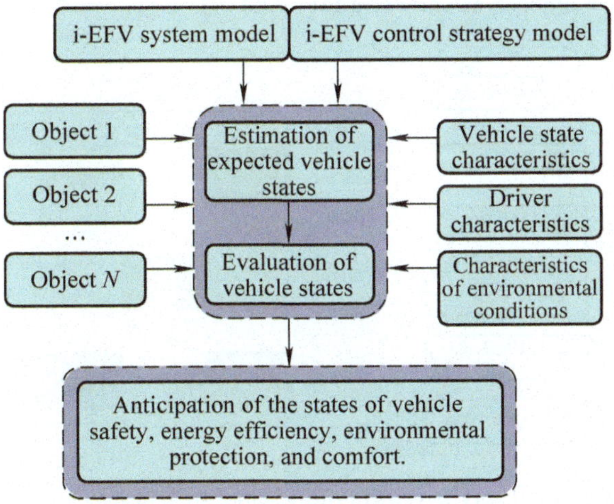

important to carry out such research to effectively identify the vehicle and environmental conditions, accurately estimate the vehicle's current and future driving safety and improve the economy of system operation.

4.2.4 Key Technologies for Multi-source Information Fusion System

As a new concept vehicle that both acts as a complex electromechanical coupled system and has the characteristics of operation nodes in transportation network, energy network and IoV, the information of i-EFV is multi-source, heterogeneous and variable in scale. Multi-source means that there are many types of sensors, such as vision sensor, radar sensor and V2X sensor etc.; heterogeneous means that the structures of input data of different sensors are not exactly the same, including target point data, map data, GNSS coordinates etc.; variable scale means that for certain same data structures, the resolution and spatio-temporal references of different sensors are not the same. For example, the measurement accuracies of vision radar and millimeter-wave radar are different and their spatio-temporal references are different as their placement positions and working clocks are different, as shown in Fig. 4.7 [19].

The information fusion of i-EFV involves two dimensions: single-vehicle multi-sensor information fusion and multi-vehicle multi-sensor coordinated positioning.

Both use the object-level information fusion perception method and the key technologies used mainly include: spatio-temporal reference alignment technology, multi-object association technology, parameter fusion technology, multi-object tracking and prediction technology and technology of classification and identification of objects, as shown in Fig. 4.8 [19].

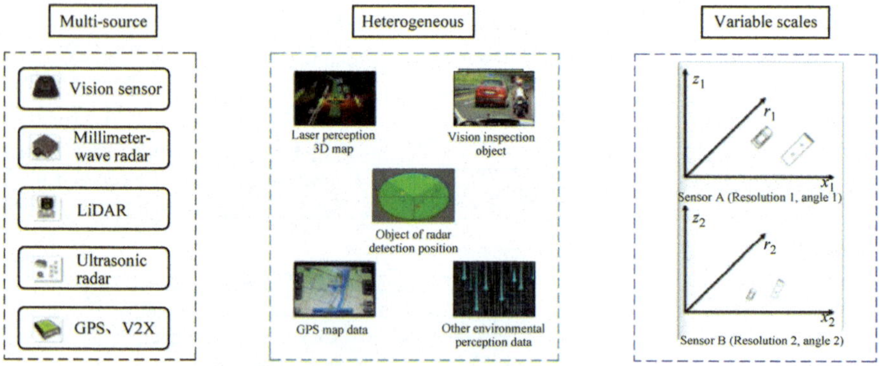

Fig. 4.7 Characteristics of i-EFV information fusion data

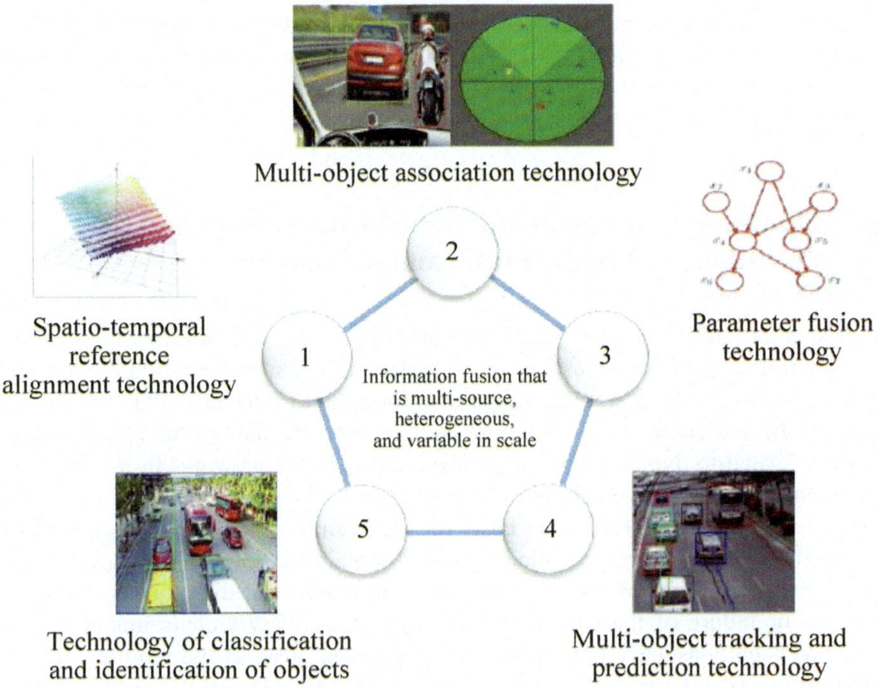

Fig. 4.8 Key technologies for i-EFV information fusion

(1) Spatio-temporal reference alignment technology: Because different sensors are installed at different locations and have different clock references, spatial rotation transform and time extrapolation are needed to achieve spatio-temporal reference alignment before object-level fusion.
(2) Multi-object association technology: There are multiple identical objects in the sensing of IV multi-sensor environment, which require multi-object association through spatio-temporal distance and trajectory correlation to ensure subsequent smooth object-level fusion.
(3) Parameter fusion technology: for a parameter of the object, there are observations from multiple sensors. For example, millimeter-wave radars and videos can all obtain the distance of the object. The fusion method such as maximum likelihood estimation can improve the estimation accuracy of the object parameter.
(4) Multi-object tracking and prediction technology: After obtaining the point traces of the object, the trajectory of the object can be tracked and predicted through the equation of motion, so as to obtain the traffic situation map about the local environment, from which the travelable area of this vehicle can be further estimated.
(5) Object classification and identification technology: the object association by multiple sensors can help obtain multiple features of the object and greatly

improve the identification of the object. For example, videos can only be recognized by images, but by fusing the radar, videos can also obtain the information on speed and RCS of the object. The combination of these information can greatly improve the object identification rate.

4.3 Traffic Environment and Vehicle State Sensing Technology Based on Information Fusion

I-EFV is an intelligent agent that integrates perception, decision-making, control, actuation and other functions. As the vehicles are becoming more intelligent, the vehicle needs to obtain a wider range of information to detect gradually from the state of the vehicle to the surrounding environment and then to the relevant information related to dynamic operation of the entire traffic network. In the functions mentioned above, the vehicle operating state and driving environment perception system provides the information basis for the realization of other functions of i-EFV. The ability of the perception system to obtain the surrounding environment and vehicle state information in real time and stably and directly determine the success or failure of the subsequent decision. The i-EFV's environment perception has been constantly improving and progressing to meet the requirements for sensing range and sensing accuracy, therefore the related technology has moved from single vehicle single sensor perception of the position of surrounding vehicles and obstacles to single vehicle multi-sensor fusion perception and then to the use of vehicle-to-vehicle/vehicle-to-infrastructure communication technology to achieve multi-vehicle multi-sensor coordinated perception.

The framework of the environment and vehicle state perception system based on multi-source information fusion is shown in Fig. 4.9. The system mainly uses multi-sensor fusion technology and the method of multi-source and multi-feature fusion. According to the characteristics of environment perception during actual driving, the order of actuation of sub-algorithm modules and the information flow of intermediate identification results are reasonably arranged. The optimization of the algorithm enables the system to meet the requirements of real-timeness while satisfying the needs of lateral and longitudinal environment perception.

The system is divided into three parts: information acquisition, information processing and information output. To realize the intended function of the system, it's necessary to obtain real-time and accurate driving environment information and vehicle state information in the first place and this system adopts monocular vision and LiDAR to obtain the environment information during driving.

The part of information processing mainly uses lane detection and obstacle identification algorithms to identify the lane lines and other vehicles running on the road during driving, divides the lanes to which the vehicles driving ahead belong and selects the object vehicle to be used for forward collision warning or as the car-following target of the cruise control system based on the lane and vehicle identification results. At the same time, based on the information on vehicle states and

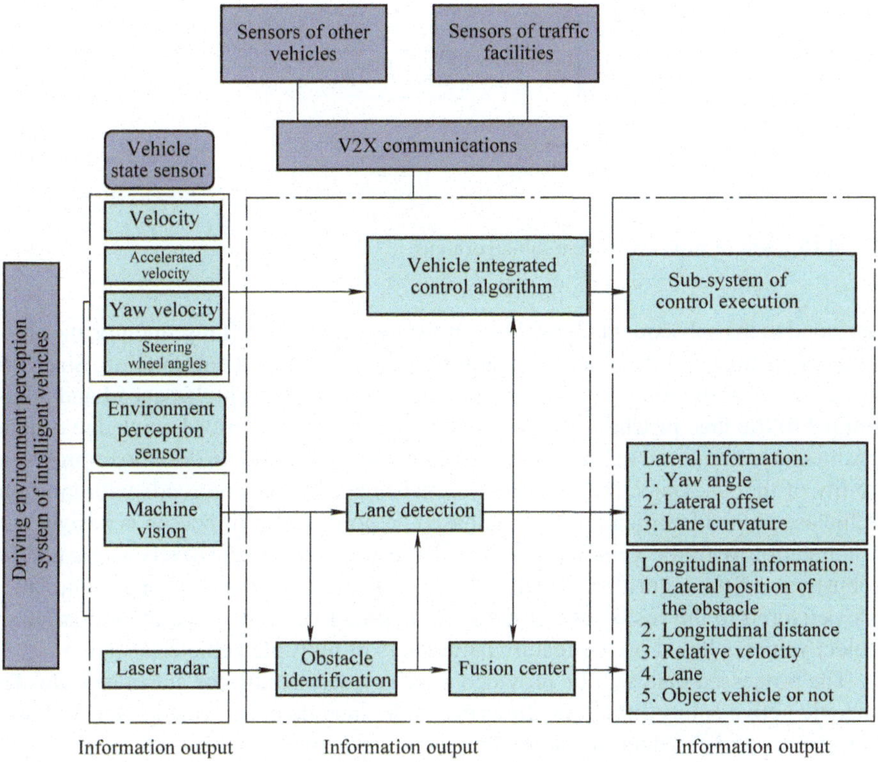

Fig. 4.9 General framework of the environment and vehicle state perception system

the lane and vehicle identification results, combined with the lane-keeping control algorithms, the lateral control command is generated for lane keeping. The characteristics of this system are mainly reflected in the fusion of the lateral and longitudinal environmental information as well as the intermediate identification results. In the existing research on environment sensing technologies, the vehicle and lane identification is often separated without comprehensive consideration due to the different needs of longitudinal and lateral active safety functions. This system is positioned to sense both lateral and horizontal environmental information of intelligent vehicles, which makes the designed system applicable to a variety of active safety functions of intelligent vehicles on the one hand and improves the performance of the system and the effect of environmental perception by controlling the actuation process of sub-algorithms and the information flow within the system on the other hand. The process of the fusion of system sub-algorithms is shown in Fig. 4.10.

Within the overall algorithm framework of the driving environment perception system, the data preparation of the system is first completed by pre-processing the images. The images enhanced with vehicle and lane line features, together with the radar data and the vehicle state information, constitute the data source of the system

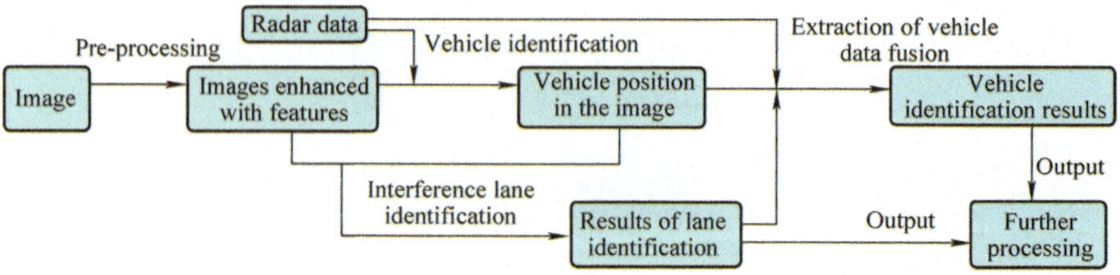

Fig. 4.10 Flow of fusion of system sub-algorithms

provided to the subsequent algorithms for further processing. Subsequently, the algorithms fuse the radar data and image information for vehicle identification. Since the system aims at the identification and tracking of multiple vehicles ahead and is not limited to the lane in which the ego vehicle is running, the prior knowledge of lane position is not required in the process of vehicle identification. In order to improve the ability of the lane detection algorithm to resist interferences on roads with multiple vehicles, a lane identification method based on point set optimization is designed in combination with the vehicle identification results, which effectively improves the robustness of the algorithm. Finally, the identification results of vehicles and lanes are delivered to the fusion center at the same time to complete the identification of object vehicles based on the feature parameters of multiple vehicles ahead.

The processed data will be provided to subsequent control or alarm modules for specific active safety functions. In terms of the lane detection results, the lane line parameters and the data on the relative position between the vehicle and the lane will be directly output to the functional modules of the lateral active safety system, where the lane information is the input parameter to the controller in the lane keeping system. In terms of longitudinal active safety applications, the vehicle identification results are not directly used as input, but instead, the information on position, relative speed and lane to which the vehicle belongs is extracted in the fusion center by using the lane and vehicle identification results to decide whether the vehicle is the object vehicle or not. Finally, the characteristic data of the vehicle is used as input parameters of the longitudinal system. The development process shown in Fig. 4.11 is used in the system.

In order to improve the accuracy and real-timeness of the vehicle identification algorithm, the overall structure of the algorithm is shown in Fig. 4.12 by fusing the obstacle data acquired by LiDAR and the visual information to detect the vehicles ahead.

The algorithm consists of four main modules. The pre-processing module is responsible for processing the original images captured by the camera to obtain the images needed for subsequent identification, including: the shadow images under the vehicle, the extracted horizontal-edge images and the vertical-edge images. The tracking module tracks these objects based on the previous round of vehicle identification results. If it's confirmed that the object still exists, then the object is marked as the vehicle detected in the current round; if the object is confirmed to have disappeared, then the object is removed from the tracking profile. The use of vehicle

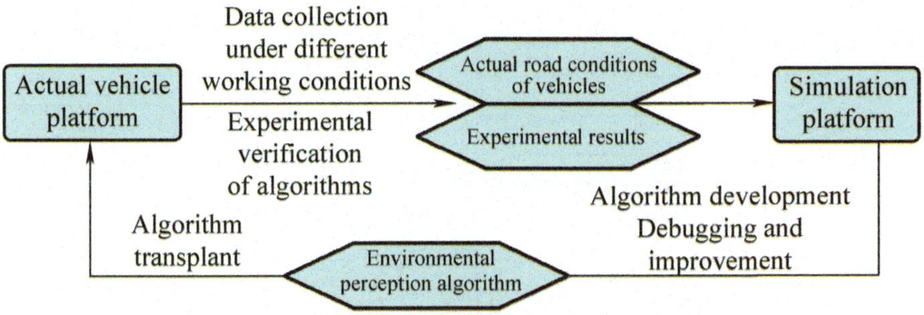

Fig. 4.11 Development process of the environment perception system

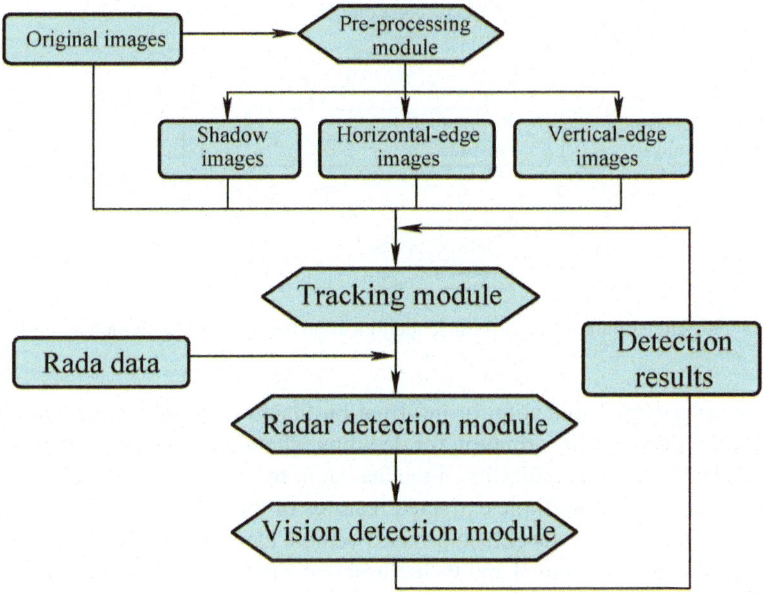

Fig. 4.12 Diagram of structure of vehicle identification algorithm

tracking technology can reduce the actuation time of the algorithm. The radar detection module compares the radar data with the vehicle position detected by the tracking module. For the object newly detected by the radar, the radar detection module is responsible for confirming the object in the image. If the object is confirmed, it will be added to the current round of identification result. The vision detection module, on the other hand, performs supplementary detection of the images. For the image areas already detected by the tracking module and the radar detection module, the vision module does not repeat the tests, which greatly reduces the scope of images to be detected.

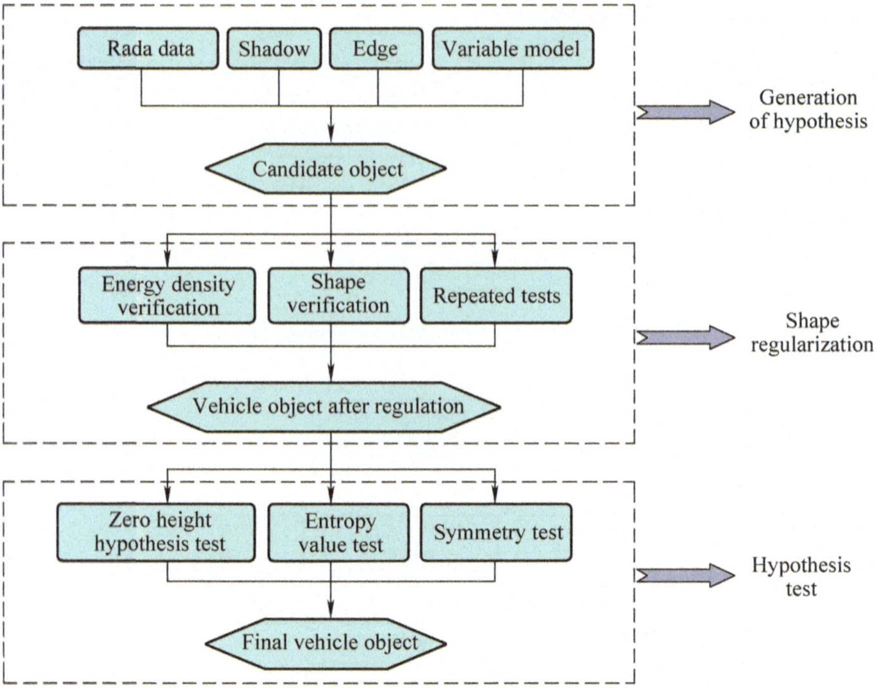

Fig. 4.13 Structure of algorithm of vehicle identification decision based on multi-level fusion

As an important application of information fusion technology, detection fusion can provide redundant information for deciding whether a candidate object is a real object and enhance the credibility of the decision. In order to more effectively make comprehensive use of multiple extracted features of the vehicle in the image as well as the radar data, a multi-level fusion-based vehicle identification decision algorithm using fuzzy logic is designed and its hierarchical structure is shown in Fig. 4.13.

The fusion decision algorithm is divided into three levels. The first level is about the generation of hypothesis of candidate vehicle objects. In this part, through comprehensive analysis of radar data, vehicle shadow features and vehicle edge features, the possible areas in which the vehicle may occupy that fit the variable model are extracted as the candidate objects for vehicle identification; due to the interference of road shadows, roadside railings and viaducts over the road, the boundary of the vehicle-occupied areas may be inaccurately defined among the candidate objects identified in the first level. Therefore, the shape regularization of the candidate objects is carried out after the first level. The accuracy of vehicle boundary definition is improved through energy density verification, shape verification and repeated tests. The accuracy improvement is very useful for subsequent hypothesis testing. The shape regularization hierarchy does not play a decisive role in the decision on vehicle candidate objects, but instead only serves for the subsequent tests. For those vehicle objects generated by hypothesis, some of them have very high confidence level and can be directly used as obtained vehicle results, while some of them do not have enough confidence level and need further hypothesis verification before they can be used as identification results. The zero height hypothesis test, entropy value test and symmetry test are used at the hypothesis testing level to verify the vehicle-occupied areas that have been shape-regularized.

4.4 Spatial Synchronization of Multi-source Sensor Information

Using a single sensor for obstacle identification in the driving area in front of the intelligent vehicle often leads to the problem of high false detection rate and omission rate. The fusion of the driving environment information obtained by radar and machine vision that is applied in obstacle avoidance can greatly improve the system's obstacle identification accuracy. However, the data obtained by the two sensors must be spatially synchronized to make the information obtained by radar match that obtained by machine vision, otherwise the accuracy of the fusion results will be affected.

This section proposes a "four-point" calibration method for spatial synchronization of LiDAR and camera, which has three advantages compared with existing synchronization methods: first, the method only needs to use the 10 m area in front of the vehicle to perform calibration and synchronization, which can meet the requirements of future industrialization where the calibration of sensors needs to be done in a small workshop; second, only the horizontal angle of the optical axis of the two sensors and the pitch angle of the camera need to be calibrated, which requires fewer parameters to be calibrated, and the calibration method is simple and easy to implement with high accuracy; third, during the driving of the intelligent vehicle, this method can use the assumption of lane parallelism to make real-time compensation for the change of the camera pitch angle, thus improving the synchronization method's practicality.

4.4.1 Establishing the Relationship of Multi-coordinate System Fusion

Before the spatial synchronization of the information obtained by radar and machine vision, the relationship between coordinate systems of the two sensors should be established first. We define the radar's coordinate system O_L-X_L-Y_L: the origin of coordinates is the projected point of the onboard radar installed on the ground and the direction *of* Y_L axis is the direction of the radar's central axis (i.e. optical axis); the camera's coordinate system O_C-X_C-Y_C: the origin of coordinates is the projected point of the camera on the ground and the direction of Y_C axis is the direction of

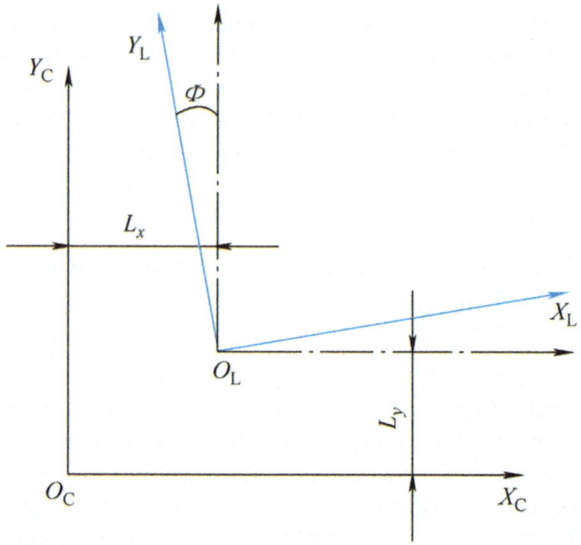

Fig. 4.14 Relationship between positions of the radar's coordinate system and the camera's coordinate system

the camera's optical axis; the image's coordinate system O_I-X_I-Y_I: the origin of coordinates is the central pixel point of the image captured by the camera and the direction *of* Y_I axis is the direction along the height of the image.

The relationship between the coordinate systems of radar and camera is shown in Fig. 4.14:

where L_x is the lateral offset of the two coordinate systems, L_y is the longitudinal offset of the two coordinate systems and Φ is the horizontal angle between the Y axes of the two coordinate systems.

The conversion between the radar's coordinate system and the camera's coordinate system is shown in Eqs. (4.1) and (4.2).

$$\begin{cases} X_C = X_L \cos\Phi - Y_L \sin\Phi + L_x \\ Y_C = Y_L / \cos\Phi + (X_C - L_x)\tan\Phi + L_y \end{cases} \tag{4.1}$$

$$\begin{cases} Y_L = \left(Y_C - L_y\right)\cos\Phi - (X_C - L_x)\sin\Phi \\ X_L = (X_C - L_x)/\cos\Phi + Y_L \tan\Phi \end{cases} \tag{4.2}$$

Equation (4.1) shows the conversion from the radar's coordinate system to the camera's coordinate system and Eq. (4.2) shows the inverse conversion relationship. The translation relations L_x and L_y of the two coordinate systems can be measured and obtained and the horizontal angle Φ between the Y axes of the two coordinate systems needs to be obtained by calibration. Since this conversion only involves the translation and rotation between the two coordinate systems, the derivation of the conversion equation will not be discussed.

The coordinate conversion relationship between the camera's coordinate system and the image's coordinate system is shown in Eqs. (4.3) and (4.4):

$$\begin{cases} P_y = \dfrac{h(H\tan\gamma_0 + 2P_y\tan\alpha_0)}{H - 2\tan\gamma_0\tan\alpha_0 P_y} \\ P_x = \dfrac{2P_x\tan\beta_0\sqrt{h^2+P_y^2}\cos\left(\arctan\frac{2p_y\tan\alpha_0}{H}\right)}{W} \end{cases} \tag{4.3}$$

$$\begin{cases} p_y = \dfrac{P_yH - Hh\tan\gamma_0}{2P_y\tan\gamma_0\tan\alpha_0 + 2h\tan\alpha_0} \\ p_x = \dfrac{P_xW}{2\tan\beta_0\sqrt{h^2+P_y^2}\cos\left(\arctan\frac{2P_y\tan\alpha_0}{H}\right)} \end{cases} \tag{4.4}$$

where (P_x, P_y) are the coordinates of any point in the camera's coordinate system, (p_x, p_y) are the coordinates of the corresponding point in the image's coordinate system. H is the height of the image (in pixel), W is the width of the image (in pixel), h is the installation height of the camera, $2\beta_0$ is the horizontal field of view of the camera, $2\alpha_0$ is the vertical field of view of the camera and γ_0 is the pitch angle of the camera. Equation (4.3) shows the conversion from the image's coordinate system to the camera's coordinate system and Eq. (4.4) is the inverse mapping equation. The camera parameters used in this system are: $H = 480$; $W = 640$; $h = 1.195$ m; $2\beta_0 = 42.5°$; $2\alpha_0 = 32.4°$. The pitch angle γ_0 needs to be calibrated.

The derivation of Eqs. (4.3) and (4.4) will be discussed in detail in the next subsection.

4.4.2 Improved Equation for Conversion Between the Coordinates of Image Coordinate System and the Camera's Coordinate System

The equation for conversion between the image's coordinate system and the camera's coordinate system derived in Refs. [20, 21] has been improved by removing the hypothesis of constant distance from the optical center to the imaging plane of the image, which further improves the accuracy of the coordinate conversion. The specific derivation process is described below.

The definition of the coordinate system is consistent with the what has been discussed in previous subsection. The imaging model is first derived for the Y axis direction, as shown in Fig. 4.15.

In Fig. 4.15, O is the optical center of the camera. Rectangle *abcd* is the imaging plane of the image with W and H as width and height respectively. The road surface area that rectangle *abcd* corresponds to is trapezoid *ABCD*. OO_C refers to the installation height of the camera h. The coordinates of any point P on the road surface

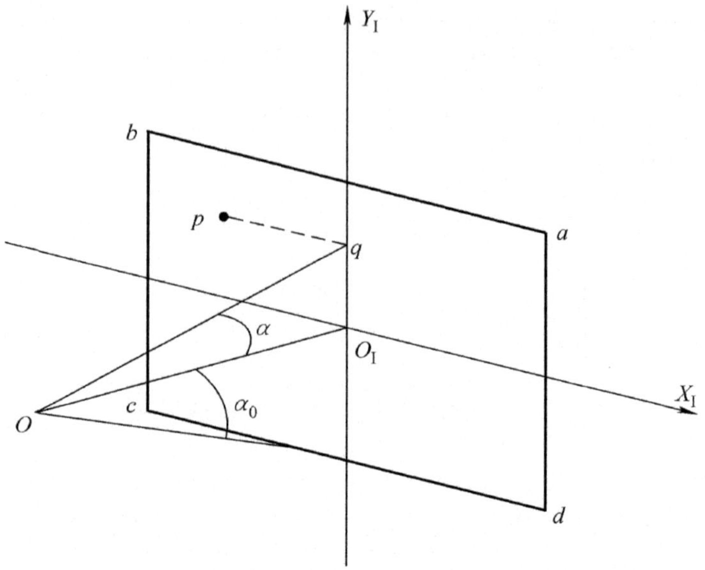

a) The image's coordinate system

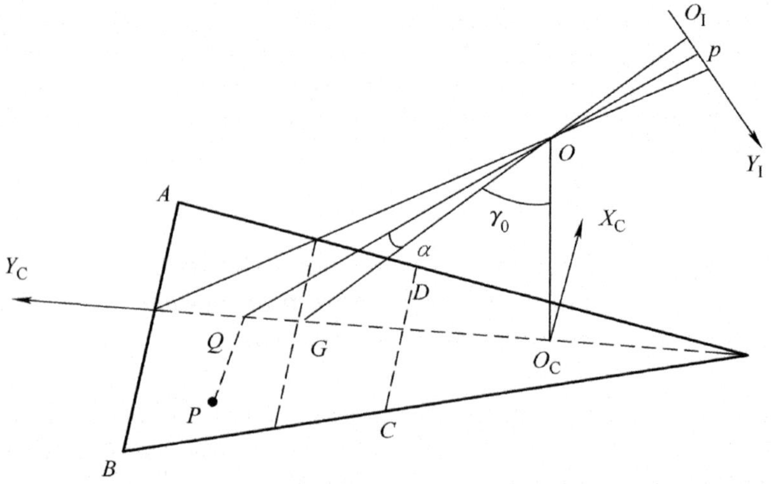

b) The camera's coordinate system

Fig. 4.15 Imaging model in the Y axis direction

are (P_x, P_y). The corresponding point of P in the image is $p\ (p_x, p_y)$. The process of deriving the equation for conversion between two coordinate systems is equivalent to deriving the correspondence between (P_x, P_y) and (p_x, p_y).

Based on the geometric relations shown in Fig. 4.15, the following equation can be derived:

$$\tan\alpha = \frac{qO_{\mathrm{I}}}{OO_{\mathrm{I}}} = \frac{p_y}{OO_{\mathrm{I}}} \tag{4.5}$$

$$\tan\alpha_0 = \frac{H/2}{OO_{\mathrm{I}}} \tag{4.6}$$

$$\alpha = \arctan\frac{2P_y\tan\alpha_0}{H} \tag{4.7}$$

$$\begin{aligned} P_y &= QO_{\mathrm{C}} \\ &= h\tan(\alpha+\gamma_0) \\ &= h\frac{\tan\alpha+\tan\gamma_0}{1-\tan\alpha\tan\gamma_0} \end{aligned} \tag{4.8}$$

By substituting Eq. (4.7) into Eq. (4.8), we can obtain the relationship between the vertical coordinates of the two coordinate systems, as shown in Eq. (4.9).

$$\begin{cases} P_y = \dfrac{h\left(H\tan\gamma_0 + 2P_y\tan\alpha_0\right)}{H - 2\tan\gamma_0\tan\alpha_0 P_y} \\ P_y = \dfrac{P_yH - Hh\tan\gamma_0}{2P_y\tan\gamma_0\tan\alpha_0 + 2h\tan\alpha_0} \end{cases} \tag{4.9}$$

Then we derive the relationship between the horizontal coordinates of the two coordinate systems, as shown in Fig. 4.16.

Based on the geometric relations in Fig. 4.16, the following equation can be derived:

$$\tan\beta_0 = \frac{W/2}{OO_{\mathrm{I}}} \tag{4.10}$$

$$\begin{aligned} \tan\beta_2 &= \frac{W/2}{Qq} \\ &= \tan\beta_0\frac{OO_{\mathrm{I}}}{Oq} \\ &= \tan\beta_0\cos\alpha \\ &= \tan\beta_0\cos\left(\arctan\frac{2P_y\tan\alpha_0}{H}\right) \end{aligned} \tag{4.11}$$

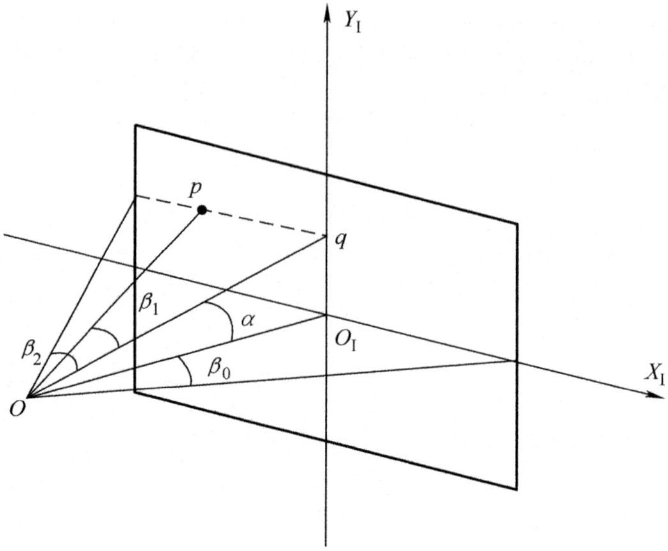

a) The image's coordinate system

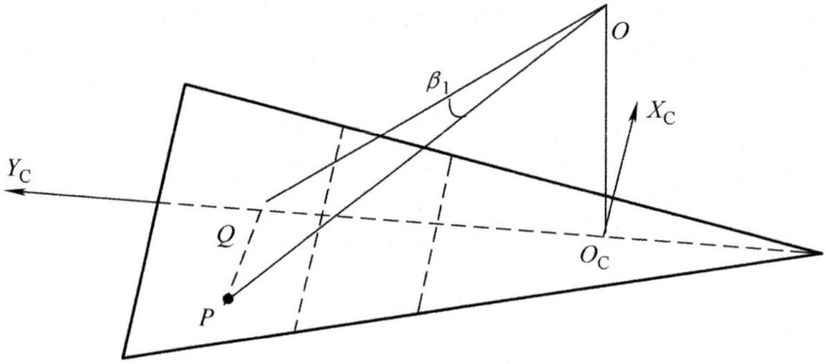

b) The camera's coordinate system

Fig. 4.16 Imaging model in the X axis direction

$$\tan \beta_1 = \frac{pq}{Oq} = \frac{2p_x \tan \alpha_0}{W} \tag{4.12}$$

The following relationship is shown in Fig. 4.16b:

$$OQ = \sqrt{OO_C^2 + QO_C^2} = \sqrt{h^2 + P_y^2} \tag{4.13}$$

$$P_x = PQ = OQ \tan \beta_1 \tag{4.14}$$

By substituting Eqs. (4.12) and (4.13) into Eq. (4.14), we can obtain the equation for conversion between the horizontal coordinates of the two coordinate systems, as shown in Eq. (4.15).

$$\begin{cases} P_x = \dfrac{2p_x \tan\beta_0 \sqrt{h^2 + P_y^2}\cos\left(\arctan\frac{2p_y \tan\alpha_0}{H}\right)}{W} \\ p_x = \dfrac{P_x W}{2\tan\beta_0 \sqrt{h^2 + P_y^2}\cos\left(\arctan\frac{2p_y \tan\alpha_0}{H}\right)} \end{cases} \tag{4.15}$$

Equations (4.9) and (4.15) together form the equation for conversion between the coordinates of the camera's coordinate system and the image's coordinate system.

4.4.3 Spatial Synchronization Calibration Method for Sensors

To complete the spatial synchronization of both LiDAR and camera sensors, four calibration points are selected in the area 10 m in front of the vehicle and arranged as shown in Fig. 4.17.

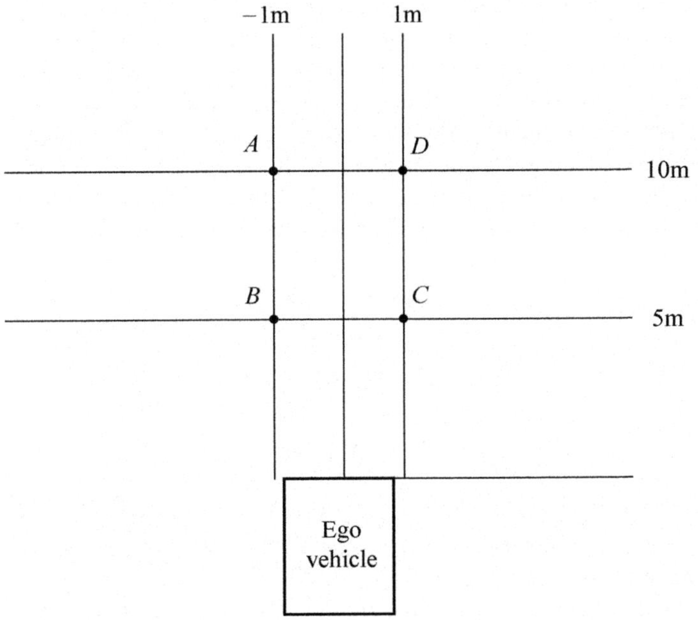

Fig. 4.17 Selection of calibration points

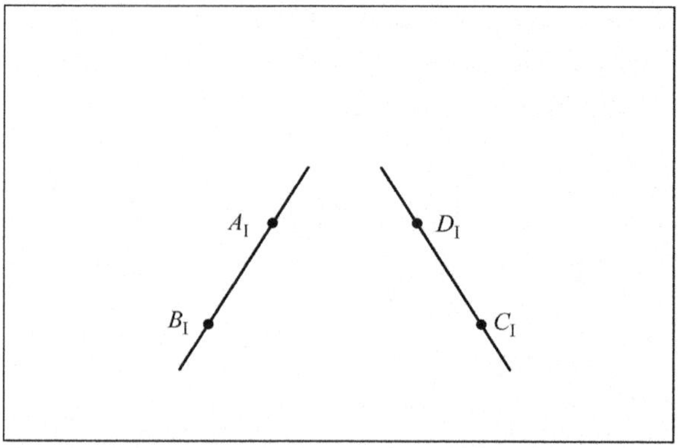

Fig. 4.18 Corresponding points of the calibration points in the image

The corresponding points A_{I}, B_{I}, C_{I} and D_{I} of the calibration points A, B, C and D in the image are shown in Fig. 4.18.

After converting the coordinates of the image's coordinate system of A_{I}, B_{I}, C_{I} and D_{I} to the coordinates of the camera's coordinate system using Eqs. (4.9) and (4.15), line segment $A_{\mathrm{I}}B_{\mathrm{I}}$ should be parallel to line segment $C_{\mathrm{I}}D_{\mathrm{I}}$. If there is no horizontal angle between the Y axes of the radar's coordinate system and the camera's coordinate system (that is, the horizontal angle is zero), these two line segments should be parallel to the Y axis of the camera's coordinate system. With a given range of possible variations of the camera's pitch angle, the pitch angle value is changed in certain stepsizes within this range and when line segment $A_{\mathrm{I}}B_{\mathrm{I}}$ is closest to being parallel to line segment $C_{\mathrm{I}}D_{\mathrm{I}}$ at a certain pitch angle, this certain pitch angle is the expected one. Similarly, given the range of possible variations of the horizontal angle between the vertical axes of the radar's coordinate system and the camera's coordinate system, the horizontal angle value is changed in certain stepsizes within this range and when at a certain horizontal angle, line segment $A_{\mathrm{I}}B_{\mathrm{I}}$, line segment $C_{\mathrm{I}}D_{\mathrm{I}}$ are closest to being parallel to the camera's optical axis. then this certain horizontal angle is the expected one. We define S_{pitch} as the indicator function for solving the pitch angle.

$$S_{\mathrm{pitch}} = (k_1 - k_2)^2 \tag{4.16}$$

$$k_1 = (X_A - X_B)/(Y_A - Y_B) \tag{4.17}$$

$$k_2 = (X_C - X_D)/(Y_C - Y_D) \tag{4.18}$$

where (X_A, Y_A), (X_B, Y_B), (X_C, Y_C), (X_D, Y_D) are the coordinates of the camera's coordinate system obtained by converting the coordinates of image coordinate system of A_{I}, B_{I}, C_{I} and D_{I} using Eqs. (4.9) and (4.15).

We define S_{yaw} as the indicator function to solve the horizontal angle,

$$S_{\mathrm{yaw}} = \left(X'_A - X'_B\right)^2 + \left(X'_C - X'_D\right)^2 \tag{4.19}$$

where X'_A, X'_B, X'_C and X'_D are the coordinate values obtained by rotating (X_A, Y_A), (X_B, Y_B), (X_C, Y_C) and (X_D, Y_D) using Eq. (5.2). The solution of the horizontal angle needs to use the results of the calibration of the pitch angle, i.e. (X_A, Y_A), (X_B, Y_B), (X_C, Y_C) and (X_D, Y_D) used here is to substitute the pitch angle obtained after calibration into Eqs. (4.9) and (4.15) and then to convert the coordinates of the image's coordinate system of A_{I}, B_{I}, C_{I} and D_{I}.

Solving the pitch angle and the horizontal angle means solving the pitch angle value and horizontal angle value that minimize the indicator functions S_{pitch} and S_{yaw} within the given range of pitch angles and horizontal angles.

Bumps and other factors during the travel of the intelligent vehicle lead to changes in the camera's pitch angle. The assumption of parallel lane lines can be used to compensate for changes in the pitch angle in real time, that is, to identify the lane lines first and then use the camera pitch angle value that makes two lane lines parallel to correct the initially calibrated pitch angle, so as to improve the accuracy of dynamic spatial synchronization of the radar and machine vision fusion system.

4.4.4 Experimental Verification of the Spatial Synchronization Method

To verify the accuracy of the spatial synchronization method, 25 locations were selected in the area of 10–100 m in front of the vehicle and obstacles were placed at these locations for static verification experiments. The acquired radar data as well as the camera detection data are shown in Table 4.1.

In the above table, (X_{L}, Y_{L}) is the obstacle position data acquired from the radar, ($X_{\mathrm{L(I\text{-}L)}}$, $Y_{\mathrm{L(I\text{-}L)}}$) is the position of the obstacle in the radar's coordinate system obtained by converting the coordinates of the obstacle in the image's coordinate system using Eqs. (4.2), (4.9) and (4.15) and X_{error} and Y_{error} are the error values of spatial synchronization. In the range from 10 to 50 m, the lateral deviation of spatial synchronization is within 0.25 m and the longitudinal deviation is within 2.30 m. What's behind the large deviation in spatial synchronization at longer distances is that the further away the image is, the greater the distance between pixels. The conversion of the same image error into the radar's coordinate system will lead to increases in distance error.

In the practical application of machine vision, obstacles beyond a certain distance are small in the image, which makes identification difficult. The more appropriate applicable distance is usually within 60 m (for cameras with ordinary focal length range). The static verification experiments have proved that the spatial synchronization method proposed in this chapter can meet the synchronization accuracy requirements of both sensors within the actual application distance of intelligent vehicles.

In order to further verify the effects of practical application of the spatial synchronization method, several dynamic experiments have been conducted. The roads used

Table 4.1 Static verification experimental data on spatial sensor synchronization

No	Y_L/m	X_L/m	$Y_{L(I-L)}$/m	$X_{L(I-L)}$/m	Y_{error}/m	X_{error}/m
1	9.6	0	9.58	−0.16	−0.02	−0.16
2	9.6	−0.96	9.52	−1.07	−0.08	−0.11
3	9.65	−1.76	9.46	−1.95	−0.19	−0.19
4	9.6	0.8	9.63	0.89	0.03	0.09
5	9.6	1.76	9.70	1.80	0.10	0.04
6	29.75	0	29.24	−0.09	−0.51	−0.09
7	29.75	−0.96	29.31	−1.09	−0.44	−0.13
8	29.75	−1.92	28.44	−1.93	−1.31	−0.01
9	29.73	0.8	29.17	0.92	−0.56	0.12
10	29.73	1.6	29.10	1.81	−0.63	0.21
11	49.65	−0.16	47.52	−0.07	−2.13	0.09
12	49.65	−1.12	47.59	−1.08	−2.06	0.04
13	49.7	−1.92	47.64	−1.79	−2.06	0.13
14	49.7	0.64	47.46	0.88	−2.24	0.24
15	49.7	1.6	47.40	1.71	−2.30	0.11
16	69.65	−0.16	63.68	−0.14	−5.97	0.02
17	69.7	−1.12	63.75	−1.09	−5.95	0.03
18	69.65	−1.92	63.79	−1.72	−5.86	0.20
19	69.7	0.64	73.49	0.94	3.79	0.30
20	69.65	1.6	73.42	1.94	3.77	0.34
21	99.7	−0.16	95.69	−0.16	−4.01	0
22	99.9	−1.28	95.75	−1.10	−4.15	0.18
23	99.7	−1.92	95.80	−1.80	−3.90	0.12
24	99.7	0.48	95.61	0.90	−4.09	0.42
25	99.7	1.44	86.88	1.56	−12.8	0.12

in the experiments are highways and city roads with good road conditions and the speed of the ego vehicle is between 50 km/h and 80 km/h. Some of the experimental results are extracted and shown in Fig. 4.19.

The black line in the figure is the result after projecting the vehicle position detected by the radar into the image. Based on the dynamic experimental results, the spatial synchronization method proposed in this section can better match the spatial information obtained by the two sensors, which can help with information fusion-based obstacle identification in the next step.

Since this chapter uses the road plane assumption commonly used by monocular vision systems, the conditions of this assumption are no longer met in the case of

a) Working condition 1 b) Working condition 2

Fig. 4.19 Dynamic experimental results of spatial sensor synchronization

an uphill or downhill road section, and errors are introduced to the spatial synchronization of the two sensors. For the next step, we should consider how to reduce the effects of changes in road slope.

4.5 Detection Object Fusion Decision Based on Multi-source Information

In order to improve the accuracy and real-timeness of the vehicle identification algorithm, we try to fuse the obstacle data acquired by LiDAR and the visual information to detect the vehicles driving ahead in this section. The overall structure of the algorithm as shown in Fig. 4.20.

The algorithm consists of four main modules. The pre-processing module is responsible for processing the original images captured by the camera to obtain the images needed for subsequent identification, which include the shadow images under the vehicle, the extracted horizontal-edge images and the vertical-edge images. The tracking module tracks these objects based on the previous round of vehicle identification results. If it's confirmed the object still exists, then the object is marked as the vehicle detected in the current round; if it's confirmed the object has disappeared, then the object is removed from the tracking profile. The use of vehicle tracking technology can reduce the actuation time of the algorithm. The radar detection module compares the radar data with the vehicle position detected by the tracking module. For the object newly detected by the radar, the radar detection module is responsible for confirming the object in the image. If the object is confirmed, it will be added to the current round of identification results. The vision detection module, on the other hand, performs supplementary detection of the images. For the image areas already detected by the tracking module and the radar detection module, the vision module does not repeat the tests, which greatly reduces the range of images to be detected.

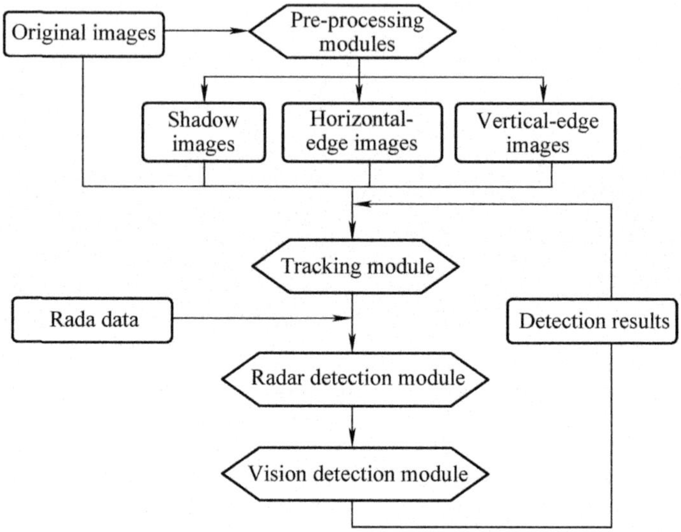

Fig. 4.20 Structure of the vehicle identification algorithm

In order to reduce the false detection rate of the vision module, strict confirmation conditions are adopted.

As an important application of information fusion technology, detection fusion can provide redundant information for deciding whether a candidate object is a real object and enhance the credibility of the decision. In order to more effectively make comprehensive use of the multiple extracted features of the vehicle in the image as well as the radar data, a multi-level fusion-based vehicle identification decision algorithm using fuzzy logic is designed in this section.

The fusion decision algorithm is divided into three levels. The first level is about the generation of hypothesis of candidate vehicle objects. In this part, through comprehensive analysis of radar data, vehicle shadow features and vehicle edge features, the possible areas that vehicles may occupy conforming to variable models are extracted as the candidate objects for vehicle identification; due to the interference of road shadows, roadside railings and viaducts over the road, the boundary of the vehicle-occupied areas may be inaccurately defined among the candidate objects identified in the first level. Therefore, the shape regularization of the candidate objects is carried out after the first level. The accuracy of vehicle boundary definition is improved through energy density verification, shape verification and repeated tests. The accuracy improvement is very useful for subsequent hypothesis testing. The shape regularization hierarchy does not play a decisive role in the decision on candidate objects, but instead only serves for the subsequent tests. For those vehicle objects generated by hypothesis, some of them have very high confidence level and can be directly used as obtained vehicle results, while some of them do not have enough confidence level and need further hypothesis verification before they can be used as identification results. The zero height hypothesis test, entropy value test and symmetry test are used at the hypothesis testing level to verify the vehicle-occupied areas that have been shape-regularized.

Due to the imprecision in vehicle features, fuzzy logic is introduced in the fusion algorithm to describe the vehicle features, that is, the weight value and membership function of the object credibility are set for each feature. The target is considered to exist only when the calculated object credibility exceeds the set threshold value. Since the credibility of radar data and vehicle shadow features is higher than that of vehicle edge information, the weight value of radar data W_{LiDAR} and the weight value of vehicle shadow features W_{shade} are set as 1.2 and the weight values of both longitudinal edge features of the vehicle W_{edge} are equal to 0.8; the membership functions of radar data, shadow features and edge features M_{LiDAR}, M_{shade} and M_{edge} are calculated, as shown in Eqs. (4.20)–(4.22).

$$M_{\text{lidar}} = \begin{cases} 0 & radar\ doesn't\ detect\ object \\ 1 & radar\ detect\ object \end{cases} \tag{4.20}$$

$$M_{\text{shade}} = \frac{N_{\text{shade}}}{\frac{3}{4} W_{\text{model}}} \tag{4.21}$$

$$M_{\text{edge}} = \frac{N_{\text{edge}}}{\frac{3}{4} H_{\text{model}}} \tag{4.22}$$

where W_{model} and H_{model} are the width and height of the vehicle planned using the variable model and N_{shade} and N_{edge} are the numbers of shadow feature pixels and edge feature pixels detected in the planning area of the variable model. In terms of the membership of radar data, the membership value is set as 1 when radar detects the object. In membership calculation, the membership value is set as 1 when the membership function is greater than 1. In the vehicle identification algorithm, the vehicle is not identified as a vehicle object when it's detected for the first time and its object identity is determined and included in the tracking profile only when the vehicle is detected for the second time. When a vehicle in the tracking profile is not detected, the vehicle is not immediately removed from the file, but instead, the location of the vehicle is continuously tracked and predicted until the vehicle is not identified 10 times in a row and then the vehicle is removed from the file. For vehicles that are being tracked, the object decision conditions can be reduced appropriately to speed up the identification process and a tracking compensation term W_{track} is added, so as to increase the credibility of object decision when the vehicle object is kept in the tracking profile and in this case the membership value is set as 0.5.

The total object presence credibility W_{hypo} is calculated at the hypothesis generation level, as shown in Eq. (4.23).

$$W_{\text{hypo}} = W_{\text{lidar}} M_{\text{lidar}} + W_{\text{shade}} M_{\text{shade}} + W_{\text{edge}} M_{\text{edge}} + W_{\text{track}} \tag{4.23}$$

It's assumed that two thresholds T_{hypo1} and T_{hypo2} exist, with $0 < T_{\text{hypo1}} < T_{\text{hypo2}} < 4.5$. When W_{hypo} is greater than T_{hypo2}, the object exists and no subsequent hypothesis testing is required; when W_{hypo} is less than T_{hypo2} and is greater than T_{hypo1}, the object exists but subsequent hypothesis testing is required to confirm the object; when W_{hypo} is less than T_{hypo1}, the object does not exist.

The work done at the shape regularization level does not affect the results of object decision. For the objects that require hypothesis testing, there are three test terms, namely, zero height hypothesis test, entropy value test and symmetry test, whose weights W_{zero}, W_{entropy} and W_{sym} are all set as 1. Their membership functions are calculated, as shown in Eqs. (4.24)–(4.26).

$$M_{\text{zero}} = \frac{Z_{\text{index}}}{0.2} \tag{4.24}$$

$$M_{\text{entropy}} = \frac{E(l)}{5} \tag{4.25}$$

$$M_{\text{sym}} = \frac{S_{\text{total}}}{0.9} M_{\text{entropy}} \tag{4.26}$$

Similarly, when the membership value is calculated to be over 1, then the membership value is set as 1. It should be noted that the entropy membership is added to the calculation of the symmetry feature membership because the image area is homogeneous for many backgrounds, resulting in a high symmetry. However, as the object vehicle is not the real one, and the addition of the entropy membership reduces the negative effect of the background on the symmetry calculation.

At the hypothesis testing level, the total confidence level of the object W_{vali} is calculated, as shown in Eq. (4.27).

$$W_{\text{vali}} = W_{\text{zero}} M_{\text{zero}} + W_{\text{entropy}} M_{\text{entropy}} + W_{\text{sym}} M_{\text{sym}} \tag{4.27}$$

Set T_{vali} as the threshold to pass the hypothesis test, with $0 < \text{T}_{\text{vali}} < 3$. When W_{vali} is greater than T_{vali}, it means that the object has passed the hypothesis test and the object exists; otherwise, the object does not exist.

4.6 Establishing the Set of Vehicle Parameter Features Based on Information Fusion

The selection of dangerous objects or car-following objects from multiple vehicle objects driving ahead that have been identified is the basis for the subsequent development of longitudinal active safety systems such as adaptive cruise control and autonomous driving. However, the current research on object vehicle identification methods has not sufficiently integrated the lateral and longitudinal perception information, resulting in insufficient stability of object vehicle identification and tracking as well as the algorithm's untimely response to the object vehicle when it changes lanes. To this end, an object vehicle identification method based on multi-feature fusion is proposed based on the previously studied vehicle identification [22] and lane identification methods.

Compared with existing object vehicle identification methods, the method designed in this section does not solely rely on individual vehicle parameters for object vehicle identification and decision, but rather combines multiple vehicle parameter features obtained from machine vision and radar data. This processing method has two advantages: first, the redundancy of decision information is used to improve the stability and accuracy of object vehicle identification and tracking; second, because of the fusion of lateral and longitudinal environmental sensing information, the algorithm can respond to the changes in object vehicle state when the object vehicle cuts in or leaves the lane.

Based on the vehicle data information obtained by machine vision and LiDAR, the lateral distance, longitudinal distance and relative speed between the ego vehicle and the preceding vehicle can be obtained. The radar also provides the In-Lane Probability data of the preceding vehicle. Therefore, these four feature quantities are selected from the vehicle data feature set, which are denoted as $D_{\mathrm{H}i}$ (relative lateral distance), $D_{\mathrm{L}i}$ (relative longitudinal distance), V_i (relative speed) and P_i (in-lane probability); i = 1, …, n, and n is the total number of vehicles identified in each round of detection. There will be two problems if the raw data such as the relative distance between the preceding vehicle and the ego vehicle are directly input as feature quantities: (1) the data of several feature quantities varies widely and are not uniformly quantified within a range of values, which is not conducive to the subsequent neural network training; (2) when the human driver performs the object vehicle identification, it's actually the result of comparing the candidate vehicles obtained in each round of detection and is not a single decision of whether a vehicle is an object or not. If the raw data is directly used as feature quantities, then the comparison part cannot be reflected. To solve these two problems, when we were selecting feature quantities, instead of simply substituting the original data into the selection process, we introduced comparison results of relative distance and other data as feature quantities based on each round of detection identification results. The four feature quantities are quantified uniformly into the [0, 1] value domain space.

(1) Lateral distance indicator

For the vehicles identified by radar and machine vision, the lateral distance between the ego vehicle and the preceding vehicle can be obtained from radar data and image data. However, since the lateral distance of the object provided by LiDAR in object detection is calculated by using the multiple tail reflection points of the scanned objects based on the algorithm set in the radar, the lateral distance value is usually inaccurate considering the width of the preceding vehicle. The lane marking information is also not taken into account in the calculation of the lateral distance data given by the radar. Therefore, for the selection of the feature quantity of lateral distance, the image information is used, that is, the distance between the bottom midpoint of the vehicle identified in the image and the lane centerline is calculated. Let the number of vehicles detected in each round be n. For each vehicle, the original lateral distance calculated from the image information is $D_{\mathrm{HI}i}$ and the smallest lateral distance among n vehicles is D_{HImin}. The longitudinal distance and relative speed of the object vehicle with the smallest lateral distance are denoted as D_{LImin} and V_{Imin}. The lateral distance of a vehicle after comparative quantification is $D_{\mathrm{H}i}$, which is calculated as shown in Eq. (4.28).

$$\begin{aligned} D_{\mathrm{H}i} &= \frac{D_{\mathrm{HImin}}}{D_{\mathrm{HI}i}} \frac{4 - D_{\mathrm{HI}i}}{4} \\ if\ \ & D_{\mathrm{H}i} < 0,\ D_{\mathrm{HI}i} = 0 \end{aligned} \tag{4.28}$$

$D_{\mathrm{H}i}$ represents the danger level of the object in terms of lateral distance and the larger the lateral distance value, the higher the danger level. In the process of comparative quantification of lateral distance, firstly, the comparison in the same round of detection is considered, that is, the closer the object is to the minimum lateral distance, the greater its lateral distance indicator is; the absolute lateral distance of the vehicle is also considered, that is, the smaller the absolute lateral distance value is, the higher the danger level is. If the lateral distance $D_{\mathrm{HI}i}$ exceeds 4 m, the lateral distance indicator $D_{\mathrm{H}i}$ is denoted as 0. Through Eq. (4.28), it's possible to quantify and compare the original lateral distance $D_{\mathrm{HI}i}$ as the lateral distance indicator $D_{\mathrm{H}i}$ and the range of values of this indicator is [0, 1].

(2) Longitudinal distance and relative speed indicators

As the vehicle longitudinal data acquired by LiDAR is more accurate and stable, the radar data is used to extract the longitudinal distance and relative speed in the feature set. When the radar misses the detection and the vehicle that has been missed is identified by image supplementation, the longitudinal distance and relative speed are offered by the image information because the vehicle data cannot be given by the radar. The longitudinal distance can be obtained by directly calculating the distance between the bottom midpoint of the vehicle identified in the image and the vehicle. The relative speed is calculated using the longitudinal distance data from this frame and the last three consecutive frames. Three speed values are obtained and the relative speed values are obtained using the filter [0.25, 0.25, 0.5].

The original longitudinal distance is denoted as $D_{\mathrm{LI}i}$, and the longitudinal distance indicator $D_{\mathrm{L}i}$ after comparative quantification is calculated as shown in Eq. (4.29).

$$\begin{aligned} D_{\mathrm{L}i} &= \left(\frac{D_{\mathrm{LImin}} - D_{\mathrm{LI}i}}{20} + 1\right)/2 \\ if\ \ & D_{\mathrm{L}i} < 0,\ \ D_{\mathrm{L}i} = 0 \\ if\ \ & D_{\mathrm{L}i} > 1,\ \ D_{\mathrm{L}i} = 1 \end{aligned} \tag{4.29}$$

The longitudinal distance indicator is compared to the object vehicle with the smallest lateral distance.

The original relative speed is denoted as $V_{\mathrm{I}i}$ and the relative speed indicator V_i is calculated as shown in Eq. (4.30).

$$
\begin{aligned}
V_i &= \left(\frac{V_{\text{Imin}} - V_{\text{I}i}}{10} + 1\right)/2 \\
if \;\; V_i &< 0, \;\; V_i = 0 \\
if \;\; V_i &< 1, \;\; V_i = 1
\end{aligned}
\tag{4.30}
$$

The relative speed indicator is also compared to the object vehicle with the smallest lateral distance.

After comparative quantification, the relative danger level of multiple vehicles identified in the same round of detection can be introduced into the feature set and the feature quantities are quantified uniformly into the [0, 1] space.

(3) In-lane probability

The radar provides "In-Lane Probability" data for detected vehicle objects, which represents the likelihood of the vehicle object being in the same lane as the ego vehicle and is therefore included in the vehicle feature set as a basis for selection of dangerous objects. When the radar misses the detection and the vehicle that has been missed is identified by image supplementation, this data is obtained by dividing the width of the vehicle in the lane identified in the image by the width of the vehicle.

4.7 Extraction and Processing of Vehicle Feature Data Based on Information Fusion

After the LiDAR data and the images captured by the camera are obtained, the feature set of the vehicle is established based on multiple information sources. However, instead of directly using vehicle feature quantities in the selection of object vehicles, the feature data extraction and pre-processing are carried out first, because: first, under certain working conditions, the vehicle feature data cannot be directly obtained and can only be acquired with some extraction algorithms. For example, when the radar has not directly detected the object vehicle and the location of the vehicle is identified through image supplementation, the longitudinal data of the vehicle can not be provided by LiDAR and can only be obtained through the monocular vision-based depth information extraction algorithm; second, certain vehicle feature data that can be directly obtained is not accurate enough or does not meet the actual road conditions, we need to make these features more applicable to the subsequent object vehicle identification through certain pre-processing algorithms; third, in order to improve the accuracy and stability of vehicle identification, the vehicle feature data and vehicle location information need to be tracked and predicted so that the algorithm can operate undisturbed and stably in the case of occasional miss detection of objects by the sensor. Based on the above considerations, the feature data in the vehicle feature set are fused and extracted with optimized processing in the first place.

4.7.1 Monocular Vision Ranging

Radar has obvious advantages in the detection of longitudinal distance and speed of the object with high detection accuracy. However, although radar data is superior to vision sensor data in the process of fusing and extracting longitudinal information of the vehicle object, the radar's relatively small viewing angle for detection easily leads to miss detection of vehicles. In this case, for those vehicles that rely on machine vision to supplement the identification, the acquisition of longitudinal information must be performed by machine vision. In order to acquire the depth information from the monocular camera, an algorithm is designed in this section to calculate the dynamic pitch angle of the camera in real time based on parallel constraints of the road boundary, so as to monitor the distance between the ego vehicle and the preceding vehicle using monocular vision in real time.

The process of solving the distance between the ego vehicle and the preceding vehicle is as follows: first, the location of the vehicle is identified in the image and the boundary of the vehicle is marked with a rectangular box; then the bottom midpoint of the rectangular box boundary of vehicle is taken and the image coordinates of the midpoint are transformed into the coordinates of the world coordinate system using Eq. (4.3). The value of the vertical coordinate of the world coordinate system is taken as the distance between the object vehicle and the ego vehicle.

There are two main factors that affect the accuracy of the algorithm results: one is the accuracy of the measured camera height and the other is the accuracy of the measured the camera pitch angle. We fixed the camera pitch angle at 89.7° and changed the height of the camera without changing other parameters and the obtained distance change data are shown in Table 4.2.

We then fixed the camera height at 1.21 m, changed the camera pitch angle without changing other parameters. The distance change data obtained are shown in Table 4.3.

When the camera height changes within the range of 20 cm, the distance change is 4.32 m, when the camera pitch angle changes within the range of 0.4°, the distance change is 5.44 m. Thus, it can be seen that the camera height has little effect on the solution of distance, while the pitch angle plays a key role in the solution of distance

Table 4.2 Measured distance varies with height

Height/m	1.11	1.16	1.21	1.26	1.31
Distance/m	27.89	28.97	30.05	31.13	32.22

Table 4.3 Measured distance varies with the pitch angle

Pitch angle/(°)	89.5	89.6	89.7	89.8	89.9
Distance/m	27.58	28.76	30.05	31.47	33.02

and largely restricts the accuracy of solving the distance between the ego vehicle and the preceding vehicle.

In order to obtain the range of changes of the camera pitch angle due to vehicle bumps in the actual road environment, lane line identification, vehicle identification and pitch angle detection were performed by driving the vehicle on the 4th Ring Road in Beijing and some of the extracted raw data records are shown in Table 4.4.

In Table 4.4, rows of data refer to the straight-line equation of the lane line obtained from the road image by the machine vision identification algorithm and the pitch angle refers to the actual pitch angle value of the camera obtained using the real-time detection algorithm. As shown from the data in the table, when the vehicle is running on a flat road, the camera pitch angle changes from 90.4° to 90.9° and for the same point on the road image, its actual distance changes from 32.91 to 43.11 m based on the solution, representing a change of 10 m. This proves that the real-time solution of the camera pitch angle is very important for the monocular vision ranging algorithm.

In the actual road environment, the lane markings are similar to parallel lines. We use the machine vision algorithm to obtain the lane markings in the image plane and then use Eq. (4.3) to calculate the straight line in the road plane that corresponds to the lane markings in the image. Theoretically, the calculated lane markings should be parallel to each other. However, due to the pitch motion of the vehicle during driving, the actual pitch angle of the camera will change. Therefore, the road lane lines calculated by using the camera pitch angle calibrated in the initialization will no longer be parallel. The algorithm proposed in this section is to use the constraints of parallel road lane lines to solve the camera pitch angle dynamically.

The algorithm for solving the actual pitch angle of the camera is derived as below. Define S_{dyn} as the indicator function for the dynamic pitch angle solution.

$$S_{\text{dyn}} = (k_{\text{L1}} - k_{\text{L2}})^2 \tag{4.31}$$

In Eq. (4.31), k_{L1} is the slope of the lane marking on the left side of the road in the world coordinate system and k_{L2} is the slope of the lane marking on the right side of the road in the world coordinate system. The following is an example of the slope solution process based on the lane marking on the left side of the road.

Table 4.4 Changes in pitch angle in the actual road environment

Left slope	Left intercept	Right slope	Right intercept	Pitch angle/(°)
0.3640	50.0000	0.6249	320.000	90.6000
0.3640	50.0000	0.6249	319.000	90.7000
0.3640	50.0000	0.6249	319.000	90.4000
0.3640	51.0000	0.5543	298.000	90.7000
0.3640	51.0000	0.5317	289.000	90.8000
0.3640	50.0000	0.5543	294.000	90.9000
0.3640	50.0000	0.6009	311.000	90.7000

Firstly, the two lane lines in the image are identified using the lane marking identification algorithm and two points P_a' and P_b' on the identified lane markings on the left side of the road are taken and their coordinates are (u_a, v_a) and (u_b, v_b) respectively. The corresponding points of P_a' and P_b' in the road plane are P_a and P_b, whose coordinates are (X_a, Y_a) and (X_b, Y_b) respectively. The slope of the lane marking on the left side of the road in the world coordinate system k_{L1} is calculated by the equation:

$$k_{L1} = \frac{X_b - X_a}{Y_b - Y_a} \tag{4.32}$$

On the condition that other camera-related parameters are fixed and only the pitch angle changes, X_a, Y_a, X_b, and Y_b can be seen as the function of the pitch angle γ_0. Thus, the slope of the lane marking on the left side of the road k_{L1} is the function of the variable pitch angle γ_0. Similarly, the slope of the lane marking on the right side of the road k_{L2} is also the function of the variable pitch angle γ_0. The solution process of k_{L2} can be obtained similarly from the solution process of k_{L1}.

Therefore, the function S_{dyn} can be considered as the function of the single-variable pitch angle, with:

$$S_{dyn} = f(\gamma_0) \tag{4.33}$$

In the process of deriving the function S_{dyn} with respect to the pitch angle γ_0, the value of γ_0 when the derivative is 0 is the actual pitch angle sought. In practice, the pitch angle can be solved discretely using the method proposed in Sect. 4.2.3.

To verify the accuracy of the monocular ranging algorithm, static experiments were conducted on a real-life vehicle and the experimental data are shown in Table 4.5.

The experimental results show that the real-time ranging algorithm after pitch angle correction can keep the error under 2% within the range of 20–60 m. The error is slightly larger at 5.82% when the actual distance reaches 10 m. In the range of 10–50 m, the difference between the actual distance and the measured distance is within 1 m, indicating that the algorithm proposed in this section can meet the accuracy

Table 4.5 Static experimental data on monocular vision ranging algorithm

Actual distance/m	Monocular vision ranging/m	Error rate/%
10	10.58	5.82
20	20.31	1.53
30	29.55	1.49
40	40.58	1.46
50	49.86	0.28
60	58.81	1.98

requirements of vehicle identification and distance calculation using machine vision in the usual range of distance measurement (10–60 m).

4.7.2 DCF Correction for In-Lane Probability

The fact that radar usually can not react to the cut-in vehicles in time implies the following situations when it's shown in the in-lane probability data: when there are two vehicles, the longitudinally distant vehicle is in the middle of the lane and the longitudinally close vehicle has just cut into the lane. The values of two vehicles' probability of being in the same lane may differ a lot based on the radar data, with the longitudinally distant vehicle's probability of being in the same lane close to 1, while the same probability of the longitudinally close vehicle close to 0. This data does not match the real-life condition and the danger level of the vehicle object.

Inspired by the DDF (Distance Discount Factor) method [23], a distance compensation factor (DCF) was designed to take into account the distance to the vehicle object in order to correct the in-lane probability data included in the radar data to make radar data better match the real-life road condition.

First, multiple vehicle objects detected by the radar in the same round of detection are compared and the object with the highest in-lane probability is selected, with its in-lane probability, lateral distance and longitudinal distance being recorded as $P_{\max}$, D_{Hmax} and D_{Lmax} respectively. The in-lane probability in the original data is denoted as $P_{\mathrm{I}i}$ and the in-lane probability after DCF correction is denoted as P_i. The DCF correction equation is shown in Eq. (4.34) as below.

$$
\begin{aligned}
&P_i = P_{\mathrm{I}i} + (P_{\max} - P_{\mathrm{I}i})\sqrt{\left(\frac{D_{\mathrm{Hmax}}}{D_{\mathrm{HI}i}}\right)^2 + \left(\frac{D_{\mathrm{Lmax}} - D_{\mathrm{LI}i}}{D_{\mathrm{Lmax}}}\right)^2}\\
&\text{if } D_{\mathrm{LI}i} \geq D_{\mathrm{Lmax}},\ P_i = P_{\mathrm{I}i}\\
&\text{if } D_{\mathrm{HI}i} \leq D_{\mathrm{Hmax}},\ P_i = P_{\max}\\
&\text{if } \left(\frac{D_{\mathrm{Hmax}}}{D_{\mathrm{HI}i}}\right)^2 + \left(\frac{D_{\mathrm{Lmax}} - D_{\mathrm{LI}i}}{D_{\mathrm{Lmax}}}\right)^2 > 1,\ P_i = P_{\max}
\end{aligned}
\tag{4.34}
$$

Compared to DDF, DCF has the following features:

(1) DDF is a one-dimensional correction factor and DCF uses a two-dimensional correction factor that integrates the influencing factors of lateral and longitudinal distances;
(2) DDF is a discount factor, that is, the adoption of DDF correction can reduce the importance of long-distance object. DCF is a compensation factor. The highest value of the in-lane probability in the original radar data is usually accurate, but the radar cannot reflect the cut-in vehicles in time, making the value of the in-lane probability of cut-in vehicles very low. DCF considers the distance, and compensates and revises the object's in-lane probability data;

(3) After the DCF correction is carried out, the consistency between the in-lane probability in the radar data and the real road condition can be improved, and the corrected data can be used as a feature to better reflect the degree of danger of the vehicle.

4.7.3 Predictive Tracking Using Kalman Filtering

In the process of vehicle recognition by fusing the information obtained by machine visual and LiDAR, the missing detection of vehicles may occur. To improve the accuracy and stability of vehicle identification, the vehicle detected initially is not identified as a vehicle object. Only when the vehicle is detected for twice in a row can it be identified as a object vehicle and be included in the tracking profile. When the vehicle in the tracking profile is not identified, it will not be deleted from the file immediately. Instead, the system continues to perform predictive tracking of its location until it fails to be detected 10 times in a row when it will be deleted from the profile.

Kalman filter can effectively track and predict the position and velocity of vehicles, so in this section, the Kalman filter method is adopted to track the vehicle objects. In this section, it's assumed that the change of relative longitudinal acceleration and relative lateral velocity between the ego vehicle and the preceding vehicle is stable, so the third-order Kalman filter is used to track and predict the longitudinal data, and the second-order Kalman filter is used to track and predict the lateral data.

The state equation of the longitudinal Kalman filter is shown in Eq. (4.35), and the observation equation is shown in Eq. (4.36).

$$s(t + \Delta t) = As(t) + Bw(t) \tag{4.35}$$

$$z(t) = Hs(t) + v(t) \tag{4.36}$$

where

$$s(t) = [d_{\mathrm{L}}, v_{\mathrm{L}}, a_{\mathrm{L}}]^{\mathrm{T}}, z(t) = [d_{\mathrm{L}}, v_{\mathrm{L}}]^{\mathrm{T}}$$

$$A = \begin{bmatrix} 1 & \Delta t & \dfrac{\Delta t^2}{2} \\ 0 & 1 & \Delta t \\ 0 & 0 & 1 \end{bmatrix}, B = \left[\frac{\Delta t^3}{6}, \frac{\Delta t^2}{2}, \Delta t\right], H = \begin{bmatrix} 1 & 0 & 0 \\ 0 & 1 & 0 \end{bmatrix}$$

d_{L} refers to the relative longitudinal distance between the preceding vehicle and the ego vehicle, v_{L} refers to the relative longitudinal velocity between the preceding vehicle and the ego vehicle, a_{L} refers to the relative longitudinal acceleration between the preceding vehicle and the ego vehicle, and Δt refers to the time interval between two rounds of detection. The system time interval used in this chapter is 0.1 s.

The form of the state equation and observation equation of the lateral Kalman filter is the same as that of the longitudinal Kalman filter, but their state variables, observation variables, system matrix and observation matrix are different.

$$s(t) = [d_{\mathrm{H}}, v_{\mathrm{H}}]^{\mathrm{T}}, z(t) = [d_{\mathrm{H}}]$$

$$A = \begin{bmatrix} 1 & \Delta t \\ 0 & 1 \end{bmatrix}, B = \begin{bmatrix} \frac{\Delta t^2}{2}, \Delta t \end{bmatrix}, H = [1 \quad 0]$$

d_{H} refers to the relative lateral distance between the preceding vehicle and the ego vehicle, v_{H} refers to the relative lateral speed between the preceding vehicle and the ego vehicle.

To verify the tracking prediction effect of Kalman filter, the longitudinal distance, lateral distance and relative longitudinal velocity of obstacles detected by radar were processed using the Kalman filter, as shown in Fig. 4.21.

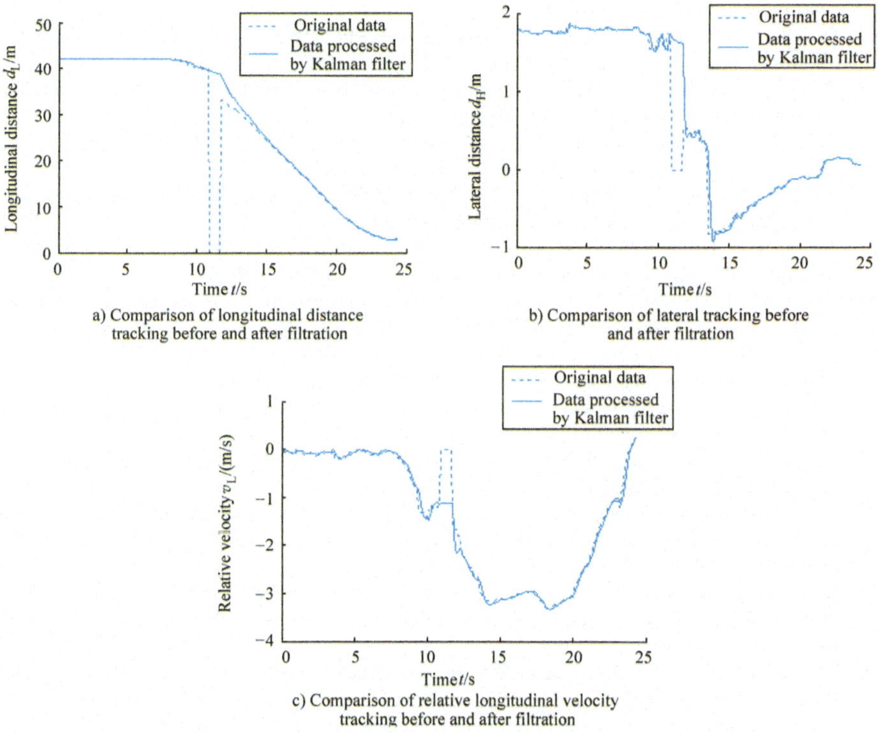

Fig. 4.21 Comparison of vehicle tracking data before and after Kalman filtration

It's shown in the figure that the object is lost for a period in the original data, that is, a missing detection occurred. Through Kalman tracking prediction, the data of the vehicle can be better estimated, thereby improving the stability of vehicle recognition and reducing the rate of missing detection. The data processed using Kalman filter can also be better used as the feature for object vehicle identification.

References

1. Mahler R (2007) Statistical multisource-multitarget information fusion. Artech House, Norwood
2. Bleiholder J, Naumann F (2009) Data fusion. ACM Comput Surv 41(1):1–41
3. Alessandretti G, Broggi A, Cerri P (2007) Vehicle and guard rail detection using radar and vision data fusion. IEEE Trans Intell Transp Syst 8(1):95–105
4. Chongzhao H, Hongyan Z, Zhansheng D (2006) Multi-source information fusion. Tsinghua University Press
5. Li Q, Chen L, Li M et al (2014) A sensor-fusion drivable-region and lane-detection system for autonomous vehicle navigation in challenging road scenarios. IEEE Trans Veh Technol 63(2):540–555
6. de Croon G (2017) Self-supervised learning: when is fusion of the primary and secondary sensor cue useful? arXiv:1709.08126
7. Sukkarieh S, Nebot EM, Durrant-Whyte HF (1999) A high integrity IMU/GPS navigation loop for autonomous land vehicle applications. IEEE Trans Robot Autom 15(3):572–578
8. Caron F, Duflos E, Pomorski D et al (2006) GPS/IMU data fusion using multisensor Kalman filtering: introduction of contextual aspects. Inf Fusion 7(2):221–230
9. Wendel J, Meister O, Schlaile C et al (2006) An integrated GPS/MEMS-IMU navigation system for an autonomous helicopter. Aerosp Sci Technol 10(6):527–533
10. Yanhua Z, Tijing C, Zhuopeng Y (2009) Implementation of MEMS-IMU/GPS integrated navigation system. J Chin Inertial Technol 17(5):552–556
11. Newman P, Ho K (2005) SLAM-loop closing with visually salient features. In: IEEE international conference on robotics and automation. IEEE, pp 635–642
12. Danelljan M, Meneghetti G, Khan FS et al (2016) A probabilistic framework for color-based point set registration. In: Computer vision and pattern identification. IEEE, pp 1818–1826
13. Leonard J, How J, Teller S, et al (2009) A perception-driven autonomous urban vehicle. In: The DARPA urban challenge. Springer, Berlin, Heidelberg, pp 727–774
14. Huang AS, Antone M, Olson E et al (2010) A high-rate, heterogeneous data set from the DARPA urban challenge. Int J Robot Res 29(13):1595–1601
15. Rauskolb FW, Kai B, Lipski C et al (2014) Caroline: an autonomously driving vehicle for urban environments. Journal of Field Robotics 25(9):674–724
16. Steux B, Laurgeau C, Salesse L, et al (2002) Fade: A vehicle detection and tracking system featuring monocular color vision and radar data fusion. In: 2002 IEEE intelligent vehicle symposium, vol 2. IEEE, pp 632–639
17. Duraisamy B, Schwarz T, Wöhler C (2013) Track level fusion algorithms for automotive safety applications. In: International conference on signal processing image processing & pattern identification. IEEE, pp 179–184
18. Zhangsong S (2010) Theory and method of target tracking and data fusion. National Defense Industry Press, Beijing
19. Ning S (2018) Research on environment perception technology for intelligent vehicle based on multi-source information fusion. Jiangsu University, Zhenjiang
20. Hattori H (2000) Stereo for 2D visual navigation. In: Proceedings of IEEE intelligent vehicles symposium, Dearborn, MI, USA. IEEE Press, pp 31–38

21. Franke U, Gavrila D, Gorzig S et al (1998) Autonomous driving goes downtown. IEEE Intell Syst Appl 13(6):40–48
22. Bensrhair A, Bertozzi A, Broggi A et al (2002) Stereo vision-based feature extraction for vehicle detection. In: Proceedings of IEEE intelligent vehicle symposium, Paris, France: IEEE Press, pp 465–470
23. Bertozzi M, Broggi A, Fascioli A et al (2000) Stereo vision-based vehicle detection. In: Proceedings of IEEE intelligent vehicles symposium, Dearborn, MI, USA. IEEE Press, pp 39–44

Chapter 5
Cooperative Control Technologies for i-EFV

The optimally integrated vehicle control is the basic problem to be solved by the key i-EFV coordinated control technologies. Exploring the mechanism of multi-physical process coupling in complex electromechanical systems and the problem of multi-system coordinated control in the "human-vehicle-infrastructure" environment is the basis for achieving the performance and operational goals of i-EFV. The multi-objective and multi-system coordinated dynamics control of i-EFV achieves the integrated and optimized performance of i-EFV by optimizing the safety, comfort, energy saving and environmental protection objectives and coordinating the vehicle driving, brake, steering, suspension and other systems.

5.1 Integrated and Coordinated Vehicle Chassis Control System

In the era of rapidly changing Internet information and electronic technologies, a variety of electronic control technologies have been developed and widely used in vehicle chassis, bringing disruptive changes to the development of vehicle technologies. The change of the vehicle chassis system began in its previous individual system or parts working independently to achieve network integration, so that the development of integrated and coordinated vehicle chassis control system continues. The vehicle chassis system is composed of several subsystems, including steering, suspension, driving, brake and other key subsystems that involves nonlinear dynamics characteristics of complex electromechanical systems coupled with each other. Each subsystem controlled respectively changes the entire vehicle system's dynamics performance. There are links between the subsystems and their synergy requires the integrated control of the chassis [1]. The electronic chassis control systems of modern vehicles have improved the vehicle dynamics by eventually implementing

K. Li et al., *The Intelligent Environment Friendly Vehicle*, Key Technologies on New Energy Vehicles,
https://doi.org/10.1007/978-981-19-4851-0_5

active control of the longitudinal, lateral, and vertical forces on the tyres, respectively. However, due to the complex coupling characteristics of the vehicle motion in different directions and forces of different directions acting on tyres, the use of different electronic chassis control systems to control tyre forces respectively inevitably leads to conflicts between control objectives and control actuation. For the conflicts or complementarity between individual electronic chassis control systems, it's important to implement coordinated control for multiple individual systems, eliminate the conflicts between different subsystems and explore the complementarity potentials between different subsystems to improve the comprehensive performance of vehicle dynamics.

Integrated Chassis Control (ICC), or Global Chassis Control (GCC), combines the control capabilities of different subsystems and works in concert with the vehicle's longitudinal, lateral and vertical dynamics or two of them to provide integrated control of longitudinal forces, lateral forces and vertical forces on tyres. The integrated control of chassis system can effectively improve vehicle dynamics, handling stability, safety and environmental protection by studying the methods for coordinated control of vehicle chassis subsystems. Due to the significant advantages of ICC in improving vehicle stability and controllability, it has received extensive attention in current research and has become the development direction of today's automotive electronic chassis control system [2–6].

However, the simple superposition of subsystems according to different control objectives is not enough to optimize the comprehensive performance of the vehicle. Therefore, only when the traditional vehicle chassis system becomes intelligent, connected and modular, can the integrated chassis control be actually realized. The vehicle dynamics integrated control system can integrate and control two or more subsystems in the vehicle chassis system. In the previous research, the concept of "integration" is limited to the linear superposition and simple combination of several subsystems, but with the increasing requirements of the performance vehicle system dynamics control, the integrated control is moving towaords multi-level, multi-dimensional and all-round coordinated control of chassis subsystems [7].

Based on current research, the integrated control structures of vehicle chassis systems can be divided into two main categories, bottom-up and top-down structures, when analyzed in terms of information and control structures [8].

(1) Bottom-up structure

In the bottom-up structure, each component forms a separate control subsystem mainly using parallel control mode and can communicate with each other via the vehicle body bus to achieve overall coordination. This structure is mostly based on the superposition of multiple subsystems at the bottom layer of the vehicle chassis system. In the subsystem controller design process, by simply integrating other subsystems' information, the information sharing among sensors and related hardware as well as integrated control at low levels and in small scale can be achieved. Therefore, it's also known as the system superposition type integrated vehicle chassis

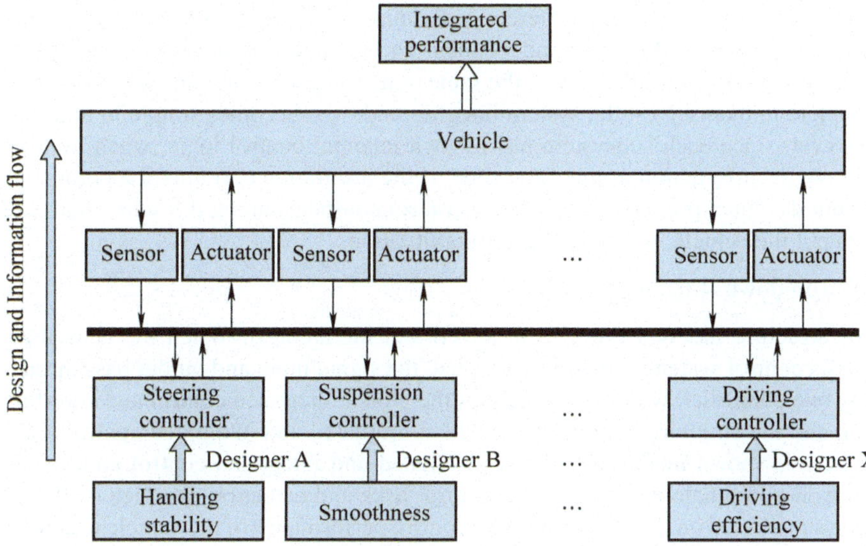

Fig. 5.1 System superposition type integrated vehicle chassis control system

control system, as shown in Fig. 5.1. Although this structure has limits in integrated control, it can fully ensure the reliability of components and the ease of implementation of the vehicle system integration.

Figure 5.2 shows the system superposition type integrated coordinated control architecture for vehicle stability.

The system superposition type stability control system has kept the control modules of individual subsystems. The main task of the upper-level controller is

Fig. 5.2 Structure of the system superposition type stability control system [9]

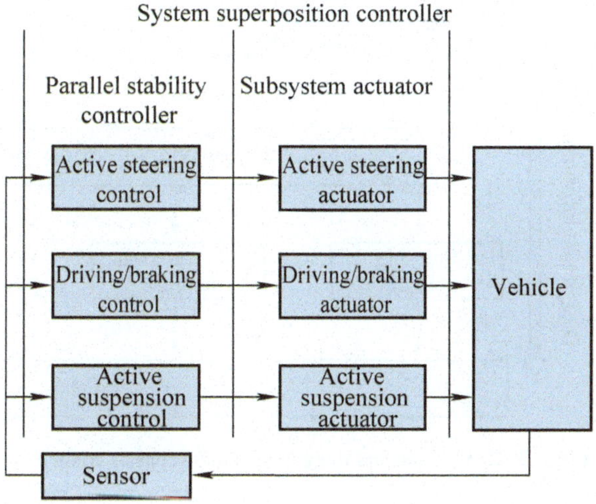

to coordinate the action of different individual subsystems. Since the individual subsystems contain their own independent and complete architectures, the system boasts good safety redundancy. At the same time, since the individual subsystems are complete in themselves, the system integration design becomes simple in this case. However, since each subsystem has its own inherent control logic, when multiple subsystems take action at the same time, if the arbitration coordination commands are not designed properly, it can cause control conflicts among the subsystems and even put the vehicle in a more dangerous situation.

(2) Top-down structure

This structure has two types: centralized and hierarchical. When the centralized chassis control system structure is adopted, the signal input and output pass through only one controller, which truly realizes the global integrated control and coordination, and the structure is therefore called coordinated and integrated control architecture, as shown in Fig. 5.3. The coordinated and integrated control architecture can well coordinate and assign the control tasks and actuation parameters of each subsystem based on the comprehensive control performance of the vehicle to achieve the optimal control effect. However, with the increasing number of chassis subsystems, the control system becomes more and more complex and can easily lead to a "dimensional disaster" [10]. In addition, the huge computational and storage load also puts higher demands on the control system hardware. The higher the integration level, the lower the changeability. Due to the limitations of current technologies, there are few vehicle model products based on this architecture, but as high performance on-board computing platform technologies develop, the centralized coordinated control of electrical and electronic architecture is the future of development.

The control structure of the stability control system of the vehicle that adopts coordinated and integrated control architecture is shown in Fig. 5.4.

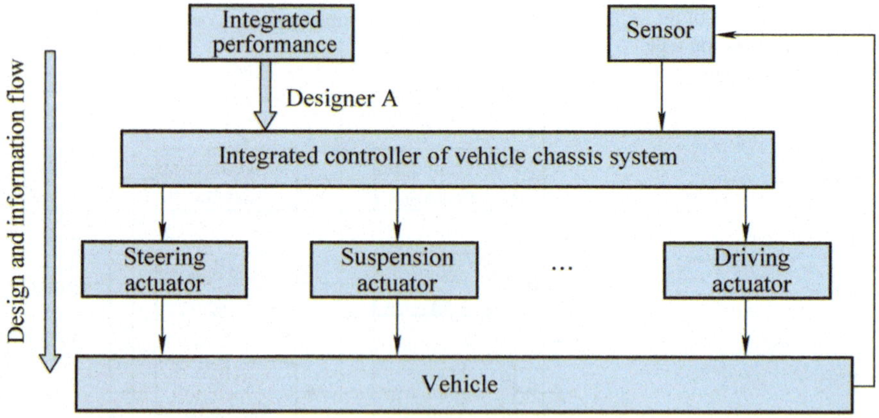

Fig. 5.3 Centralized and integrated vehicle chassis control system

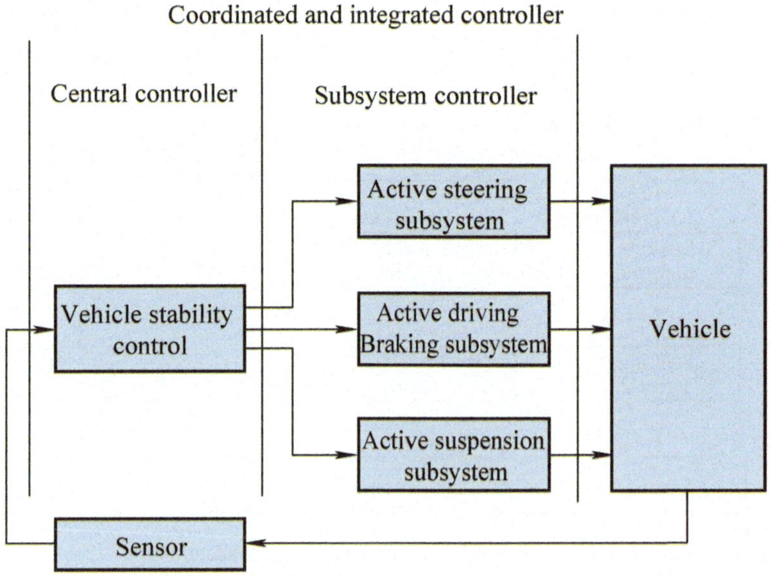

Fig. 5.4 Structure of the stability control system of the vehicle that adopts coordinated and integrated control architecture [9]

In the coordinated and integrated stability control system, the subsystems do not have the control modules of the original systems, but only keep the function of their actuator. The task of vehicle stability control is performed by the upper-level stability control system. The upper-level stability control system calculates the expected direct yaw torque and the expected brake/driving force based on the comparison of the expected motion state and the actual motion state and distributes the actuation commands among the subsystems through the mid-level coordinated control system. This type of control system is better coordinated and can more reasonably optimize the vehicle performance.

Since the stability coordinated control structure of the vehicle that adopts coordinated and integrated control architecture can better coordinate vehicle subsystems to allow them to play a greater role in ensuring the vehicle stability. Therefore, the stability control method applied in the vehicle that adopts coordinated and integrated control architecture has become an important topic in the field of vehicle stability coordinated control.

The most widely used and well-studied integrated control system of the vehicle that adopts the top-down structure nowadays is the hierarchical coordinated control structure, as shown in Fig. 5.5. This control structure can decompose the complex global multi-system chassis integration problem into relatively simple local subsystem control tasks and then design the upper-level controller to carry out the coordinated control among different local subsystems.

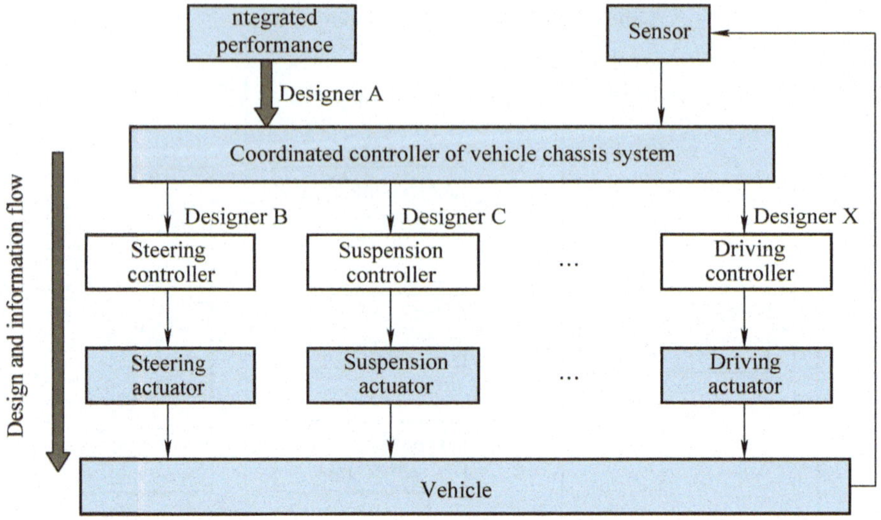

Fig. 5.5 Hierarchical and integrated vehicle chassis control architecture

The current research on GCC mainly focuses on vehicle's longitudinal, lateral and vertical dynamics and explores further on the longitudinal forces, lateral forces and vertical forces on tyres or two of them, so as to improve vehicle driving/brake performance and maneuvering stability performance. Figure 5.6 shows the hierarchical control structure of the vehicle stability coordinated control system.

Fig. 5.6 Hierarchical control structure of the stability coordinated control system

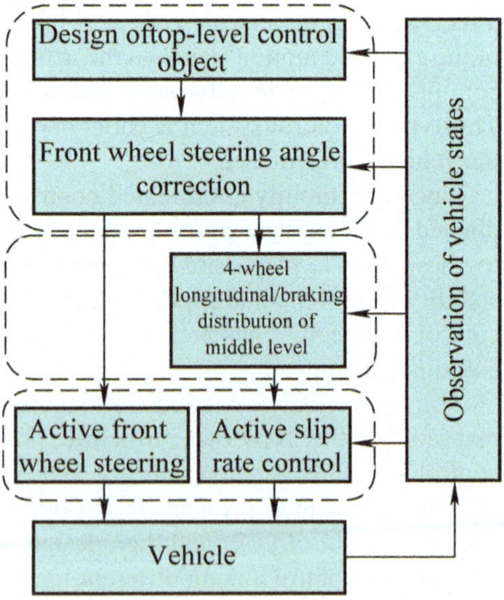

The hierarchical control structure of vehicle stability coordinated control system is divided into top, middle and bottom levels. The top level of the system is used to design the control objectives to meet the driver's requirements, the direct yaw torque M_z and front wheel steering angle correction value $\Delta\delta_f$ that meet the stability requirements are calculated with corresponding calculation methods. The middle level of the system optmally distributes the direct yaw torque calculated by the top level into the longitudinal driving/braking force of four wheels. The bottom level of the system consists of the active front wheel steering actuator and the active slip rate control subsystem. The active front wheel steering actuator directly executes the front wheel steering angle correction value $\Delta\delta_f$ calculated by the top level of the system. The active slip rate control subsystem first converts the four-wheel longitudinal driving/brake force F_{xi} calculated by the middle level into the target slip rate λ_d and controls the target slip rate by controlling the engine torque and brake pressure of the brake system. The top, middle and bottom levels work in coordination to achieve optimal control of vehicle stability and maneuverability.

In summary, based on current research, the existing chassis system integration usually adopts a top-down structure, and its design process is a process of specifying an abstract control objective to each different actuator. This indicates that current research on integrated chassis control reflects the idea of top-level design and has put forward various ideas on issues such as effective coordination among subsystems and framework designs, but a sound theory and system have not been formed from a methodological perspective. there are following aspects that need further research:

(1) vehicle chassis is a complex and large hierachical system, the comprehensive performance of the vehicle relies on the performance of subsystems and each subsystem is controlled by the top-level subsystem while dominating the bottom-level subsystem. Therefore, it's necessary to design the relationship among the subsystems of each level to ensure that the work of each subsystem is fully integrated with the improvement of the overall performance of the vehicle to maximize its effectiveness.

Like all large systems, solving such integrated control-related theoretical and technical problems requires, on the one hand, an in-depth analysis of the interrelationship and mechanism of action of the subsystems of the controlled object, andon the other hand, this complex large system problem needs to by synthesized into a system coordination solution mechanism problem. The design of the control system is then required, which including information integration of multiple sensors at the bottom level to ensure system reliability; the integration of the middle-level monitoring system (environmental model, evaluation unit, control law generation unit); the overall integration of the top level by integrating various decisions distributed in different locations and functions into a common coordinated framework, so that the various components of the system can effectively support and cooperate with each other under a unified goal.

(2) The vehicle chassis is composed of several subsystems, involving many key components which are responsible for different functions or structures, interacting and coupling with each other. How to seek coordination between performance and structure and optimization is a concern for engineers. At present, most of the research on the synergy among subsystems is still preliminary, mainly focusing on system decoupling methods and control, and there is a lack of systematic analysis of and methodological research on theories and technologies related to the mutual cooperation, coordination and negotiation of subsystems, which barely considers the mutual matching among multiple subsystems and between the subsustem and vehicle performance in a comprehensive way. The adopted models, control theories, methods and control strategies are not very ideal in practical applications and have not addressed the problem of effective control and coordinated framework design for such large systems, which can barely help achieve the coordination of the system and meet the requirements of substantial improvement of the comprehensive performance of the vehicle.
(3) The key subsystems of vehicle chassis (such as steering, suspension, driving and brake etc.) are closely related to the control of tyre forces, so the study of the coupling relationship between the different directions of tyre force is the core of the study of integrated control of chassis system dynamics. Most of the current studies use simple and isolated models for research on integration, which have certain limitations. Therefore, in order to improve the dynamics of vehicle systems, it's necessary to solve theoretical and technical problems encountered in the design of integrated chassis systems in macro and micro perspectives and from top to bottom, so as to more effectively apply advanced and practical control theories and methods in the integrated systems of key chassis components, especially to seek a breakthrough in the depth of system control theory research and collaborative framework design and to develop vehicle chassis integrated control technologies, in an effort to make substantial progress in the research on theories, methods and applications related to vehicle integrated control. Therefore, it has not only laid theoretical foundation for the practical applications of chassis system integration, but also accumulated valuable experience in industrialization.

Therefore, the integrated system in the future must focus more on the top-level design and the research on the coordinated control theories and methods on the bottom level. We need to conduct top-level design and planning of methods and technologies of coordination and cooperation among subsystems, comprehensively consider the lower-level subsystem control performance requirements and the optimization of matching of the overall performance objectives of the vehicle, so as to truly solve the problem of effective control and coordinated framework design of the large chassis system and carry out in-depth research on model solving and coordinated control of multiple subsystems, including dynamic modeling, real-time acquisition of data and expert system inference and decision-making, in an effort to conduct intelligent negotiations and cooperation among subsystems and to realize substantial improvement in subsystem control while effectively improving the comprehensive performance of vehicles.

5.2 Coorperative Control Methods for Vehicle Chassis System Based on Top-Level Design

Top-level design is a core research idea and concept in systems science that comes from system control and management and has become widely used in various areas of systematology [11]. Everything in the universe is inextricably linked to its surroundings and each object is both a constituent element of a macro system and a complex of micro systems. Top-level design is an integrated and managed combination of elements based on the mechanisms of action of macro functions and micro elements [12]. From the persective of systematology, top-level design is the process of unified planning and macro layout of system hierarchy and constituent elements to achieve overall goal optimization. In the process of design, it's required that the functions and structures of each level comply with the macro and top-level requirements and the interaction model reflecting the interaction between its own functions and environmental parameters is constructed. In coordinated systems, top-level design is mainly a process of resolving internal conflicts and contradictions through integrated planning, adjustment and laying out the internal resources of the system in a coherent manner [13].

Top-level design is a new design method since the emergence of systems science that can greatly improve the efficiency of the system. Its essence is a design idea from top level to bottom level, with a holistic and big-picture view, optimizing various resources and elements within the system, designing strategies and planning ideas to achieve consistent goals, functional coordination, scientific structures, unified standards and resource sharing. The characteristics of top-level design from a theoretical point of view are reflected in the overall big-picture view of coordination, rigorous logic and resolute and strong implementation. From the perspective of information fusion, top-level design aims to integrate the functions and information resources of subsystems in accordance with the control objectives and control strategies of the system and in response to specific performance requirements, so as to solve the problems of poor information communication, inefficient cooperation and low-level information sharing among subsystems. As an advanced design concept and approach, top-level design has been widely used in the fields of intelligent robotics, aerospace control, system engineering, economic management and military industry. Applied in the field of the coordinated control of vehicle chassis systems, it can not only ensure the realization of the function of each subsystem, but also improve the overall performance of the vehicle system and solve the problems and technical bottlenecks encountered in the integration process, namely

$$F > \sum F_i (i = 1, 2, \ldots, n)$$

where F_i is the subsystem performance and F is the overall system performance of the vehicle.

In the process of top-level design, the first thing to complete is the system hierarchy design, followed by the design of functions. The design of large complex system hierarchy often uses the "top-down, level-by-level decomposition" method which divides the whole system into top, middle and bottom levels. Each level is designed with relatively independent functionalized sub-modules, each of which contains three parts: input and output, logical inference, and storage and memory. The decomposed sub-modules can also be called sub-systems. In the top-down modeling process, in addition to gauranteeing the functionality of subsystems, the interaction and coordination among subsystems should be studied. Only when individual subsystems contain their own specific functional attributes can the overall functionality and goals of the system be achieved.

On the other hand, since the birth of mankind, coordinated behavior has been formed in the operation of large social systems and has advanced along with the development of human civilization, but people are accustomed to its daily applications without knowing its fundamental mechanism. In the age of technological innovation and information explosion in the twenty-first century, people have found that each large system is composed of individuals and the only way to make the whole system to work properly and achieve expected effect is to make individuals cooperate. Therefore, from academic research to theoretical applications, a new wave of research on cooperation was launched and the utility and deeper meaning of "synergy" have been constantly explored. With the development of network technology, coordinated technology has gradually matured, with the typical representative being the relationship network model, which is a simple but widely-used protocol. This model mainly solves the problems of task distribution, resource conflict and knowledge conflict and has been applied in the cooperation of aircrafts and other carriers and multi-computer cooperation.

Cooperation is the process or ability of different subsystems within a large complex system to cooperate and negotiate with each other, exert a positive influence on the external environment, strengthen the overall system and meet the needs of each subsystem. Synergy refers to the mechanisms and characteristics of mutual cooperation and negotiation among different subsystems. The systematic study of such mechanisms and characteristics leads to a scientific theory called synergetics.

Synergetics is an emerging discipline founded by the famous scientist Hermann Haken in 1970s and has gradually become an important branch of systems science through the long-term development and integration of multiple disciplines. After continuous development and enrichment theories, synergestics has gradually become a comprehensive discipline known as "synergetics science" and is widely used. It extracts and studies the common features from different things, explores the synergy of various complex systems from disorder to coherence, studies the structure and function of large complex systems at the macro level and has led to the discovery of the control principles and management science of large complex systems.

Synergistic theories and methods have practical values in research and applications in many areas such as astronomy, geography, physics, chemistry, biology, sociology, economic management and engineering sciences etc. For example, it's impossible to predict the behavior of each individual in sociology, but the coordinated nature of individual behavior is used to explore the macro patterns of social group behaviors. Similarly, applying this approach and concept in the field of engineering can solve a series of complex system coordinated control problems and build a coordinated organizational framework to meet the overall performance requirements based on the implementation of the function of different subsystems.

Top-level design is a process of solving problems from abstract to concrete, from qualitative to quantitative, decomposing and combining problems from top to bottom, simulating human cognition and thinking process. Large complex systems consist of several subsystems with different structures or functions that need coordination and cooperation, which fuly reflects the role of synergy. It can be seen that these two research theories are not independent of each other, but rather closely related.

Based on the above theoretical research and applications, it's not difficult to find that the top-level design theory can give full play to the respective advantages of the internal resources of the integrated system in the top-down integration process. If combined with the synergy theory, it can solve the inevitable problems of parameter uncertainty, mutual interference of sensors and actuators, big data and sudden changes of system states that are often faced in the research of large complex systems and will surely become a hot topic for the research on complex system integration and control theory technology in the field of science and engineering in the future.

Therefore, the problem of combining top-level design ideas and synergy theory has attracted many researchers engaged in process control, computer applications and systems science to conduct research from different perspectives of system modeling, analysis and application. Certain research progress and results have been achieved. The practical application results show that the application of system synergy theory related to top-level design to study these large complex systems improves the control effect while ensuring the efficiency and safety of the system compared with the traditional methods.

For intelligent vehicles, the characteristics of their chassis systems in terms of coordinated control include:

(1) Individual subsystems contain autonomous attributes. Subsystems have moved from passively waiting for work commands from top-level subsystems to actively sensing the performance requirements of their surroundings for the corresponding subsystems. The communication and interaction to achieve theses subsystems' respective control priorities and authority have been established. Subsystems cooperate and control online in real time, synthesize problems, analyze causes and coordinate solutions to avoid redundant procedures and improve processing efficiency.
(2) The improvement of the control efficiency of large systems. The function and definition of traditional hierarchical structure of integrated chassis control is relatively simple and independent, which facilitates the design of the system, but also reduces the control efficiency. For this reason, the coordinated control of chasis system will reduce the interference of the top-level subsystem to the

lower-level subsystem, that is, after the tasks have been assigned, the subsystems can be freely combined to maximize their effectiveness through synergy.

(3) Real-timeness of control algorithm. In order to cope with the uncertainty of the external environment and the randomness of driver intention and to obtain the optimal control and accurate feedback of the overall performance of the vehicle, the coordinated control of the chassis system has also put forward more requirements for real-timeness.

In the process of using synergy theory for integrated control of chassis systems, the lack of the overall design at the top level is not allowed. Therefore, the traditional centralized control system has been turned into the designer and supervisor at the top level, so that more control is transferred to the coordinated system. The coordinated system also has the rights of task assigning and decision-making and can take over the management of lower-level subsystems when its intervention is required in specific situations. Cooperative control of the top-level design is to determine the working condition of the system based on the vehicle and environment state, determine the responsibilities and authority of the subsystems in real time and collect the results returned by all subsystems as an evaluation reference. At the same time, the coordinated control of top-level subsystem and lower-level subsystem will be further integrated to improve the overall performance and control accuracy of the vehicle through the integration of control information and returned information on vehicle and environment state.

Based on the idea of top-level design and synergy theory, the framework of i-BEV integrated and coordinated vehicle chassis control system has been proposed, as shown in Fig. 5.7. The framework divides the integrated and coordinated vehicle chassis control system into three functional areas from the perspective of coordinated control: top level, lower level and middle coordinated level. The top level and middle level are modeled centrally based on the top-level design idea of module design. According to the functional hierarchy, the control process of coordinated vehicle chassis control system based on top-level design is divided into three stages: top-level design, coordinated optimization and autonomous control [14].

(1) Top-level design

In the top-level design stage, based on the working condition classification and by following the principles of safety, energy efficiency, environmental protection, comfort, traffic efficiency performance and overall balanced and integrated allocation, the top-level subsystem carries out comprehensive vehicle performance index determination and establishes the control task allocation model.

(2) Coordinated optimization

In the coordinated vehicle chassis control system, the following principles are adopted to solve the problem of task allocation and coordination of subsystems: based on the natural relationships between subsystems, fully consider the effect of their roles on the feedback on the overall control of the vehicle, combine the dynamics of different subsystems, dynamically coordinate the comprehensive performance of the

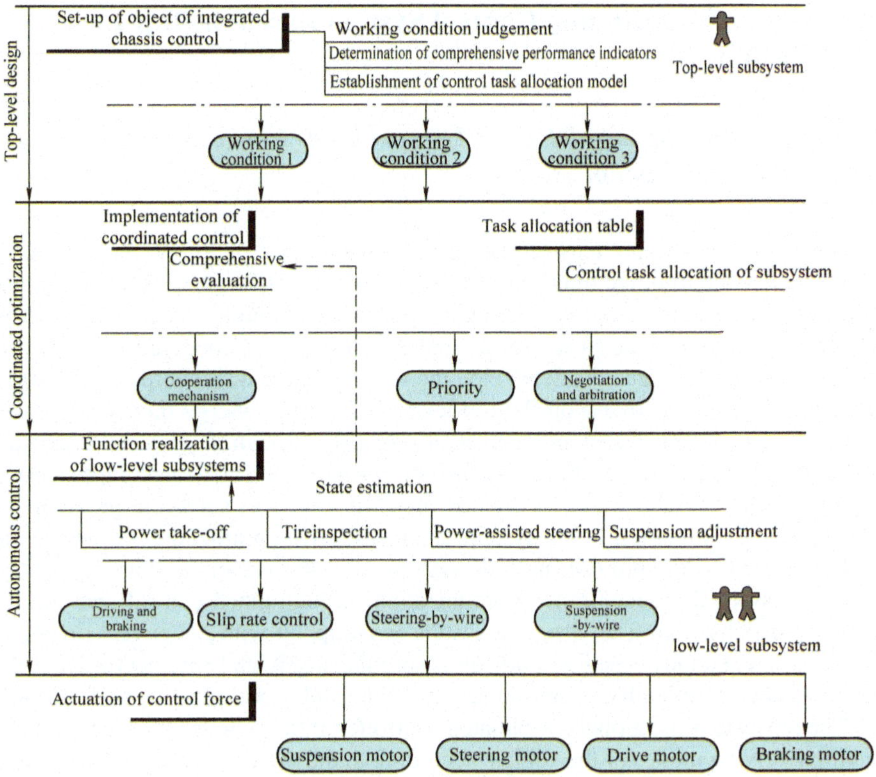

Fig. 5.7 Framework of the integrated vehicle chassis control system

vehicle such as driving safety, vehicle stability, energy efficiency and environmental protection, comfort and smoothness and driving efficiency to deal with coordination and cooperation problems, that is, whether to cooperate to promote or solve the conflicts to create a task allocation table and then establish cooperation mechanisms, set priorities and conduct negotiation and arbitration to optimize synergies.

(3) Autonomous control

After the top-level subsystem has optimized the coordinated vehicle chassis control system, the lower-level subsystems carry out assigned tasks and give feedback on state according to their own control structures and functions, i.e. adjusting the corresponding control parameters based on the comprehensive performance of the vehicle and the need for optimization of coordinated control and driving the corresponding actuators to generate operation force.

5.3 i-EFV Coordinated Control System Based on Top-Level Design

5.3.1 *Architecture of i-EFV's Hierarchical Coordinated Control System*

According to the comprehensive operational performance of i-EFV and the multi-objective cooperation, the multi-element coperation in the large system of "human-vehicle-infrastructure" nonlinear complex dynamics, the multi-agency cooperation of "driving/brake/steering/suspension," and the multi-modal cooperation of "driver/driving assistant/integrated safety" etc., the i-EFV hierarchical coordinated control structure is proposed based on the top-level design and synergy theory, as shown in Fig. 5.8. The control system receives information through the bus and integrates the acquired vehicle and traffic environment information to carry out comprehensive control of the vehicle, which is divided into four levels to solve the problems of multi-mode coordination of the system, multi-objective optimization of the vehicle and dynamic coordination of multiple systems. For the vehicle system, its control mode is essentially decentralized optimization, decentralized control and coordination compensation. The structure system is equipped with better real-timeness, robustness and control security and can ensure the steady and safe operation of the vehicle system, although it compromises part of the global optimization performance.

The top level is the global coordinated controller of the vehicle which coordinates multiple driving modes of the vehicle according to the driver's operation, actual and expected vehicle/environmental state and mainly addresses the problem of "driver/driving assistant/integrated safety" multi-mode control coordination.

The middle level deals with the "human-vehicle-infrastructure" multi-objective coordinated optimization, which contains a number of sub-optimization controllers based on the vehicle's working mode. Based on the vehicle state, environmental information and driver's operation, characteristics information and "safety, comfort, energy efficiency and environmental protection" multi-objectives, the middle level conducts coordinated optimization. Based on the idea of hierarchical control, we do not consider the dynamic response characteristics of the system and the impact they cause, but only consider the steady-state characteristics of the system, and integrate safety, comfort, energy efficiency and environmental protection for multi-objective steady-state optimization.

There are often mutual constraints and coupling relationships among different types of multiple performance objectives of i-EFV such as safety, comfort, energy efficiency and environmental protection. Based on full consideration of the driver's behavioral characteristics, the real-time state of the vehicle, changes in the road environment and the predictive control theory, in order to overcome the problem of the nonlinear coupling among multiple systems and objectives, simplify the control system design and make sure the control is real-time, we have designed the "human-vehicle-infrastucture" large-system multi-objective coordinated control

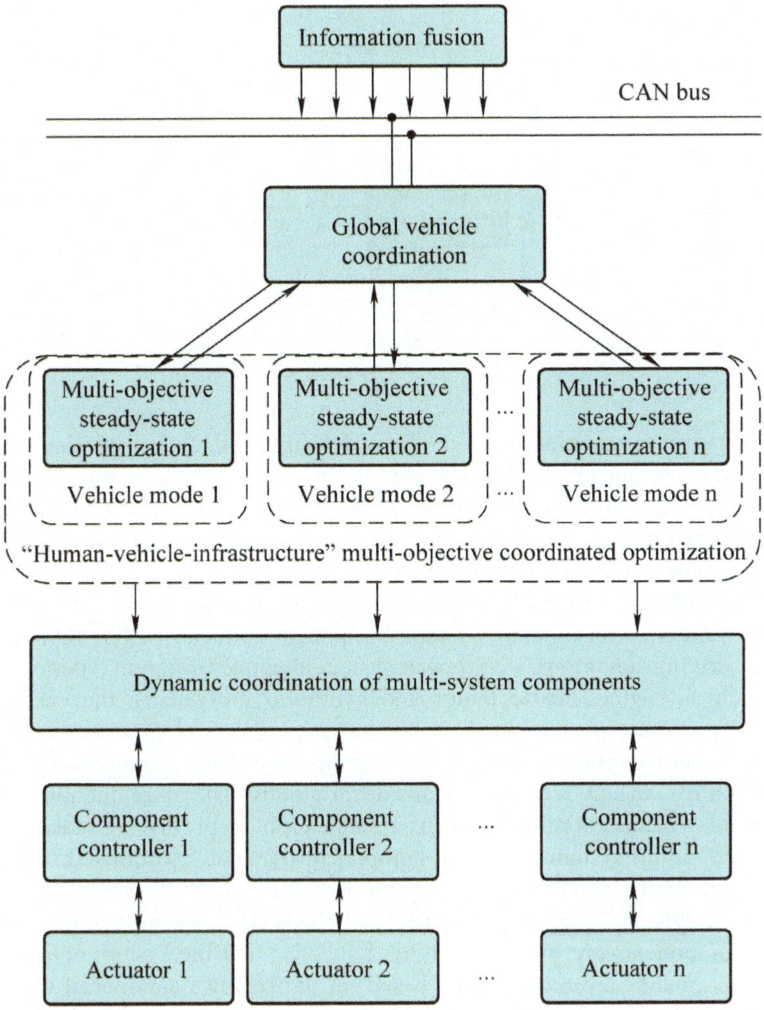

Fig. 5.8 Architecture of i-EFV hierachical coordinated control system

system structure, as shown in Fig. 5.9, to achieve the optimized coordinated control in the local time domain of the vehicle system.

As shown in Fig. 5.9, a multi-objective (safety, comfort, energy efficiency and environmental protection) optimization function is developed based on the driver characteristics, current vehicle/environment state and vehicle motion reference targets, followed by the real-time optimization carried out through the "human-vehicle-infrastructure" coupling prediction model and rolling optimization. The research focuses on: (1) method for mathematical quantification of performance indicators such as safety, energy efficiency, environmental protection and comfort

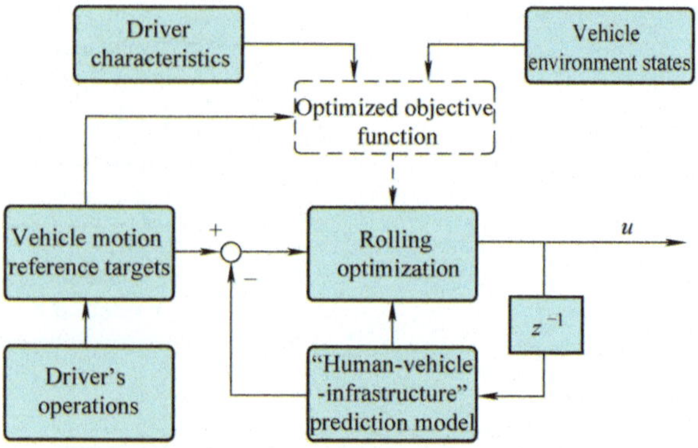

Fig. 5.9 Structure of multi-objective coordinated optimization of "human-vehicle-infrastructure" system

that meet driver expectations; (2) coordinated design methods considering multiple-performance optimization objectives and multi-performance objective coordinated control optimization methods; (3) multi-objective optimal control methods under different constraints and solutions to real-timeness, robustness and stability problems of multi-performance objective coordinated control.

The lower level of i-EFV hierarchical coordinated control system deals with the dynamic coordination of multi-system components. For an i-EFV vehicle under complex driving conditions, as its power system integrates different types of subsystems such as engine, eletric motor and hydraulic subsystems, the energy flow transmission of the system based on this structure shows obvious nonlinear characteristics and uncertainty. The internal dynamic response characteristics and the cost of control objectives of subsystems differ greatly. Therefore, the multi-system component dynamic coordination focuses on the problem of "driving/brake/steering/suspension" multi-system and multi-component dynamic coordinated control. To address the above problems, the overall structure of multi-system and multi-component dynamic coordinated control is proposed, as shown in Fig. 5.10. The system components are numbered as i_1, i_2, … i_m, and the system components's control commands are redistributed based on the steady-state control commands for "human-vehicle-infrastructure" multi-objective coordinated optimization while considering the system dynamic response characteristics and the impact of the control costs in order to achieve a fast and economic response and the steady-state optimization objectives of the system. Firstly, based on the steady-state objectives of different system components, design the integrated response speed and economic indicator of the dynamic process, and then perform comprehensive optimization to obtain the expected control quantity uij for component ij (j = 1, 2, … m). Subsequently, in order to meet the fast response requirements of the system, a two-degree-of-freedom model matching control algorithm based on the component specification model Gij using feedforward Fij plus feedback Cij, with final output to the component controller and actuators. The key research contents include: (1) the method for coordinated design of performance indicators such as safety and economy of coupled nonlinear "mechanical/electric/hydraulic" system; (2) the method for comprehensive

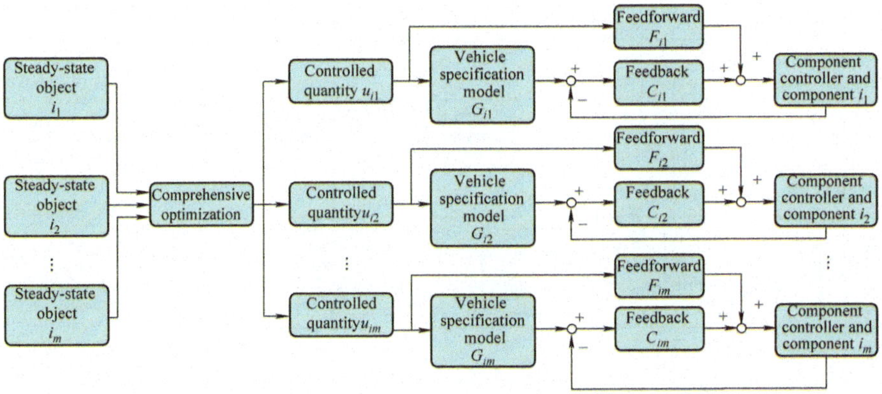

Fig. 5.10 Dynamic coordination of multi-system components

performance optimization of "mechanical/electric/hydraulic" integrated system and nonlinear coupled dynamic coordinated control method for uncertain system; (3) the jitter, robustness and real-timeness problems of the nonlinear dynamic coordinated optimization control method.

As a typical application of i-EFV, the vehicle hierarchical control system structure of i-HEV is shown in Fig. 5.11.

The vehicle hierarchical control system of i-HEV is divided into four levels. The top level is the intelligent assisted driving control target setting level that can identify the intelligent assisted driving sub-mode according to the state of the ego vehicle and the preceding vehicle and the cruise speed input by the driver and sets the expected longitudinal acceleration of the vehicle in various modes; the upper level is the vehicle driving/brake torque setting level that sets the demanded torque in manual driving mode and automated driving mode respectively and discriminates and coordinates different demanded torque for the vehicle driving mode according to the driving operation and finally sets the expected driving/brake torque for the vehicle; the middle level is the powertrain torque distribution level, which identifies the power system operating modes and sets the expected torque commands for different powertrain components based on the powertrain component state information and the expected driving/brake torque; the lower level is the powertrain torque coordinated control level, which identifies the torque of engine and electric driving, the motor and EVB system and ultimately determines the torque commands for the engine and driving motor and the brake pressure commands for the EVB system.

In Fig. 5.11, α is the accelerator pedal travel, β is the brake pedal travel, P_g is the current gear position of the transmission, V is the velocity of the ego vehicle, P_A is the position of the intelligent assisted driving function switch, V_r is the relative vehicle velocity, D_r is the relative inter-vehicle distance, P_s is the position of the cruise speed setting switch, a_x is the ego vehicle's longitudinal acceleration, a_{xd} is the expected acceleration, T_{dd} is the expected driving torque in manual mode, T_{ba} is the expected brake torque in manual mode, T_{da} is the expected driving torque in

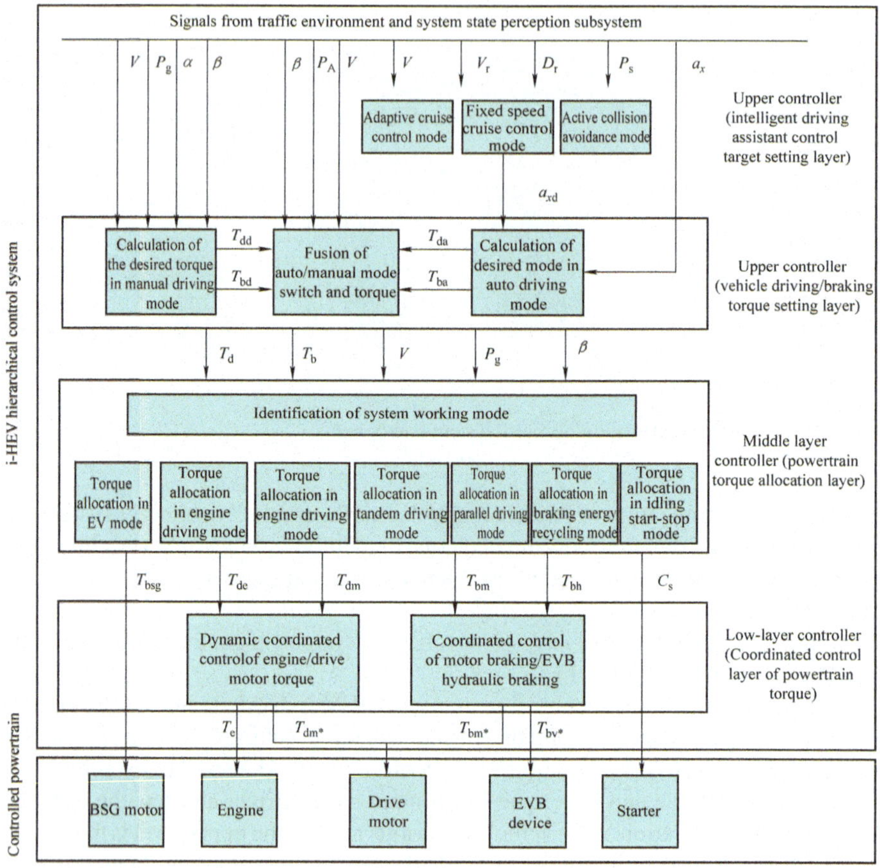

Fig. 5.11 Structure diagram of vehicle hierachical control system

auto mode, T_{ba} is the expected brake torque in auto mode, T_d is the expected driving torque for the vehicle, T_b is the expected brake torque for the vehicle, T_{de} is the expected engine driving torque, T_{dm} is the expected motor driving torque, T_{bm} is the expected motor brake torque, T_{bh} is the expected hydraulic brake torque, T_{bsg} is BSG motor torque command, T_e is the engine torque command, T_{dm*} is the motor driving torque command, T_{bm*} is the motor brake torque command, T_{bv*} is the EVB brake torque command and C_s is the starter start command.

As another typical application of i-EFV, the overall architecture of the driving force hierarchy control system for all-wheel independent electric driving vehicles is shown in Fig. 5.12.

In Fig. 5.12, δ is the steering wheel angle input by the driver, θ_α is the accelerator pedal opening by the driver, λ_d, β_d is the expected yaw rate and sideslip angle calculated by the reference model, γ is the measured signal from the vehicle yaw rate sensor, β is the sideslip angle estimated by the observer, $\hat{v}$ is the longitudinal vehicle velocity estimated by the observer, $\hat{\mu}$ is the road friction coefficient estimated

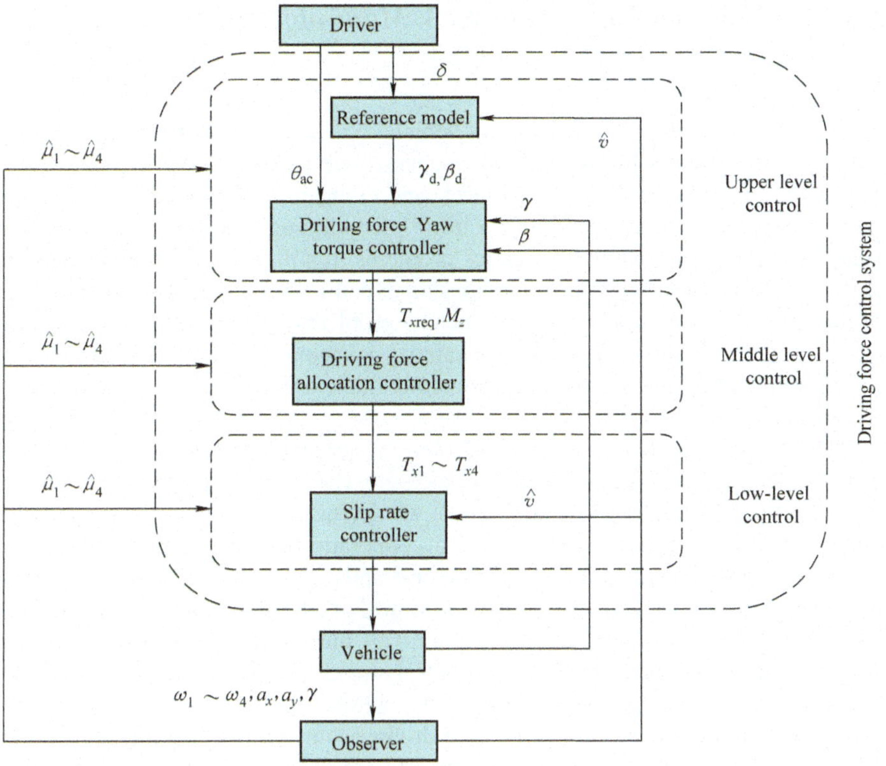

Fig. 5.12 Overall structure of the driving force control system for four-wheel independent electric driving vehicles

by the observer, a_x and a_y are the longitudinal acceleration and lateral acceleration measured by the vehicle accelerometer, ω is the wheel speed of each wheel and T_x is the torque of each drive motor.

The all-wheel independent electric driving vehicle's driving force control system consists of three levels, of which the top level is the driving force and yaw torque setting level, the middle level is the multi-objective optimized driving force distribution level and the lower level is the slip rate control level. The driving force and yaw torque setting level is located in the top level of the entire four-wheel independent electric driving vehicle's driving force control system and its main function is to output the driver's expected longitudinal driving torque and stabilize the yaw rate of the vehicle. The multi-objective optimized driving force distribution level assigns the yaw torque value M_z from the top level and the driver's expected longitudinal torque T_{req} to each wheel. The input to the slip rate control level is the expected and target slip rate of the wheels and the output is the driving torque that controls wheels. The main function of the slip rate control level is to control the drive motor's torque to avoid the slipping of the wheels and to prevent the vehicle's lateral stability from being compromised by the sharp drop in lateral tyre forces during the slipping process.

5.3.2 Integrated Control of i-EFV Multi-objective and Multi-system Synergy

The control principles and decisions of vehicle dynamics mainly focus on the separate study of different important performances, such as vehicle driving safety, fuel economy, emission performance control, and control of driving noise and vibration performance related to driving comfort etc. Although these methods can well solve the relevant individual dynamics problems, it's difficult to comprehensively and synergistically solve the comprehensive dynamics problems with the presence of the coupling of complex systems, such as problems of multi-objective cooperation of driver, vehicle and road driving safety, fuel economy and riding comfort and the problems of multi-system cooperation of vehicle driving, brake, suspension and steering etc.

The i-EFV has characteristics including coupled controlled objects, various control functions and complex vehicle objects etc. How to make the "human-vehicle-infrastructure" of the i-EFV as an organic whole and realize the optimal dynamics control of multi-objective and multi-system cooperation by considering the dynamics of complex system in the multi-motion process and multi-system coupling is another important research topic on the dynamics control of i-EFV integrated with complex electromechanical system. Integrated control of multi-objective and multi-system cooperation dynamics means that the control strategy for i-EFV needs to consider not only the multi-objective coordination, including driving safety, energy efficiency and environmental protection and driving comfort, but also the cooperation of multiple systems such as vehicle driving, suspension, steering and brake systems. The integrated control of i-EFV multi-objective and multi-system cooperation has the following problems:

(1) "Driver/driver assistant/integrated safety" multi-mode cooperation

The i-EFV integrated control requires different working modes such as normal driving, longitudinal/later driver assistant and "human-vehicle-infrastructure" integrated safety to be realized based on driver demand, vehicle state and traffic environment information. The above-mentioned system control modes often trigger several modes to work at the same time under certain emergency conditions. Therefore, how to monitor and coordinate multiple vehicle modes such as "driver/driver assistant/integrated safety" in real time to achieve multi-mode cooperation is one of the problems that need to be solved for i-EFV control cooperation.

(2) "Human-vehicle-infrastructure" multi-objective cooperation

The integrated control of i-EFV requires not only the safety, comfort, energy efficiency and environmental protection of the vehicle to be realized, but also the coordinated optimization of i-EFV, driving environment and pedestrians. The integrated control not only involves the vehicle dynamics system, but also takes into account driving habit and behavioral characteristics of the driver as well as the road conditions and traffic environment where the vehicle is running. Therefore, the integrated i-EFV control process is about taking the "human-vehicle-infrastructure" generalized mechanical dynamics system composed of driver-vehicle-road traffic environment as the controlled object and carrying out the coordinated optimization control.

(3) "Driving/braking/steering/suspension" multi-system cooperation

i-EFV integrates complex electromechanical coupling system, including driving, braking, steering, suspension and other mechanical, electronic and hydraulic multi-systems and each sub-system contains multiple actuating components. Different subcomponents have different system operating ranges, response characteristics and control costs. Therefore, the integrated i-EFV control requires dynamic coordination of the complex "machine-electric-hydraulic" coupled system composed of multiple systems and components to achieve the fastest integrated system response and optimal economy.

5.4 Typical Applications of i-EFV Coordinated Control Technology

5.4.1 Intelligent-Assisted Driving Coordinated Control Technology for HEVs

5.4.1.1 Strategies for Multi-objective Coordinated Human–Machine Co-driving Multi-mode Switching Control

•

(1) Strategy for coordination and control of driver operation and intelligent-assisted driving switching

In the real-life vehicle operation, intelligent assisted driving control and driver control require frequent and repeated switching control. In the process of switching, firstly, it's necessary to design strategies for coordination and control of driver operation and intelligent assisted driving intentions to ensure smooth and reasonable switching between driver operation and intelligent assisted driving in the actual driving; secondly, in the process of switching, it's necessary to coordinate and control each powertrain components (including engine, drive motor, generator and EVB) to achieve smooth changes in the output control during dynamic switching and to ensure that vehicle can accelerate according to the expected demand. The coordinated control strategies involve the strategy for coordination and control of intentions to switch between driver operation and intelligent-assisted driving and the strategy

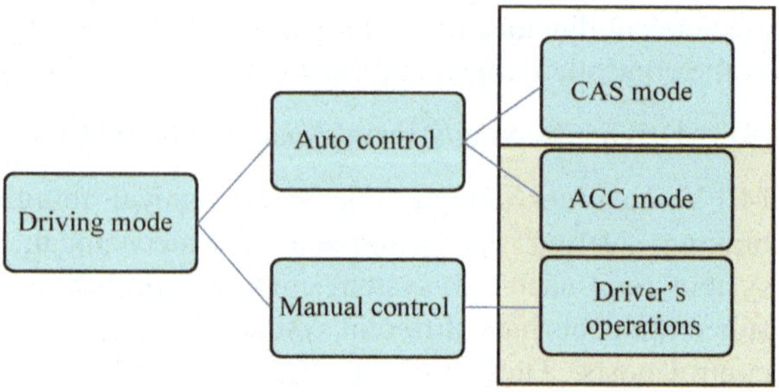

Fig. 5.13 Division of driver operation and intelligent-assisted driving intentions

for coordination and control of driving force to switch between driver operation and intelligent-assisted driving, which are analyzed in detail as follows.

(a) Strategy for coordination and control of intentions to switch between driver operation and intelligent-assisted driving

In general, vehicle driving is divided into two modes: manual (driver-operated) and automated (intelligent-assisted driving). The intelligent-assisted driving is divided into two subsystems: adaptive cruise control (ACC) and active collision avoidance (CAS). Therefore, the automated mode is divided into two modes: adaptive cruise control and active collision avoidance. The driving mode of the vehicle is divided, as shown in Fig. 5.13.

As shown in Fig. 5.13, driver operation, adaptive cruise control (ACC) and active collision avoidance system (CAS) constitute the three major operating modes during vehicle driving. Driver operation and adaptive cruise control are independent and have the same priority and the driver can operate on his own or enter adaptive cruise control mode after making certain settings and operations. The active collision avoidance system has the highest priority. Once entering into this working mode, driver operation and adaptive cruise control are automatically disabled and the vehicle enters into the forced braking working state until exiting the active collision avoidance mode.

To correspond to the actual vehicle operation modes, vehicle operation modes that can be used for vehicle control are designed based on the vehicle driving modes. The set of vehicle operation modes is defined as $\Omega = \{N_1, N_2, \ldots, N_p\}$. Each N_i corresponds to a vehicle driving mode, where $N_i = \{X_1, X_2, X_3, X_4\}$. x_i are all Boolean variables; 1 represents true and 0 false. x_i corresponds to a certain vehicle state and its truth table is shown in Table 5.1.

Table 5.1 x_i truth table

Variable	True (1)	False (0)
x_1	Driving operation	Automated operation
x_2	Driven	Non-driven
x_3	Sliding brake	Non-sliding brake
x_4	Forced braking	Non-forced braking

Based on the actual vehicle operation, we can get the vehicle operation mode $\Omega = \{N_1, N_2, \ldots, N_p\}$, where the truth values of N_i are listed as follows:

$$
\begin{aligned}
N_1 &= \{0, 0, 0, 0\} \\
N_2 &= \{1, 1, 0, 0\} \\
N_3 &= \{1, 0, 1, 0\} \\
N_3 &= \{1, 0, 0, 1\}
\end{aligned}
$$

To calculate the vehicle operation mode in real time, define the function:

$$
\Gamma(x) = \begin{cases} 1 & x \geq 0 \\ 0 & x < 0 \end{cases} \tag{5.1}
$$

Then x_i are calculated respectively, as shown in Eq. (5.2).

$$
\begin{aligned}
x_1 &= \Gamma(\text{CAS\_ Disable} - 1)\{\Gamma[\Gamma(\text{APS} - 8) + \Gamma(\text{BPS} - 1) + \Gamma(10 - V) \\
&\quad + \Gamma(\text{Key\_ Disable} - 1)]\} \\
x_2 &= \Gamma(\text{CAS\_ Disable} - 1)[1 - \Gamma(x_1 - 1)]\Gamma(\text{ACC\_ trq} - 25) \\
x_3 &= \Gamma(\text{CAS\_ Disable} - 1)[1 - \Gamma(x_1 - 1)]\Gamma(15 - \text{ACC\_ trq})\Gamma(90 + \text{ACC\_ trq}) \\
x_4 &= \Gamma\{[1 - \Gamma(\text{CAS\_ Disable} - 1)] \\
&\quad + [1 - \Gamma(110 + \text{ACC\_ trq})][1 - \Gamma(x_1 - 1)]\}
\end{aligned} \tag{5.2}
$$

When *CAS_Disable* is 1, it means that the active collision avoidance function is cancelled and 0 means that the active collision avoidance function is triggered; *APS*represents the driver's accelerator pedal; *BPS* represents the brake pedal signal, 1 means the pedal is depressed and 0 means the pedal is not depressed; *ACC_trq* represents the driving torque calculated in ACC mode; *V* represents the current vehicle velocity; when *Key_Disable* is 1, it means the ACC function is off, 0 means the ACC function is on.

(b) Strategy for coordination and control of driving force to switch between driver operation and intelligent-assisted driving

In the process of swtiching from driver operation to intelligent-assisted driving, due to the possible prominent s in driving tasks, there may be large sudden changes in the driving force before and after the mode switches and the vehicle acceleration will change drastically, which will greatly affect the driving comfort. On the other hand, as the driver may feel quite differently, it may lead to frequent driver intervention in the vehicle system. In addition, due to sudden changes in system driving force, for the vehicle transmission system, auto parts may produce a large impact, affecting the life of the transmission system. Therefore, during the mode switching process, the vehicle power needs continuous and smooth transition. For this reason, this study proposes a a method for coordination and control of driving force featuring the variable reference curve.

Suppose that the vehicle operation mode is manual control mode when the driver's demanded torque is *Driver_trq*. After switching to the vehicle automated control mode, the demand drive torque calculated at this time is *Auto_trq*. In order to ensure the system response speed, define the total time of mode switching as T_0. At this time, the torque needs to smoothly change from the driver-controlled driving torque *Driver_trq* to the driving torque *Auto_trq* calculated in automated mode. Let the torque switching process change curve be f, then the function curve f is a function of time and torque *Driver_trq*. Also, since the demanded torque calculated in automated mode changes in real time during the mode switching process, the torque swiching process change curve f is also related to the value of this change.

Let the torque calculated in automated mode at the beginning of the mode switching process be *Auto_trq*(0), and after the time of t passes, the torque calculated in automated mode is *Auto_trq*(t) and the change curve from 0 to t is f_t, then at the time of $t + dT$, the change curve f' should meet the following conditions:

$$\begin{aligned} f'(t) &= f(t) \\ f'(T) &= Auto\_Trq(t + \mathrm{d}T) \end{aligned} \tag{5.3}$$

If the mode switching process curve is simplified to a straight line, the driving torque change curve during the mode switching process will be as shown in Fig. 5.14. Discretizing this continuous process, let the step size be δ, then the driving torque value at the moment of $(k + 1)\delta$ is:

$$f(k+1) = \frac{Auto\_Trq(k) - f(k)}{T - k\delta} \cdot \delta + f(k) \tag{5.4}$$

(2) Expected vehicle acceleration coordinated control strategy based on the characteristics of the hybrid driving system

After the expected acceleration has been calculated based on the driver's car-following characteristics, the actually controlled acceleration needs to be calculated based on the expected acceleration, which will be further used to calculate the expected driving torque in order to overcome the external interventions and changes in the vehicle state during the actual vehicle operation. For the hybrid system, the characteristics of its driving system engine and motor differ greatly. In the process of different parts working and mode switching, there may be great changes in expected acceleration, resulting in the system responding slower. Also, the vehicle can not quickly follow the preceding vehicle, causing the vehicle to vehicle distance to increase or decrease sharply. At the same time, it will lead to frequent driver interventions, thus compromising driving comfort.

In order to resolve the differences between the dynamic characteristics of the switching between hybrid driving system engine and motor and the power transmission characteristics, the expected vehicle acceleration coordinated control strategy

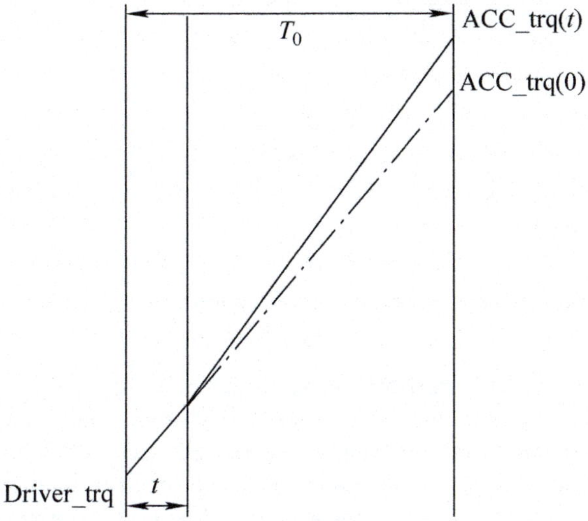

Fig. 5.14 Diagram of driving torque change during mode switching process

based on the characteristics of the hybrid driving system is proposed here. It is used to solve the drastic changes in the characteristics of controlled objects caused by the mode switching among the hybrid system engine-only driving, motor-only driving and engine/motor co-driving in the driving process.

According to the characteristics of the hybrid system, the control system is adjusted in real time according to the switching between the engine and motor. Here, we propose a model matching controller based on real-time vehicle state feedback with a two-degree-of-freedom structure for mode switching to achieve coordinated control of the expected acceleration calculation and hybrid driving power calculation, improve the control system performance and meet the vehicle's need for rapid reponse.

The structure of the two-degree-of-freedom controller based on mode switching is shown in Fig. 5.15, where the feedback on current vehicle driving state is given by the vehicle system and the model switching supervisor decides which controller should be used at the moment.

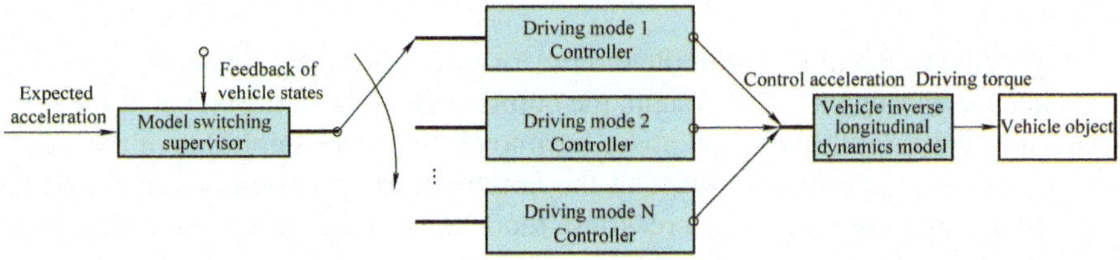

Fig. 5.15 Two-degree-of-freedom controller structure based on mode switching

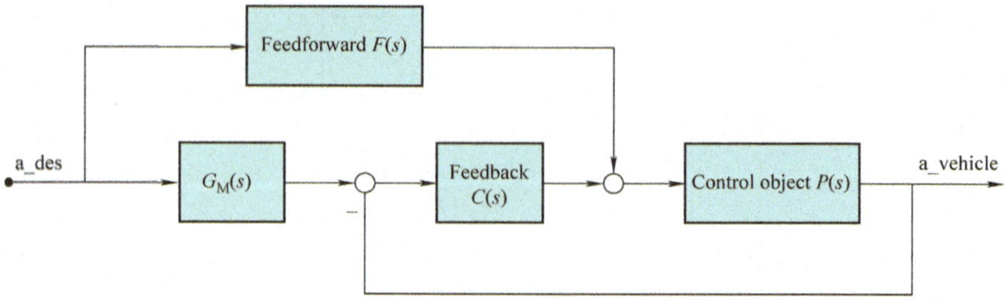

Fig. 5.16 Two-degree-of-freedom structure model for matching of controllers

Since the system characteristics of the vehicle are different in different driving states and especially in the modes of engine-only driving and motor-only driving, the system characteristics differ greatly, the controller set will be divided into three sub-controllers according to the characteristics of the controlled objects: engine-only driving, motor-only driving and engine/motor co-driving etc. The sub-controller used is determined by the current system driving state that has been reported.

For each sub-controller in the corresponding controller set, a two-degree-of-freedom structure model is used to match the controllers. The structure of the model is shown in Fig. 5.16.

Where $P(s)$ is the characteristics of transfer between the actual response vehicle's acceleration and the expected acceleration.

The transfer function of the model for controller matching is:

$$G_M(s) = \frac{1}{s+1} \tag{5.5}$$

The feedforward transfer function is:

$$F(s) = \frac{G_M(s)}{P(s)} \tag{5.6}$$

The feedback uses PI control and $P(s)$ is the vehicle object transfer function. The vehicle object transfer function can be obtained by using the method of model identification. $P(s)$ will identify the vehicle transfer function based on three different driving modes, i.e. the engine-only driving, motor-only driving and hybrid driving modes.

According to the model identification theory, after a single-frequency sinusoidal signal passes through a linear system, the output remains as a sine signal of the same frequency, but its amplitude and phase have to change. The amplitude of the output sine signal is equal to the product of the amplitude of the original signal and the value of the amplitude frequency response function of the linear system at this single frequency; the phase of the output sine signal lags behind the phase of the original signal, and the lag value is the value of the phase frequency response function of the linear system at this single frequency.

Based on the above principles, respectively, for the engine-only driving, motor-only driving, and the engine/motor co-driving modes, take the vehicle working speed range of 18–90 km/h and set a working point for every 9 km/h. In each working point and under each working condition, use the established active collision avoidance vehicle dynamics system model, input sine signals of different frequencies of the vehicle object, record the output signal of the lower controlled object, analyze the amplitude and phase relationship of the input and output signals, obtain the frequency response characteristics of the controlled vehicle object and then use the Levy method to obtain the transfer function model of the controlled vehicle.

For engine-only driving and motor-only driving models, the identification of transfer function models is shown in the equation and the obtained identification diagram is shown in Figs. 5.17 and 5.18.

$$S_e = \frac{2.139}{s + 2.526} \tag{5.7}$$

$$S_M = \frac{95.72}{s + 99.69} \tag{5.8}$$

In the hybrid driving process, since the engine responds much slower than motor, the system transfer characteristics of the hybrid driving process are basically the same as that of engine-only driving mode. Therefore, the transfer characteristics of the vehicle in the engine-only driving, motor-only driving and the hybrid driving modes can be obtained, as shown in Eqs. (5.9)–(5.11).

$$S_e = \frac{2.139}{s + 2.526} \tag{5.9}$$

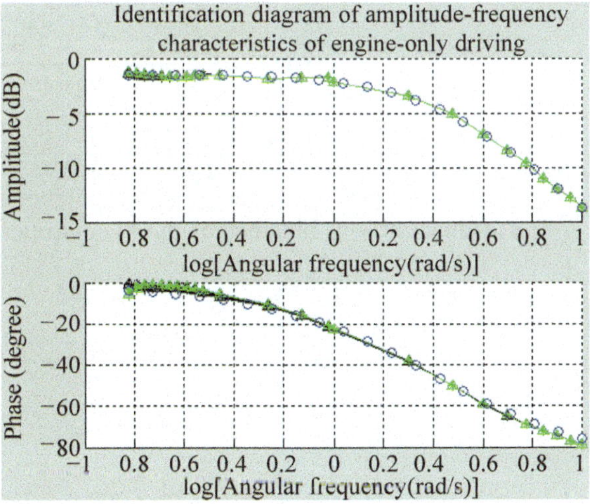

Fig. 5.17 Identification results of the Engine-only driving model

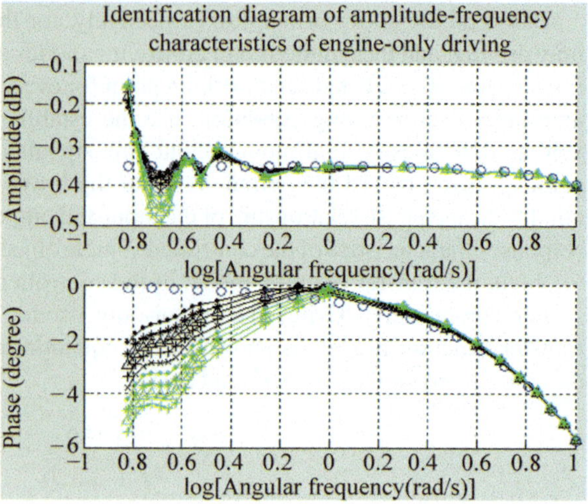

Fig. 5.18 Identification results of the driving motor-only driving model

$$S_{\mathrm{M}} = \frac{95.71}{s + 99.69} \tag{5.10}$$

$$S_{\mathrm{h}} = \frac{2.139}{s + 2.526} \tag{5.11}$$

Based on the vehicle transfer function characteristics obtained in different driving modes, the transfer functions of the feedforward F of the two-degree-of-freedom structure model matching controller can be obtained, as shown in Eqs. (5.12)–(5.14) respectively.

$$F_{\mathrm{e}}(s) = \frac{G_{\mathrm{M}}(s)}{P_{\mathrm{e}}(s)} = \frac{s + 2.526}{2.139s + 2.139} \tag{5.12}$$

$$F_{\mathrm{M}}(s) = \frac{G_{\mathrm{M}}(s)}{P_{\mathrm{M}}(s)} = \frac{s + 99.69}{95.71s + 95.71} \tag{5.13}$$

$$F_{\mathrm{h}}(s) = \frac{G_{\mathrm{M}}(s)}{P_{\mathrm{h}}(s)} = \frac{s + 2.526}{2.139s + 2.139} \tag{5.14}$$

The feedback $C(s)$ is obtained by the PI controller design method.

5.4.1.2 Multi-system Coordinated Control Technology for Intelligent-Assisted Driving Mode Switching

(1) Coordinated control technology for engine and drive motor in the intelligent-assisted driving state

In the intelligent-assisted driving state, the torque of the engine and the drive motor needs to be controlled in real time according to the relative vehicle velocity and the inter-vehicle distance collected by the radar and the vehicle state information collected by the vehicle state sensor, so that the vehicle always maintains better fuel economy and better safety.

Specifically, when the engine and the drive motor work steadily in the steady intelligent-assisted driving state, the control system keeps the economy as its main control objective while ensuring the safety; in process of mode switching of the intelligent -assisted driving, the vehicle state has undergone certain changes. At this point, priority should be given to ensuring safety and this is achieved through the faster system response. Therefore, in process of mode switching of intelligent-assisted driving, the driving torque of the engine and the drive motor should be distributed in a comprehensive and coordinated manner to ensure the response speed of the system and improve vehicle safety.

As shown in Fig. 5.19, there are four intelligent-assisted driving modes: fixed-speed cruise, car following, approaching and active collision avoidance. Because the vehicle needs greater braking deceleration in the active collision avoidance mode, which is a strong brake mode, only the coordinated control of the engine and the drive motor in the driving state is considered, and the control of the vehicle in the active collision avoidance mode is not considered for the time being.

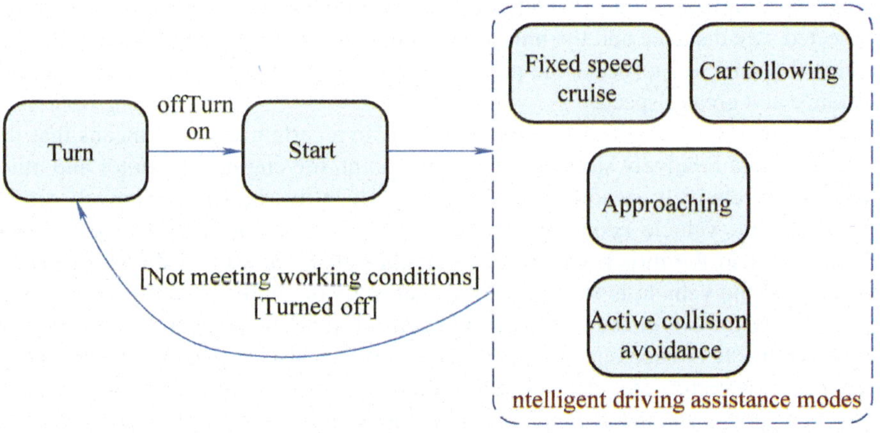

Fig. 5.19 Intelligent-assisted driving control mode

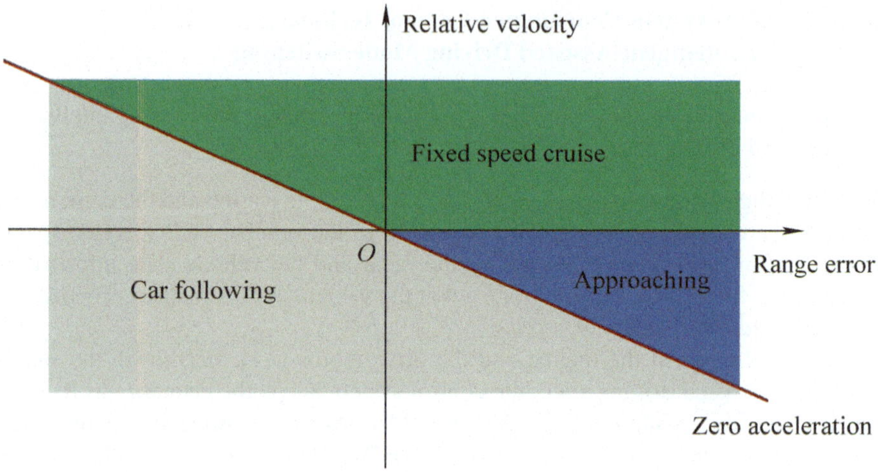

Fig. 5.20 Intelligent-assisted driving mode switching conditions

Fixed-speed cruise, car following and approaching modes are switched according to the pattern shown in Fig. 5.20. The distance error in the figure is obtained by subtracting the safe distance from the actual inter-vehicle distance (vehicle velocity $\times$ time distance $+$ constant), and the relative vehicle velocity is the velocity of the preceding vehicle minus the ego vehicle's velocity. The fuchsia zero acceleration line is the line consisting of distance error and relative vehicle velocity that makes the expected acceleration calculated by the inter-vehicle distance keeping control algorithm zero. The cyan area of fixed-speed cruise is the working condition in which the car drives at the speed set by the driver, the gray area of car following is the working condition in which the vehicle follows the preceding vehicle at the expected safe distance and the blue area of approaching is the working condition in which the vehicle approches the preceding vehicle at a lower speed from a greater distance at a greater speed.

As shown in Fig. 5.20, when the vehicle is in a certain mode, it means that the vehicle is in a relatively steady state. At this point, the engine, the motor and other parts work relatively steadily and the control quantity change is relatively small. At this point, vehicle economy should be set as the main control objective and the engine and the motor should be comprehensively controlled. Mode switching means that the vehicle is switching from one state to another and the vehicle is in an unsteady transition state in which the control quantity of each component may vary greatly. It is necessary to carry out comprehensive and coordinated control, ensure the system's adequate response, aim at vehicle safety and ensure adequate system response characteristics. The following is an introduction to the coordinated control of the driving torque of the engine and the drive motor in the steady mode of intelligent-assisted driving and in the process of switching of intelligent-assisted driving mode respectively.

In the steady intelligent-assisted driving mode, the engine and the drive motor work relatively steadily and the vehicle works steadily, and the control system ensures safety with economy as its main control objective. In order to meet the requirements of real-time vehicle control, the coordinated control strategy for the engine and the motor based on the optimal engine working range is used. According to the diagram of characteristics of fuel consumption based on the optimal engine working range, we have four modes: motor-only driving, engine-driving with power generation, generator-only driving and engine/drive motor co-driving, as shown in Fig. 5.21.

In the process of mode switching of the intelligent-assisted driving, as the vehicle is switching to a different mode, priority should be given to ensuring the fast response of the system to improve vehicle safety. Therefore, a transition mode is needed in addition to the normal steady intelligent-assisted driving mode. The complete working modes of intelligent-assisted driving are shown in Fig. 5.22.

As shown in Fig. 5.22, in addition to the fixed-speed cruise, approaching and car following modes which are divided according to the current working states of the ego vehicle and the preceding vehicle, the intelligent-assisted driving is also divided into fixed-speed cruise-approaching transition mode, fixed-speed cruise-car following transition mode and approaching-car following transition mode based on different transition processes. In the transition mode, as mentioned earlier, the torques of the

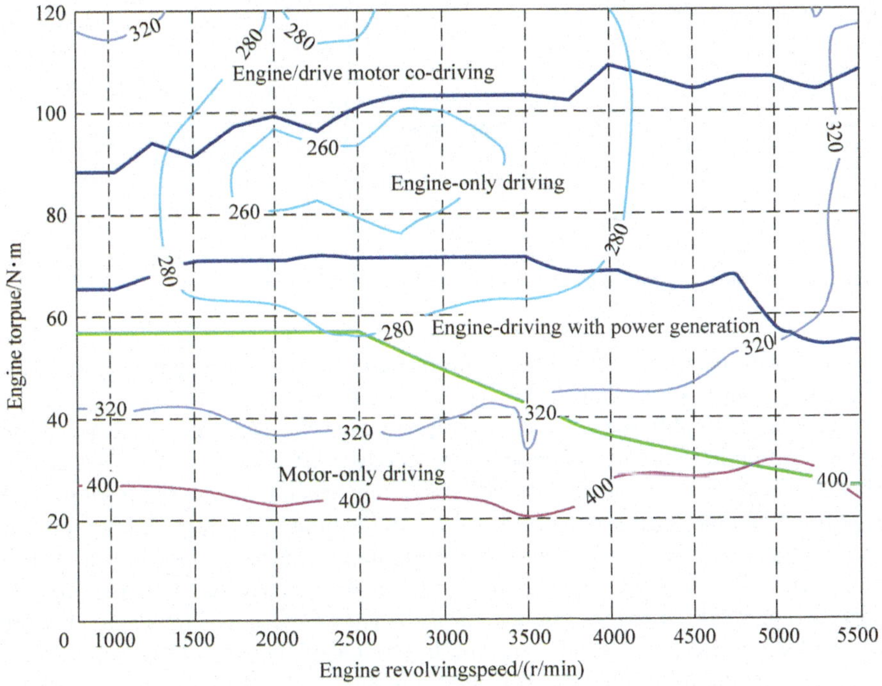

Fig. 5.21 Coordinated control strategy for engine and motor based on the optimal engine working range

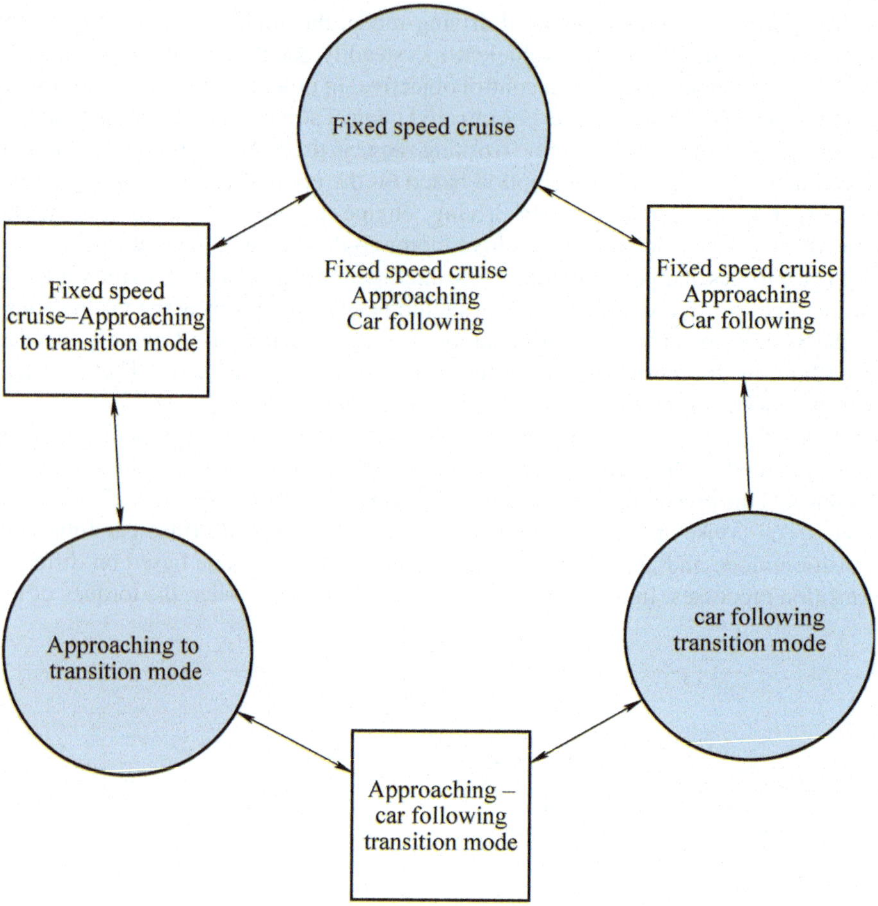

Fig. 5.22 Division of working modes of intelligent-assisted driving

engine and the drive motor need to be coordinated and controlled separately to ensure a faster response of the vehicle.

In order to simplify the control and help the control system operate faster, in the process of mode switching of the intelligent-assisted driving, the motor's torque is changed first to meet the need of the control system because the motor responses fast. Assuming that the fixed-speed cruise mode is switching to the approaching mode at this moment, and the system enters the fixed-speed cruise—approaching transition mode. If we set the time it takes to transition from fixed speed cruise mode to approaching mode as $T_{CC\_Approach}$, the expected torque calculated in the previous fixed-speed cruise mode as *CC_trq*, the torque calculated in current approaching mode as *Approach_trq*, and the expected torque calculated by the control system since the beginning of the transition mode as *Req_trq*, then:

$$Req\_trq = Approach\_trq \tag{5.15}$$

Discretize this continuous process, let the step size be δ, then the expected torque of the control system at moment $k\delta$ is $Req\_trq(k)$ and the torque of the engine and the drive motor calculated during the transitional process is $T_{engine}(k)$ and $T_{motor}(k)$ respectively. Let:

$$\begin{aligned}\Delta(k+1) &= Req\_trq(k+1) - Req\_trq(k)\\ Req\_trq(k+1) &= T_{\text{engine}}(k+1) + T_{\text{motor}}(k+1)\\ T_{\text{engine}}(k+1) &= T_{\text{engine}}(k) + \alpha_1(k+1)\Delta(k+1)\\ T_{\text{motor}}(k+1) &= T_{\text{motor}}(k) + [1-\alpha_1(k+1)]\Delta(k+1)\end{aligned} \tag{5.16}$$

The torques of the engine and the drive motor based on the optimal engine working range at each moment are $T^{opt}_{engine}(k)$ and $T^{opt}_{motor}(k)$ respectivley, then:

$$\begin{aligned}\Delta(k+1) &= Req\_trq(k+1) - Req\_trq(k)\\ \text{Req\_ trq}(k+1) &= T^{\text{opt}}_{\text{engine}}(k+1) + T^{\text{opt}}_{\text{motor}}(k+1)\\ T^{\text{opt}}_{\text{engine}}(k+1) &= T^{\text{opt}}_{\text{engine}}(k) + \lambda_1(k+1)\Delta(k+1)\\ T^{\text{opt}}_{\text{motor}}(k+1) &= T^{\text{opt}}_{\text{motor}}(k) + [1-\lambda_1(k+1)\Delta(k+1)]\end{aligned} \tag{5.17}$$

To achieve mode switching during the transitional process, $\alpha_1(k)$ changes according to a certain pattern, but the following condition needs to be met at the end of fixed-speed cruise-approaching transition mode, i.e. at the moment of $T_{CC\ to\ approach}$:

$$\alpha_1\left(\frac{T_{\text{CC\_ Apprpach}}}{\delta}\right) = \lambda_1\left(\frac{T_{\text{CC\_ Apprpach}}}{\delta}\right) \tag{5.18}$$

to finally achieve the economy of the control system after the end of the dynamic mode-switching process.

In order to simplify the control system, assume that the transitional process $\alpha_1(k)$ changes in the form of a straight line and the goal of this change is to achieve the regular coefficient $\lambda_1(k)$ based on the economic distribution at the end of the transition mode. Then the fixed-speed cruise-approaching transition mode change curve $\alpha_1(k)$ is shown in Fig. 5.23. At moment of $(k+1)\delta$, the value of $\alpha_1(k+1)$ is:

$$\alpha_1(k+1) = \frac{\lambda_1(k) - \alpha_1(k)}{T - k\delta}\delta + \alpha_1(k) \tag{5.19}$$

As the corresponding vehicle states are different as the vehicle is switched into different modes, so the parameters of different mode-switching processes are also different. After various experiments are conducted, the parameters are designed as follows:

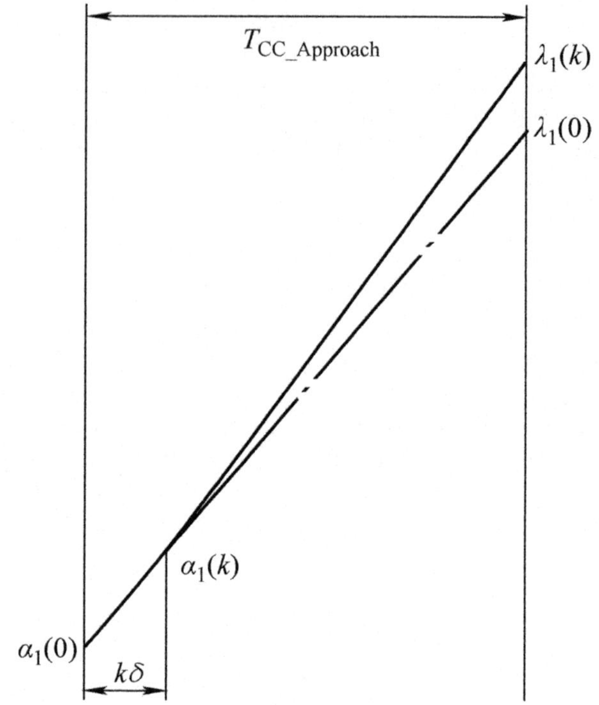

Fig. 5.23 Fixed-speed cruise-approaching transition mode change curve $\alpha_1(k)$

$$
\begin{aligned}
T_{CC\_Approach} &= 2s \\
T_{Apprpach\_Dis} &= 1s \\
T_{CC\_Dis} &= 1.5s
\end{aligned}
\tag{5.20}
$$

2. Coordinated control technology for motor brake and electro-hydraulic brake system in the intelligent-assisted driving brake state

For intelligent hybrid vehicles, the brake system consists of two sets of brake systems: motor brake system and electro-hydraulic brake system. The dynamic response characteristics of these two sets of brake systems are not the same. In general, the motor responds very faster and the hydraulic brake responds slower than the motor due to its need for pressure building and transmission. In addition, motor brake also involves energy regeneration. Therefore, in the intelligent-assisted driving brake state, the motor brake torque and electro-hydraulic brake torque need to be coordinated and controlled. On the one hand, we need to ensure the fastest response of the brake torque, so that the brake torque changes continuously and smoothly. On the other hand, we need to ensure that the motor regenerates the most energy, so that the economy of the vehicle is maximized.

Without considering the vehicle stability, the cletric motor regenerative energy model is mathematically modeled to determine the brake torque distribution for the motor as well as the electro-hydraulic system in real time. Let the motor brake torque be T_m, the motor rotational speed be n, the motor efficiency be $\eta(T_m, n)$, and the mechanical transmission efficiency be η_b. In order to get the maximum brake regenerative energy, we need to maximize the following objective function.

$$\max|P_\mathrm{c}| = \frac{\pi}{30}(-T_\mathrm{m})\eta_\mathrm{b}\eta(T_\mathrm{m}, n) \tag{5.21}$$

$$\begin{aligned} &\text{s.t. } T^*_\mathrm{mmax} \le T_\mathrm{m} \le 0 \\ &0 \le -T_\mathrm{m} i_\mathrm{m} i_\mathrm{f} \le mgz' R_\mathrm{w} \\ &0 \le \frac{-T_\mathrm{m} i_\mathrm{m} i_\mathrm{f}}{mg\left(\gamma_\mathrm{f} + z'\frac{h_\mathrm{g}}{L}\right) R_\mathrm{w}} \le \mu_\mathrm{max} \end{aligned} \tag{5.22}$$

In the objective function equation above, the factors that affect the amount of regenerative energy are motor brake torque T_m, motor efficiency $\eta(T_m, n)$, and battery charging efficiency η_b. The brake regenerative energy is mainly determined by two factors: motor brake torque T_m and charging efficiency $\eta(T_m, n)$ η_b, which are analyzed separately below.

The motor brake torque T_m is limited by the maximum brake torque of the motor, the maximum brake torque of the motor needs to be calculated in real time and is determined by the following equation.

$$\begin{aligned} T^*_\mathrm{mmax} &= T_\mathrm{mmax} W_\mathrm{SOC} W_v \\ T_\mathrm{mmax} &= \begin{cases} T_\mathrm{mn}, & n_\mathrm{m} \le n_\mathrm{mn}; \\ \frac{9550 P_\mathrm{mn}}{n_\mathrm{mn}}, & 0.8 \le \mathrm{SOC} \le 0.9 \\ 0, & v \ge 30\,\mathrm{km/h} \end{cases} \\ W_\mathrm{SOC} &= \begin{cases} 1, & \mathrm{SOC} < 0.8 \\ 10(0.9 - \mathrm{SOC}) & 0.8 \le \mathrm{SOC} < 0.9 \\ 0, & 0.9 \le \mathrm{SOC} \le 1 \end{cases} \\ W_v &= \begin{cases} 0, & 0 \le v < 20\,\mathrm{km/h} \\ \frac{v-20}{10}, & 20\,\mathrm{km/h} \le v < 30\,\mathrm{km/h} \\ 1, & v \ge 30\,\mathrm{km/h} \end{cases} \end{aligned} \tag{5.23}$$

Based on the above equation, the curve of $T^*_\mathrm{m\,max}$ is drawn, as shown in Fig. 5.24.

The charging efficiency $\eta_b\eta(T_m, n)$ varies with the motor regeneative torque and rotational speed. The motor charging efficiency and battery charging efficiency diagrams are shown in Figs. 5.25 and 5.26 respectively.

Through numerical simulation, it can be found that increasing the motor regenerative brake torque can increase the brake regenerative energy under sufficient road friction conditions. Therefore, the motor brake torque that can maximize brake regenerative energy is shown as below:

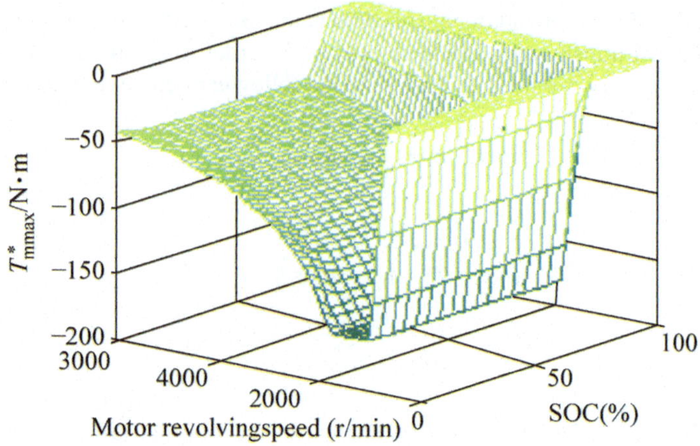

Fig. 5.24 Maximum motor brake torque change curve $T^*_{\rm m\,max}$

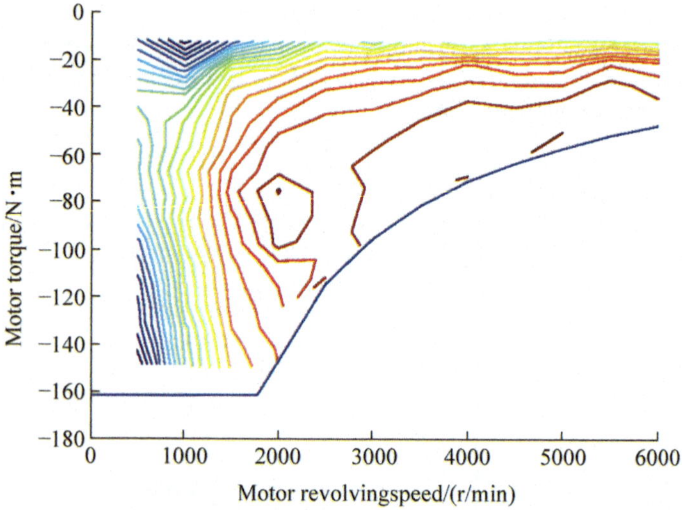

Fig. 5.25 Motor charging efficiency diagram

$$T_{\rm mopt} = -\min\left(-T^*_{\rm mmax}, \frac{F' R_{\rm w}}{i_{\rm m} i_{\rm f}}, \frac{\mu_{\max}\left(mg\gamma_{\rm f} + F'\frac{h_{\rm g}}{L}\right)R_{\rm w}}{i_{\rm m} i_{\rm f}}\right) \tag{5.24}$$

Based on the above analysis, while maintaining the stability of the vehicle, priority shoule be given to using the motor for brake, which can maximize the brake regenerative energy and improve the economy of the vehicle. However, when the required brake torque of the vehicle is relatively large and the motor-only brake cannot meet

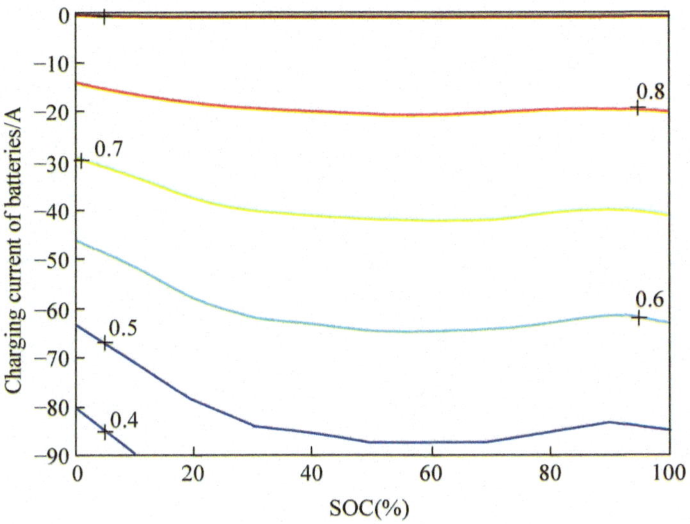

Fig. 5.26 Battery charging efficiency diagram

the brake demand of the vehicle, the electro-hydraulic brake system is needed for compensation.

In the process of coordinated controlled of the motor and the electro-hydraulic brake system, the following two requirements need to be met:

(a) Considering safety issues, the sum of the brake torques of the motor and the electro-hydraulic brake system should meet the entire vehicle's brake torque demand.
(b) considering the issue of system response accuracy, the motor has relatively fast response and high precision while the traditional electro-hydraulic brake system has relatively slow response and low precision. Therefore,we need to coordinate the motor and the hydraulic brake system to achieve accurate control of the total brake torque.

In order to meet the above two requirements, the motor and the electro-hydraulic brake system is coordinatedly controlled: (1) the electro-hydraulic brake system responds in advance to meet the expected brake torque following; (2) when the motor cannot fully meet the brake demand, the electro-hydraulic system provides the steady-state part of the brake force, and the motor precisely provides the remaining brake force through closed-loop control.

The target strategy is shown in Fig. 5.27:

The diagram of improved basic logic is shown in Fig. 5.28.

The logical relationship is analyzed as follows:

1. As the motor braking process is a process of power generation to charge the battery. When the battery SOC is greater than the charge saturation value, that is, when the motor brake is not allowed, the required brake torque is provided entirely

Fig. 5.27 Target strategy chart

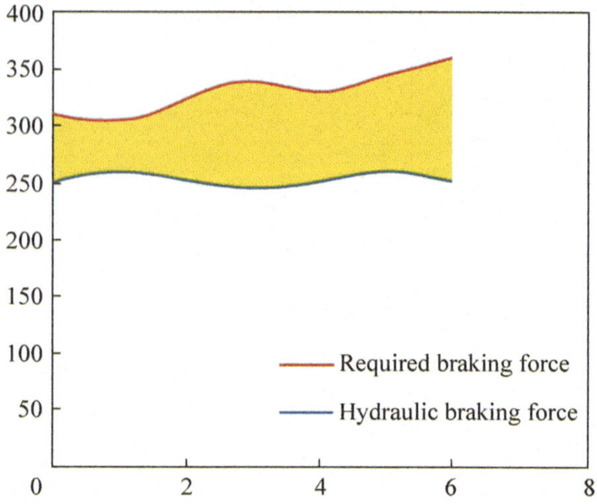

by the electro-hydraulic brake system. In this case, SOC < SOCSlipingChrgShop (charge saturation value).

2. In order to make the motor to continue provide the total brake force needed even if the *Tdes* increases after the EBS is turned on to respond. The EBS should be turned on in advance, i.e. before *Tdes* reaches *T_MP_Max*. Therefore, the EBS turn-on point set in this logic is formulated as:

$$T_{\text{des}} + t \times \mathrm{d}T > T_{\text{MP\_ Max}} \tag{5.25}$$

where t is the delay in EBS turn-on response, which varies according to the value of the command to turn on the EBS. dT is the real-time derivative of the expected torque, which is used to predict how *Tdes* changes.

3. After the conditions are met, turn on EBS and give the initial command value. The value is selected based on how EBS is turned on and is expected to be larger for fast response. The expected wheel torque value provided by the initially selected EBS is 1500 N m.
4. During EBS working process, the CAN bus can receive real-time feedback on pressure value signasl and convert it into torque value and is expressed as *T_EBS_Act* in the logic. When the following formula is satisfied:

$$T_{\text{EBS\_ Act}} > 0 \tag{5.26}$$

The torque command value will be changed to 0.5*Tdes* and the command should be added only once. This is because if the command value of the turn-on point is kept at 1500 N m, there will be too much torque provided by the EBS and this will make it impossible to make full use of the motor brake force.

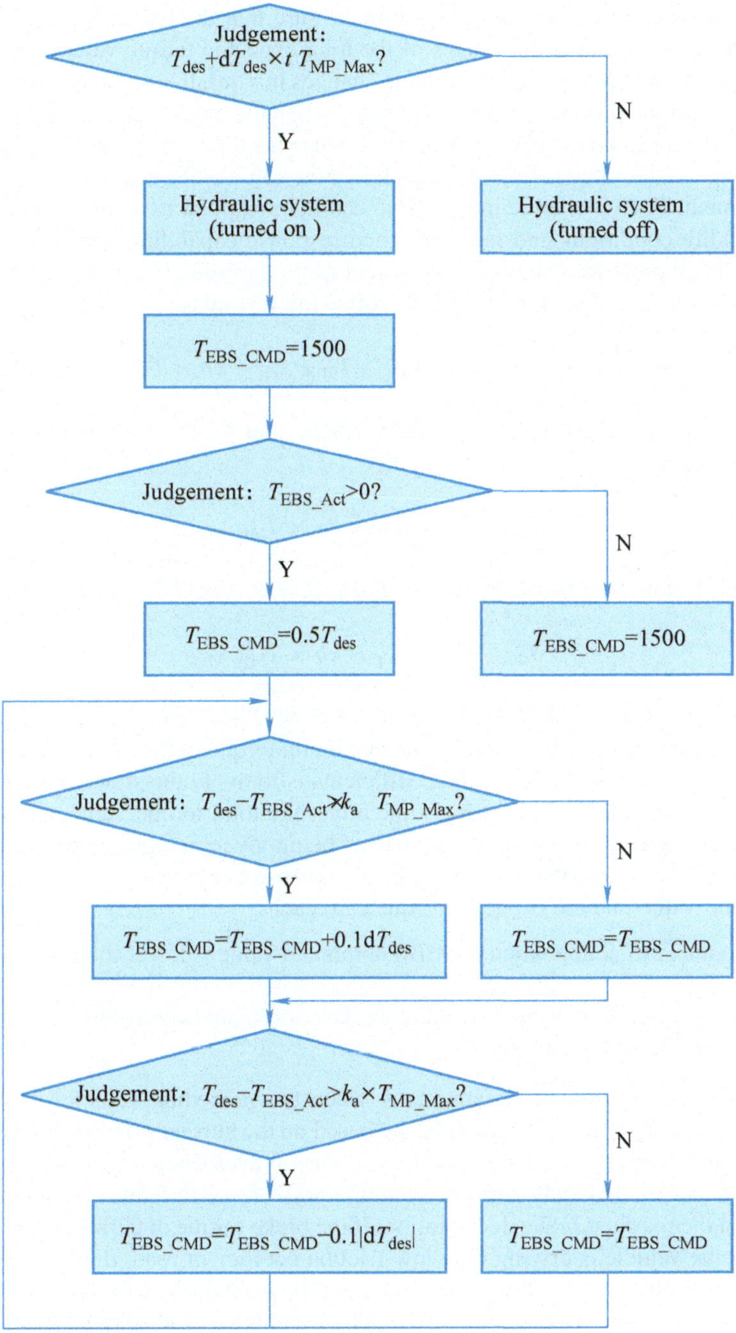

Fig. 5.28 New logic diagram

5. When the EBS comes into play, to make sure that the EBS command value is stabilized within certain ranges of the total expected torque values, changes in the motor torque are used to regulate changes in *Tdes*. It's set in the logic that the EBS command value remains unchanged when the motor value changes within a certain range and then changes stepwise when it's out of the range. The size of the change in the value is related to the derivative value of the expected torque at that moment. Such a change in the value can help respond to a variety of complex working conditions and facilitate good response capabilities. By detecting the feedback pressure value of EBS in real time, a mini-cycle is formed when the conditions are met, as shown in the following formulas:

$$ka \times T_{\text{MP\_ Max}} \leq T_{\text{des}} - T_{\text{EBS\_ Act}} \leq kb \times T_{\text{MP\_ Max}} \tag{5.27}$$

At this point, *T_EBS_CMD_T_EBS_CMD*, that is, the EBS command value remains unchanged

$$T_{\text{des}} - T_{\text{EBS\_ Act}} < ka \times T_{\text{MP\_ Max}} \tag{5.28}$$

At this point, $T\_EBS\_CMD = T\_EBS\_CMD - c^{*}|dT|$.

$$T_{\text{des}} - T_{\text{EBS\_ Act}} > kb \times T_{\text{MP\_ Max}} \tag{5.29}$$

At this point, $T\_EBS\_CMD = T\_EBS\_CMD + c^{*}|dT|$.

The coefficients ka\kb in front of the maximum torque of the motor are adjustable values. The purpose of adding the coefficients is to avoid the slow response of the EBS to the command value in real time affect the total torque value, and to make sure the command values are in the state of being given in advance to make up for its unsatisfactory speed and accuracy. c is its growth coefficient and depends on the maximum value that the expected torque can reach.

6. Stop the EBS action when the EBS command value is lower than 0.

(3) Motor brake and electro-hydraulic brake coordinated control technology in the vehicle's anti-lock brake state.

In the intelligent-assisted driving state, when the preceding vehicle or obstacles detected are very close to the vehicle and based on the current safe time distance, the vehicle immediately enters the automatic collision avoidance state, and the vehicle brake system immediately generates enough brake force to reduce vehicle velocity based on the required brake deceleration. If the brake torque of current vehicle is too large or the vehicle is driving on a low-friction coefficient road, the wheels will be locked, thus affecting the brake efficiency and brake stability of the vehicle. At the same time, since the brake system of the vehicle contains a motor brake system and a hydraulic brake system, it's necessary to design a complete set of technologies for the coordinated control of motor brake and electro-hydraulic brake under the anti-lock brake condition of the vehicle to ensure the brake efficiency and brake stability. The

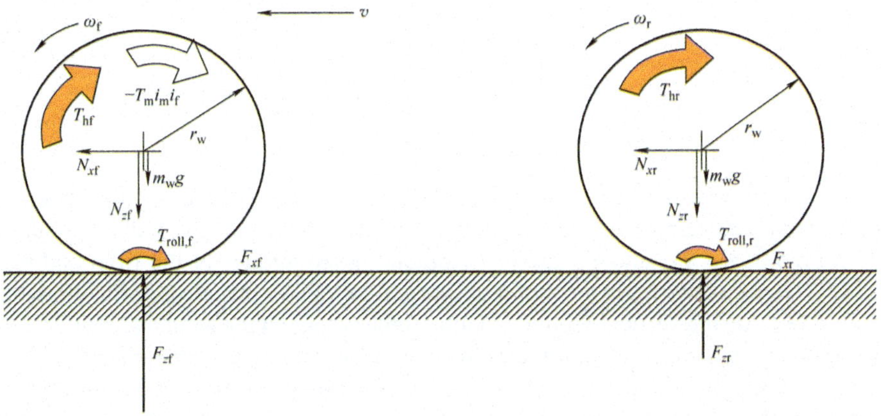

Fig. 5.29 Analysis of the force acting on the front and rear wheels of front-axle electric drive vehicle

coordinated control strategy proposed in this paper will be analyzed in detail in the next step.

The intelligent hybrid vehicle studied in this section is a front-wheel drive vehicle and only the front wheels are equipped with the motor brake system. Therefore, the analysis of the force acting on the front and rear wheels when the vehicle brakes is shown in Fig. 5.29. When the vehicle brakes, the front wheels act with the feedback brake torque of the motor and the friction brake torque of the hydraulic system, and the rear wheels only act with the hydraulic brake torque. The ground acts on the wheels with tangential brake force to produce the required brake strength.

The following state equation for the front wheel slip rate can be obtained based on the analysis:

$$\dot{\lambda}_{\mathrm{f}} = g_{\mathrm{f}} + b_{\mathrm{m}} u_{\mathrm{m}} + b_{\mathrm{hf}} u_{\mathrm{hf}} \tag{5.30}$$

where

$$g_{\mathrm{f}} = \frac{r_{\mathrm{w}}}{J_{\mathrm{f}} v}\left[T_{\mathrm{roll,f}} - F_{z\mathrm{f}} \mu_{\mathrm{f}} r_{\mathrm{w}}\right] - \frac{(1-\lambda_{\mathrm{f}})}{m v}\left[F_{\mathrm{aero}} + F_{z\mathrm{f}} \mu_{\mathrm{f}} + F_{z\mathrm{f}} \mu_{\mathrm{r}}\right] \tag{5.31}$$

$$b_{\mathrm{m}} = \frac{r_{\mathrm{w}}(1+e_{\mathrm{m}}) i_{\mathrm{m}} i_{\mathrm{f}}}{J_{\mathrm{f}}} \tag{5.32}$$

$$b_{\mathrm{hf}} = \frac{r_{\mathrm{w}}(1+e_{\mathrm{hf}})}{J_{\mathrm{f}}} \tag{5.33}$$

u_{m} and u_{hf} refer to the force controlled by motor brake and hydraulic brake and are defined as

$$u_{\mathrm{m}} = \frac{-T'_{\mathrm{m}}}{v} \tag{5.34}$$

$$u_{\mathrm{hf}} = \frac{T'_{\mathrm{hf}}}{v} \tag{5.35}$$

where m is the vehicle mass, r_{w} is the wheel radius, ω_{f} and ω_{r} are the angular velocities of the front and rear wheels respectively, ω_{v} is the vehicle velocity v divided by the wheel radius r_{w}. T_{hf} and T_{hr} are the hydraulic brake torques of the front and rear wheels respectively, T_{m} is the motor torque and its value is negative during braking. F_{zf} and F_{zr} are the vertical ground reaction force acting on the front and rear wheels respectively. i_{m} is the transmission ratio from the motor to the main reducer, i_{f} is the transmission ratio of the main reducer. J_{f} is the rotational inertia of the front wheels, including the wheels on both sides of the front axles, the rotor of the drive motor and the transmission parts etc., J_{r} is the rotational inertia of the rear-axle wheels. $T_{\mathrm{roll,f}}$ and $T_{\mathrm{roll,r}}$ are the rolling resistance torques acting on the front and rear axle wheels respectively. μ_{f} and μ_{r} are the brake force coefficients of the front and rear wheels respectively.

In the electric vehicle brake system studied in this section, T_{hf}, T_{h} and T_{m} are generated by controlling the relevant components through the hydraulic regulating unit controller and the motor controller and errors between their actual and demanded values are inevitable. Let the actuation errors of T_{hf}, T_{hr} and T_{m} be e_{hf}, e_{hr} and e_{m} respectively, then we have

$$T_{\mathrm{hf}} = (1 + e_{\mathrm{hf}})T'_{\mathrm{hf}} \tag{5.36}$$

$$T_{\mathrm{hr}} = (1 + e_{\mathrm{hr}})T'_{\mathrm{hr}} \tag{5.37}$$

$$T_{\mathrm{m}} = (1 + e_{\mathrm{m}})T'_{\mathrm{m}} \tag{5.38}$$

where T'_{hf} and T'_{hr} are the required hydraulic brake torque of the front and rear wheels respectively and T'_{m} is the demanded torque of the motor. It is necessary to design $\mathrm{u_m}$ and $\mathrm{u_{hf}}$ to converge λ_{f} to the target value λ'_{f}.

As shown in the expression of g_{f}, both the state equations of the front and rear wheel slip rate shown in Eq. (5.30) are strongly nonlinear and the general linear control method cannot solve the problem. Also, both motor brake torque and hydraulic brake torque have actuation errors and e_{hf} and e_{m} in Eqs. (5.32) and (5.33) cannot be obtained in real time, that is, both b_{m} and b_{hf} have uncertainty. However, by optimizing the controller design, the actuation errors of each brake torque can be controlled within a certain range, that is, b_{m} and b_{hf} satisfy

$$0 < b_{\mathrm{mmin}} \le b_{\mathrm{m}} \le b_{\mathrm{mmax}} \tag{5.39}$$

$$0 < b_{\mathrm{hfmin}} \le b_{\mathrm{hf}} \le b_{\mathrm{hfmax}} \tag{5.40}$$

where $b_{\mathrm{m\,min}}$ and $b_{\mathrm{m\,max}}$ are the lower and upper limits of b_{m} and $b_{\mathrm{hf\,min}}$ and $b_{\mathrm{hf\,max}}$ are the lower and upper limits of b_{hf}.

For the first-order nonlinear system shown in Eq. (5.30), the control variables u_{m} and u_{hf} need to be designed so that the front wheel slip rate λ_{f} converges to the required value for top level control and distribution. If the use of either u_{m} or u_{hf} can simplify the problem to a control problem of a single-input nonlinear system, the existing nonlinear control methods can be applied. However, in practice, since u_{m} and u_{hf} represent the motor brake and the hydraulic brake respectively with their own advantages, they need to be combined to achieve the overall optimal effect. Motor brake allows for energy regeneration, converting the kinetic energy of the vehicle into electrical energy for later use and the motor responds fast and accurately and is easy to measure, allowing for better slip rate control. In addition, the use of motor brake reduces the wear of brake friction components in hydraulic brake and thus reduces the likelihood of heat fade. However, due to the limited capacity of motor brake, hydraulic brake is required to supplement at high brake intensities. Therefore, the motor brake weighting coefficient α_{m} is introduced to regulate the steady-state proportional relationship between the motor brake and the hydraulic brake in the front wheel slip rate control. The value of α_{m} is set within the range of

$$0 \le \alpha_{\mathrm{m}} \le 1 \tag{5.41}$$

When $\alpha_{\mathrm{m}} = 1$, the front wheel brake force is generated entirely by the motor brake and the hydraulic system does not generate brake pressure in the front wheel cylinders; when $\alpha_{\mathrm{m}} = 0$, the front wheel brake force is generated entirely by the hydraulic brake and there is no energy regeneration during braking; when $0 < \alpha_{\mathrm{m}} < 1$, the front wheel braking force is generated by the motor brake and hydraulic brake together, part of the kinetic energy used during braking is regenerated by the motor and stored in the battery and part is dissipated in the air through the friction between the brake caliper and the brake disc. To design slip rate electro-hydraulic composite controller in the Eq. (5.30) as

$$u_{\mathrm{m}} = \hat{b}_{\mathrm{m}}^{-1}\left[\alpha_{\mathrm{m}}\hat{u}_{\mathrm{f}} - k_{\mathrm{m}}\mathrm{sgn}(s_{\mathrm{f}})\right] \tag{5.42}$$

$$u_{\mathrm{hf}} = \hat{b}_{\mathrm{hf}}^{-1}\left[(1-\alpha_{\mathrm{m}})\hat{u}_{\mathrm{f}} - k_{\mathrm{hf}}\mathrm{sgn}(s_{\mathrm{f}})\right] \tag{5.43}$$

where

$$\hat{u}_{\mathrm{f}} = \hat{\lambda}_{\mathrm{f}}' - \hat{g}_{\mathrm{f}} \tag{5.44}$$

$\hat{b}_{\mathrm{m}}$, $\hat{b}_{\mathrm{hf}}$ and $\hat{g}_{\mathrm{f}}$ are estimates of b_{m}, b_{hf} and g_{f} respectively and

$$\hat{b}_{\mathrm{m}} = (b_{\mathrm{mmin}}b_{\mathrm{mmax}})^{\frac{1}{2}} \tag{5.45}$$

$$\hat{b}_{\mathrm{hf}} = (b_{\mathrm{hfmin}} b_{\mathrm{hfmax}})^{\frac{1}{2}} \tag{5.46}$$

$$\left|g_{\mathrm{f}} - \hat{g}_{\mathrm{r}}\right| \le E_{\mathrm{gr}} \tag{5.47}$$

s_{f} is the sliding mode switching function, defined as

$$s_{\mathrm{f}} = \lambda_{\mathrm{f}} - \lambda'_{\mathrm{f}} \tag{5.48}$$

where λ'_{f} is the target front wheel slip rate obtained by the top control based on the current driver brake demand, vehicle electrical system state and road surface state. It can be seen that when the system is on the sliding mode surface $s_{\mathrm{f}} = 0$, there is

$$\lambda_{\mathrm{f}} = \lambda'_{f} \tag{5.49}$$

sgn is a symbolic function, defined as

$$\operatorname{sgn}(x) = \begin{cases} 1, & x > 0 \\ -1, & x < 0 \end{cases} \tag{5.50}$$

k_{m} and k_{hf} are the term-switching gain parameters. When k_{m} and k_{hf} meet certain conditions, the system tends to the sliding mode surface $s_{\mathrm{f}} = 0$ at the specified speed, so that the actual slip rate follows the target value.

It can be proven that if the term-switching gain k_{m} and k_{hf} satisfies the conditions

$$k_{\mathrm{m}} \ge \alpha_{\mathrm{m}}\left[(\beta_{\mathrm{m}} - 1)\left|\hat{u}_{\mathrm{f}}\right| + \beta_{\mathrm{m}}\left(E_{\mathrm{gf}} + \eta\right)\right] \tag{5.51}$$

$$k_{\mathrm{hf}} \ge (1 - \alpha_{\mathrm{m}})\left[(\beta_{\mathrm{hf}} - 1)\left|\hat{u}_{\mathrm{f}}\right| + \beta_{\mathrm{hf}}\left(E_{\mathrm{gr}} + \eta\right)\right] \tag{5.52}$$

then the accessibility condition can be satisfied.

$$\frac{1}{2}\frac{\mathrm{d}}{\mathrm{d}t}s_{\mathrm{f}}^2 \le -v_{\mathrm{s}}|s_{\mathrm{f}}| \tag{5.53}$$

That is, the system converges to the sliding mode surface $s_{\mathrm{f}} = 0$ at least at the rate of v_{s} ($v_{\mathrm{s}} > 0$) where.

$$\beta_{\mathrm{m}} = \left(\frac{b_{\mathrm{mmax}}}{b_{\mathrm{mmin}}}\right)^{\frac{1}{2}} \tag{5.54}$$

$$\beta_{\mathrm{hf}} = \left(\frac{b_{\mathrm{hfmax}}}{b_{\mathrm{hfmin}}}\right)^{\frac{1}{2}} \tag{5.55}$$

Due to the inevitable errors between the actually controlled system and the mathematical equations described for it, there are always errors and delays in the measurement of the feedback signal. Also, due to the influence of the system inertia and other factors, the use of sliding mode variable structure control often causes the system to cross back and forth near the sliding mode switching surface, resulting in jitter. In order to weaken the jitter, the integral saturation function isat is introduced instead of the symbolic function. isat is defined as

$$\mathrm{isat}(x, \phi, k_{\mathrm{i}}) = \begin{cases} 1, & x \geq \phi \\ \frac{x}{\phi} + \frac{k_{\mathrm{i}}}{\phi} \int x\mathrm{d}t, & -\phi < x < \phi \\ -1, & x \leq -\phi \end{cases} \tag{5.56}$$

where

$$\phi > 0 \tag{5.57}$$

The integral saturation function is substituted in (2.42) (2.43) to get

$$u_{\mathrm{m}} = \hat{b}_{\mathrm{m}}^{-1}\left[\alpha_{\mathrm{m}}\hat{u}_{\mathrm{f}} - k_{\mathrm{m}}\mathrm{isat}(s_{\mathrm{f}}, \phi_{\mathrm{f}}, k_{\mathrm{if}})\right] \tag{5.58}$$

$$u_{\mathrm{hf}} = \hat{b}_{\mathrm{hf}}^{-1}\left[(1 - \alpha_{\mathrm{m}})\hat{u}_{\mathrm{f}} - k_{\mathrm{hf}}\mathrm{isat}(s_{\mathrm{f}}, \phi_{\mathrm{f}}, k_{\mathrm{if}})\right] \tag{5.59}$$

Figure 5.30 shows the front wheel slip rate control in the emergency brake condition. In emergency brake, the target slip rate of the front wheels is the optimal slip rate of the road, so as to reduce the brake distance and to ensure that the wheels have sufficient lateral force. In Fig. 5.30, the vehicle electrical system allows for brake energy regeneration, but the control does not pursue the maximum efficiency of brake energy regeneration and instead focuses on stabilizing the front wheel slip rate at the target value by coordinating the motor brake with the hydraulic brake. The hydraulic brake provides the steady-state part of the brake demand to compensate for the limited motor brake power, while the motor brake provides the dynamic adjustment function of the brake demand. Therefore, we make use of the rapid and accurate motor response to effectively adjust the front wheel slip rate.

5.4.2 Hybrid Multi-system Energy Management and Coordinated Control Technology

In this section, by taking the NEV equipped with the series-parallel hybrid power system as an example, we will discuss the designed RCP control plan for the multi-energy management control of the entire vehicle, as shown in Fig. 5.31, which is divided into several parts such as power demand calculation, vehicle operation

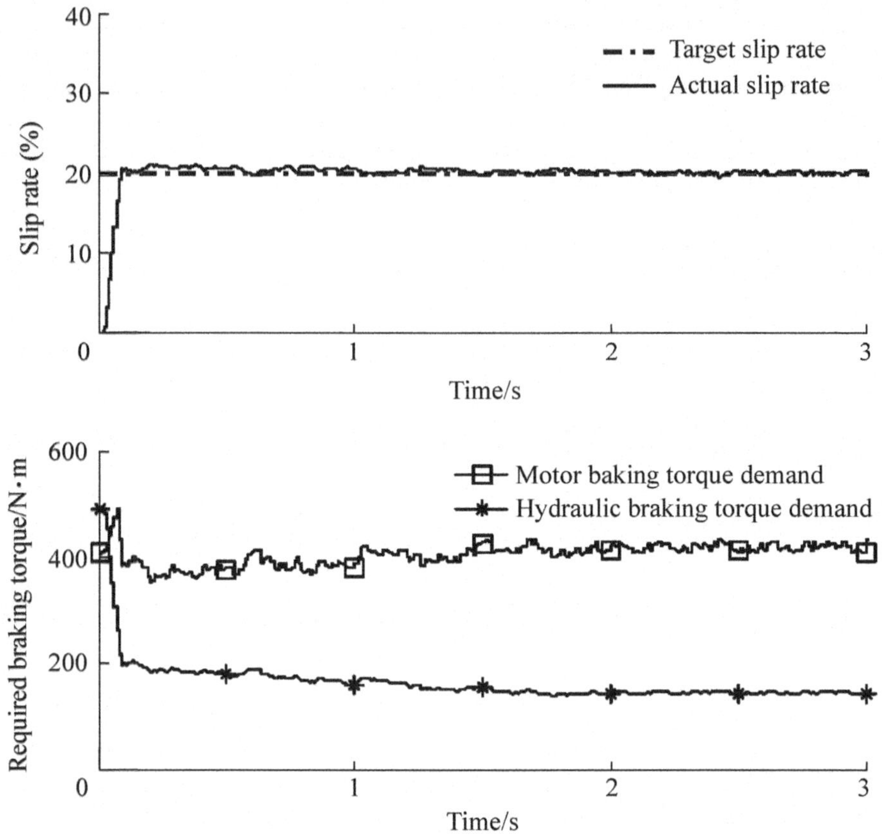

Fig. 5.30 Front wheel slip rate control during emergency brake (allowing for brake energy regeneration)

state identification, power distribution control, dynamic coordinated control, fault diagnosis and power limitation etc.

5.4.2.1 Power Demand Calculation

Power demand calculation is the calculation of the required driving power or brake power of the power system in the current state based on the driver's operating commands and the vehicle state information, which is used as the condition for inputs to the power distribution control. The required driving torque is shown in Fig. 5.32.

The definition curved surface of demanded driving torque in Fig. 5.32 is obtained by the "nine-point definition method", which requires to determine the demanded torque values for the nine key points shown in Table 5.2.

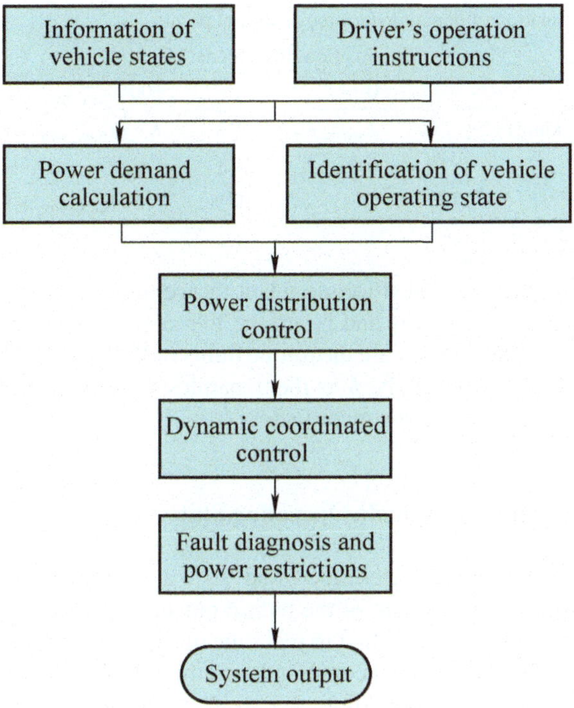

Fig. 5.31 RCP control plan

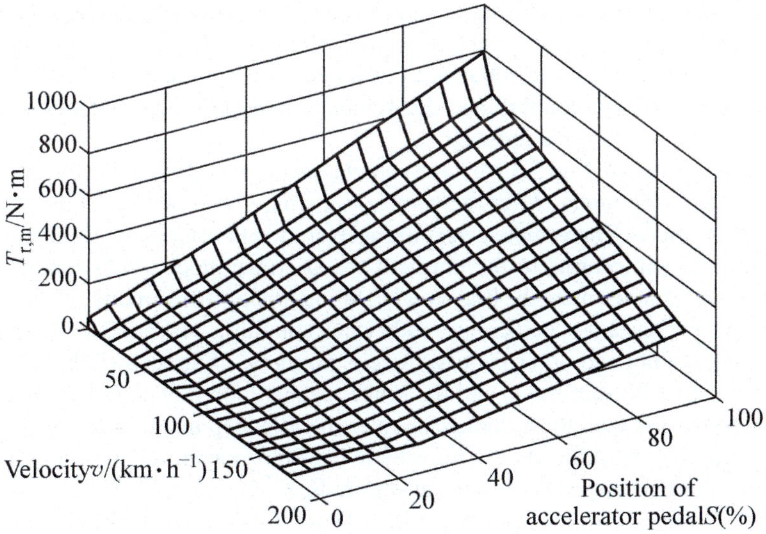

Fig. 5.32 Definition curved surface of demanded driving torque

Table 5.2 Determination of demanded torque at special points

		Position of gas pedal (%)		
		0	30	100
Vehicle velocity (km/h)	0	T_{start_PM}	C2	$K \cdot (T_{\max\_eng} \times i_1 + T_{PM\_0})$
	10	0	C3	C4
	175	C1	0	$K \cdot (T_{\max\_eng} \times i_4 + T_{PM\_175})$

where T_{start_PM} is the demanded torque when the vehicle starts slowly, T_{max_eng} is the maximum engine torque, i_1 and i_4 are the first and fourth gear ratios of AMT respectively, T_{PM_0} and T_{PM_175} are the torque limits of PM when the vehicle velocity is at 0 and 175 km/h respectively. K is the dynamics adjustment factor, C1, C2, C3 and C4 are the smoothness adjustment factors.

5.4.2.2 Identification of Vehicle Operating State

First of all, it's necessary to determine in advance a variety of operating states of the vehicle based on the structure of the hybrid power system and the current operating state of the vehicle is identified in real time during the operation of the vehicle according to the driver's operating commands and the vehicle's current state information, which is also used as the condition for inputs to the power distribution control. According to the power system structure of the series-parallel hybrid vehicle and based on the differences in the working states of the engine and the two motors, the vehicle operating conditions can be divided into the following different operating modes. The operation mode switching logic is shown in Fig. 5.33. The arrow points to the target state from the initial state of mode switching and the number on the arrow is the condition number and the vehicle operation mode switches when the corresponding condition is satisfied.

It can be seen from Fig. 5.33 that the vehicle can be started from the idle stop mode or the vehicle stop mode. After the vehicle is started, the vehicle is in the constant-speed cruise mode. The vehicle enters the acceleration booster mode when condition 1 is satisfied, and returns to the constant-speed cruise mode when condition 2 is satisfied; the vehicle enters the brake regeneration mode when condition 3 is satisfied and returns to constant-speed cruise mode when condition 4 is satisfied. Considering the driving feeling and other factors, there is no mode of direct switching between the acceleration booster mode and the brake regeneration mode. When condition 5 or condition 12 is met, the vehicle can enter the idle stop mode or the vehicle stop mode from any operating state. The vehicle can start operating with a normal start or try an emergency start if the normal start fails.

The switching conditions shown in Fig. 5.33 are still designed based on the driver's current operating commands (e.g. the position of accelerator pedal and brake pedal etc.), component state information (SOC and engine speed etc.) and vehicle state information (vehicle velocity and gear etc.).

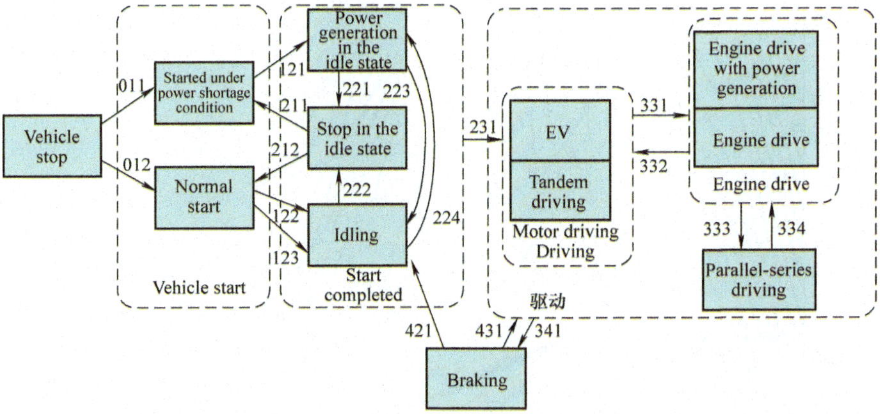

Fig. 5.33 The operation mode switching logic for series-parallel hybrid vehicle

5.4.2.3 Power Distribution Control

Based on the vehicle power allocation strategy, the target driving torque of the engine and the motor and the target output power of the battery are determined in real time based on the power demand obtained in real time and the vehicle's current operating state. In order to make the components operate efficiently during vehicle operation, the working zones of these componenets need to be divided first according to the optimized ICE curve power allocation strategy and the efficiency characteristics of the power components.

Based on the rated torque line of the main motor, an efficient operating line of the main motor is defined based on the Map diagram of the main motor's efficiency to determine the working range of the engine and the main motor, which is denoted as T_{PM_Hieff}, as shown in Fig. 5.34.

In order to make the engine work in the area with good fuel economy and avoid the working conditions with poor fuel economy such as low speed and small load, this section divides the engine's fuel consumption rate Map diagram into three areas, as shown in Fig. 5.21. The area between curve T_{eng_EcoUp} and curve $T_{eng\_EcoDown}$ is the economic working area of the engine, which is the target working area of the engine, the area below curve $T_{eng\_EcoDown}$ is the low load area and the area above curve T_{eng_EcoUp} is the booster area. Since the output power of the engine is coupled to the main motor at the rear end of the transmission, the current gear in the AMT must also be taken into account when determining the working states of the engine and the main motor. The efficient operating line of the main motor is converted into the engine end in different gears N to obtain curve $T_{PM\_Hieff}(N)$ shown in Fig. 5.35, with $N = 1, 2, 3, 4$.

Based on the battery's charging and discharging efficiency curves, the battery's working area is divided into four parts: prohibited working area, low power area, high efficiency working area and high power area, as shown in Fig. 5.36. The battery

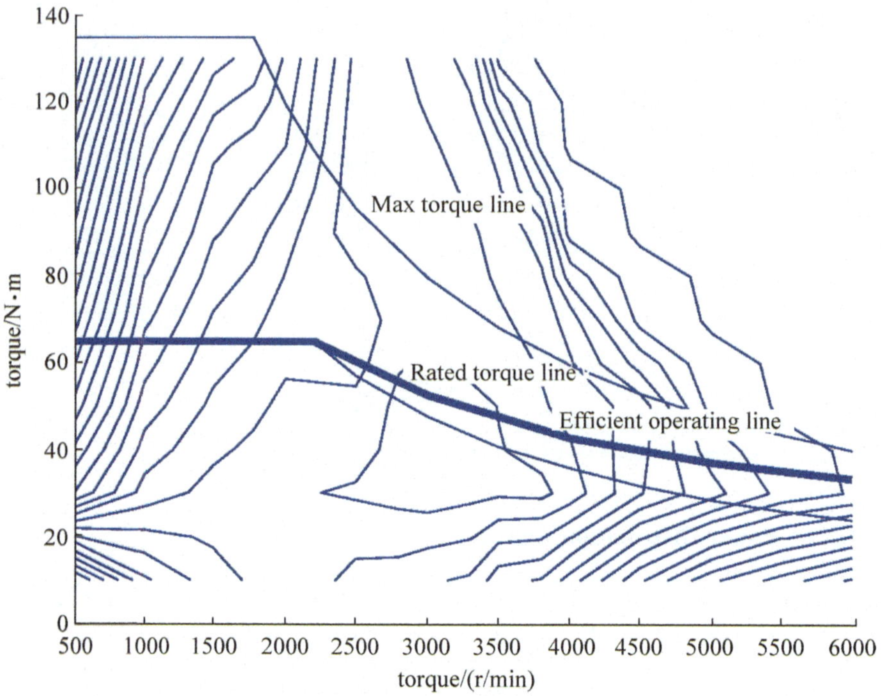

Fig. 5.34 Efficient operating lines of the main motor

state of charge (SOC) value Scmin ~ 1 is the area where the battery works normally. If SOC is below Scmin, it indicates that the battery is seriously undercharged and continued discharging will affect battery life, so the battery should be charged in time.

Based on the efficiency optimization of different components, the obtained power allocation strategy is shown in Table 5.3.

5.4.2.4 AMT Dynamic Coordinated Control

According to the characteristics of the hybrid power system, we have studied the dynamic coordinated control of the motor torque during the AMT gearshift and adjusted the target driving torque of the motor. In a conventional vehicle, the engine is the only power source and the power is transmitted to the driving axle through the transmission and the main reducer. During vehicle gear shift, the gears of the transmission are in neutral for a short time and power is inevitably be interrupted. The special structure of the hybrid vehicle that has been studied can solve this problem: since the power of the main motor is coupled at the output shaft of the transmission, the driving axle can get a certain driving torque from the main motor in any gear of the AMT, thus ensuring that the power is not interrupted.

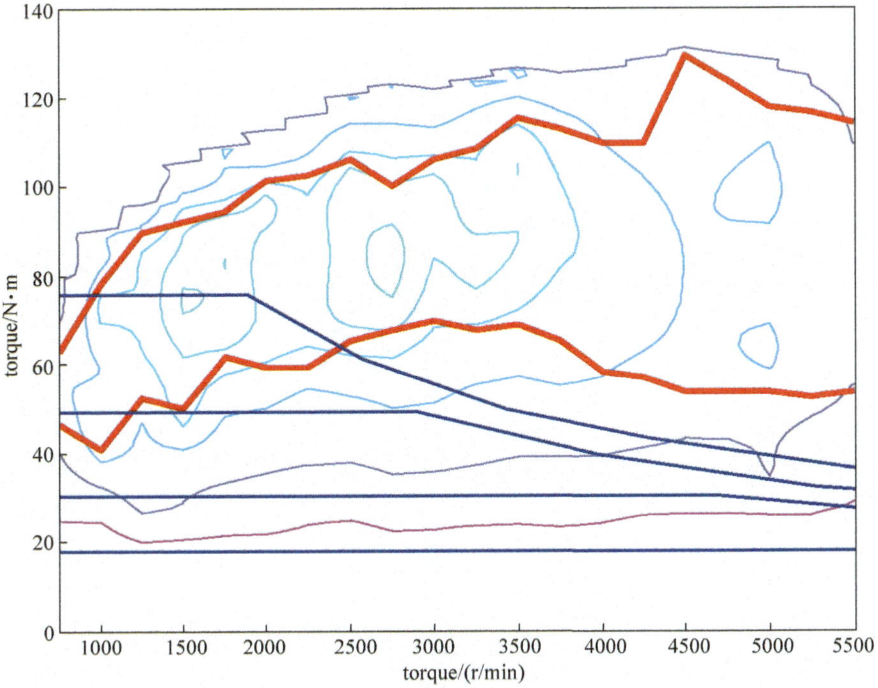

Fig. 5.35 Optimization of the working areas of the main motor and the engine

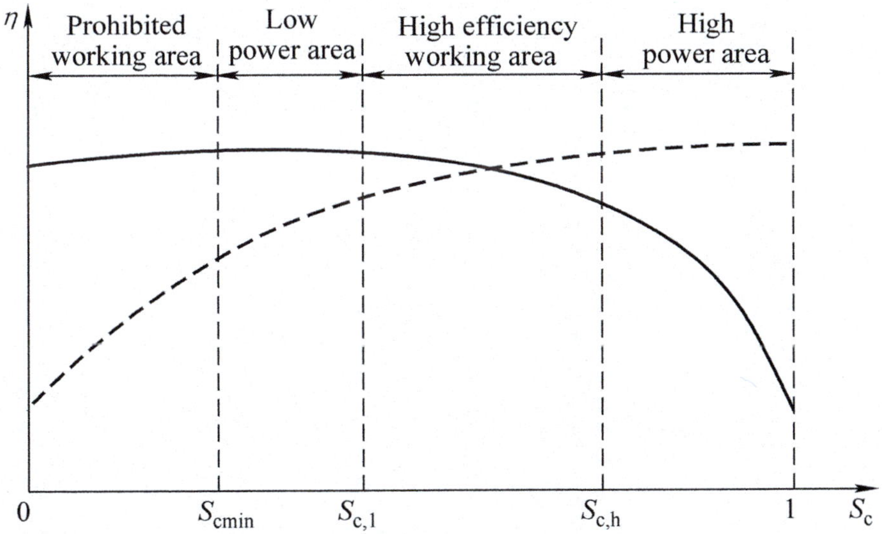

Fig. 5.36 The battery's working areas

Table 5.3 Torque distribution in different driving modes

	Battery electric driving	Series driving	Engine driving	Engine driving with power generation	Hybrid driving
Condition	$T_{r,e} < T_{m,e}(N)$ and $S_c \geq S_{c,l}$	$T_{r,e} < T_{m,e}(N)$ and $S_{cmin} \leq S_c < S_{c,l}$	$T_{e,d} \leq T_{r,e} < T_{e,u}$ and $T_{r,e} \geq T_{m,e}(N)$ and $S_c \geq S_{c,l}$	$(T_{m,e}(N) \leq T_{r,e} < T_{e,d}$ and $S_c < S_{c,l})$ or $S_c < S_{cmin}$	$T_{r,e} \geq T_{e,u}$
T_e	0	$i_{g,e} W_{SOC} T_{g,r}$	$T_{r,e}$	$T_{e,d}$	$\min(T_{emax}, T_{r,e} - T_m / i_g)$
T_m	$T_{r,m}$	$T_{r,m}$	0	0	$\min(T_{mmax}, (T_{r,e} - T_{e,u}) i_g)$
T_g	0	$-W_{SOC} T_{g,r}$	0	$-W_{SO}C\min(T_{g,r}, (T_{e,d} - T_{r,e}) / i_{g,e})$	0

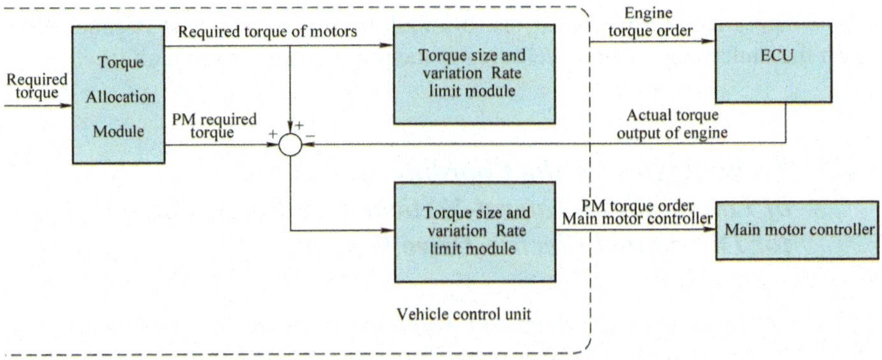

Fig. 5.37 Diagram of coordinated control of gearshift process

Due to slower response the engine, the actual output torque of the engine is introduced as the feedback quantity in order to ensure that the main motor can make up for the driver's demanded torque during the gearshift process. The diagram of the coordinated control of the gearshift process is shown in Fig. 5.37.

5.4.2.5 Fault Diagnosis and Power Limitations

According to the current fault state of the components transmitted by different controllers through CAN bus, the severity of the fault occurring in the vehicle is determined and the work load of the powertrain components is adjusted in real time according to the fault level or the power system work is stopped when a serious fault occurs.

The fault diagnosis function of the vehicle control unit is to use the state information of different powertrain components transmitted by the CAN communication network and various self-collected vehicle state information to conduct real-time detection of the operating state of the vehicle and different powertrain components, diagnose whether the hybrid powertrain system or component is malfunctioning and provide a basis for the power limitation control of the vehicle.

There are two types of fault diagnosis methods currently used: (1) system fault diagnosis based on the powertain fault codes and fault level codes transmitted by different powertrain control units through the CAN bus; (2) fault diagnosis based on the processing of powertrain state information transmitted by different powertrain control units through the CAN bus.

In Method 1, the vehicle control unit only needs to classify the system fault levels based on different component fault levels and generate system fault level codes to report errors to the system.

In Method 2, the corresponding diagnostic signals and diagnostic methods need to be selected according to the specific fault state and whether the fault has occurred will be determined through analysis. If a fault occurs, a fault code is generated to report the fault to the system according to the pre-defined system fault level.

5.4.3 Technologies for the Coordinated Control of Longitudinal-Lateral-Vertical Tyre Force for Distributed Electric-Drive Vehicle

With the progress and development of automotive technologies, a wide variety of vehicle chassis electronic control technologies have been widely studied and applied. These chassis electronic control systems implement active control over the longitudinal, lateral and vertical forces of tyres respectively, to improve vehicle dynamics and maneuvering stability. However, due to the complex coupling characteristics of vehicle motion and tyre forces in all directions, using different chassis electronic control systems to control a single aspect of tyre forces inevitably leads to conflicts between control objectives and control actuation. To address the conflicts or complementarity between single chassis electronic control systems, we need to implement the coordinated control of multiple single systems, eliminate the conflicts between subsystems and explore the complementarity between different subsystems, which is important for improving the comprehensive performance of vehicle dynamics. The coordinated vehicle stability control technology integrates the resources of different subsystems of a vehicle to coordinate and optimize the vehicle performance by coordinating the control of two or more of the steering system, suspension system and brake/driving system. Compared with traditional vehicle stability control technology, the coordinated vehicle stability control technology can not only ensure that the original performance of subsystems remain unchanged, but also optimize the stability and dynamics of the vehicle to the largest extent due to the integrated coordination of the working ranges and working states of different subsystems.

Distributed electric-drive vehicles are equipped with independent electric drive systems, independent steering systems and active suspension systems with multiple controlled objects. In order to improve the vehicle dynamics, it's necessary to clarify the specific dynamics requirements during vehicle driving and then determine the control objectives of different actuator systems and finally control different actuators to achieve the expected dynamics response. Based on this idea, we've designed a hierarchical system for the coordinated control of longitudinal-lateral-vertical tyre forces for distributed electric-drive vehicles and the structure of the system is shown in Fig. 5.38.

In this hierarchical control system, the top level determines the expected driving state of the vehicle based on driver input, vehicle state parameters and road information and sets control objectives such as expected combined force and torque; the middle level establishes objective functions and constraints to optimize the distribution of expected combined force and torque of longitudinal, lateral and vertical tyre forces generated by different wheels; the lower level precisely controls the motor

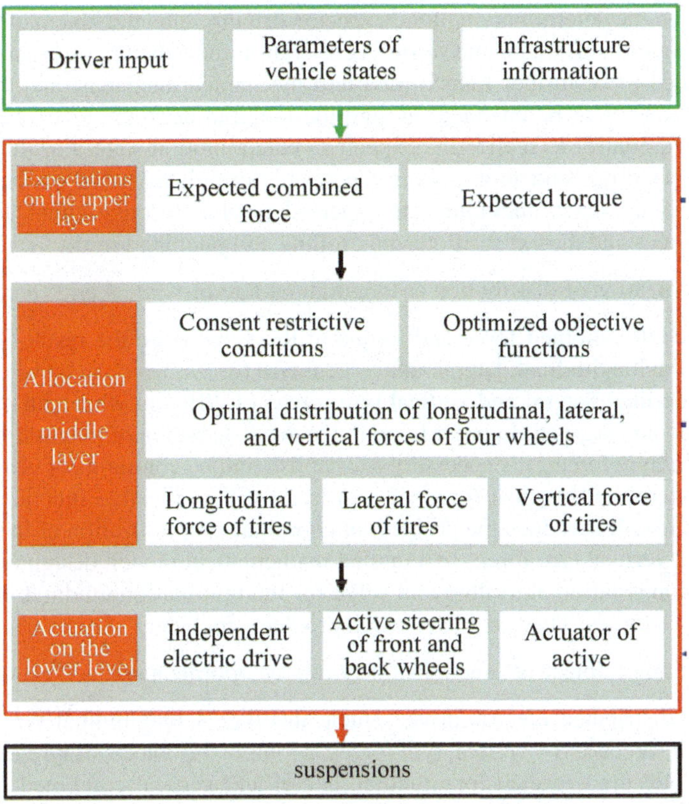

Fig. 5.38 Hierarchical system for the coordinated control of longitudinal-lateral-vertical tyre forces

driving system, front/rear wheel active steering system and active suspension system to optimize the allocated tyre forces. The specific research problems that need to be solved for the coordinated control of longitudinal, lateral and vertical tyre forces of distributed electric-drive vehicles using this hierarchical control system include:

(1) Formulation of vehicle control objectives

The ultimate goal of the system of the coordinated control of longitudinal-lateral-vertical tyre forces for distributed electric-drive vehicle is to implement the coordinated control of different subsystems and improve the vehicle dynamics performance. For the vehicle in motion, the expected longitudinal, lateral and vertical driving states of the vehicle should be clarified first, including the expected longitudinal acceleration to be achieved, steering based on the expected driving trajectory, vertical acceleration of smaller vehicle body and reasonable vehicle body yawing/rolling/pitching motion attitudes. For this purpose, the expected driving states of the vehicle, including longitudinal, lateral and vertical acceleration requirements as well as yawing, rolling and pitching motion requirements are determined first based on driver input, vehicle state parameters and road information.

Based on the determination of the expected driving state, the kinematic demands such as longitudinal, lateral and vertical acceleration and yawing, rolling and pitching motion need to be further transformed into mechanical demands, and the vehicle force state needs to be changed through the longitudinal-transverse-vertical force coordinated control to realize the expected vehicle motion state. For this purpose, the corresponding control objectives need to be formulated based on the expected vehicle driving states, including the expected combined longitudinal, lateral and vertical forces and the expected yawing, rolling and pitching torque.

(2) Optimization of distribution of longitudinal-lateral-vertical tyre forces

The expected combined force and torque of the vehicle are the result of the joint action of the longitudinal, lateral and vertical tyre forces, which must be distributed into longitudinal, lateral and vertical tyre forces of different wheels in order to be controlled and executed. In order to optimize the distribution of longitudinal, lateral and vertical tyre forces, it's necessary to establish unified contraints of vehicle longitudinal, lateral and vertical forces, design reasonable objective functions for tyre force optimization, reduce the problem of coordinated optimization of longitudinal, lateral and vertical tyre forces to a unified mathematical model, design an effective optimization solution algorithm and solve for the optimal longitudinal, lateral and vertical tyre forces. The specific research problems include:

① Selection of objective functions for tyre force optimization

Since the distributed electric-drive vehicle involves 4 longitudinal tyre forces, 4 lateral tyre forces and 4 vertical tyre forces, a total of 12 objects of tyre force distribution, while the expected longitudinal, lateral and vertical combined forces and yawing, rolling and pitching torque of the vehicle provide only 6 constraint equations for the tyre forces of different wheels, the tyre force variables therefore have sufficient degrees of freedom. Without any constraints on the tyre forces, any set of expected combined force or torque can be achieved by an endless number of tyre force distribution plans. However, different tyre force distribution plans correspond to different driving, steering and suspension system actuation states and thus correspond to different states of vehicle dynamics, from which a unique tyre force distribution plan must be selected to achieve optimal vehicle dynamics. Based on this consideration, in order to determine the optimal tyre force allocation plan, a reasonable objective function for tyre force optimization must be designed to effectively evaluate the effect of optimal tyre force allocation.

② Determination of tyre force constraints

The tyre force of different wheels is not merely a mathematical variable and its range of values is limited by the actual physical system. First, since the tyre force arises from relative motion between the tyre and the ground, the size of the tyre force of different wheels is limited by the maximum friction between the ground and the tyre, that is, the tyre force of different wheels should satisfy the constraints of tyre friction limits of different wheels. Second, the actuation of the tyre force is realized by the driving, steering and suspension systems and the range of values and dynamic rates of change of the tyre force are also limited by the actuation range of corresponding actuators, that is, the tyre force of different wheels should meet the constraints of the actuator system. In addition, the tyre force of different wheels should also satisfy the expected combined force and torque constraints pointed out in research problem (1). In conclusion, the optimal distribution of tyre force of different wheels must take into account the expected combined force and torque constraints, the constraints of tyre friction limits and the actuator system.

③ Optimization of problem-solving algorithm

On the basis of solving problems ① and ②, the tire force optimization objective function and the tire force constraint conditions constitute a complete tire force optimization distribution problem. Therefore, an effective algorithm must be designed to obtain the optimal solution of the optimization problem to enable the optimal distribution of tire forces.

In short, since the optimal distribution of longitudinal-lateral-vertical tire forces involves many optimization variables, many constraints and is highly non-linear, how to design a reasonable optimization objective function and effective optimization algorithm to achieve the purpose is the focus and difficulty of research.

(3) Precise actuator control

After the optimal distribution values of longitudinal, lateral and vertical forces of each wheel are obtained, the obtained values must be transformed into control objectives to accurately control the motor drive, active steering and active suspension systems to achieve the optimal distribution values of tyre forces.

References

1. Fan Y, Yi L (2005) Automobile system dynamics. China Machine Press, Beijing
2. Kawakami H, Sato H, Tabata M et al (1992) Development of Integrated system between active control suspension, active 4WS TRC and ABS. SAE Trans 101(6):326–333
3. Wallentowitz H (1990) Scope for the integration of powertrain and chassis control systems: traction control-all-wheel drive-active suspension. In: Proceedings of the international congress on transportation electronics, pp 439–453
4. Zhenfu C (2006) Status and trends of the vehicle chasses control technology. Automot Eng 28(2):105–113
5. Gordon T, Howell M, Brandao F (2003) Integrated control methodologies for road vehicles. Veh Syst Dyn 40(1–3):157–190
6. Fan Y, Daofei L (2008) Review on integrated vehicle dynamics control. Trans Chin Soc Agric Mach 39(6):1–7
7. Huang C, Chen L, Chao-Chuii Y (2013) Non-line modeling and control of semi-active suspensions with variable dampiiig. Veh Syst Dyn 51(10):1568–1587
8. Semmler SJ, Rieth PE (2006) Global chassis control-the networked chassis. In: SAE paper, 2006-01-1954
9. Van Zanten AT (2002) Evolution of electronic control systems for improving the vehicle dynamic behavior. In: The 6th international symposium on advanced vehicle control, Hiroshima, Japan, 2002

10. Shu-en Z (2010) Research of vehicle chassis integrated control based on multi-model intelligent hierarchical control. Chongqing University, Chongqing
11. Llewelyn AI (1989) Review of CAD/CAM. Comput Aided Des 21(5):297–302
12. Mantyla M (1990) A modeling system for top-down design of assembled products. IBM J Res Dev 34(5):636–659
13. Sutherland IE (1964) Sketch pad a man-machine graphical communication system. In: Proceedings of the SHARE design automation workshop. ACM, pp 6329–6346
14. Chen H (2014) Study of the cooperative chassis control system theory and technology based on the top level design. Jiangsu University, Zhenjiang

Chapter 6
Implementation of Intelligent Electric Vehicle Energy-Efficiency Control Based on Structure Sharing

Intelligent electric vehicles using the structure-sharing architecture have shared sensor information, including information issued by the forward radar, which can be used for both active safety control and energy-efficiency control of electric vehicles, which makes it possible to make full use of system information resources to improve the energy utilization of vehicles. This chapter digs into the intelligent energy-efficiency control methods for electric vehicles based on the preceding vehicle motion information obtained through sensor sharing: relying on the perception of the preceding vehicle motion information by radar and other sensors to conduct real-time assessment of vehicle collision risk and safety situation and dividing into four traffic scenarios based on the relative motion between the ego vehicle and the preceding vehicle, which corresponds to four working modes; by dynamically optimizing the torque of the vehicle drive motor in different modes and increasing the motor brake, the energy consumption of the electric vehicle is reduced on the condition that safety is guaranteed and driving intentions have been considered. In order to analyze and verify the energy-efficiency effect of the designed energy-efficiency controller, a simulation platform has been established for the development and debugging of the energy-efficiency controller and the evaluation of the strategy simulation according to the structure characteristics of the intelligent electric vehicle power system and the vehicle information flow and energy flow transfer characteristics, and the simulation and analysis of the energy-efficiency effect of the energy-efficiency controller under typical working conditions has been carried out. Finally, a real-life vehicle comparison experiment has been conducted on battery electric city buses.

K. Li et al., *The Intelligent Environment Friendly Vehicle*,
Key Technologies on New Energy Vehicles,
https://doi.org/10.1007/978-981-19-4851-0_6

6.1 Overall Architecture of Intelligent Energy-Efficiency Control System

6.1.1 Principles of Control System Design

Existing researches on energy-efficiency control of electric vehicles show that optimizing the vehicle driving speed and motor torque output at the same time can reduce the energy consumption of the vehicle [1]. The proposed Intelligent Energy-saving Control (IEC) system can further optimize the motor torque based on the relative motion between the two vehicles while ensuring the driving safety, and therefore has the potential to save energy further after the introduction of the preceding vehicle motion information. The IEC design is based on the following principles:

(1) Ensure safety while considering driving intentions

The conventional energy-efficiency driving controller is mostly realized by directly limiting the maximum motor output torque or by reducing the motor output torque in a fixed proportion; since the motion state of the preceding vehicle is not considered, the maximum output torque cannot be dynamically adjusted; this will easily cause the vehicle to be underpowered under some extreme working conditions such as sharp acceleration, which to a certain extent affects the realization of the driving intentions. IEC can ensure no collision with the preceding vehicle by considering the motion state of the preceding vehicle in the first place, and secondly, it also needs to identify and judge the driver's intentions according to the human–machine interface and pedal information to reduce the interference of the control system in the driving intentions.

(2) Coordination of dynamics and economy

IEC requires energy-efficient driving by dynamically optimizing drive motor torque and vehicle velocity while ensuring vehicle driving safety and taking into consideration the driving intentions. During the actual operation of electric vehicles, the power and economy of the vehicle are closely related to the drive motor torque output and the two are susceptible to certain conflicts. In order to achieve more energy saving, it's necessary to limit the maximum output torque and torque change rate of the motor, which will inevitably lead to a decrease in vehicle dynamics and the acceleration performance of the vehicle to be affected. Similarly, in order to ensure the vehicle dynamics, it's necessary to reduce the limitation to motor torque, which will result in low energy efficiency. For this reason, it's necessary to distinguish different situations, distinguish different energy-efficiency modes according to different driving scenarios and adopt different optimization strategies for the electric motor torque in different modes to achieve the coordination of dynamics and economy.

(3) Real-time optimization of motor torque

For the problem of multi-objective optimal control of vehicle longitudinal motion, the current mainstream approach is to apply the model predictive control theory and adopt rolling optimization for the comprehensive optimization of system safety, comfort and economy. However, this method is difficult to guarantee the real-timeness of the control due to its large computational volume. The rule-based torque optimization control method has the advantages of good real-timeness and practicality and is more widely used. However, the design of the rules relies on engineering experience and it's difficult to guarantee the optimality of the control. For this reason, it's necessary to combine the respective advantages of these two methods and design a practical torque optimization control strategy to achieve real-time optimization of motor torque.

6.1.2 Control System Architecture

Considering the above principles in IEC design, this chapter proposes the energy-efficiency control architecture of "hierarchical control and mode-by-mode optimization." The basic idea of energy efficiency is: to evaluate the driving safety situation based on the perception of the relative speed and distance of the ego vehicle to the preceding vehicle by radar and other sensors, and develop four different scenarios based on the relative motion between two vehicles that correspond to four energy-efficiency modes accordingly. The energy-efficiency modes are switched in real time according to the changes in driving scenarios and the motor driving torque is dynamically optimized in different modes to reduce energy consumption and increase brake regenerative energy, so as to realize energy-efficiency control. Based on the idea of hierarchical control, the IEC system architecture is designed, as shown in Fig. 6.1 [2]. The control system is divided into three levels: the scenario analysis level (top level), the mode decision and switching level (middle level) and the motor torque control level (bottom level) respectively.

(1) Scenario analysis level

The main task of the scenario analysis level is to identify the vehicle states and driving intentions and evaluate the safety situation according to the motion state of the preceding vehicle. It includes three functional modules: the driver intention identification module, the system state identification module and the safety situation evaluation module. The main function of the driver intention identification module is to identify driving intentions based on IEC switch, accelerator pedal opening, brake pedal opening, current gear, steering wheel angle and directional light signal etc.; the main function of the system state identification module is to determine whether the IEC function needs to be turned on based on information such as vehicle velocity, yaw rate, system fault state and relative distance and velocity between the ego vehicle and the preceding vehicle; the main function of the safety situation evaluation module is to analyze the scenarios based on the information on the relative motion between the ego vehicle and the preceding vehicle, determine the maximum working distance

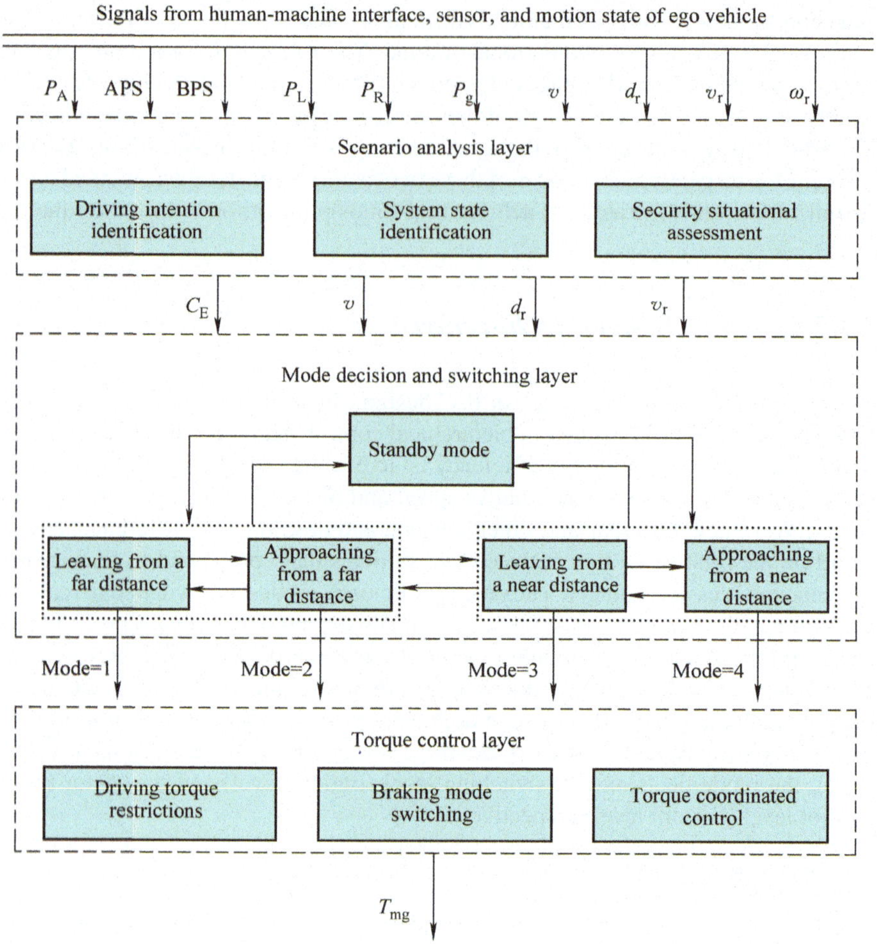

Fig. 6.1 Diagram of general architecture of IEC system

and safety distance of the energy-efficiency controller and determine the IEC "on" and "standby" state.

(2) Mode decision and switching level

The main task of the mode decision and switching level is to divide the energy-efficiency modes according to different scenarios and to carry out mode switching control. According to the relative distance and speed of the ego vehicle to the preceding vehicle acquired by the forward radar and other sensors, the driving scenarios are divided into four modes according to the risk of collision: leaving from a far distance, approaching from a far distance, leaving from a near distance and approaching from a near distance, which correspond to four energy efficiency modes that are switched based on the vehicle motion state and the relative motion between the ego vehicle and the preceding vehicle.

(3) Motor torque control level

The motor torque control level, also known as the lower control level, is the core of energy-efficiency control and includes three functional modules: the driving torque limiting module, the brake mode switching module and the torque coordinated control module, whose main task is to control the motor driving torque and brake torque in different modes in real time and ensure a smooth transition of the output torque during mode switching. The task of the driving torque limiting module is to optimize the motor output driving torque by proportional limiting and peak reduction; the task of the brake mode switching module is to implement motor brake in advance when the driver intends to slow down or when the risk of collision with the preceding vehicle increases; the task of the torque coordinated control module is to achieve seamless and smooth transition of the motor output torque by increasing the buffer and limiting the torque change rate when different energy-efficiency modes are switched, so as to ensure driving comfort and smooth vehicle velocity.

In Fig. 6.1, P_{A} represents the position of the IEC switch, APS represents the travel of the accelerator pedal, BPS represents the travel of the brake pedal, P_{g} represents the current gear of the transmission, P_{L} and P_{R} represent the position of the left-turn and right-turn light switches, v represents the vehicle velocity, v_{r} represents the relative vehicle velocity, d_{r} represents the relative inter-vehicle distance, ω_{r} represents the yaw rate, C_{E} represents the "on" state of IEC and T_{mg} represents the output torque of the drive motor.

6.2 Scenario Analysis-Based Energy-Efficiency Mode Decision and Switching

6.2.1 Driving Scenario Classification Based on Safety Situation Assessment

Driving safety situation assessment is both the primary prerequisite for ensuring driving safety and the main basis for scenario classification. The main indicators to assess the safety and collision risk of car-following motion are include Time to Collision (TTC), Time Headway (THW), the driver's expected early warning time and deceleration etc. The vehicle safe distance models established using the corresponding indicators are mainly primary models based on the safe time distance (TTC or THW) and secondary models based on the expected deceleration.

The safe distance model based on the TTC is as follows:

$$d_{\mathrm{s1}} = \mathrm{TTC} \cdot v_{\mathrm{r}} \tag{6.1}$$

Considering the driver's expected early warning time and brake deceleration, the safe distance model is established as follows:

$$d_{s2} = t_N v_r + \frac{v_r^2}{2a_0} \tag{6.2}$$

where t_N is the proximity collision early warning time (or the driver's expected reaction time), v_r is the relative vehicle velocity, a_0 is the set deceleration, and $v_r^2/2a_0$ is the minimum brake distance required to avoid a collision.

Meanwhile, considering the emergency brake of the preceding vehicle, the safe distance between the two vehicles is modeled as follows:

$$d_{s3} = \frac{v^2}{2a_N} \tag{6.3}$$

where v is the vehicle velocity and a_N is the expected deceleration from a near distance based on driving habits.

In order to take into account the driving safety at different relative vehicle velocities and at different vehicle velocities, the larger of the above three models has been selected as the safe distance d_s for the scenario to be discussed in this chapter.

$$d_s = \max(d_{s1}, d_{s2}, d_{s3}) \tag{6.4}$$

The maximum working distance d_m of IEC is determined in the similar way. The maximum working distance is the distance at which the IEC works. Firstly, the distance should be within the detection range of radar and other sensors; secondly, different working distances can be obtained by adjusting the driver's expected early warning time and expected deceleration according to the driver's habits.

By taking into account the forward radar detection range, the safe time distance between the two vehicles and driving expectation, the maximum working distance of the IEC system is modeled as follows:

$$\begin{cases} d_m = \max\left(\mathrm{TTC} \cdot v_r, t_F \cdot v_r + \dfrac{v_r^2}{2a_0}, \dfrac{v^2}{2a_F}\right) \\ d_m \le r \end{cases} \tag{6.5}$$

where t_F is the expected early warning time from a far distance based on driving habit, a_F is the expected deceleration from a far distance based on driving habit, and r is the maximum detection range of the forward radar.

The safe distance d_s and maximum working distance d_m obtained according to the above formula help us to develop four scenarios based on the relative motion between the two vehicles as well as the levels of the risk of collision, as shown in Fig. 6.2.

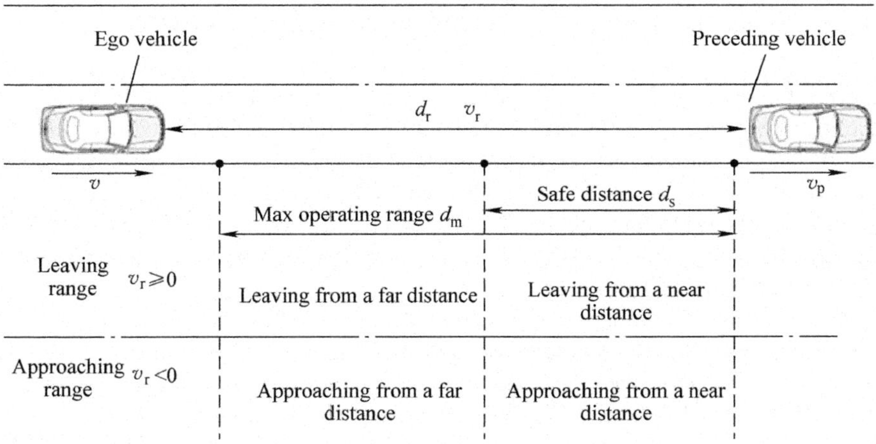

Fig. 6.2 Division of driving scenarios based on safety situation assessment

(1) "Leaving from a far distance" scenario: the relative distance between the two vehicles is greater than the safe distance and the velocity of the preceding vehicle is greater than the ego vehicle's velocity. In this scenario, the ego vehicle drives away from the preceding vehicle beyond range of the safe distance. Unless the preceding vehicle suddenly brakes emergently, there is basically no risk of collision between the two vehicles.
(2) "Approaching from a far distance" scenario: the relative distance between the two vehicles is greater than the safe distance and the velocity of the preceding vehicle is less than the ego vehicle's velocity. In this scenario, the ego vehicle approaches the preceding vehicle beyond the safe distance and the two vehicles have lower risk of collision because the relative distance is greater than the safe distance.
(3) "Leaving from a near distance" scenario: the relative distance between the two vehicles is less than the safe distance and the velocity of the preceding vehicle is greater than the ego vehicle's velocity. In this scenario, the ego vehicle drives away from the preceding vehicle within the safe distance and the two vehicles have the risk of collision if the preceding vehicle brakes emergently, as the relative distance is less than the safe distance, so the risk of collision increases.
(4) "Approaching from a near distance" scenario: The relative distance between the two vehicles is less than the safe distance and the velocity of the preceding vehicle is less than the ego vehicle's velocity. In this scenario, the vehicle approaches the preceding vehicle within the safe distance and the two vehicle have the greatest risk of collision because the relative distance is small and is shrinking.

6.2.2 Scenario Change-Based Mode Switching Control

During the driving process, the scenario changes dynamically with the change in the velocity of the ego vehicle and the relative motion between the ego vehicle and the preceding vehicle. Based on the division of scenarios, four energy-efficiency modes have been developed based on the size of the risk of collision between the ego vehicle and the preceding vehicle in four scenarios. The correspondence between the energy-efficiency modes and the relative motion scenarios after the system is turned on (=1), as shown in Table 6.1.

Under real-life complex traffic conditions, the motion state of the preceding vehicle is dynamic and time-varying and there are cut-in and cut-out situations, which will bring about frequent changes in the scenarios based on the relative motion between the vehicle the preceding vehicle and easily cause frequent switching of the IEC energy-efficiency modes. Too frequent switching of the energy-efficiency modes will not only affect the stability of the control system, but also affect the smoothness of the vehicle and the driver's feeling. For this reason, it's necessary to design a switching buffer in the mode switching process to reduce the frequent switching of modes.

The mode switching buffer includes a distance buffer and a speed buffer. The distance buffer $[d_1\ d_2]$ is increased when switching between the "far distance" modes (mode 1 and mode 2) and the "near distance" modes (energy-efficiency mode 3 and energy-efficiency mode 4), that is, switching from the "near distance" mode to the "far distance" mode when $d_r > d_s + d_1$; when $d_r < d_s - d_2$, we need to switch from the "far distance" mode to the "near distance" mode. When switching between "far distance" modes (mode 1 and mode 3) and "near distance" modes (mode 2 and mode 4), we need to increase the speed buffer $[v_1\ v_2]$, that is, switching from the leaving mode to the approaching mode when $v_r > v_1$ and switching from the leaving mode to the approaching mode when $v_r < v_2$. The buffer parameters d_1, d_2, v_1 v_2 can be obtained by calibration and debugging.

Based on the above analysis, the logical relationship of mode switching has been developed, as shown in Fig. 6.3.

Table 6.1 Correspondence between energy-efficiency modes and scenarios

Energy-efficiency mode	Relative motion scenario	Differentiation condition
Mode 1	Leaving from a far distance	$d_m \geq d_r > d_s$ and $v_r > 0$
Mode 2	Approaching from a far distance	$d_m \geq d_r > d_s$ and $v_r \leq 0$
Mode 3	Leaving from a near distance	$0 < d_r \leq d_s$ and $v_r > 0$
Mode 4	Approaching from a near distance	$0 < d_r \leq d_s$ and $v_r \leq 0$
The standby mode	–	$d_r > d_m$ or $v = 0$

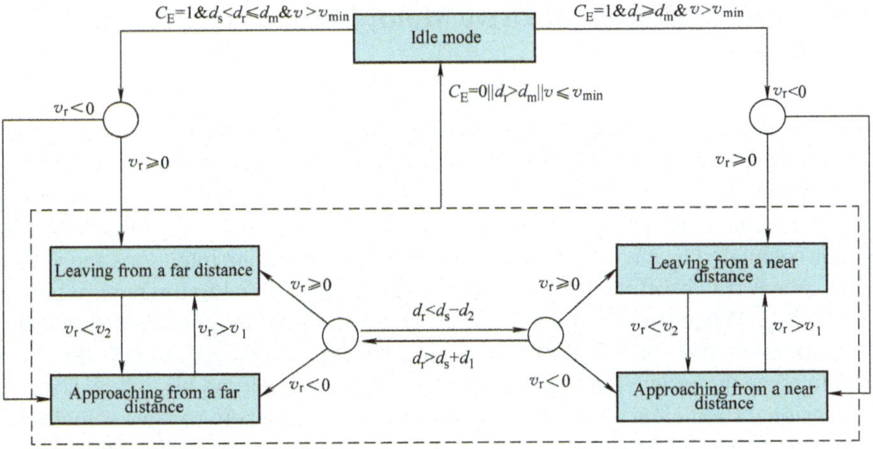

Fig. 6.3 Diagram of mode switching logic

6.2.3 *Driving Intentions-Based System Turn-On Control*

The overriding principle in the design of the IEC system is to meet the driver's intentions and ensure the driving safety of the vehicle. Therefore, whether IEC is on or off depends on the driver's intentions and the state of the vehicle in motion.

Driver's intentions are mainly reflected in the human–machine interface and driving operation, including the position of IEC switch, the position of directional light switches, transmission gear, travel of the brake pedal, travel of the accelerator pedal and steering wheel angle. To meet driving intentions, the IEC is turned off when its switch is off, a non-forward gear is selected and the travel of the accelerator pedal is greater than a set threshold. The vehicle motion state includes the vehicle velocity, yaw rate and system fault state. To ensure driving safety and not interfere with driver operation, the IEC is turned off when the vehicle is stopped, making a turn or changing lanes (directional light switches are turned on or yaw rate is greater than the threshold value) or when a system fault occurs.

Based on the above analysis, the control strategy for IEC system turn-on and turn-off has been established:

$$C_E = \begin{cases} 1, & (P_A = 1 \ \& \ \alpha < \alpha_t \ \& \ P_L = 0 \ \& \ P_R = 0 \ \& \ P_g > 1 \ \& \ \omega_r < \omega_0 \ \& \ d_r > 0) \\ 0, & others \end{cases} \tag{6.6}$$

where α_t and ω_0 are the set accelerator pedal opening and yaw rate thresholds respectively.

6.3 Optimal Control of Drive Motor Torque in Different Modes

The torque optimization control strategy is the core of IEC. How to reasonably optimize the motor torque in different modes while ensuring safety and taking into account the driving intentions is the focus of research in this section. The torque optimization strategy proposed in this section is shown in Fig. 6.4. T_{req} is the demanded torque calculated by the vehicle controller based on the driving intentions and vehicle state. When IEC is off, the motor output torque T_{cmd} is equal to the required motor torque T_{req}. When IEC is on, the required motor torque is output to the motor after the correction and optimization by IEC. Coefficient K is the key to IEC design. K is defined as the torque optimization coefficient in different modes, which is a two-dimensional coefficient table related to the relative distance between two vehicles and ranges from 0 to 1. It mainly includes the following coefficients: driving torque optimization coefficient K_{FD} of the "far distance" scenario, driving torque optimization coefficient K_{ND} of the "near distance" scenario, brake torque optimization coefficient K_{AB} of the "approaching" scenario, brake torque optimization coefficient K_{LB} of the "leaving" scenario, coefficient K_{TC} of the "approaching" scenario and safe time distance coefficient K_{TS} of the "leaving" scenario etc.

In order to design the above optimization coefficients, this section uses the advantages of Model Predictive Control (MPC) method in multi-objective optimization control to calculate, extract and fit the optimization coefficient table in different modes. In the application of MPC in the car-following motion control, because the

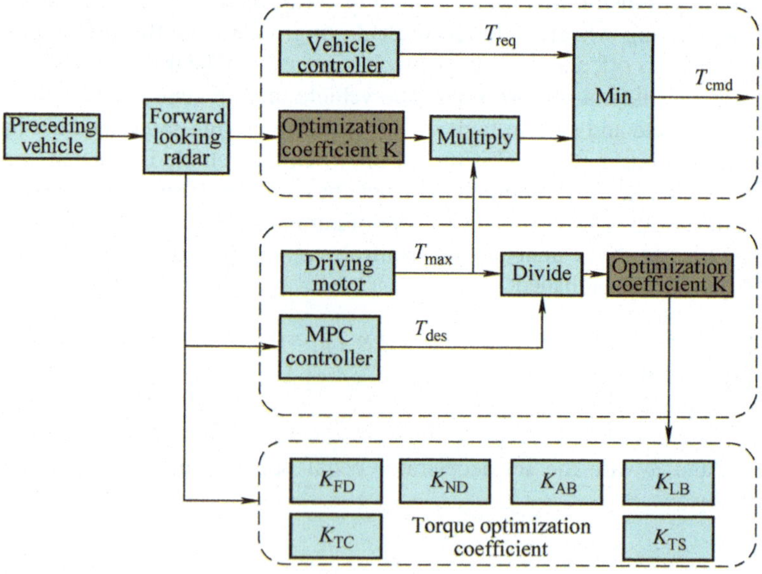

Fig. 6.4 Diagram of the drive motor torque optimization control strategy

design adds the vehicle acceleration as one of the optimization objectives while considering the car-following safety while, a better balance between safety and energy efficiency is achieved. However, the real-time performance of MPC control is poor due to the large amount of calculation, but its control output under the same working condition can be used as a reference for the design of the torque optimization coefficient table. The correspondence between the motion state of the two vehicles and the motor torque is calculated by MPC under typical following conditions, and a parameter table is obtained. The correspondence between the relative motion of the vehicles and K under different modes is analyzed and the torque optimization coefficient table in four modes is extracted. The method of online table search is used to optimize the motor output torque of the vehicle in different energy-efficiency modes, thus ensuring the real-time control.

6.3.1 Rules for Drive Motor Torque Optimization in Different Modes

The motor torque control strategy is a strategy to optimize the motor driving torque and brake torque for each energy-efficiency mode to reduce energy consumption. In order to ensure driving safety and to take into account driving intentions, a "mode-based optimization" strategy is used with different torque optimization rules for different energy-efficiency modes.

(1) "Leaving from a far distance" mode (energy-efficiency mode 1)

In the "leaving from a far distance" mode, the driving is relatively safe and the risk of collision is small. In order to avoid unnecessary high acceleration, the maximum driving torque of the motor is partially limited and the motor torque is optimized by using the "far distance" scenario's driving torque optimization coefficient K_{FD}; meanwhile, according to the deceleration intention of the driver and the safety situation, the electric motor brake torque is optimized by using the brake torque optimization coefficient K_{LB}, which appropriately increases the motor brake. The motor torque optimization rules are:

$$T_{m\_1} = \begin{cases} \min\left(T_{max} K_{FD}, T_{req}\right), & T_{req} > 0 \\ \left(T_{min} - T_{req}\right) K_{LB} + T_{req}, & T_{req} \leq 0 \end{cases} \tag{6.7}$$

where T_{m_1} is the motor torque command in IEC energy-efficiency mode 1 and T_{max} and T_{min} are the maximum driving torque and maximum brake torque of the motor respectively, depending on the motor rotational speed and the battery pack charge etc.

(2) "Approaching from a far distance" mode (energy-efficiency mode 2)

In the "approaching from a far distance" mode, the vehicle is approaching the preceding vehicle and the risk of collision between the two vehicles is increased.

In addition to optimizing the output torque of the motor through the "far distance" scenario's driving torque optimization coefficient K_{FD}, the rate of change of the output torque needs to be limited to avoid rapid acceleration to reduce energy consumption. At the same time, when the originally demanded torque is less than 0, the motor brake torque is optimized by the brake torque optimization coefficient K_{AB} of the "approaching" scenario to increase the motor brake according to the driver's deceleration intention and safety situation. The motor torque optimization rules are:

$$T_{m\_2} = \begin{cases} \min(T_{max} \cdot K_{FD} \cdot K_{TC},\ T_{req}) \& \frac{d_T}{d_t} \le a_2, & T_{req} > 0 \\ (T_{min} - T_{req})K_{AB} + T_{req}, & T_{req} \le 0 \end{cases} \quad (6.8)$$

where T_{m_2} is the motor torque command in IEC energy-efficiency mode 2, K_{TC} is the safe time distance coefficient defined to quantitatively evaluate the safety situation of the "approaching" scenario and a_2 is the torque change rate limit.

(3) Leaving from a near distance mode (energy-efficiency mode 3)

In the "leaving from a near distance" mode, the relative distance between two vehicles is less than the safe distance and the risk of collision between two vehicles is further increased. The "near distance" scenario's driving torque optimization coefficient K_{ND} is used for optimization; the change rate of motor driving torque is also limited. Meanwhile, when the originally demanded torque is less than 0, based on the driver's deceleration intention, part of the motor's brake force is actively applied and the motor brake torque is optimized using the brake torque optimization coefficient K_{LB} to achieve more regenerative brake energy. The motor torque optimization rules are:

$$T_{m\_3} = \begin{cases} \min(T_{max} \cdot K_{ND} \cdot K_{TS}, T_{req}) \& \frac{d_T}{d_t} \le a_2, & T_{req} > 0 \\ (T_{min} - T_{req})K_{LB} + T_{req}, & T_{req} \le 0 \end{cases} \quad (6.9)$$

where T_{m_3} is the motor torque command in IEC energy-efficiency mode 3, K_{TS} is the safe time distance coefficient defined in order to quantitatively assess the safety situation of the "leaving" scenario and is a fitting function with TTS (Time to Safe) as the independent variable, where TTS indicates the time it takes for the vehicle to maintain its current relative speed until to a safe distance in the "leaving from a near distance" scenario, i.e. $\text{TTS} = (d_s - d_r)/v_r$.

(4) "Approaching from a near distance" mode (energy-efficiency mode)

In the "approaching from a near distance" mode, the relative distance between the two vehicles is less than the safe distance and as the two vehicles are approaching each other, the risk of collision is the greatest. In this case, to ensure driving safety, we need to directly cut off the motor drive and enter motor brake mode. When the originally demanded torque is greater than 0, it indicates that the driver has not yet had time to brake or the electric motor brake has not yet taken effect. Therefore, we should enter the motor brake in advance and use the "approaching" scenario's safe time distance coefficient K_{TC} to optimize the motor brake torque to increase the motor brake. When the originally demanded torque is less than 0, the motor brake torque is optimized using the brake torque optimization coefficient K_{AB} based on the safety situation and the driver's deceleration intention and the motor brake torque is actively increased to improve the energy regeneration rate. The motor torque optimization rules are:

$$T_{m\_4} = \begin{cases} T_{min} - T_{min} K_{TC}, & T_{req} > 0 \\ (T_{min} - T_{req}) K_{AB} + T_{req}, & T_{req} \leq 0 \end{cases} \tag{6.10}$$

where T_{m_4} is the motor torque command in IEC energy-efficiency mode 4, and the safe time distance coefficient K_{TC} of the "approaching" scenario is a fitting function with the safe time distance (TTC) as the independent variable, reflecting different degrees of increased brake energy regeneration in different safety situations.

(5) The standby mode

In the standby mode, the system does not limit or optimize the driving or brake torque of the vehicle, while only outputting the originally demanded torque calculated by the vehicle controller:

$$T_{m\_0} = T_{req} \tag{6.11}$$

where T_{m_0} is the motor torque command of the IEC system in the standby mode and T_{req} is the originally demanded torque of the motor calculated by the vehicle control strategy.

6.3.2 MPC-Based Longitudinal Car-Following Motion Control Algorithm

After the development of the motor torque optimization rules for each mode is completed, we need to design the series of optimization coefficients included in the rules. The design of torque optimization coefficients is the core of the realization of energy efficiency and the key to balancing the multiple objectives of safety and energy efficiency. To this end, a longitudinal car-following motion control system based on model predictive control has been designed to input the typical motion conditions of the preceding vehicle and output the controlled motor torque that is based on car-following safety and energy efficiency, which provides the data basis for the design of motor torque optimization coefficients.

6.3.2.1 Modeling of Longitudinal Car-Following System Dynamics

Suppose that only the longitudinal driving conditions of the vehicle are considered and steering, lane changing and overtaking etc. are ignored, the steering system is not considered and the yaw motion is ignored, and a simplified vehicle longitudinal dynamics model has been established:

$$\begin{cases} F_t = \dfrac{T_{mg} i_g}{r} - (F_r + F_a + F_{aero}) \\ F_r = mgf_r \\ F_a = \delta ma \\ F_{aero} = C_d A v^2 / 21.15 \end{cases} \tag{6.12}$$

where F_t is the vehicle's longitudinal driving force, T_{mg} is the drive motor torque, F_r is the rolling resistance, F_a is the acceleration resistance, F_{aero} is the air resistance, m is the vehicle mass, f_r is the rolling resistance coefficient, δ is the rotational mass conversion coefficient, a is the vehicle straight driving acceleration, C_d is the air resistance coefficient, A is the windward area, i_g is the gear ratio of the main reducer and v is the vehicle velocity.

By using the frequency response method to identify the longitudinal dynamics of the vehicle, the following transfer function is obtained [3]:

$$a = \frac{K_g}{T_g s + 1} a_{des} \tag{6.13}$$

where a is the actual acceleration of the vehicle, a_{des} is the expected acceleration of the vehicle, K_g is the system gain; T_g is the system delay. The system delay is relatively small compared to the conventional vehicle.

The longitudinal dynamics model of the car-following system couples the vehicle longitudinal dynamics and the inter-vehicle kinematics. A unified longitudinal dynamics model of the car-following system can be established by integrating the inter-vehicle kinematics model with the vehicle longitudinal dynamics model.

The kinematic relationship between the ego vehicle and the preceding vehicle is:

$$\begin{cases} d_r = s_p - s_f \\ v_r = v_p - v \end{cases} \tag{6.14}$$

where d_r is the relative distance and v_r is the relative speed between the two vehicles, s_p is the displacement and v_p is the velocity of the preceding vehicle, s_f is the displacement and v is the velocity of the ego vehicle.

Combining Eqs. (6.13) and (6.15), we get:

$$\begin{cases} \dot{d}_r = v_r + a_p T - aT \\ \dot{v}_r = a_p - a \\ \dot{a} = K_g u / T_g - a / T_g \end{cases} \tag{6.15}$$

where a_p is the acceleration of the preceding vehicle and T is the sampling step size.

By taking the system state variable $\mathbf{x} = [d_r, v_r, a]^T$, the control variable $u = a_{des}$ and the longitudinal car-following dynamics is modeled as:

$$\dot{x} = A_0 x + B_0 u + G_0 \eta \tag{6.16}$$

where $\mathbf{A}_0 = \begin{bmatrix} 0 & 1 & -T \\ 0 & 0 & -1 \\ 0 & 0 & -1/T_g \end{bmatrix}$, $\mathbf{B}_0 = \begin{bmatrix} 0 \\ 0 \\ K_g/T_g \end{bmatrix}$ and $\mathbf{G}_0 = \begin{bmatrix} 1 \\ 1 \\ 0 \end{bmatrix}$ are the coefficient matrices; $\eta = a_p$ is the input disturbance.

Furthermore, by discretizing Eq. (6.16), the discrete system state space equation is obtained as:

$$x(k+1) = Ax(k) + Bu(k) + G\eta(k) \tag{6.17}$$

where $\mathbf{A} = \mathbf{I} + T \cdot \mathbf{A}_0, \mathbf{B} = T \cdot \mathbf{B}_0$ and $\mathbf{G} = T \cdot \mathbf{G}_0$ are the coefficient matrices of the discrete state equations and $\eta(k)$ is the input disturbance at the current moment.

If the output disturbance is ignored, then the system output is:

$$y(k) = C \cdot x(k) \tag{6.18}$$

where $\mathbf{C}$ is the output coefficient matrix.

6.3.2.2 Cost Function and Constraint Design for Performance Indicators

The goals of the design of vehicle longitudinal car-following control algorithm are safety, energy efficiency and comfort. During the actual car-following process, the driver only responds to a large tracking error, so a 2-norm is used to quantify each indicator [4]. The two vehicles' relative distance errors and speed errors are used as the safety indicator. If we set w_d and w_v as the weight coefficients of distance errors and speed errors respectively, then the safety indicators are:

$$L_t = w_d(d_r - d_{des})^2 + w_v v_r^2 \tag{6.19}$$

where d_{des} is the expected safe distance, $d_{des} = \tau \cdot v + d_0$, τ is the safe time distance, d_0 is the safe distance during parking.

Related studies have shown that the instantaneous energy consumption of electric vehicles increases as acceleration increases [5], so the acceleration is used to quantify the energy-efficiency indicator of electric vehicles. Since the actual acceleration of the vehicle with the ACC system depends on the expected acceleration, the energy-efficiency indicator is designed as:

$$L_e = w_u u^2 \tag{6.20}$$

The comfort indicator relies on the driver's feelings and usually needs to be in line with the driver's expected car-following distance and makes sure the longitudinal acceleration is within the driver's acceptable range. Therefore, a method of designing constraints on acceleration and state variables is used:

$$\begin{cases} u_{\min} \le u \le u_{\max} \\ \mathrm{x}_{\min} \le \mathrm{x} \le \mathrm{x}_{\max} \end{cases} \tag{6.21}$$

where $\mathbf{x_{min}} = \begin{bmatrix} d_{\mathrm{rmin}} \\ v_{\mathrm{r\,min}} \\ a_{\min} \end{bmatrix}$ and $\mathbf{x_{max}} = \begin{bmatrix} d_{\mathrm{r\,max}} \\ v_{\mathrm{r\,max}} \\ a_{\max} \end{bmatrix}$ are the minimum and maximum values of different parameters of the state variables respectively.

The comprehensive indicators are linearly weighted and summed to transform the multi-objective problem into a single-objective problem that involves different indicators. Based on the summation of different quantitative indicators, the following cost function is obtained:

$$L = \mathrm{x}^{\mathrm{T}} \cdot w_{\mathrm{x2}} \cdot \mathrm{x} + w_{\mathrm{x1}} \cdot \mathrm{x} + w_{\mathrm{u}} \cdot u^2 + c \tag{6.22}$$

where w_{x2} is the weighting factor of the quadratic term, w_{x1} is the weighting factor of the one-order term and c is a constant.

6.3.2.3 Optimal Solution of the System Prediction Problem

First, the prediction model of the car-following motion system is established. Assuming that the current moment is k and i is the predictive time increment, then we have:

$$\mathrm{x}(k+i) = A\mathrm{x}(k+i-1) + Bu(k+i-1) + G\eta(k+i-1) \tag{6.23}$$

Without loss of generality, we assume that the length of the predictive time domain is P, then:

$$\mathrm{X} = A_P\mathrm{x}(k) + B_P U + G_P H \tag{6.24}$$

where,

$$\mathbf{A}_P = \begin{bmatrix} \mathbf{A} \\ \mathbf{A}^2 \\ \vdots \\ \mathbf{A}^P \end{bmatrix}, \quad \mathbf{B_P} = \begin{bmatrix} \mathbf{A}^0\mathbf{B} & 0 & \cdots & 0 \\ \mathbf{A}^1\mathbf{B} & \mathbf{A}^0\mathbf{B} & \cdots & 0 \\ \vdots & \vdots & \vdots & \vdots \\ \mathbf{A}^{P-1}\mathbf{B} & \mathbf{A}^{P-2}\mathbf{B} & \cdots & \mathbf{A}^0\mathbf{B} \end{bmatrix}, \quad \mathbf{U} = \begin{bmatrix} u(k|k) \\ u(k+1|k) \\ \vdots \\ u(k+P-1|k) \end{bmatrix},$$

$$\mathbf{G}_P = \begin{bmatrix} \mathbf{A}^0\mathbf{G} & 0 & \cdots & 0 \\ \mathbf{A}^1\mathbf{G} & \mathbf{A}^0\mathbf{G} & \cdots & 0 \\ \vdots & \vdots & \vdots & \vdots \\ \mathbf{A}^{P-1}\mathbf{G} & \mathbf{A}^{P-2}\mathbf{G} & \cdots & \mathbf{A}^0\mathbf{G} \end{bmatrix}, \quad \mathbf{H} = \begin{bmatrix} \eta(k|k) \\ \eta(k+1|k) \\ \vdots \\ \eta(k+P-1|k) \end{bmatrix}$$

Second, by linearly weighting all cost functions in the predictive time domain and transforming it into a predictive type, we get:

$$J = X^{\mathrm{T}} W_{X2} X + W_{X1} X + U^{\mathrm{T}} R U + C \tag{6.25}$$

where,

$$\mathbf{W}_{X2} = \begin{bmatrix} w_{x2}(k+1|k) & & 0 \\ & \ddots & \\ 0 & & w_{x2}(k+\mathrm{P}|k) \end{bmatrix}, \quad \mathbf{R} = \begin{bmatrix} w_{\mathrm{u}}(k+1|k) & & 0 \\ & \ddots & \\ 0 & & w_{\mathrm{u}}(k+\mathrm{P}|k) \end{bmatrix},$$

$$\mathbf{W}_{X1} = \begin{bmatrix} w_{x1}(k+1|k) & \cdots & w_{x1}(k+\mathrm{P}|k) \end{bmatrix}, \quad \mathbf{C} = \sum_{i=1}^{P} c(k+\mathrm{i}|k)$$

Again, by extending all constraints to the entire predictive time domain and transforming it into a predictive type, then we get:

$$\begin{cases} u_{\min} \le u(k+i/k) \le u_{\max} \\ x_{\min} \le x(k+i/k) \le x_{\max} \end{cases}, \quad i = 0 : P-1 \tag{6.26}$$

Finally, Eq. (9.24) is substituted into Eq. (9.25) to establish the quadratic programming type of the prediction optimization problem, as shown in Eq. (6.27) and the effective set method is chosen to solve it.

$$J = \frac{1}{2} U^{\mathrm{T}} \Phi U + f^{\mathrm{T}} U + C_{\mathrm{final}} \tag{6.27}$$

where $\mathbf{\Phi} = 2\left(\mathbf{B}_p^{\mathrm{T}} \cdot \mathbf{W}_{x2} \cdot \mathbf{B}_P + \mathbf{R}\right)$, $\mathbf{f} = 2\left(\mathbf{x}(k)^{\mathrm{T}} \cdot \mathbf{A}_{\mathbf{P}}^{\mathrm{T}} \cdot \mathbf{W}_{X2} \cdot \mathbf{B}_P + \frac{1}{2}\mathbf{W}_{X1} \cdot \mathbf{B}_P\right)^{\mathrm{T}}$ and C_{final} are constants.

6.3.3 Extraction and Fitting of the Motor Torque Optimization Coefficient Table

Based on the model predictive control algorithm, the Simulink model of the controller is established and integrated in the platform for simulation of the electric vehicle following motion. Based on the relative motion between the ego vehicle and the

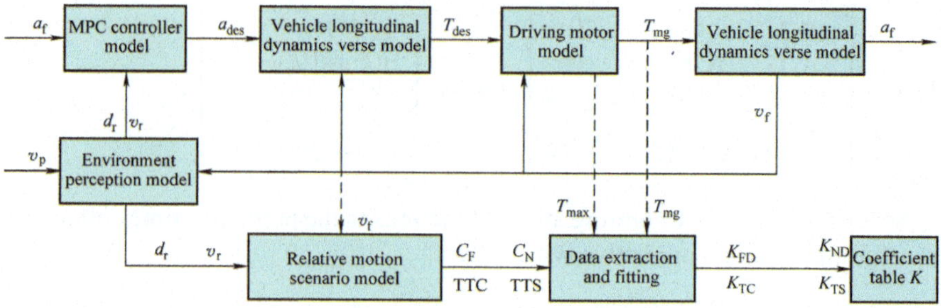

Fig. 6.5 Extraction of motor torque optimization coefficients for electric vehicle following motion based on MPC

preceding vehicle, the relative motion scenario model concerning the two vehicles is established. We then input the typical motion conditions of the preceding vehicle and output the controlled motor torque that is based on the car-following safety and energy efficiency. We then calculate the motor torque by using MPC under typical car-following conditions and the maximum torque that the motor can provide at this time, analyze the correspondence between the relative motion between the two vehicles and the motor torque optimization coefficient in different modes and extract and fit the torque optimization coefficient table for each mode, as shown in Fig. 6.5 [6].

In Fig. 6.5, to quantify the proximity of the vehicle to the preceding vehicle during the car-following motion, the "far distance" scenario's proximity indicator C_F and the "near distance" scenario's proximity indicator C_N are defined respectively, where:

$$C_F = \frac{d_r - d_s}{d_m - d_s} \tag{6.28}$$

$$C_N = \frac{d_r}{d_s} \tag{6.29}$$

The designed driving torque optimization coefficients K_{FD} and K_{ND} are functions that are close to exponential functions, with the following correspondence relationships:

$$K_{FD} = f_1(C_F) \tag{6.30}$$

$$K_{ND} = f_2(C_N) \tag{6.31}$$

when $C_N = 1$, that is, the relative distance between the two vehicles is equal to the safe distance, there's need for mode switching at this time. Let $K_{ND} = K_{FD}$ to guarantee a seamless and smooth transition of torque during the mode switching process.

The designed safe time distance coefficients K_{TC} and K_{TS} as functions of TTC and TTS have the following correspondence relationships:

$$K_{TC} = f_3(TTC) \tag{6.32}$$

$$K_{TS} = f_4(TTS) \tag{6.33}$$

In addition, in order to take into account the driver's intentions, the brake torque optimization coefficients are designed to be piecewise linear functions of the brake pedal opening BPS respectively, which reflects the increase in motor brake with the increase in brake demand, where $K_{AB} > K_{LB}$, which indicates that the brake torque optimization coefficient of the two vehicles that are approaching each other should be greater than the brake torque coefficient of the two vehicles are moving away from each other.

We then run the Simulink model and output the demanded motor torque T_{des}, maximum driving torque T_{max} and maximum brake torque T_{min} that are optimized offline by the MPC controller and then fit them according to the following equation:

$$f_1(x) = \begin{cases} y = \dfrac{T_{des}}{T_{max}} \\ x = \dfrac{d_r - d_s}{d_m - d_s} \end{cases} \tag{6.34}$$

$$f_2(x) = \begin{cases} y = \dfrac{T_{des}}{T_{max}} \\ x = \dfrac{d_r}{d_s} \end{cases} \tag{6.35}$$

$$f_3(x) = \begin{cases} y = \dfrac{T_{des}}{T_{max}, f_1(x)} \\ x = \dfrac{d_r}{v_r} \end{cases} \tag{6.36}$$

$$f_4(x) = \begin{cases} y = \frac{T_{des}}{T_{max} f_2(x)} \\ x = \frac{d_s - d_r}{v_r} \end{cases} \tag{6.37}$$

For better visual representation and experimental debugging, the approach of fitting a piecewise linear function is adopted in this section and the fitted coefficients are shown in Fig. 6.6.

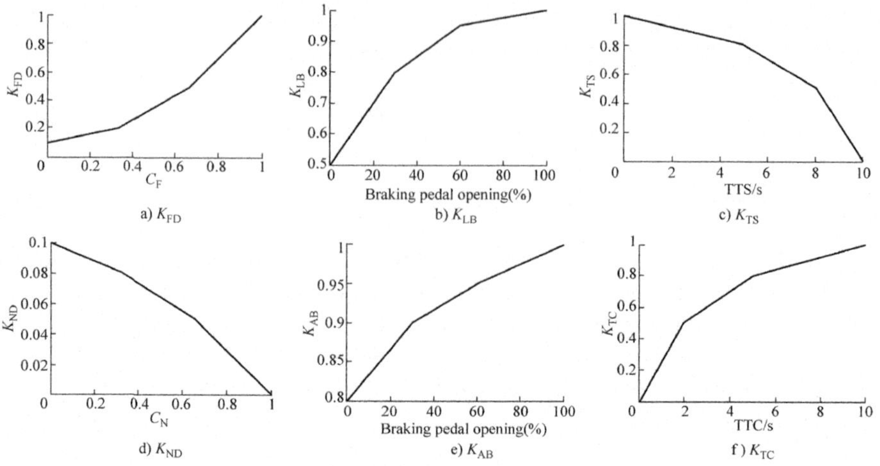

Fig. 6.6 Fitted motor torque optimization coefficients

6.4 Simulation Analysis of EV Intelligent Energy-Efficiency Control

Based on the designed control algorithm, the IEC forward simulation platform for development, debugging of the energy-efficiency controller and strategy simulation evaluation is established. Based on this simulation platform, the typical extreme working conditions of China Urban Driving Condition (CUDC) and sharp acceleration-sharp deceleration of the preceding vehicle are used to follow the preceding vehicle according to the designed longitudinal following model, so as to compare and analyze the energy-efficiency effects before and after the IEC is turned on.

6.4.1 Simulation Platform Design

The IEC simulation platform is mainly used for development, debugging and simulation analysis of longitudinal driver-assistant system control strategy for electric vehicles and should meet the following basic requirements:

(1) The basic structure of the simulation platform should be consistent with the real-life vehicles, so that the simulation model can better reflect the real-life vehicle's longitudinal dynamics characteristics;
(2) The simulation platform consists of different component modules, which allows for the addition or removal of system components as needed and can also be quickly transferred to other types of vehicle simulation models;

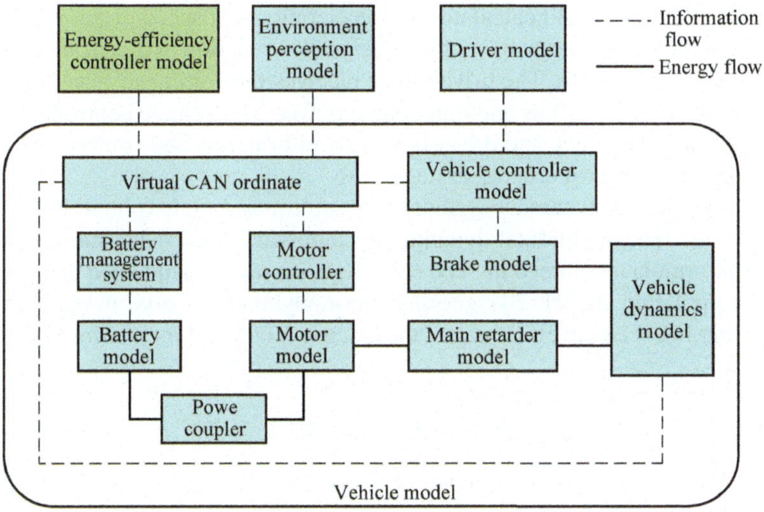

Fig. 6.7 Structure of forward simulation platform for EV intelligent energy-efficiency control system

(3) The main structure, information flow and energy flow transfer of the simulation platform are consistent with the real-life vehicles, which can be used for rapid development, performance debugging and strategy optimization of the control system.

Based on the structural characteristics of information sharing among environmental perception systems in intelligent electric vehicles, the IEC hierarchical modular forward simulation platform is designed with components of the real-life vehicle as unit modules and the information flow of the model is designed according to the flow in the real-life vehicle control system and components. The overall structure is shown in Fig. 6.7.

As shown in Fig. 6.7, the established simulation platform mainly consists of the IEC model, the environmental perception model (radar model), the driver model, and the vehicle model. The IEC model is the core of the simulation platform and the key module for verification of the energy-efficiency control strategy, which mainly contains three sub-modules: mode division, mode switching and torque optimization control. The environmental perception model mainly simulates the function of environmental perception sensors to obtain the information on the traffic environment and the relative motion between vehicles. The driver model is used to simulate the driver's operation and output information related to accelerator pedal and brake pedal opening etc. The vehicle model contains the vehicle controller, the battery and its management system, the drive motor and its controller, the power coupler, the main reducer, the brake and other component models and vehicle dynamics models. The vehicle dynamics model mainly deals with the longitudinal dynamics characteristics and simulates the relationship between the longitudinal driving force and resistance during vehicle motion and calculates the acceleration and speed of the vehicle. The information flow and energy flow of the designed simulation platform are consistent with the real-life vehicle. The drive motor receives the command from the vehicle controller through CAN bus and outputs the power. The vehicle controller receives the vehicle state feedback and driver operation information and controls the vehicle motion based on its own control strategy.

The longitudinal dynamics system, drive motor and high-voltage battery pack are modeled. The longitudinal dynamics system is modeled using the longitudinal dynamics formula mentioned in Sect. 6.3. The drive motor is modeled using a quasi-static interpolation model. By ignoring thermodynamic effects, it establishes the relationship between the efficiency of the electric motor and its rotational speed and torque.

$$\begin{aligned} \eta_{\mathrm{mg}} &= E_{\mathrm{eff}}\left(\omega_{\mathrm{mg}}, T_{\mathrm{mg}}\right) \\ T_{\mathrm{mg\_max}} &= T_{\mathrm{torque}}\left(\omega_{\mathrm{mg}}\right) \end{aligned} \tag{6.38}$$

where η_{mg} is the efficiency of the motor/generator, which is related to its current rotational speed ω_{mg} and torque T_{mg}. $T_{\mathrm{mg\_max}}$ is the maximum torque at the speed of ω_{mg}. The actual power required by the motor is shown in Eq. (6.39) and is directly related to efficiency η_{mg}. When the generator works, the power is negative, which corresponds to $j = 1$ in the following equation; when the motor works, the power is positive, which corresponds to $j = -1$ in the following equation.

$$P_{\mathrm{mg}} = T_{\mathrm{mg}}\omega_{\mathrm{mg}}\eta_{\mathrm{mg}}^{j} \tag{6.39}$$

The model of the battery mainly describes the battery SOC changes, as shown in Eq. (6.40):

$$\mathrm{S}_{\mathrm{soc}}(k+1) = \mathrm{S}_{\mathrm{soc}}(k) - \frac{P_{\mathrm{battery}}(k)}{E_{\mathrm{battery}}}\Delta t(k) \tag{6.40}$$

where $E_{battery}$ is the total battery energy and P_{battery} is the battery power, which turns positive when the battery is discharged and turns negative when the battery is charged.

The environmental perception model contains the simulation of the relative distance between the ego vehicle and the preceding vehicle and the relative speed of the two vehicles. The in-vehicle bus delay is not considered and the state information of each component is sent and received regularly. The final and complete forward simulation platform built for the energy-efficiency control of intelligent electric vehicles is shown in Fig. 6.8.

The simulation platform takes a city's electric buses as the object of application and the vehicle longitudinal dynamics parameters and parameters of key components involved are shown in Table 6.2. The external characteristic curve of the drive motor is shown in Fig. 6.9.

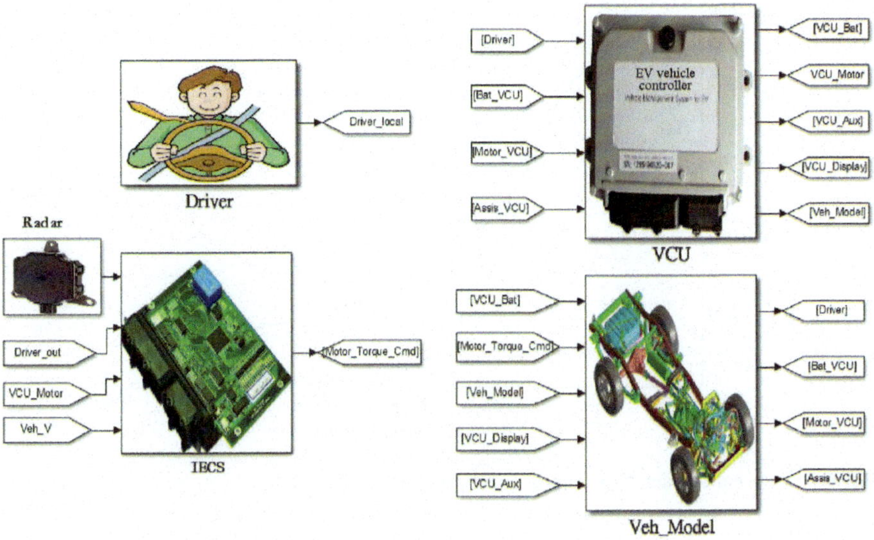

Fig. 6.8 Simulink simulation model of intelligent energy-efficiency control system for electric vehicles

Table 6.2 Parameters of vehicle dynamics and key components

Parameter type	Parameter name	Value
Vehicle longitudinal dynamics	Vehicle mass (kg)	17,000
	Tyre rolling radius (m)	0.52
	Ratio of rotational speed and deceleration	6.17
	Windward area (m^2)	6.93
	Air resistance coefficient	0.32
	Rolling resistance coefficient	0.011
	Mechanical transmission efficiency	0.95
Motor	Maximum torque (N.m)	2100
	Maximum rotational speed (rpm)	3000
Battery	Maximum capacity (A.h)	100
	Initial SOC value (%)	70

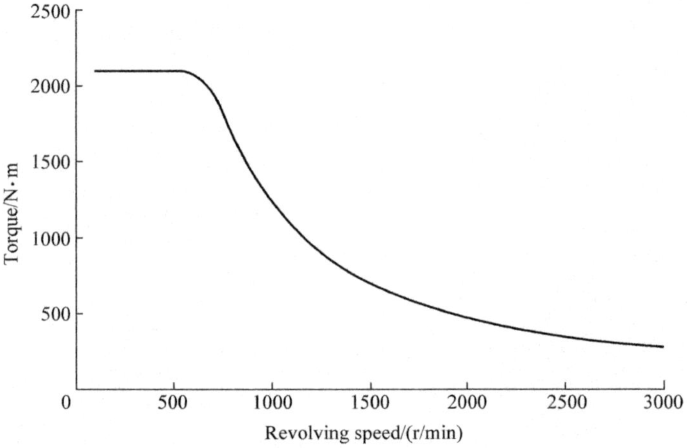

Fig. 6.9 External characteristic curve of drive motor

6.4.2 Simulation Plan Design

The designed IEC can assist energy-efficient driving. In order to analyze and verify its energy-efficiency effects, the paper conducts simulation analysis from two aspects:

(1) Using the preceding vehicle duty cycle and typical extreme conditions, let the same driver drive the ego vehicle to follow the preceding vehicle and compare the car-following situation and energy consumption before and after IEC is turned on;
(2) Under the CUDC, adjust the driver model parameters to simulate multiple drivers driving the ego vehicle to follow the preceding vehicle and compare the car-following safety and distribution of energy consumption before and after IEC is turned on.

For this purpose, the typical extreme conditions of the preceding vehicle is designed. The conditions contain the preceding vehicle with sharp acceleration, constant speed and sharp deceleration and are used to simulate and analyze the characteristics of response of vehicle components, working state and vehicle velocity change under rapid changes of preceding vehicle velocity. The preceding vehicle's working conditions are designed to accelerate from standstill with constant acceleration of 1.5 m/s^2 for 10 s, then maintain a constant speed for 20 s and then decelerate with constant brake deceleration of -1.5 m/s^2 until the vehicle stops, as shown in Fig. 6.10.

In order to comprehensively compare the car-following errors and energy consumption of electric vehicles before and after IEC is turned on, China Urban Driving Condition (CUDC) is used to simulate and evaluate the system. This is shown in Fig. 6.11.

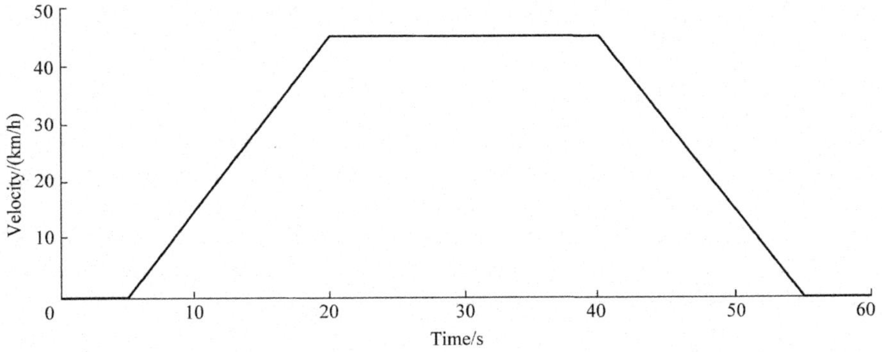

Fig. 6.10 Typical extreme working conditions of the preceding vehicle

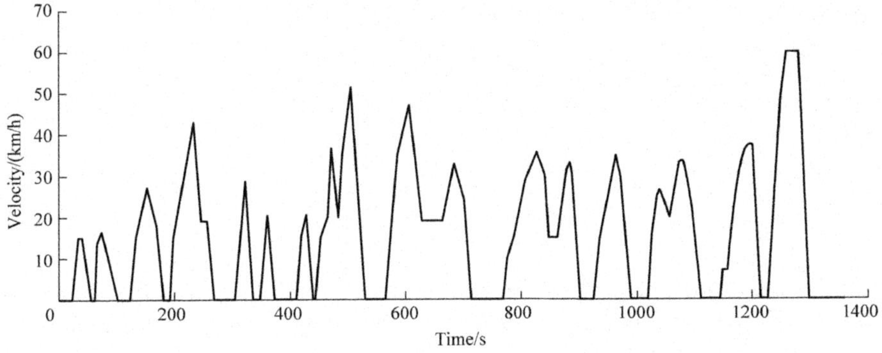

Fig. 6.11 CUDC

In order to make a comprehensive analysis and comparison of simulation results, it's necessary to quantitatively define the indicators of car-following errors, comfort and economy.

(1) Car-following error indicator

For vehicle longitudinal motion, the car-following error is defined as the relative velocity and safe distance that the vehicle maintains with the preceding vehicle in the process of car following. The following equation is used to design the car-following error indicator $S_{index.}$

$$S_{\text{index}} = \frac{1}{N} \sum_{k=1}^{N} \left[\left| \frac{\Delta d(k)}{K_{\text{D}}} \right| + |\Delta v(k)| \right] \tag{6.41}$$

where N is all the sampled data points of the duty cycle, K_{D} is the relative distance error weighting coefficient, $\Delta d(k)$ is the error between the actual inter-vehicle distance and the expected safe distance at moment k and $\Delta v(k)$ is the relative vehicle velocity at moment k.

(2) Driving comfort indicator

The driving comfort indicator is characterized by the magnitude of longitudinal acceleration during car following. The smaller the acceleration amplitude, the smoother the driving and the more the driving comfort. Conversely, the larger the acceleration amplitude, the less the driving comfort. In this paper, the following formula is used to evaluate whether the system meets the basic requirements of driver for driving comfort from the perspective of meeting driver's comfort needs.

$$a_{\min} \leq a \leq a_{\max} \tag{6.42}$$

where a is the acceleration of the vehicle during driving, $a_{\min}$ is the minimum acceleration constraint and $a_{\max}$ is maximum acceleration constraint.

(3) Economic indicators

The economic indicator is expressed in terms of the energy consumption per 100 km of the high-voltage battery pack:

$$E_{\text{index}} = \frac{s}{100} \int_0^t U_b I_b \mathrm{d}t \tag{6.43}$$

where E_{index} is the value of energy consumption per 100 km, s is the number of kilometers driven, U_b is the voltage of the high-voltage battery pack and I_b is the current of high-voltage battery pack.

The driver model uses the most common PID model, with the errors in the expected safe distance as the controller input and the acceleration or brake pedal as the output. In this case, the primary model for the expected safe distance is chosen, as shown in the following equation.

$$d_{\text{des}} = \tau_{\text{des}} v + d_0 \tag{6.44}$$

where τ_{des} is the expected safe car-following time distance and d_0 is the safe distance when the vehicle is stationary.

6.4.3 *Simulation Analysis of Vehicle Energy-Efficiency Control Effects*

6.4.3.1 Overall Simulation Results

Figure 6.12 shows the 90 sets of simulation results obtained for different driver parameters changes under CUDC conditions. Figure 6.12a shows the comparison of distribution of energy consumption and car-following errors before and after IEC is turned on and Fig. 6.12b shows the distribution of energy-efficiency rates after IEC is turned on and the distribution of car-following errors for different car-following distance weighting factors K. The statistics of the simulation results of these 90 sets are shown in Table 6.3.

The statistics of the simulation results in Table 6.3 show that the energy consumption per 100 km of the vehicle before IEC is turned on under CUDC ranges between 64 and 70 kWh/100 km and the energy consumption per 100 km of the vehicle after IEC is turned on ranges between 52.5 and 66.5 kWh/100 km. As shown in Fig. 6.12a, the

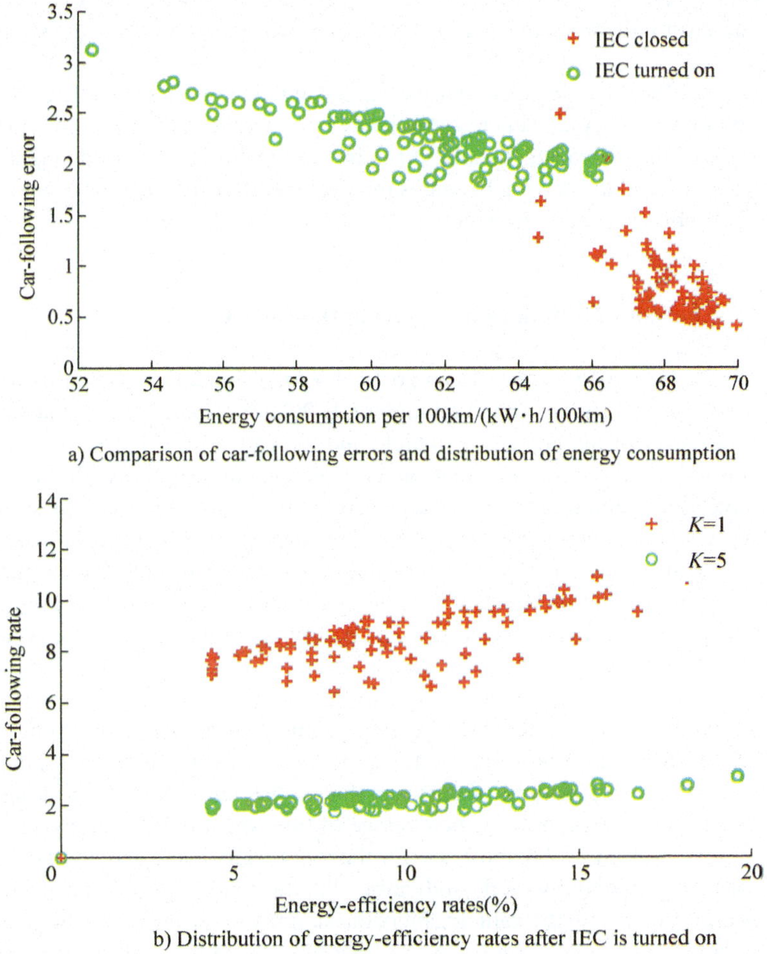

Fig. 6.12 Overall simulation results with different driving characteristics

Table 6.3 Statistics of overall simulation results

IEC 状态 IEC state	百公里能耗 (kWh/100 km) Energy consumption per 100 km (kWh/100 km)			能量节约率 (%) Energy-efficiency rate (%)		
	平均值 Avg.	最大值 Max.	最小值 Min.	平均值 Avg.	最大值 Max.	最小值 Min.
1	61.5	66.5	52.5	9.6	19	4
0	68.1	70	64			

car-following errors become larger as the energy consumption decreases and the car-following error indicator stays within a reasonable range except for some simulation results.

In Fig. 6.12b, it can be seen that the energy-efficiency rates after IEC is turned on are all above 4%. As the energy-efficiency rate increases, the car-following error increases with the increasing distance error, indicating that the car-following distance becomes larger and the driving becomes more gentle after IEC is turned on to ensure safety and achieve energy efficiency.

6.4.3.2 Analysis of Simulation Results in the Duty Cycle

Figures 6.13, 6.14, 6.15, 6.16, 6.17 and 6.18 show the simulation comparison results before and after the IEC is turned on under CUDC. Figure 6.13 shows the variation curve of velocity of the preceding vehicle and the ego vehicle before and after the IEC is turned on. It can be seen that the ego vehicle can keep up with the velocity of the preceding vehicle before and after IEC is turned on. Figure 6.14 shows the correlation curve of energy consumption before and after IEC is turned on and it can be seen that the cumulative energy consumption is reduced after IEC is turned on compared with that before IEC is turned on. Figure 6.15 shows the correlation curve of motor torque and it can be seen that the peak driving torque of the motor is partially reduced and the brake torque is increased to different degrees after IEC is turned on; Fig. 6.16 shows a section of the motor torque comparison curve extracted from Fig. 6.15 and it can be seen that the torque of the motor is limited to different degrees after IEC is turned on. Figure 6.17 shows the curve of IEC mode switching; with the changes in the relative motion between the two vehicles, the IEC working mode keeps switching. Statistically, the percentage of time spent in IEC energy-efficiency modes 1–4 is 24.2, 31.1, 7.8 and 1.9% for the whole duty cycle and another 35% of time is spent in the standby mode, indicating that during this period of time, the IEC mode does not work and the vehicle enters the standby mode due to the large relative distance between the two vehicles. No collision occurred between the two vehicles in this duty cycle and the relative distance was beyond the safe distance for 90.3% of the time, indicating that the IEC system has taken into account the car-following safety.

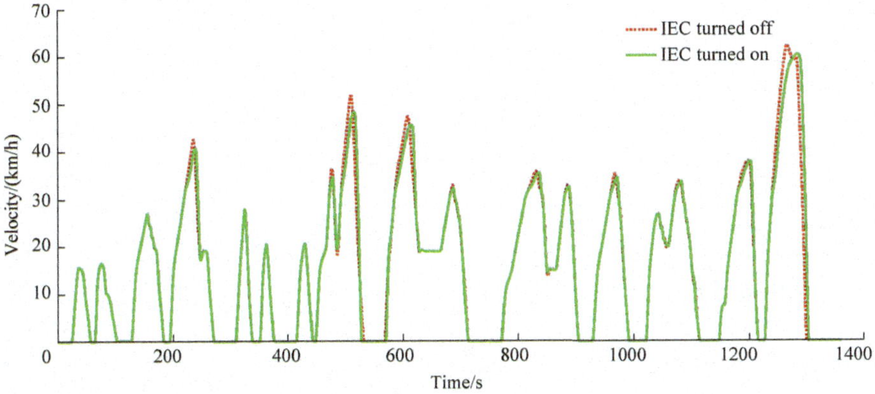

Fig. 6.13 Comparison of vehicle velocity in the duty cycle

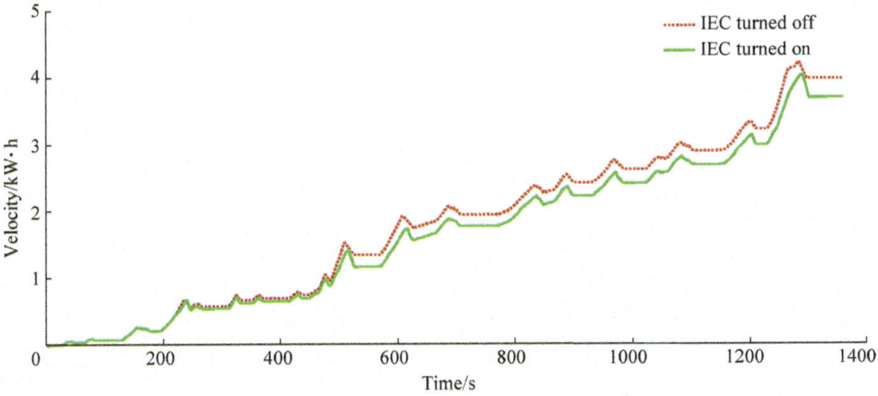

Fig. 6.14 Comparison of motor energy consumption in the duty cycle

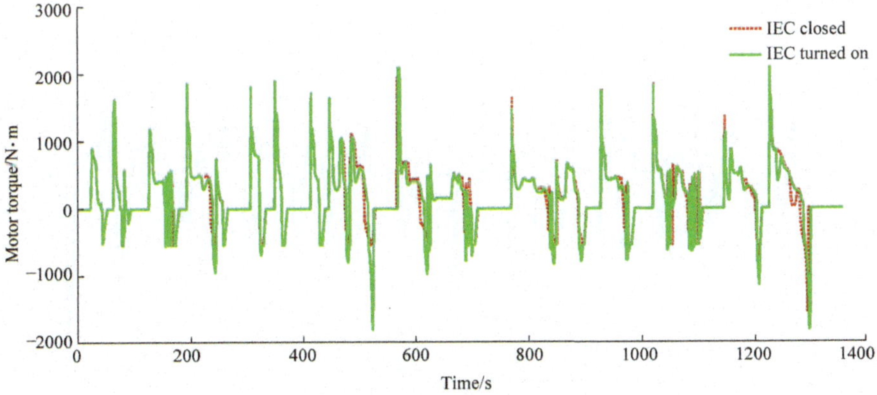

Fig. 6.15 Comparison of motor torque in the duty cycle

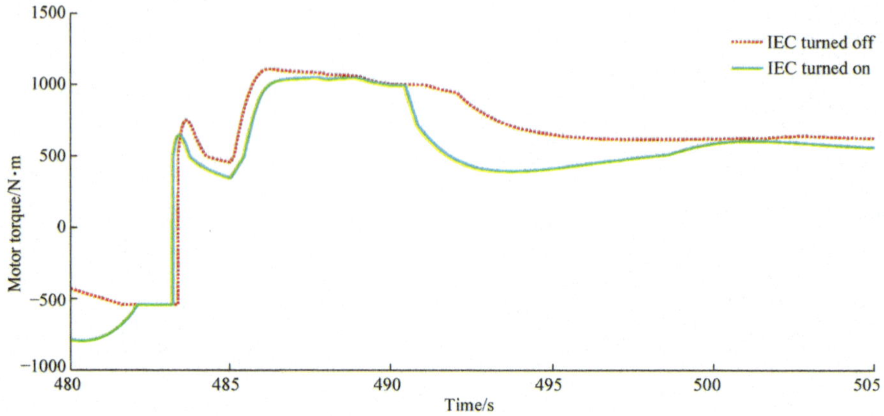

Fig. 6.16 Comparison of motor torque under some working conditions

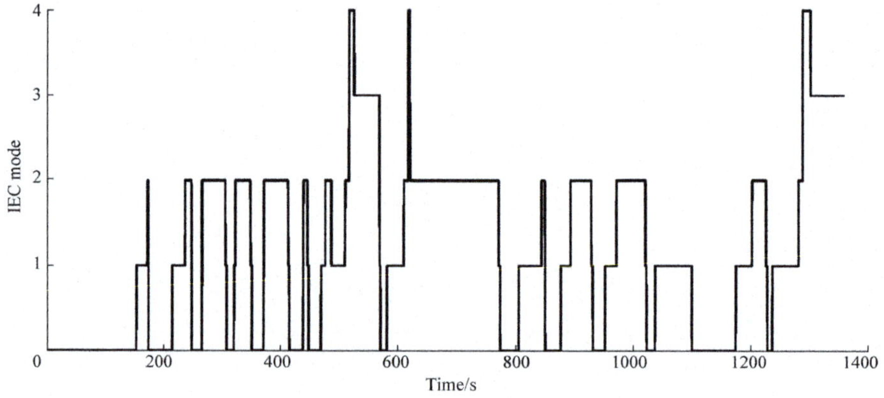

Fig. 6.17 IEC mode switching in the duty cycle

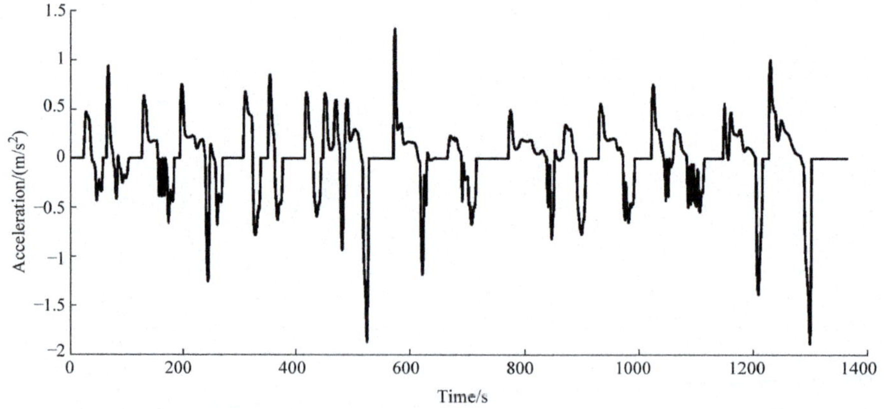

Fig. 6.18 Changes in vehicle acceleration

Table 6.4 Simulation results in CUDC

IEC state	Car-following error ($K_D = 5$)	Energy consumption per 100 km (kWh/100 km)	Energy-efficiency rate (%)
1	1.90	62.6	7.1
0	0.82	67.4	

Figure 6.18 shows the change of vehicle acceleration after the IEC is turned on and it can be seen that the vehicle acceleration is controlled between -1.9 m/s^2 and 1.3 m/s^2, showing good driving smoothness.

The statistics of car-following error and energy consumption under CUDC are shown in Table 6.4. Although the car-following error after IEC is turned on becomes larger, it's still within a reasonable range and the car-following state is maintained. The energy consumption of the vehicle after IEC is turned on is reduced by 7.1%.

6.4.3.3 Analysis of the Simulation Results Under Extreme Working Conditions

Figure 6.19 shows the simulation comparison results before and after the IEC is turned on under the sharp acceleration-sharp deceleration extreme working conditions. (a) shows the variation curve of the velocity of the preceding vehicle and the ego vehicle before and after the IEC is turned on. It can be seen that the vehicle can keep up with the velocity of the preceding vehicle before and after the IEC is turned on. Compared to before the IEC was turned on, the velocity of the ego vehicle after the IEC was turned on did not overshoot in the case of sharp acceleration, which means that the acceleration of the vehicle at high speed was limited, while the speed dropped faster in the case of sharp deceleration, which indicates that the deceleration increased after the motor brake was added. (b) shows the comparison of energy consumption before and after IEC is turned on. It can be seen that the energy consumption of the vehicle increases more slowly at sharp acceleration, while at sharp deceleration, more energy is regenerated and the system achieves more obvious energy-efficiency effects by increasing motor brake. (c) shows the comparison of the car-following error before and after the IEC is turned on. It can be seen that the car-following speed error and distance error increase after the IEC is turned on, which is due to the fact that the acceleration of the vehicle is limited after IEC is turned on and the car-following response becomes slower. (d) shows the comparison of the motor torque before and after IEC is turned on. It can be clearly seen that the motor torque is limited after the vehicle is accelerated to a certain degree and the motor brake torque is increased in the deceleration and braking stage. (e) shows the energy-efficiency mode switching after the IEC is turned on and the working mode of IEC is switched according to the relative motion between the ego vehicle and the preceding vehicle. As shown in the figure, when the vehicle accelerates to 17 s, the relative distance between the two vehicles decreases to within the IEC action distance and the system enters mode 1;

as the velocity of the ego vehicle increases, at 40 s, the difference between the speed of the two vehicles exceeds the set speed buffer and the system enters mode 2; as the vehicle keeps approaching the preceding vehicle, the relative distance between the two vehicles decreases to a safe distance. To avoid a collision, the system enters mode 4 directly. At this time, the motor brake torque is increased and the vehicle decelerates fast until the distance between the two vehicles increases to beyond the safe distance, the system enters mode 2; thus ensuring the vehicle's driving safety. The simulation results has proved the effectiveness of the mode switching control strategy.

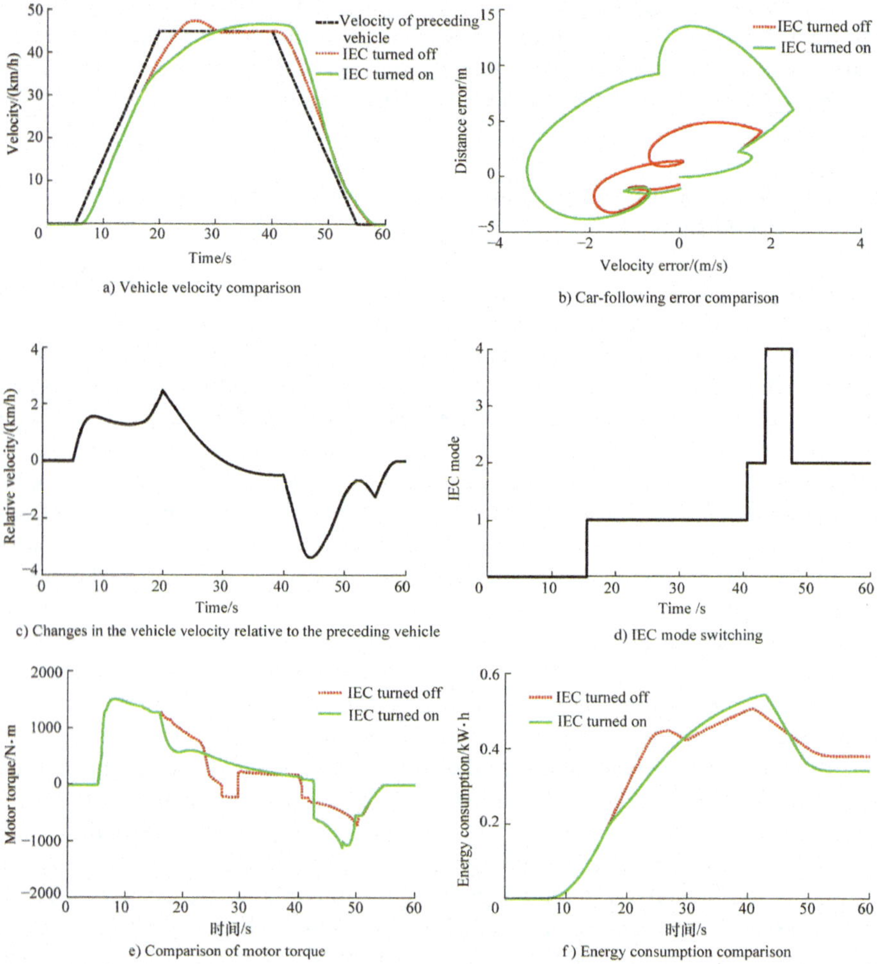

Fig. 6.19 Simulation results under extreme working conditions

Table 6.5 Comparison of energy consumption under typical extreme working conditions

IEC state	Car-following error (K_D = 5)	Energy consumption (kWh)	Energy-efficiency rate (%)
1	2.16	0.343	10.4
0	1.14	0.383	

The statistics of the car-following error and energy consumption under typical extreme working conditions are shown in Table 6.5. As shown in the table, the car-following error increases after the IEC is turned on, but It's shown in Fig. 6.19 that the vehicle maintains the car-following state and the vehicle energy consumption is reduced by 10.4% after the IEC is turned on. The simulation results, has proved the effectiveness of the torque optimization control method.

6.5 Experimental Study of Energy-Efficiency Control of Intelligent Electric Vehicles Based on Structure Sharing

In order to further verify the effectiveness of the proposed energy-efficiency control method that is applied in real-life vehicles, based on the proposed control strategies and algorithms, an intelligent energy-efficiency controller was designed and developed, an experimental data acquisition system was built and two battery electric city buses were selected as the experimental platform. Multiple sets of real-life vehicle experiments under actual dynamic traffic conditions were conducted on two typical city roads.

6.5.1 Design of Experiment Plan

6.5.1.1 General Experiment Plan

By taking into account the characteristics of the proposed energy-efficiency controller and the general process of the real-life vehicle experiment, the overall experimental plan is designed, as shown in Fig. 6.20, which mainly includes the steps such as energy-efficiency controller development, data acquisition system construction, selection of experimental sample vehicles and road conditions and experimental data processing and analysis. For the key part of data acquisition and processing, experimental verification and analysis are conducted from the following three aspects:

(1) The experiment was conducted under actual urban traffic conditions, with professional test drivers driving the vehicle through the whole road section

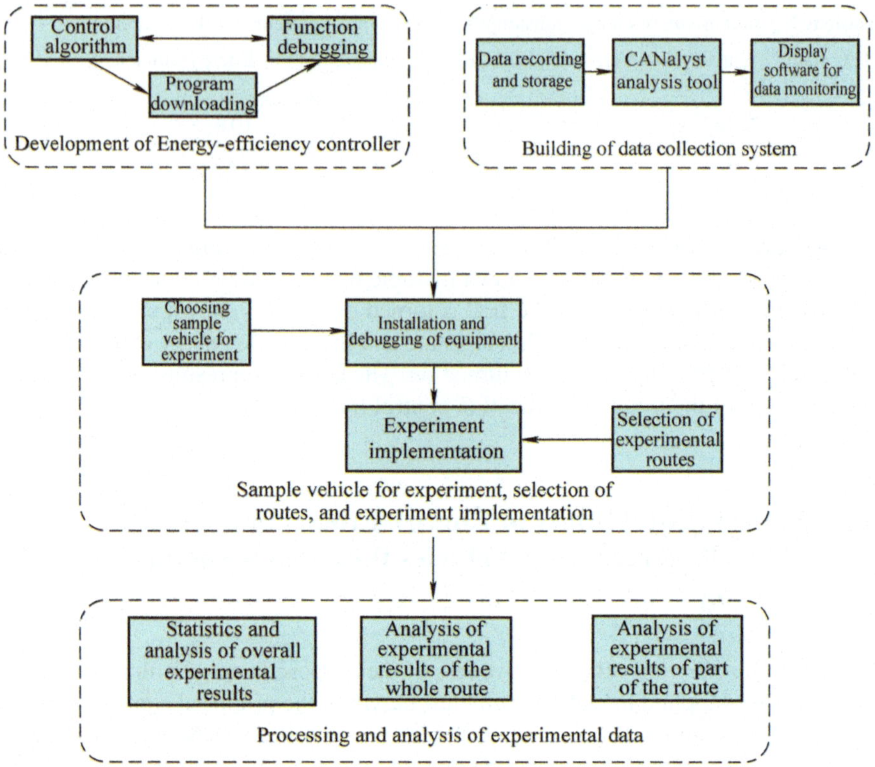

Fig. 6.20 General plan of the real-life vehicle experiments

turning on and off the IEC system respectively, while collecting information related to the working of the components under the working conditions before and after starting the IEC, including vehicle velocity, pedal, relative distance to the preceding vehicle, motor torque, IEC mode and energy consumption etc.

(2) Since the traffic environment is dynamic and time-varying, there are many factors affecting the energy consumption of electric vehicles and a single experiment is not enough to illustrate the energy efficiency state of those vehicles. In order to gather information on energy efficiency of IEC under different actual traffic conditions, two BEVs were selected for turning on and off IEC under different road conditions in multiple experiments and the average energy efficiency effects of IEC of these two vehicles in different experiments were compared and gathered separately.

(3) Due to the large amount of experimental data generated from the whole road section, observation is not easy. In order to facilitate the comparison and analysis of the workings of different components before and after IEC is turned on, the data of the typical road section was extracted for analysis, the changes in the

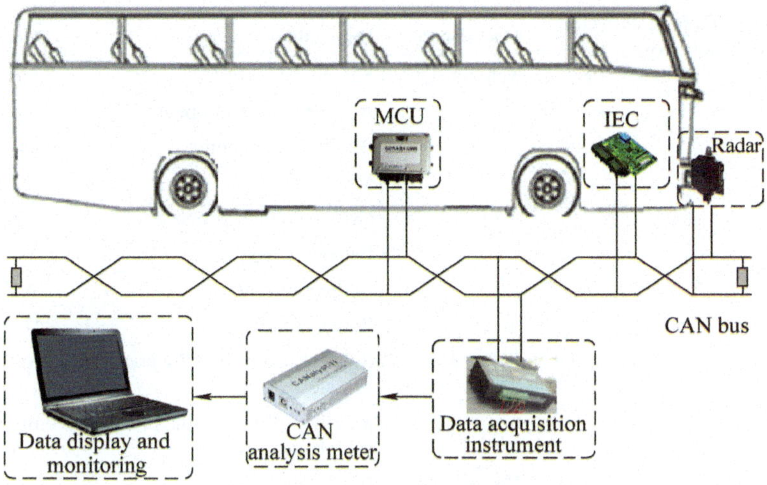

Fig. 6.21 Connection of the experimental data acquisition system

original torque and actual torque of the motor were compared and analyzed and the resulting changes in energy consumption were calculated.

6.5.1.2 Experimental Data Acquisition System

In order understand the working state of different components during the experiment, a data acquisition system was established to read the data from the vehicle control system and IEC system through the CAN analyzer. The collected data was monitored in real time through the computer interface during the experiment, as shown in Fig. 6.21.

In order to understand the energy consumption of the whole road section and the workings of different components, all the information related to the whole road section was collected during the experiment. Then the typical road section was selected for analysis as needed. The mainly collected data are shown in Table 6.6.

6.5.1.3 Sample Vehicles and Road Conditions Used in the Experiment

Compared with other models, city buses not only have higher safety requirements, but also have relatively higher energy-efficiency potential because of their large size, high energy consumption, urban use and frequent start-stops. Therefore, battery electric city buses were chosen as the experimental vehicles. In order to verify the universality of the proposed control method and the energy-efficiency effects under different road conditions, two sample vehicles A and B have been chosen for the experiment on different urban roads, both of which have the forward collision warning function and are equipped with millimeter-wave radar sensors in front of the vehicle.

Table 6.6 Table of collected real-life vehicle experiment data

Data source	Data name	Data unit
Vehicle	Vehicle velocity	m/s
	Accelerator pedal opening	%
	Brake pedal opening	%
	Yaw rate	rad/s
	Original motor torque	N·m
	Battery voltage	V
	Battery current	A
	Battery SOC	%
Radar sensor	Relative distance to the preceding vehicle	m
	Relative velocity to the preceding vehicle	m/s
IEC system	IEC mode	
	Safe time distance	s
	Motor torque command	N·m

In order to verify the energy-efficiency effect of the designed energy-efficiency controller under different urban road conditions, congested urban road conditions and general urban road conditions were selected as experimental conditions. Sample vehicle A has been selected for a section of an actual road in a city in central China. The total length of the road section is 5.8 km and the experiment is conducted in a round trip, so the total length is 11.6 km. The traffic flow of the road section is large, representing the actual working conditions of city buses. Sample vehicle B has been selected for an actual highway in an urban area in southern China and the experimental route is a closed-loop route with a total length of 15.7 km. This road section of highway includes a section of urban expressway that represents the actual traffic conditions of general urban roads.

6.5.1.4 Experimental Procedure and Main Parameter Settings

Many factors can affect the energy consumption of electric vehicles. In order to minimize the effect of factors such as traffic environment, driver's driving skills and in-vehicle electric equipment on the comparison of energy consumption, we strictly follow the principles to make sure the experimental conditions are consistent throughout the experiment. These conditions are listed as follows:

(1) The experiments were conducted under good weather conditions and the experimental roads were urban roads in plains with asphalt pavement and a slope that is negligible. For each group of comparison experiments, the experiments were conducted successively during the same time period to minimize the differences in the traffic flow of the same group of comparison experiments.

(2) The experimental vehicle was driven by professional test drivers with more than 10 years of driving experience, turning on and off the IEC system respectively to travel the same whole road section without stopping at will, while ensuring driving safety and the consistency of driving style throughout the experiment.
(3) In the comparison experiment, the loading mass, tyre pressure, SOC and other technical states of the experimental vehicle were kept the same. The air conditioning, windshield wipers, headlights and other electric equipment in the vehicle were turned off and the parameters of the IEC system were kept the same during the experiment after being set according to the driver's expectations.

Since the experiments were conducted on an open urban road and the traffic flow is dynamic and time-varying, it was difficult to achieve identical experimental conditions during the implementation of the comparison experiments. For this reason, each sample vehicle was used for several comparison experiments for us to compare the energy consumption before and after turning on the IEC system through statistical analysis of larger sample size of data. During the experiments, the two sample vehicles were subjected to a total of 20 comparison experiments with a total mileage of more than 500 km.

The main parameters of the experimental sample vehicles and IEC system are shown in Table 6.7.

Table 6.7 Main parameters of the experimental sample vehicles and IEC system

Parameter		Unit	Value
Sample vehicle A	Size	mm	8040 × 2360 × 3250
	Total mass	kg	11,480
	Motor rated power	kW	80
Sample vehicle B	Size	mm	10,490 × 2500 × 3400
	Total mass	kg	17,000
	Motor rated power	kW	100
IEC system	Collision early warning time from a far distance	s	3
	Collision early warning time from a near distance	s	1
	Desired deceleration from a far distance	m/s^2	1.5
	Desired deceleration from a near distance	m/s^2	3

6.5.2 Experimental Analysis of Energy-Efficiency Control of Vehicles in Congested Urban Road Conditions

6.5.2.1 Overall Experimental Results

Sample vehicle A was used under congested urban road conditions, with an average mileage of 11.3 km in a single experiment. 13 experiments were conducted with the IEC turned on and off respectively on the same road section and 26 sets of experimental data has been obtained, as shown in Table 6.8.

The experimental results in Table 6.8 show that the average speed of sample vehicle A is 15.7 km/h during the experiment and the average energy consumption is 5.2 kWh during the experiment that includes the average driving energy consumption of 7.9 kW, the average regenerative brake energy of 2.7 kW and the average energy consumption per 100 km of 46.2 kWh/100 km. In the 13 experiments with IEC on (IEC = 1), the average speed is 16.1 km/h, with the average energy consumption of 5 kWh and the average energy consumption per 100 km of 44.5 kWh/100 km; in the 13 experiments with IEC off (IEC = 0), the average speed is 15.2 km/h, with the average energy consumption of 5.4 kWh and the average energy consumption per 100 km of 47.8 kWh/100 km. The overall experimental results show that there is a significant reduction in the energy consumption per 100 km of the vehicle after IEC is turned on.

Figure 6.22 shows the comparison of the energy consumption per 100 km and the distribution of the energy-efficiency rates before and after the IEC is turned on in the 13 experiments. The figure shows that the energy consumption of different experiments fluctuates greatly due to the dynamic and time-varying nature of the actual traffic flow and the differences in the actual operation of the driver in different experiments, which causes the energy saving rates of different experiments to fluctuate greatly. The statistical results of the 13 experiments show that the overall energy consumption per 100 km of the sample vehicle with IEC on is significantly lower than that of the sample vehicle with IEC off. The maximum energy-efficiency rate is 15.7%, the minimum rate is 1%, with the average being 7%.

6.5.2.2 Analysis of Experimental Results for the Whole Road Section

Figures 6.23, 6.24, 6.25, 6.26, 6.27, 6.28, 6.29 and 6.30 shows a set of experimental results of sample vehicle A with IEC on during the journey along the road section. The whole distance travelled during the experiment is 11,440 m and it takes 2464 s to finish the journey. The average speed of the vehicle was 16.7 km/h and the battery SOC value decreased from 70.8 to 63.6% under the entire working conditions.

Figure 6.23 shows the changes in the velocity of the ego vehicle throughout the single experiment. Since the experimental working conditions are set as congested conditions on the main roads in urban areas with large traffic flow and the experiments are designed to simulate the actual city bus route, the vehicle stopped several times

Table 6.8 Experimental results of real-life vehicle energy consumption under congested urban road conditions

Number of experiments	IEC state	Mileage (km)	Time (s)	Average vehicle velocity (km/h)	Power consumption (kWh)	Driving energy consumption (kWh)	Regenerative energy (kWh)	Consumption per 100 km (kWh/100 km)
1	1	11.3	2419	16.8	4.44	6.78	2.33	39.25
	0	11.2	2819	14.3	4.46	6.95	2.49	39.66
2	1	11.1	2964	13.5	4.43	6.87	2.44	39.90
	0	11.3	2344	17.3	4.69	7.46	2.77	41.57
3	1	11.3	2494	16.3	4.45	6.80	2.35	39.39
	0	11.4	2514	16.3	4.59	7.24	2.65	40.31
4	1	11.2	2394	16.8	4.42	7.23	2.82	39.59
	0	11.4	2549	16.1	5.08	7.88	2.80	44.39
5	1	11.4	2159	19.0	5.03	7.65	2.62	44.09
	0	11.2	3379	11.9	5.84	8.80	2.96	52.31
6	1	11.4	2464	16.6	5.53	8.72	3.19	48.62
	0	11.3	2619	15.6	6.24	9.44	3.20	55.03
7	1	11.3	2444	16.7	4.90	7.62	2.72	43.23
	0	11.4	2559	16.1	5.58	8.86	3.28	48.77
8	1	11.4	2379	17.3	5.44	8.36	2.92	47.62
	0	11.4	2624	15.7	5.62	8.70	3.07	49.27
9	1	11.5	1879	22.1	5.54	8.73	3.19	48.04
	0	11.5	2297	18.1	6.19	9.53	3.33	53.68
10	1	11.2	3569	11.3	5.58	8.15	2.57	49.65
	0	11.2	3404	11.8	5.69	8.07	2.38	50.79

(continued)

Table 6.8 (continued)

Number of experiments	IEC state	Mileage (km)	Time (s)	Average vehicle velocity (km/h)	Power consumption (kWh)	Driving energy consumption (kWh)	Regenerative energy (kWh)	Consumption per 100 km (kWh/100 km)
11	1	11.2	3134	12.9	4.89	7.02	2.13	43.66
	0	11.2	3189	12.7	5.23	7.36	2.13	46.53
12	1	11.3	2974	13.7	5.26	7.58	2.32	46.67
	0	11.9	2754	15.6	5.64	7.77	2.13	47.23
13	1	11.4	2464	16.7	5.56	8.09	2.53	48.57
	0	11.4	2514	16.3	5.96	8.51	2.56	52.47

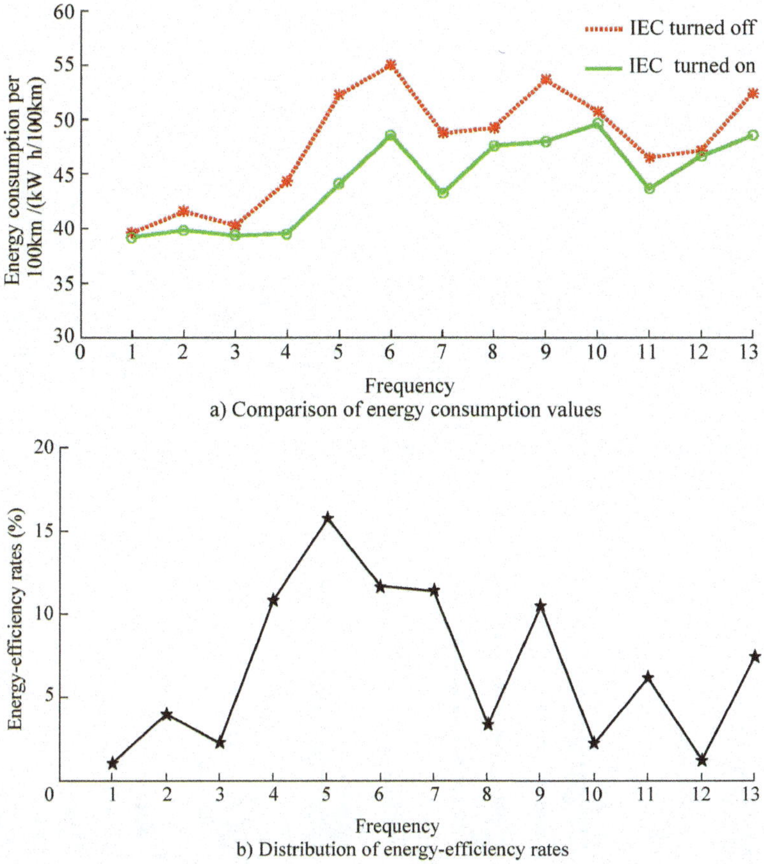

Fig. 6.22 Overall experimental results

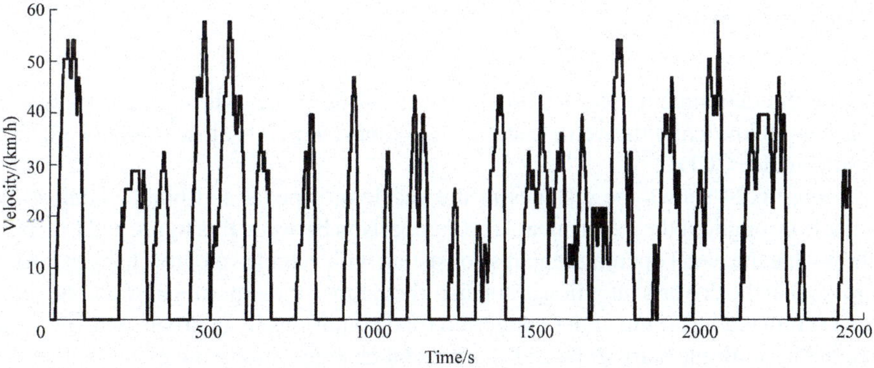

Fig. 6.23 Changes in vehicle velocity throughout the experiment

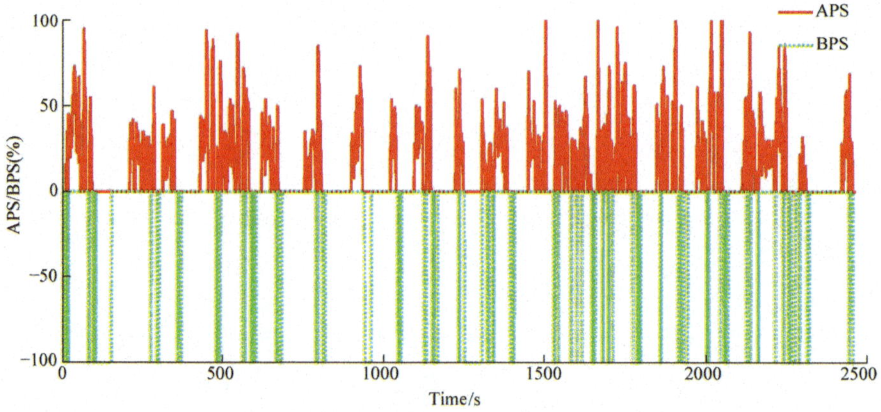

Fig. 6.24 Changes in the accelerator pedal and brake pedal throughout the experiment

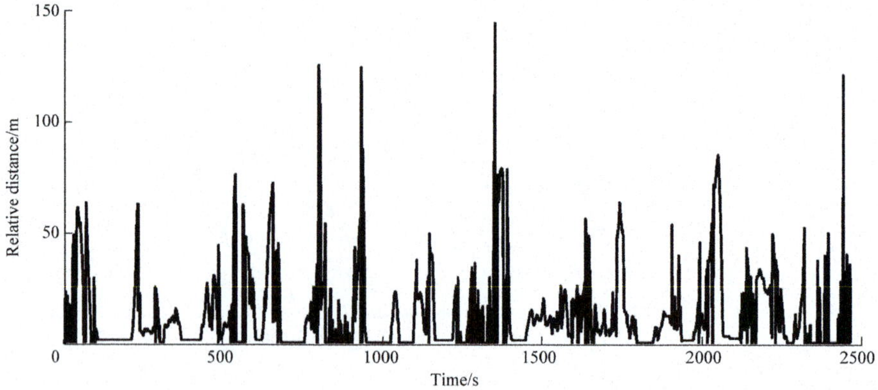

Fig. 6.25 Changes in the relative distance between the ego vehicle and the preceding vehicle throughout the experiment

during the experiment. The maximum speed of the vehicle during the experiment reached 58 km/h and the vehicle stopped 16 times, which is close to the simulation results under CUDC.

Figure 6.24 shows the changes in the accelerator pedal and brake pedal of the vehicle throughout the experiment. The comparison between this figure and Fig. 6.23 shows that the pedal opening remains consistent with vehicle velocity changes, which proves that the driver's intention is not disturbed. Since the experiment was conducted on a simulated bus route under congested road conditions in an urban area, there are inevitably multiple start-stops and sharp pedal changes during the experiment.

Figure 6.25 shows the changes in the relative distance between the ego vehicle and the preceding vehicle detected by the radar throughout the experiment. The maximum distance to the vehicle that can be detected by the radar can reach 150 m. If there is no

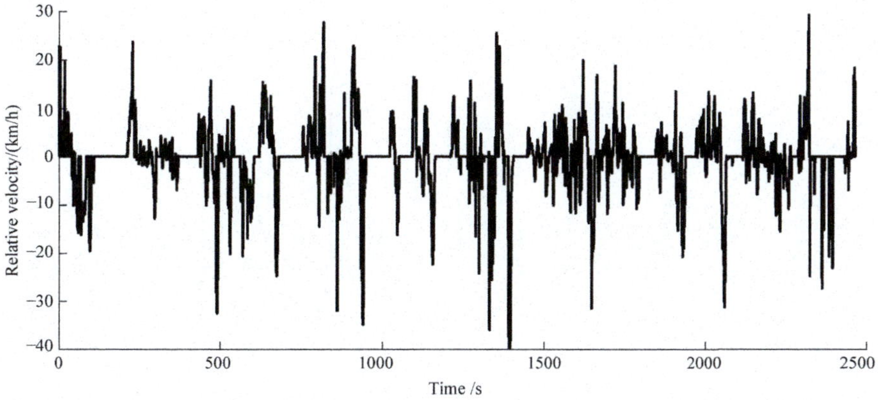

Fig. 6.26 Changes in the relative speed between the ego vehicle and the preceding vehicle throughout the experiment

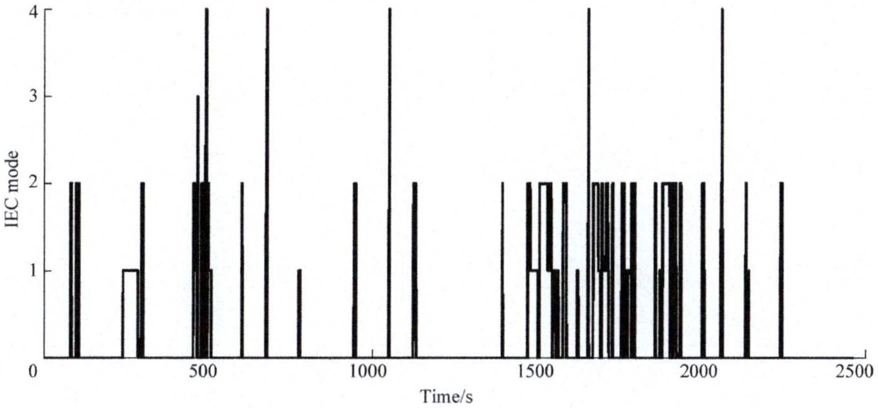

Fig. 6.27 IEC mode switching throughout the experiment

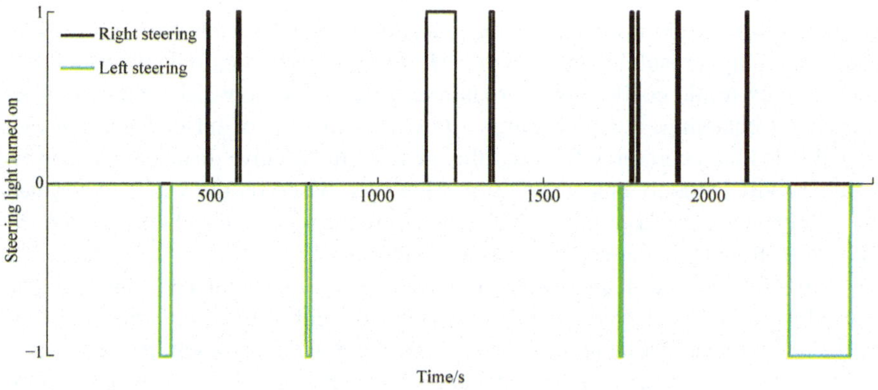

Fig. 6.28 The turning on of directional light throughout the experiment

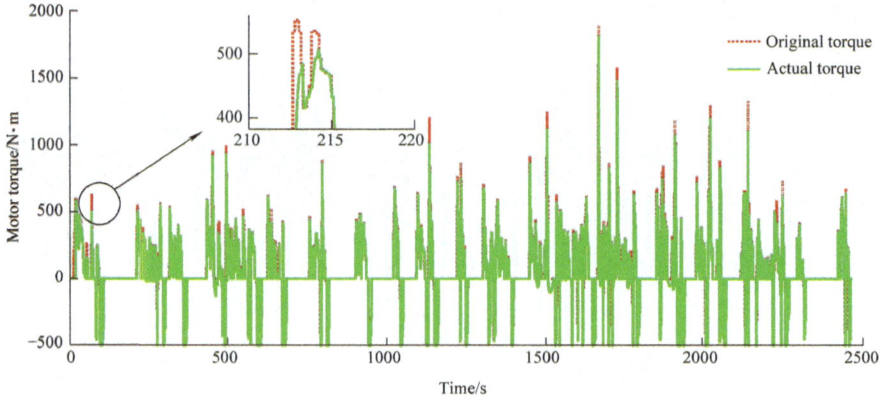

Fig. 6.29 Comparison of changes in motor torque throughout the experiment

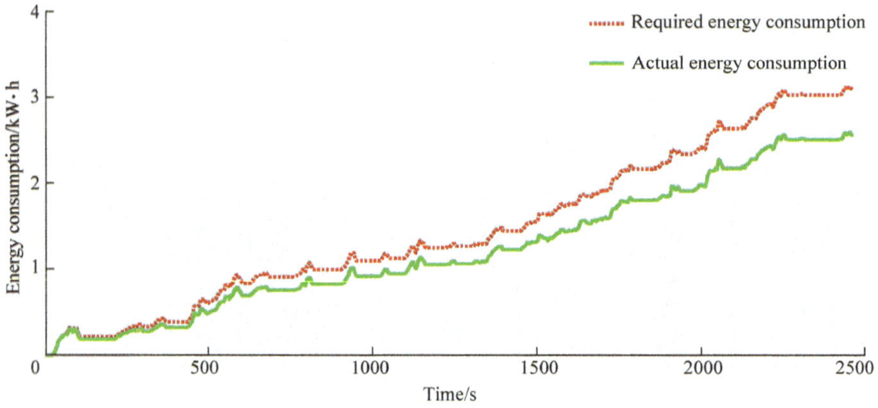

Fig. 6.30 Comparison of motor energy consumption throughout the experiment

preceding vehicle or the distance to the preceding vehicle exceeds the radar detection range, the output result is null, which is shown as 0 in the figure by default. Due to the complex traffic conditions in urban areas, the traffic flow around the vehicle is constantly changing, if a preceding vehicle cuts in or cuts out or when the radar signal is lost at a certain time point, the curve of the relative distance between two vehicles will suddenly change or jump. However, based on the whole experiment, the vehicle maintains a certain safe distance from the preceding vehicle and achieves safe car-following under congested road conditions.

Figure 6.26 shows changes in the relative speed between the ego vehicle and the preceding vehicle detected by the radar under the entire working conditions. Due to the complex and variable traffic flow throughout the experiment, the curve of the relative speed shown in the figure appears to have several sudden changes and jumps due to the cut-in or cut-out of the preceding vehicle. In general, the speed curve is consistent with the distance curve shown in Fig. 6.25 and the two are integrals and derivatives within a certain range, except, of course, when there is a cut-in or cut-out of the preceding vehicle. According to the designed IEC control strategy, the working mode of IEC switches with the changes in the relative distance and speed between the two vehicles.

Figure 6.27 shows IEC mode switching. The mode switching pattern is closely related to the changes in the relative distance shown in Fig. 6.25 and the changes in the relative speed shown in Fig. 6.26. For example, between 110 and 210 s, the IEC switches to the standby mode when the radar has not detected the preceding vehicle. Statistically, IEC remains in the standby mode for 97% of the time, which is determined by the control strategy for turning on and turning off IEC. In any case that the preceding vehicle's relative distance to the vehicle exceeds the IEC's maximum acting distance or radar detection range, the directional light is turned on or the yaw rate is greater than the set limit, the opening or rate of change of the accelerator pedal is greater than the set limit and the radar signal is temporarily lost, IEC enters the standby mode to avoid interference with the driver and ensure driving safety.

Figure 6.28 shows the situation that the directional light is turned on during the experiment. The displayed value 1 means the right-turn light is on and the displayed value -1 means the left-turn light is on. Since the experimental road conditions are main roads in urban areas in the form of a round trip route along the main road for a group of experiments, there are several lane changes or turns throughout the experiment. After the directional light is turned on, the IEC enters the standby mode, which corresponds to the IEC mode switching shown in Fig. 6.27. For example, between 2200 s and 2300 s, although the radar detected the changes in the relative speed and relative distance between the two vehicles, IEC remained in the standby mode because the left-turn light was turned on.

Figure 6.29 shows the changes in vehicle's drive motor torque under the entire working conditions, in which the dashed line represents the original torque calculated by the VCU and the solid line represents the actual torque after IEC optimization, from which it can be seen that the peak of the driving torque of the motor is significantly reduced and the brake torque of the motor is also increased during some time periods. The changes are basically consistent with the IEC mode switching shown in Fig. 6.27.

Figure 6.30 shows the comparison of the curves of energy consumption of the drive motor under the entire working conditions. The dashed line represents the required energy consumption calculated by multiplying the original motor torque by the motor speed and the solid line represents the actual energy consumption calculated by multiplying the actual motor torque by the motor speed. Given the actual traffic flow is complex and variable and the traffic conditions of different experiments vary greatly, it's therefore not convenient to compare and analyze the energy consumption curves measured by the two sets of experiments after turning on and turning off the IEC. In order to better understand the energy consumption reduction after IEC is turned on, the above-mentioned method of calculating the motor energy consumption by using the motor torque command has been adopted.

As shown in the figure, the motor energy consumption after IEC optimization is significantly reduced.

6.5.2.3 Analysis of Experimental Results for a Typical Road Section

In order to observe and analyze the state of the vehicle and the working of different components after the IEC is turned on, data from the typical road section has been taken for analysis.

Figure 6.31 shows experimental results for a road section from the experiments on sample vehicle A, which contains experimental data such as vehicle velocity, relative distance to the preceding vehicle, relative speed, the situation of directional light turned on, IEC mode and motor torque etc. In order to facilitate comparison and analysis, all the experimental measurements have been put together for analysis.

Figure 6.31a shows the changes in vehicle velocity. The figure shows that during this time period, the vehicle first decelerates and brakes to stop before starting to accelerate and there is sharp deceleration and acceleration due to the actual complex traffic flow. (b) shows the changes in vehicle accelerator pedal (APS) and brake pedal (BPS). Through comparison with (a) figure, we can see that the changes in pedal opening are consistent with the changes in vehicle velocity and the driving intention is not disturbed.

Figure 6.31c shows the relative distance between the ego vehicle and the preceding vehicle, as measured by the radar. According to the figure, the vehicle maintains a safe distance from the preceding vehicle except at 486 s when the target of radar detection was lost or cut out. Figure 6.31d shows the relative speed between the ego vehicle and the preceding vehicle and the changes in the relative speed and the changes in the relative distance shown in Fig. 6.31c are the inputs for the IEC energy-efficiency mode switching control.

Figure 6.31e shows the situation where the directional light is turned on and it can be seen that the right-turn light was turned on between 485 and 489 s, indicating turning or lane changing, which is consistent with the sudden changes shown in Fig. 6.31c and d during this time period. Figure 6.31f shows the IEC mode switching and it can be seen that the IEC system performs mode switching as the relative motion between the two vehicles and the yaw rate changes. For example, at 481 s, IEC entered energy-efficiency mode 1 in the "approaching from a far distance" scenario and started the motor brake in advance and then IEC entered the standby mode as the directional light was turned on. At 495 s, the ego vehicle had made the turn or changed lanes and the radar had detected that the preceding vehicle was approaching the vehicle from a near distance, and the ego vehicle entered energy-efficiency mode 4 and the vehicle slowed down until it stopped. Then, when the vehicle started to accelerate, based on the motion state of the preceding vehicle, for example, at 496 s, the radar had detected that the preceding vehicle was in the "leaving from a near distance" state and the vehicle entered energy-efficiency mode 3, followed by the standby mode as the accelerator pedal changed sharply.

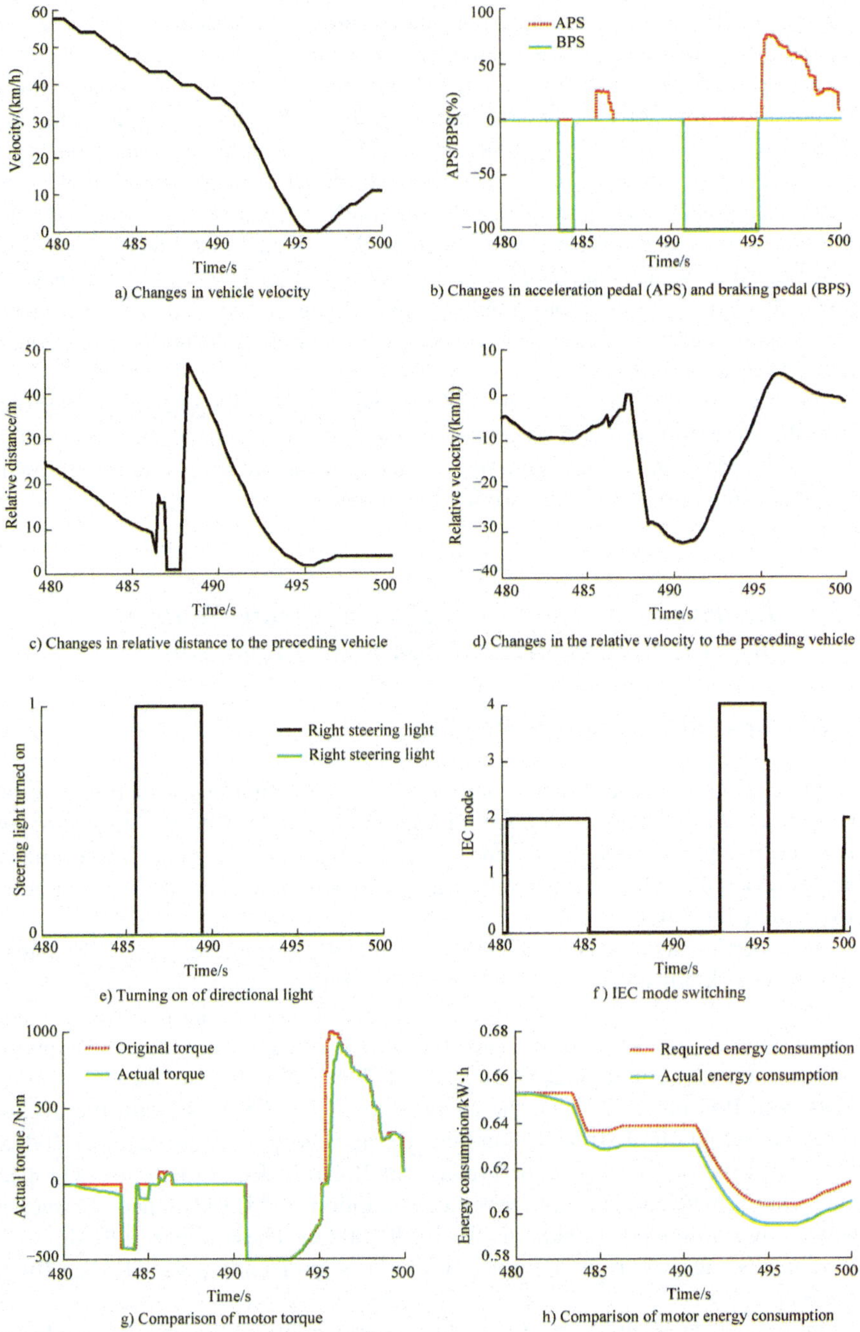

Fig. 6.31 Experimental results for a road section from the experiments on sample vehicle A under actual road conditions

Figure 6.31g shows the changes in the motor torque, the dashed line represents the original motor torque calculated by the VCU and the solid line represents the actual motor torque after IEC optimization. It's shown in the figure that the brake torque of the motor is significantly increased in advance during 481–485 s and some other time periods. At around 495 s, the working state of IEC was in mode 4 and mode 3 and the torque of the motor was optimized and reduced. IEC entered the standby mode within 3 s thereafter due to the sharp changes in the accelerator pedal until at 498 s, after buffering, as the vehicle velocity increased, the vehicle was in close proximity to the preceding vehicle and the IEC entered energy-efficiency mode 2. Figure 6.31h shows the changes in the motor energy consumption in this road section in detail. It's shown in the figure that the motor energy consumption is significantly reduced after IEC optimization.

According to the above analysis, the designed energy-efficiency controller optimizes the motor torque in real time based on the information on the preceding vehicle detected by radar and achieves energy efficiency while ensuring safe car-following and also taking into account the driving intentions.

6.5.3 Experimental Analysis of Energy-Saving Control of Vehicles in General Urban Road Conditions

6.5.3.1 Overall Experimental Results

Sample vehicle B was conducted on a general road containing a section of urban expressway with a single experiment of 15.7 km in length, with IEC turned on and off respectively. 7 experiments were conducted to obtain 14 sets of experimental data, of which the vehicle velocity and energy consumption data has been extracted, as shown in Table 6.9.

The experimental results in the table show that the average velocity of the sample vehicle is 46 km/h under actual urban traffic conditions and the average energy consumption throughout the experiment is 11.1 kWh that includes the driving energy consumption of 16.5 kW, the average regenerative brake energy of 5.3 kW and the average energy consumption per 100 km of 70.9 kWh/100 km. In the 7 experiments with IEC on (IEC = 1), the average speed is 47.5 km/h, with the average energy consumption of 10.8 kWh and the average energy consumption per 100 km of 68.7 kWh/100 km; in the 7 experiments with IEC off (IEC = 0), the average speed is 45.3 km/h, with the average energy consumption of 11.5 kWh and the average energy consumption per 100 km of 73.1 kWh/100 km. The overall results show that the vehicle's energy consumption per 100 km is significantly reduced when IEC is turned on.

Figure 6.32 shows the comparison of the energy consumption per 100 km before and after the IEC is turned on and the distribution of the energy-saving rates in the 7 experiments. The figure shows that the energy consumption of different experiments

Table 6.9 Experimental results of energy consumption of sample vehicle B under actual urban traffic conditions

Number of experiments	IEC state	Mileage (km)	Time (s)	Average vehicle velocity (km/h)	Power consumption (kWh)	Driving energy consumption (kWh)	Regenerative power (kWh)	Energy consumption per 100 km (kWh/100 km)
1	1	15.7	1115	50.7	12.27	17.78	5.51	78.08
	0	15.7	1290	43.7	13.40	19.90	6.50	85.35
2	1	15.7	1225	46.0	11.10	17.00	5.90	70.70
	0	15.7	1180	47.7	11.50	17.10	5.60	73.25
3	1	15.7	1210	46.8	9.42	14.48	5.06	59.84
	0	15.7	1210	46.7	10.80	15.90	5.20	68.79
4	1	15.7	1165	48.4	10.05	14.50	4.45	64.14
	0	15.7	1135	49.6	10.70	15.50	4.70	68.15
5	1	15.7	1170	48.3	11.50	16.50	5.00	73.07
	0	15.7	1275	44.4	11.80	17.60	5.80	75.16
6	1	15.7	1220	46.4	10.52	15.51	4.99	66.83
	0	15.7	1320	42.7	10.70	15.40	4.70	68.15
7	1	15.7	1225	46.1	10.77	16.92	6.14	68.74
	0	15.7	1325	42.5	11.46	16.60	5.20	72.96

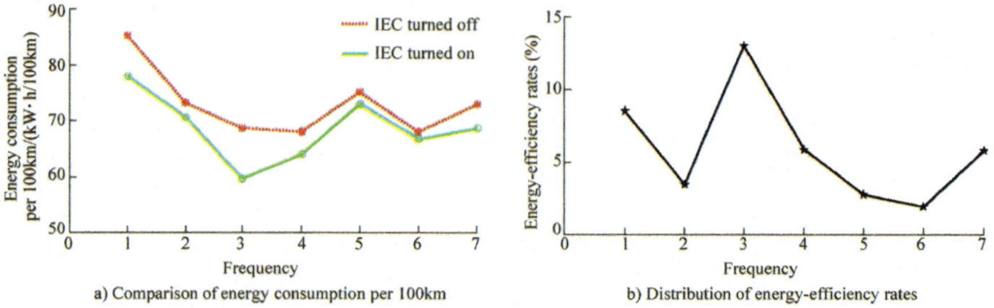

Fig. 6.32 Statistics of overall experimental results

fluctuates greatly due to the dynamic and time-varying nature of the actual traffic flow and the differences in the actual operation of the driver in different experiments, which causes the energy saving rates of different experiments to fluctuate greatly. However, the results of the 7 experiments show that the overall energy consumption per 100 km of the sample vehicle with IEC on is significantly lower than that of the sample vehicle with IEC off. The maximum energy-saving rate of the 7 experiments is 13%, the minimum rate is 1.9%, with the average being 5.9%.

6.5.3.2 Analysis of Experimental Results for the Whole Road Section

Figures 6.33, 6.34, 6.35, 6.36, 6.37 and 6.38 shows a set of experimental results of the vehicle with IEC turned on during the journey along the road section. The whole distance travelled during the experiment is 15,655 m and it takes 1225 s to finish the journey. The average speed of the vehicle is 46 km/h and the battery SOC value dropped from 89.3 to 83.1% under the entire working conditions. The measured power consumption is 11.1 kWh and the energy consumption per 100 km is 70.7 kWh/100 km.

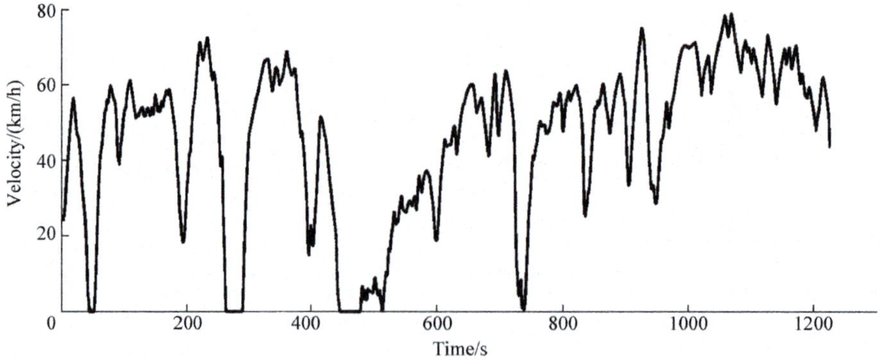

Fig. 6.33 Changes in vehicle velocity throughout the experiment

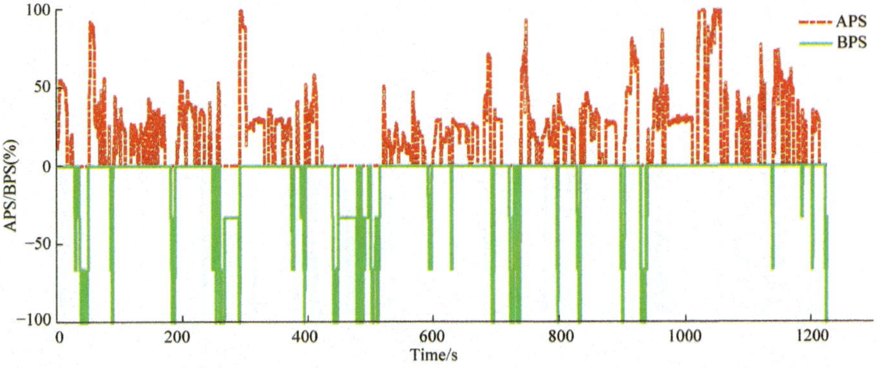

Fig. 6.34 Changes in the acceleration pedal (APS) and brake pedal (BPS) throughout the experiment

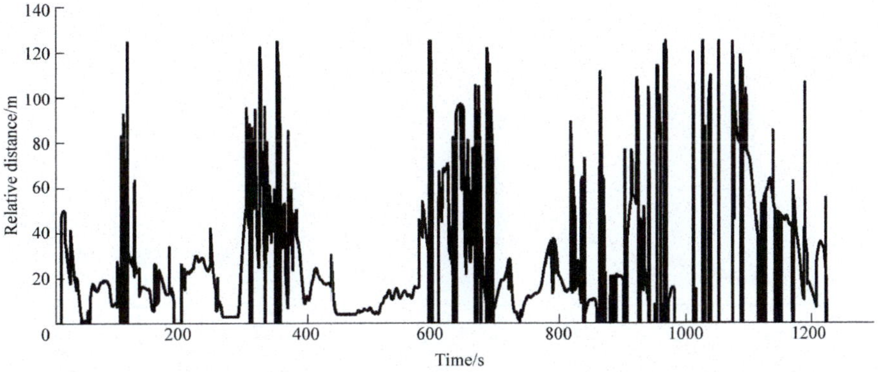

Fig. 6.35 Changes in the relative distance between the ego vehicle and the preceding vehicle throughout the experiment

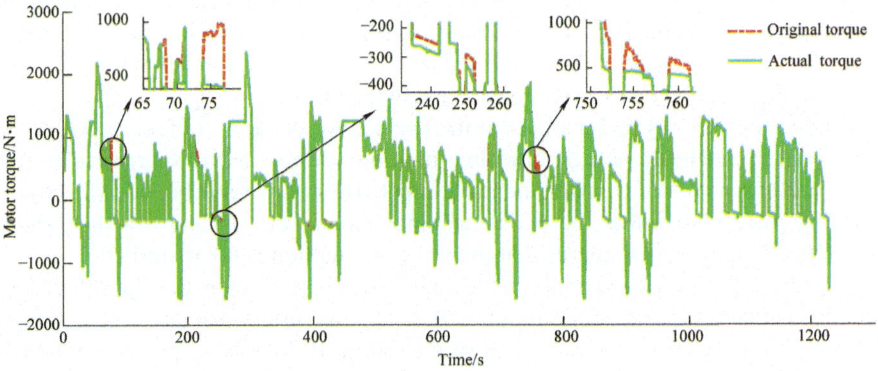

Fig. 6.36 Comparison of changes in motor torque throughout the experiment

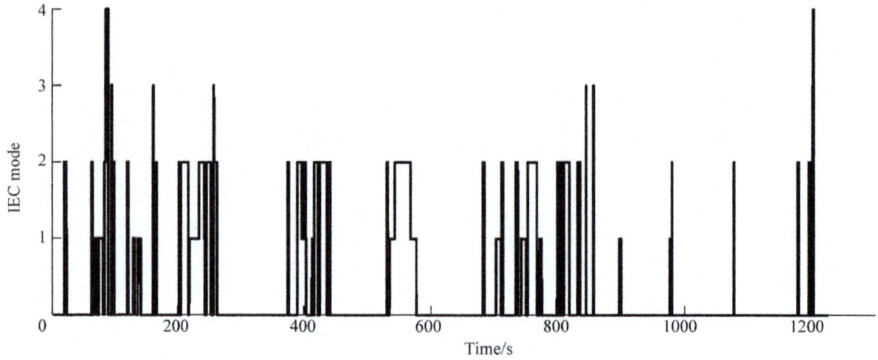

Fig. 6.37 IEC mode switching throughout the experiment

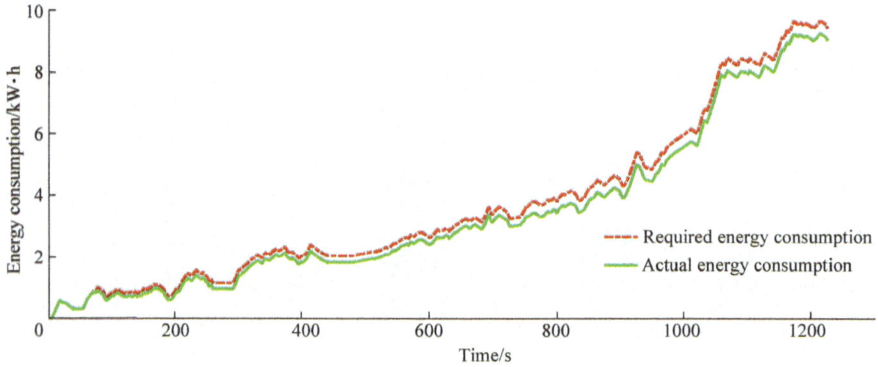

Fig. 6.38 Comparison of motor energy consumption throughout the experiment

Figure 6.33 shows the changes in the velocity of the ego vehicle throughout the experiment. Due to the high traffic flow in some road sections of the city, the vehicle stopped 5 times in some of the congested road sections. Between 1000 s and 1200 s, the vehicle travelled on the expressway at a higher speed.

Figure 6.34 shows the changes in the accelerator pedal and brake pedal of the vehicle. Affected by the actual traffic flow, there are several cases of sharp acceleration and deceleration braking during the experiment. The comparison between this figure and Fig. 6.33 shows that the pedal opening remains consistent with the vehicle velocity changes, which proves that the driver's intention is not disturbed.

Figure 6.35 shows the changes in the relative distance between the ego vehicle and the preceding vehicle detected by the radar. The maximum distance to the vehicle that can be detected by the radar can reach 130 m. If there is no preceding vehicle or the distance to the preceding vehicle exceeds the radar detection range, the output result is null, which is shown as 0 in the figure by default. Due to the dynamic and time-varying nature of the actual traffic flow, if a preceding vehicle cuts in or cuts out or when the radar signal is lost at a certain time point, the curve of the relative distance between two vehicles will suddenly jump and change. However, in general, it shows that the vehicle maintains a certain safe distance to the preceding vehicle.

Figure 6.36 shows changes in vehicle's drive motor torque, where the dashed line represents the original torque calculated by the VCU and the solid line represents the actual torque after IEC optimization. It's shown in the figure that the driving torque of the motor was significantly reduced between around 75 s and 755 s and the brake torque of the motor was increased at around 250 s.

Figure 6.37 shows IEC mode switching. The mode switching pattern is related to the changes in the vehicle velocity shown in Fig. 6.33 and the changes in the relative distance shown in Fig. 6.35. For example, IEC switches to the standby mode after 1000 s when the radar has not detected the information on the preceding vehicle. The vehicle velocity is relatively high and the motor torque is not limited during this period. The IEC mode switching shown in Fig. 6.37 is also consistent with the changes in the motor torque shown in Fig. 6.36, that is, the motor torque is optimized more significantly in the energy-efficiency mode 3 and mode 4. Statistically, IEC entered energy-efficiency mode 1–4 at 12 s, 67 s, 3 s and 2 s respectively. For the rest of the time, the vehicle was in the standby mode and remained in the standby mode for 93% of the time, which is determined by the control strategy for turning on and turning off IEC. In any case that the preceding vehicle's relative distance to the vehicle exceeds the IEC's maximum acting distance or radar detection range, the directional light is turned on or the yaw rate is greater than the set limit and the radar signal is temporarily lost, IEC enters the standby mode to avoid interference with the driver.

Figure 6.38 shows the comparison of the curves of energy consumption under the entire working condition. Since the actual traffic flow is dynamic and time-varying and the state of the preceding vehicle is also randomized in each of the experiments, it's not meaningful to compare the energy consumption curves before and after turning on IEC in a single experiment. In order to better compare the energy consumption reduction after IEC is turned on, the actual motor torque and motor speed after IEC is turned on are multiplied to calculate the actual energy consumption of the motor and the original torque and motor speed calculated by the VCU are multiplied to calculate the motor's required energy consumption. The comparison between the two is shown in Fig. 6.38. It's shown in the figure that the energy consumption of the motor after IEC optimization is significantly reduced by about 4.3%.

6.5.3.3 Analysis of Experimental Results for a Typical Road Section

The long experimental data from the whole road section makes it difficult to observe. In order to analyze the working of different components after IEC is turned on, a set of experimental data from the typical road section has been taken for analysis.

Figure 6.39 shows the experimental results of sample vehicle A on a typical road section. Figure (a) shows the changes in vehicle velocity and Figure (b) shows the changes in vehicle accelerator pedal and brake pedal. The comparison of the two figures shows that the changes in pedal opening are consistent with the changes in vehicle velocity and the driving intention is not disturbed. Figure (c) shows the relative distance between the ego vehicle and the preceding vehicle and it can been seen from the figure that the two vehicles maintain a certain safe distance. Figure (d) shows the relative speed between the ego vehicle and the preceding vehicle. If the distance value is greater than 0, it indicates that the two vehicles are leaving from each other and if the distance value is not more than 0, it indicates that the two vehicles are driving close to each other and the changes in the distance value provides the inputs for energy-efficiency mode switching. Figure (e) shows the changes in yaw rates. If the absolute value of the yaw rate is greater than the set threshold, IEC system will think that the vehicle is changing lanes or making a turn and will enter the standby mode to avoid interference with the driver. Figure (f) shows the IEC mode switching and It's shown in the figure that the IEC system performs mode switching as the relative motion between two vehicles changes and the yaw rate changes. For example, at 83 s, the yaw rate was greater than the set threshold and IEC entered the standby mode. At 88 s, the vehicle had changed lanes, as the relative distance between the two vehicles was smaller, but the relative speed was greater than 0, the system entered mode 3; as the vehicle velocity increased, the relative speed between the two vehicles turned negative at 93 s and the system entered mode 4, so that the vehicles decelerated to ensure the safe distance. Figure (g) shows the changes in motor torque, where the dashed line represents the original motor torque calculated by VCU and the solid line represents the actual motor torque after IEC optimization. It's shown in the figure that the motor driving torque is significantly reduced during 80–84 s and 100–102 s and the motor brake torque was partially increased in the time period of 93–100 s. Figure (h) shows the changes in the motor energy consumption in this road section in detail and it can be clearly seen from the figure that the motor energy consumption is reduced after IEC optimization.

6.5.4 Comparison of Experimental Results of Energy-Efficiency Control of Vehicles in Different Urban Road Conditions

In order to analyze the effects of traffic conditions on the energy-efficiency effect, the motion states and energy-efficiency rates of sample vehicle B and sample vehicle A in the experiments were calculated and compared, as shown in Table 6.10.

According to the table, sample vehicle A shows significantly better energy-efficiency effects during the experiment than sample vehicle B. In addition to the differences in the vehicle control strategy and driver's operation of the vehicle, the working conditions chosen for the experiments on the two vehicles are also one of the main factors causing this difference in energy-efficiency effects. Sample vehicle B ran on a general urban road which included a section of urban expressway at an

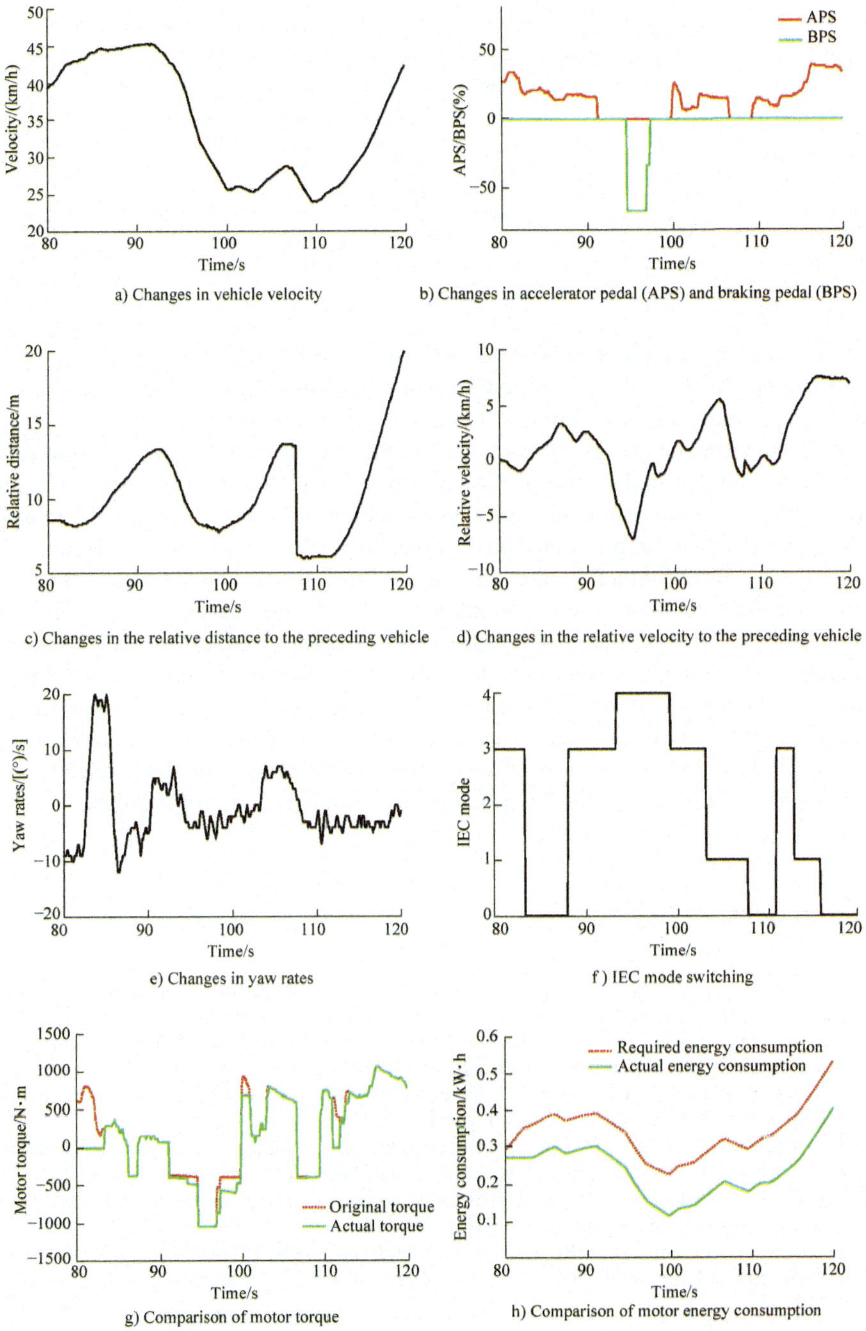

Fig. 6.39 Experimental results for a typical road section from the experiments on sample vehicle B

Table 6.10 Comparison of experimental results of sample vehicle B and sample vehicle A

Vehicle	IEC state	Average vehicle velocity (km/h)	Energy consumption per 100 km (kWh/100 km)	Energy-efficiency rate (%)		
				Max	Min	Avg
Sample vehicle A	On	16.1	44.5	15.7	1	7
	Off	15.2	47.8			
Sample vehicle B	On	47.5	68.7	13	1.9	5.9
	Off	45.3	73.1			

average velocity of 46 km/h, while sample vehicle A ran on a congested road along an urban bus route with high traffic flow and more complicated traffic conditions, at an average velocity of 15.7 km/h.

The experimental energy-efficiency effect under congested urban road conditions are better than that under general road conditions, which is also determined by the IEC strategy. On the one hand, under congested urban road conditions, the relative distance between the ego vehicle and the preceding vehicle is quite small and the IEC is turned on more frequently, the vehicle can thus receive better energy-efficiency effect. On the other hand, as traffic conditions are more time-varying under congested urban road conditions, the vehicle starts and stops more frequently and the driver's operation varies greatly each time. Once turned on, IEC optimizes the motor torque based on the actual working conditions, which corrects the aggressive operation of the driver to a certain extent, the vehicle can thus receive a better energy-efficiency effect. The experimental results of the two sample vehicles in the table have further verify the effectiveness of the proposed energy-efficiency control strategy.

References

1. Van Keulen T, De Jager B, Foster D et al (2010) Velocity trajectory optimization in Hybrid Electric trucks. In: American control conference. IEEE, pp 5074–5079
2. Xie L, Zhang D, Luo Y, Li K, Chen R (2018) Radar sharing energy-saving control strategy for intelligent hybrid electric vehicle. J Tsinghua Univ (Sci Technol) 58(3):286–291
3. Li SE, Li K, Wang J (2013) Economy-oriented vehicle adaptive cruise control with coordinating multiple objectives function. Veh Syst Dyn 51(1):1–17
4. Luo Y, Chen T, Li K (2015) Multi-objective decoupling algorithm for active distance control of intelligent hybrid electric vehicle. Mech Syst Signal Process 64–65:29–45
5. Dang R, Li S, Wang J et al (2012) Adaptive cruise control system of electric vehicles considering both economy and safety. Autom Eng 34(5):379–384
6. Xie LQ, Luo YG, Zhang DH, Rui C, Li KQ (2019) Energy-saving control strategy of intelligent electric vehicle based on preceding vehicle movement. Mech Syst Signal Process 9(130):484–501

Chapter 7
Implementation of i-HEV ACC Based on Control Collaboration

Based on the established new-concept vehicle structure and key technology system of i-EFV, this chapter focuses on the ACC of intelligent hybrid electric vehicles (i-HEVs) as a typical application, aiming to design and comprehensively implement a safe, economic and comfortable overall structure of the i-HEV ACC control system based on its multi-performance objectives and the requirements of the multi-mode coordinated control of the drive/braking system.

7.1 Structure of the i-HEV ACC Control System

According to existing studies on optimizing energy allocation of hybrid power systems, optimizing vehicle velocity and system power at the same time can further improve fuel economy. The designed i-HEV ACC can optimize velocity control in a local range when when following the preceding vehicle, which indicates a potential to improve vehicle fuel economy while balancing longitudinal driving safety and comfort. However, compared with conventional ACC, i-HEV ACC is an integration of electromechanical coupling systems such as engine, motor and battery, which represents higher complexity in terms of vehicle control target, controlled object's structure and fuel economy characteristics of the system. The design of the system mainly faces the following challenges:

(1) Multi-objective coordinated optimization

i-HEV ACC requires a coordinated optimization of multiple performance objectives including longitudinal driving safety, fuel economy of hybrid power system and longitudinal driving comfort. During the operation of the i-HEV ACC system, performances such as longitudinal driving safety, fuel economy and longitudinal driving comfort are coupled with each other, harboring contradictions and conflicts. If the ego vehicle follows the preceding vehicle that frequently accelerates and decelerates to achieve accurate car-following, it will inevitably increase the changes of the

K. Li et al., *The Intelligent Environment Friendly Vehicle*,
Key Technologies on New Energy Vehicles,
https://doi.org/10.1007/978-981-19-4851-0_7

engine and motor's torque and thus the system's fuel consumption, which improves the longitudinal driving safety while reducing the fuel economy of the vehicle. At the same time, the excessive acceleration and deceleration caused by the fast car-following demand will also increase the driver's sense of frustration and reduce the driving comfort. On the other hand, excessively pursuing fuel economy by reducing the engine and motor torque changes will increase the errors in car-following velocity and distance and compromise the car-following effect, even leading to situations such as preceding vehicle cutting-in and driver intervention, which may reduce the safety of vehicle car-following.

(2) Discreteness and nonlinearity of multiple systems

In an i-HEV ACC, engine, motor, hydraulic brake and AMT operate in a multi-mode discrete switching manner. The fuel efficiency characteristics of each subsystem are highly nonlinear, and the dynamic response characteristics of each component of the drive/braking system are greatly different. Therefore it is necessary to comprehensively consider the discretization, nonlinearity and dynamic characteristics of the above systems to design the i-HEV ACC control system.

(3) Coupling of multiple energy economies

On the one hand, since fuel and electric power systems are coupled with each other, the economy of a hybrid power system shall be measured through both the instantaneous fuel consumption of the engine and the consumption of electricity, instead of solely the former. On the other hand, the system economy is coupled with the future driving conditions of the vehicle: since the electricity consumed is mainly compensated by engine charging, the economy of a hybrid power system is not only related to the current working state of system components, but also related to the charging and discharging state of the engine and motor battery pack in the future.

The framework of the i-HEV ACC control system is designed based on the idea of hierarchical control by comprehensively considering problems such as multi-objective coordinated optimization, multi-system discretization, nonlinearity, and multi-energy economy coupling, as shown in Fig. 7.1.

As suggested in Fig. 7.1, the problem of i-HEV ACC optimal control that requires the comprehensive coordination of multiple performance objectives (safety, economy, and comfort) is decoupled into sub-problems to be addressed in a hierarchical manner, such as multi-objective steady-state optimization and multi-system dynamic coordination, etc. The layer of multi-objective steady-state optimization, based on the steady-state characteristics of each sub-system, mainly addresses the coordinated optimization issues of longitudinal driving safety, fuel economy of hybrid power system and longitudinal driving comfort. The layer of multi-system dynamic coordination, based on the difference of dynamic response characteristics of each subsystem, mainly addresses the problems of dynamic coordination control of engine, motor and EVB hydraulic braking system in the process of mode switching and joint operation. In addition, since the efficiency characteristics of fuel and electric power systems are coupled with each other, it is necessary to estimate the battery's equivalent fuel consumption factor based on the traffic environment information

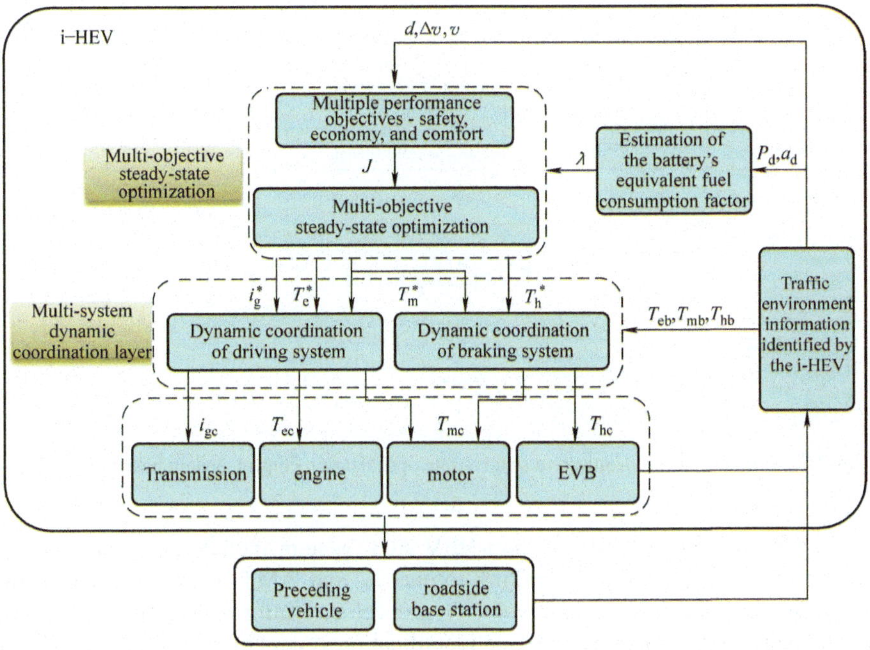

Fig. 7.1 Structure of the i-HEV ACC control system

identified by the i-HEV, so as to solve the problem of conversion between fuel and electricity's economic indicators. Detailed explanations are as followed.

(1) Multi-objective steady-state optimization

This layer ignores the dynamic response characteristics of the engine, motor, hydraulic brake and AMT systems, and only considers the steady-state efficiency characteristics of each component to calculate the steady-state optimization control instructions of each component in the process of i-HEV ACC car-following.

In order to realize the coordinated optimization of the system's longitudinal driving safety, fuel economy and longitudinal driving comfort, Nonlinear Model Predictive Control (NMPC) theory is applied to design the steady-state optimization control algorithm of the i-HEV-ACC system based on the discrete and nonlinear characteristics of the fuel efficiency and power transmission of the i-HEV-ACC system. The control system's structure is shown in Fig. 7.2.

As shown in Fig. 7.2, to simplify the difficulty of solving the optimal control of nonlinear model prediction of the i-HEV ACC, the vehicle's longitudinal acceleration is limited as the indicator of comfort [1] and designed as a constraint for the optimal control of multi-objective stead states. The driver's requirement for longitudinal driving comfort is met by limiting the vehicle's corresponding driving power and braking power, while i-HEV ACC focuses on coordinated optimal control of longitudinal driving safety and fuel economy.

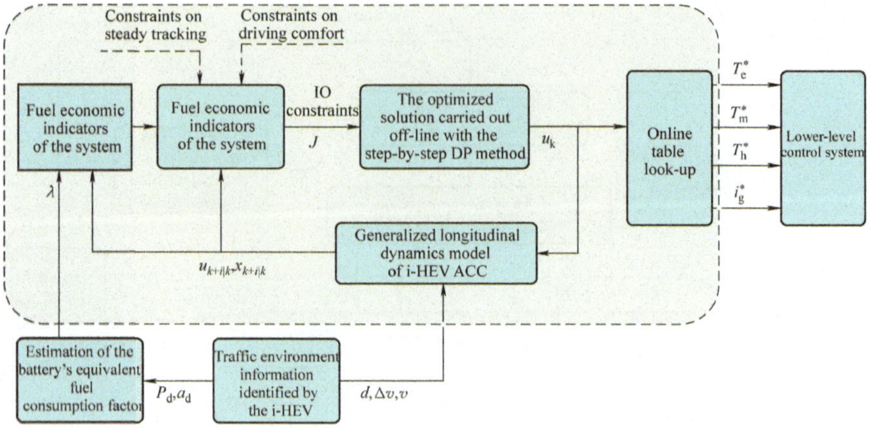

Fig. 7.2 Structure of multi-objective steady-state optimization control system based on NMPC

The controlled variables of the i-HEV ACC system include engine torque T_e, motor torque T_m, hydraulic braking torque T_h and AMT gear i_g. Based on the car-following distance, relative velocity and velocity information obtained by the vehicle and environment information identification system of i-HEV, the future state is predicted by using the established generalized longitudinal dynamics model of i-HEV ACC. Based on the estimated battery's equivalent fuel consumption factor λ and the controlled variable of vehicle prediction $u_{k+i|k}$, the fuel economy of the hybrid power system (fuel and electricity) is calculated. Considering the comfort and steady car-following constraints, the nonlinear constrained optimization objective function is further established coordinating the car-following safety indicator and the fuel economy indicator of i-HEV system within the prediction domain; in order to realize the real-time solution and control of the problem of the nonlinear model's predictive optimization, the optimized solution is carried out off-line with the step-by-step Dynamic Programming (DP) method, and the optimal control law of each component of the i-HEV ACC was obtained by online table look-up.

i-HEV ACC multi-objective steady-state optimal control based on NMPC effectively solves the problem of coordinated optimal control with the presence of the coupling and conflicts of multiple performance targets, such as safety, economy, and comfort. As shown in Fig. 7.3, assuming that the preceding vehicle runs at a uniform velocity, the iHEV ACC control process starts from the initial car-following error and gradually transitions to a stable car-following state, which provides room for the coordinated optimization of vehicle longitudinal driving safety, fuel economy and longitudinal driving comfort. Every car-following process of i-HEV ACC from unbalance to balance is a process of local coordinated optimization which is realized by applying the NMPC theory. The control process of i-HEV ACC following the movement of the preceding vehicle is described as: integrate the control of engine, motor, battery pack, AMT, and hydraulic braking system components over a period, to make the ego vehicle run at the car-following distance and velocity as expected

Fig. 7.3 i-HEV ACC car-following process

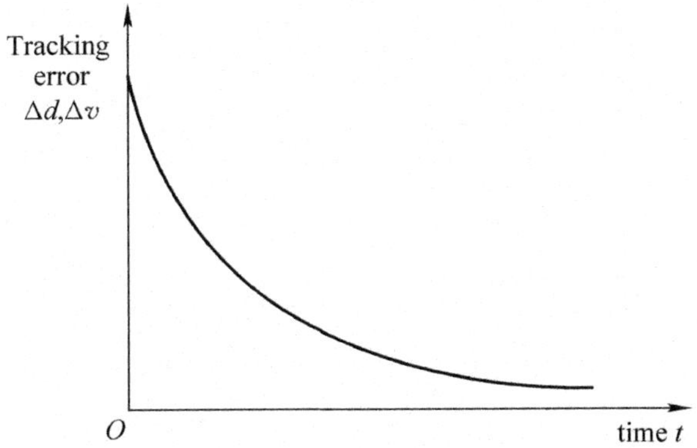

by the driver, and to realize longitudinal driving comfort while keeping the optimal fuel economy during car-following.

(2) Multi-system dynamic coordination

The driving and braking processes of i-HEV include different working modes, such as engine-only driving, motor-only driving, engine/motor co-driving, motor energy regeneration, and motor/hydraulic co-braking, etc., which requires coordinated control of various components to achieve the continuity of power output during mode switching. Engine and hydraulic brake, on the other hand, is slow in response and cannot be precisely controlled, while the motor has fast response, high control precision, and measurable torque. From the perspective of improving comprehensive control of the system performance, it's required to coordinate the control of the engine, hydraulic braking, and electrical components to achieve fast and accurate system dynamic response.

Based on the dynamic response and control precision characteristics of engine, hydraulic brake and motor, a dynamic coordinated control algorithm of drive and braking system combining open loop and closed loop is designed in this book. The engine and hydraulic brake are applied with open loop control based on specification models $G_E(s)$, $G_h(s)$ to reduce the vehicle's real-timeliness and accuracy demand for engine, hydraulic and brake response, while the motor is applied with closed-loop control by adopting the two-degree-of-freedom model matching controller with feedforward and feedback to rapidly respond to the vehicle's requirement for driving/braking force, and to improve the longitudinal driving safety when the preceding vehicle sharply accelerates or decelerates. The basic structures of dynamic coordinated control of driving and braking systems are shown in Figs. 7.4 and 7.5.

Figure 7.4 shows the dynamic coordinated control structure of the i-HEV ACC driving system, which mainly solves the problem of coordinated control of engine and motor during mode switching and joint operation. Since the engine has a slower response and the motor has a faster response, from the perspective of system safety and fast responsiveness, the torque error caused by the slow response of the engine can be compensated by the motor in real-time to realize the fast following of the

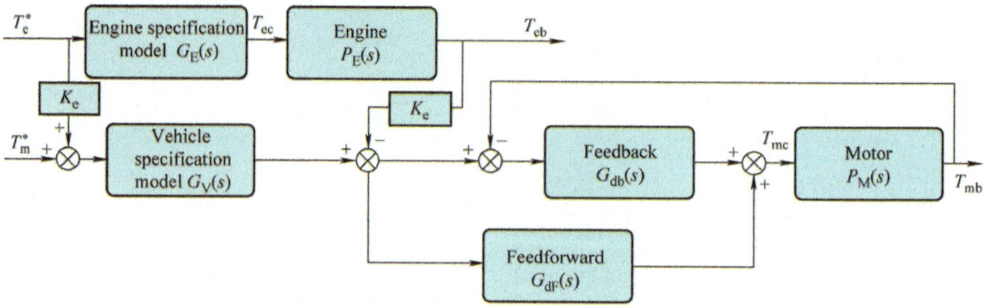

Fig. 7.4 Dynamic coordinated control of driving system

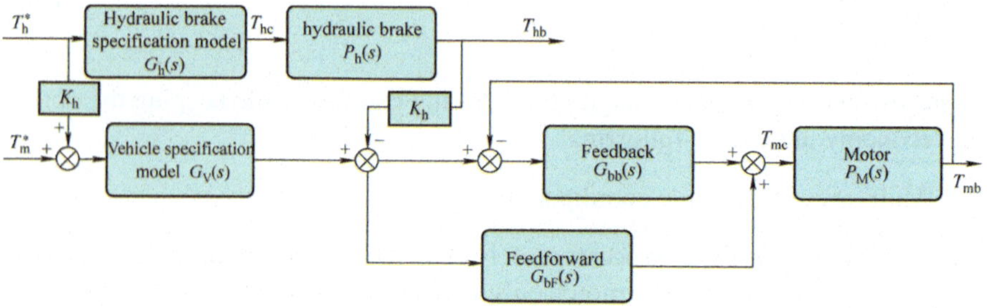

Fig. 7.5 Dynamic coordinated control of braking system

desired driving power. Because the AMT and the shift process control of the designed i-HEV are completed by the maker itself, this paper mainly studies the dynamic coordinated control algorithm of mode switching and joint operation of the engine and the motor, to meet the requirements of continuous power supply and fast following of the expected drive torque during the driving process.

Figure 7.5 shows the dynamic coordinated control structure of the i-HEV ACC braking system. Safety is the only goal during the braking process. Because the hydraulic brake has a slower response and the motor has a faster response, the dynamic coordinated control of the braking system composed of the motor and the hydraulic brake is mainly realized by coordinating the motor braking energy regeneration torque and the hydraulic brake torque to address the slow response of the hydraulic brake and meet the vehicle's demand for fast braking.

(3) Estimation of equivalent fuel consumption factor of battery

In order to establish a quantitative fuel economy indicator for the i-HEV ACC system, it is required to comprehensively consider the engine fuel consumption and electric energy consumption. The conversion method of battery's equivalent fuel consumption is adopted in this paper: the electric energy is converted into fuel consumption based on the estimated battery's equivalent fuel consumption factor. Because i-HEV integrates wireless communication systems such as GPS/GIS and wireless V2I communication, the information of traffic condition characteristics can be obtained. Therefore, it's assumed in the paper that the driving condition of the vehicle is obtained, and the i-HEV overall energy optimization allocation problem is established based on the simplified engine fuel consumption model and battery and motor efficiency model. The optimal power allocation problem is solved based on the unconstrained optimization theory to estimate the battery's equivalent fuel consumption factor λ of i-HEV ACC.

7.2 Technical Difficulties and Key Points of i-HEV ACC

Based on the characteristics of i-HEV ACC including multi-objective coordinated optimization, discretization and nonlinearity of multi-system and mutual coupling of multi-energy economies, the technical difficulties and key points involved in the i-HEV ACC control system are summarized in terms of information identification, definition of performance indicators and system target of execution, as shown in Fig. 7.6.

As shown in Fig. 7.6, i-HEV ACC has characteristics such as identification information coupling, diverse control targets, and system response difference. For the i-HEV ACC system, the collected environmental information includes the relative velocity and distance between the preceding vehicle and the ego vehicle and current

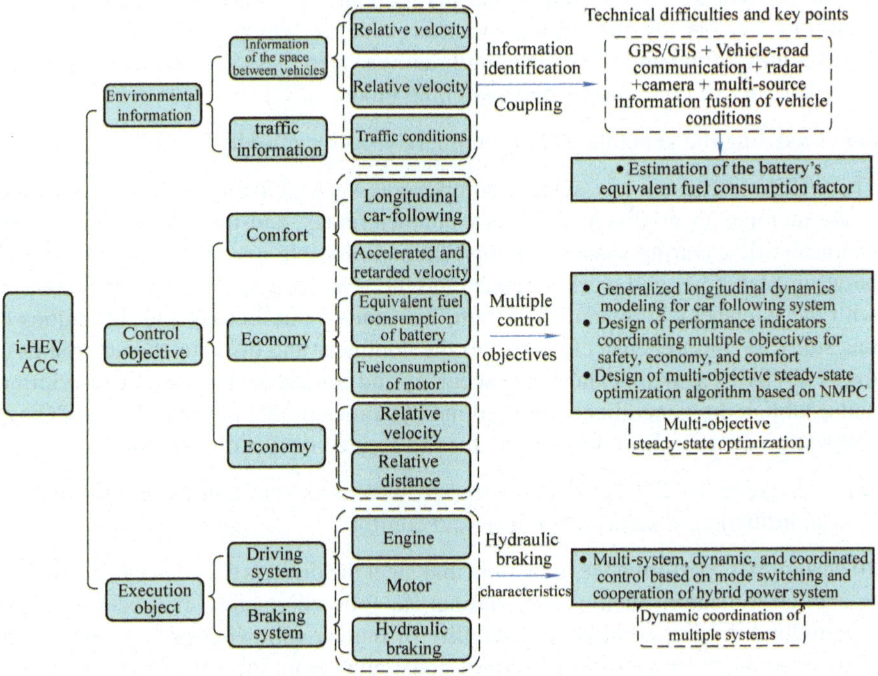

Fig. 7.6 Schematic diagram of the key technologies of the i-HEV ACC control system

traffic conditions. The goal is to achieve comprehensive longitudinal safety, economy, and comfort performance in the car-following process. The executive power components of the system mainly include the engine, motor, and hydraulic brake. The key technologies of i-HEV ACC control system design include multi-source information fusion based on GPS/GIS/V2I communication/radar/camera/vehicle state; coordination and optimization of multiple objectives including safety, economy, and comfort; dynamic coordinated control of electromechanical coupling system based on mode switching and response characteristics of the power system.

This chapter focuses on the design of the i-HEV ACC multi-objective and multi-system coordinated control algorithm and the estimation of related characteristic parameters, which mainly involve the following 5 technical difficulties or key points: (1) modeling of the i-HEV ACC generalized longitudinal dynamics system; (2) i-HEV ACC performance indicator design considering multi-objective coordination of safety, economy, and comfort; (3) design of the multi-objective steady-state optimization control algorithm of i-HEV ACC based on nonlinear model prediction theory; (4) design of the multi-system dynamic coordinated control algorithm based on the mode switching and joint operation of the hybrid power system; (5) estimation of equivalent fuel consumption characteristics of battery based on operating conditions. Points (1), (2), and (5) are the basis of the multi-objective steady-state optimal controller in point 3), which respectively involve problems such as control system modeling, control goal design, and identification of efficiency character parameters of the hybrid power system. Point (4) is about the lower-level controller of the multi-objective steady-state optimization in point (3), which carries out the coordinated control of the underlying driving/braking system based on the actual response characteristics of components and switching mode.

(1) Modeling the generalized longitudinal dynamics of i-HEV ACC

The i-HEV integrates key system components such as the engine and the drive/brake motor with nonlinear efficiency characteristics, and the AMT with discrete characteristics. During system driving/braking, multiple power components switch between each other and work together, and the vehicle dynamics system is coupled with the inter-vehicle dynamics system. It is one of the focuses and difficulties of this study to comprehensively consider the nonlinear and discrete characteristics of the i-HEV ACC system, balance the accuracy and conciseness of system description, and establish the generalized longitudinal dynamics model of the i-HEV ACC car-following system to lay a foundation for the optimal design of the control system.

(2) Designing i-HEV ACC performance indicators considering multi-objective coordination of safety, economy, and comfort

To design ACC performance indicators that meet the driver's expectation, in addition to explicit indicators such as longitudinal car-following safety, fuel economy, and longitudinal driving comfort, the stability of the car-following process also needs to be guaranteed by a proper design of constraints from the perspective of control system design. To achieve the driver's expected longitudinal driving characteristics for a specific driving distance, it is required to establish reasonable economy, safety, and comfort indicators based on the spatial domain of the vehicle's driving distance and comprehensively consider the engine fuel consumption and battery power consumption. Therefore, it's another focus and difficulty of this study to comprehensively consider multiple performances including safety, economy, and comfort when designing the objective function of the multi-objective coordinated i-HEV ACC optimal control.

(3) Designing the multi-objective steady-state optimal control algorithm of i-HEV ACC based on the nonlinear model prediction theory

Since the i-HEV ACC control system model has nonlinear and discrete characteristics, it is required to design the corresponding control algorithm based on the nonlinear model prediction theory in order to achieve accurate optimal control of the system. On the other hand, NMPC requires a large amount of calculation, making it difficult to realize the online control of the system. So far, there is no mature solution for the online rolling optimization of the nonlinear model prediction. Therefore, it is also one of the focuses and difficulties of this study to deduce the corresponding nonlinear model prediction problem based on the nonlinear and discrete characteristics of the i-HEV ACC control system model while carrying out an effective optimal solution to achieve online control.

(4) Designing the multi-system dynamic coordinated control algorithm based on mode switching and joint operation of hybrid power system

In the hybrid drive/braking system, the engine and hydraulic braking system have a slow response and low control precision, while the motor has fast response and high control precision. The process of system driving/braking contains mode switching and joint operation between the engine and motor driving, and between hydraulic braking and motor braking. It's a focus of this study to comprehensively consider the response characteristics of the engine, motor, and hydraulic braking system to realize smooth transition, and fast response of overall power during the multi-system mode switching and joint operation of those systems.

(5) Estimating the equivalent fuel consumption characteristics of the battery

The equivalent fuel consumption characteristics of the battery in a hybrid power system is an important basis for optimal energy allocation, which is highly nonlinear due to the complexity of hybrid system objects, and the estimation of characteristics is closely related to the current working scenario of the system and energy allocation control strategy adopted. It's a focus of this study to reasonably simplify the i-HEV hybrid power system and estimate the equivalent fuel consumption characteristics of the battery.

7.3 Key Technologies of the i-HEV ACC System

7.3.1 Multi-objective Steady-State Optimization

In this section, the generalized longitudinal dynamics model of the i-HEV ACC system is first established based on the i-HEV ACC system's steady-state characteristics, and the quantitative objective function and constraint control system integrating safety, economy, and comfort is designed by taking into account the system's multi-energy and control stability requirements. On this basis, the i-HEV ACC multi-objective steady-state optimization problem is derived based on the nonlinear model predictive theory, and then the real-time control law calculation method combining offline rolling optimization solution and online table look-up is designed for the problem of real-time optimization of the nonlinear model predictive problems.

7.3.1.1 Modeling the Generalized Longitudinal Dynamics of i-HEV ACC

The complex structure of the i-HEV hybrid power system is simplified based on the real-life debugging experience of the original HEV system's energy allocation control, and the overall structure of the i-HEV ACC generalized longitudinal dynamics model is established. The longitudinal dynamics model of i-HEV with hybrid characteristics is then established by taking into account characteristics of i-HEV ACC such as multi-mode discrete switching between driving/braking systems, discrete shifting of the AMT, and quadratic nonlinear wind resistance of the dynamics system. The i-HEV dynamics and inter-vehicle kinematics system are further integrated to establish the i-HEV ACC generalized longitudinal dynamics model.

(I) Overall structure of i-HEV ACC generalized longitudinal dynamics model

The i-HEV power system covered in this chapter is a series–parallel hybrid structure, as shown in Fig. 7.7. Its driving system includes an engine, a BSG, and a PM drive motor, and the braking system includes a PM motor brake and an EVB hydraulic brake. As indicated in the figure, the hybrid power system structure contains two motors, in which the generator is used to charge the battery or quickly start the engine and the driving motor is used to drive or recover braking energy. The working modes of the hybrid power system include engine-only driving, battery-only driving, engine driving + power generation, BSG power generation + PM motor driving, engine/PM motor co-driving, PM motor braking energy regeneration, and PM motor/EVB co-braking, etc.

Then the i-HEV ACC generalized longitudinal dynamics system model is established by focusing on the coordinated optimization of safety, economy, and comfort objectives based on the steady-state characteristics of the system and ignoring the dynamic response characteristics of each component. The overall structure is as

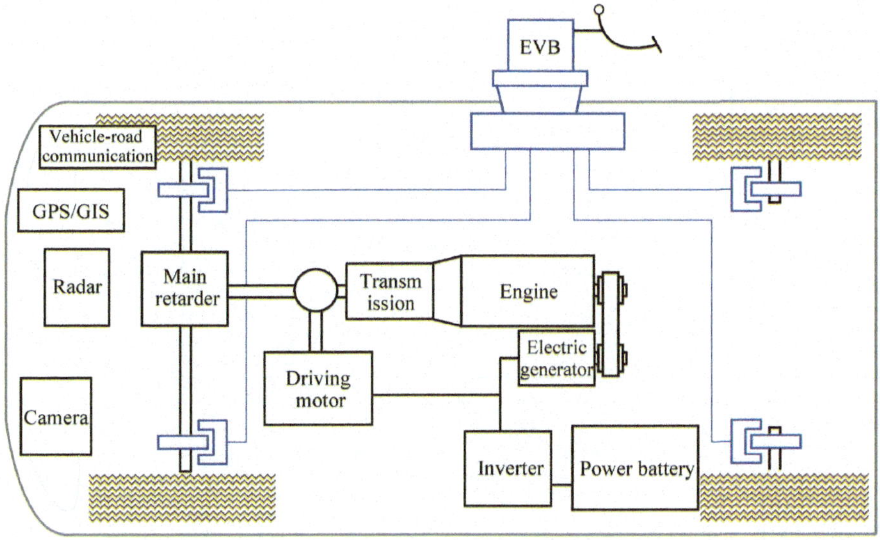

Fig. 7.7 Schematic diagram of the overall structure of the i-HEV system

shown in Fig. 7.8, in which T_e is the expected engine torque, T_m is the expected torque of the equivalent motor (EM) integrating BSG and PM drive motor, T_h is the expected torque of EVB hydraulic brake, i_g is AMT transmission ratio, F_w is vehicle air drag, F_f is rolling resistance, F_i is slope resistance, v is the velocity of the ego vehicle, a_f is the acceleration of the ego vehicle, a_p is the acceleration of the preceding vehicle, d_{des} is the safe inter-vehicle distance expected by the driver, Δd is distance error, and Δv is relative velocity.

As indicated by Fig. 7.8, the i-HEV power and transmission system is composed of the engine, the motor, the hydraulic braking and transmission system, and the relative movement relationship between the ego vehicle and the preceding vehicle is jointly expressed by the ego vehicle's velocity, acceleration and relative distance. The system input is the expected control torque of the engine, the motor and the hydraulic brake, and the transmission ratio of AMT. The torque of each component acts on the tire through the transmission system to generate the corresponding acceleration of the ego vehicle. On the other hand, the acceleration of the ego vehicle affects the velocity of the ego vehicle and the relative velocity and distance between the ego vehicle and the preceding vehicle. It can be seen that the generalized longitudinal dynamics model of i-HEV ACC can be decoupled into the longitudinal dynamics model of the i-HEV and the inter-vehicle kinematics model, while the vehicle longitudinal dynamics and the inter-vehicle kinematics are coupled to each other through the velocity and acceleration of the ego vehicle and the preceding vehicle.

The following is an analysis of the conversion relationship between the i-HEV ACC's equivalent motor EM, generator BSG and PM drive motor.

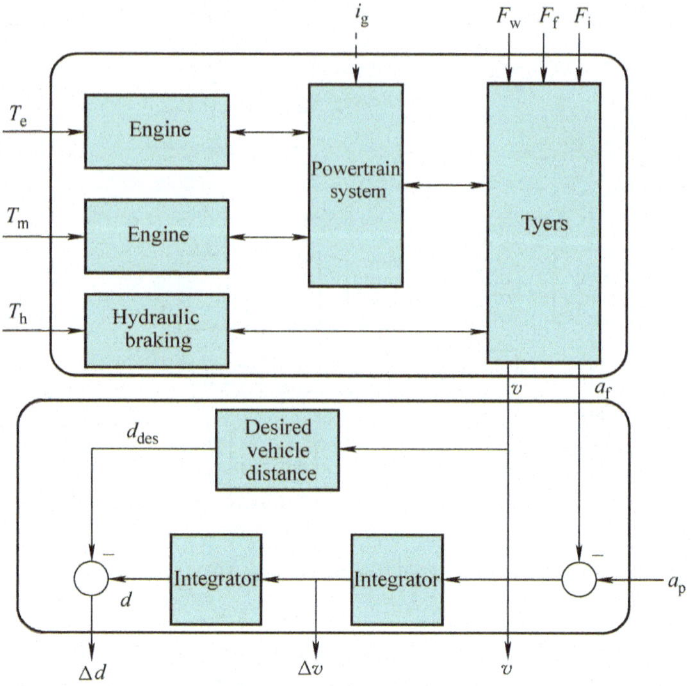

Fig. 7.8 Schematic diagram of the i-HEV ACC generalized longitudinal dynamics system model

In Fig. 7.7, the speed ratio of the engine to the BSG $i_{e\_BSG} = 0.625$, the rated torque of the BSG $T_{BSGRate} = 20$ Nm, and the corresponding rated generation torque of the engine end $T_{BSGRate}/i_{e\_BSG} = 32$ Nm. At this point, the engine is in a poor working point, which does not conform to the basic principle of hybrid power allocation for optimizing the working interval of the engine. Therefore, from the perspective of simplifying the integrated control of the i-HEV ACC system and the optimal energy allocation strategy, the serial driving form is not considered in the division of driving modes of the i-HEV power system, that is, the mode of engine power generation and drive motor driving is not considered. This basic processing method is consistent with the rule-based energy allocation strategy of real-life HEV that adopts this structure and does not affect the comprehensive control of i-HEV ACC and the optimal energy allocation of the hybrid power system. Based on the above analysis, the BSG and the PM drive motor do not work at the same time in the steady-state optimal energy allocation, so the equivalent motor EM combining the drive motor and the generator functions is designed. The equivalent motor EM has the same output transmission ratio as the PM drive motor, which is converted to the generator torque T_{BSG} pass coefficient K, as shown in Eq. (7.1).

$$
\begin{aligned}
& i_{\mathrm{m}} = i_{\mathrm{PM\_AMT}}, \quad K = \frac{i_{\mathrm{PM\_AMT}} i_{\mathrm{e\_BSG}}}{i_{\mathrm{g}}} \\
& T_{\mathrm{BSG}} = T_{\mathrm{m}} K, \quad T_{\mathrm{PM}} = T_{\mathrm{m}}
\end{aligned}
\tag{7.1}
$$

where i_{m} is the speed ratio of equivalent motor EM to the transmission output shaft, $i_{\mathrm{PM\_AMT}}$ is the speed ratio of the drive motor to the transmission output shaft, and T_{PM} is the torque of the drive motor.

(II) i-HEV longitudinal dynamics modeling

The main function of i-HEV ACC is longitudinal vehicle car-following, and its working scenarios include generally minor acceleration and deceleration movement of vehicles on the good road surface, which does not involve vehicle stability. Given the modeling requirements of the control system for the upper-level multi-objective steady-state optimization of the i-HEV ACC, the longitudinal dynamics system of the i-HEV is simplified as follows considering the accuracy and simplicity of the model:

(1) The longitudinal movement of the vehicle is considered, and the influence of lateral and vertical movement on the longitudinal dynamic characteristics is ignored;
(2) The tire sliding, spinning and the deformation of the drive shaft are ignored in good road surface and conventional driving conditions;
(3) The left and right of the vehicle are completely symmetrical, and the load transfer between front and rear axles is ignored. The longitudinal movement characteristics of the vehicle body are described by a single-wheel model;
(4) In the driving system, only steady-state characteristics are considered for the engine and the motors and the instant realization of expected torque is assumed;
(5) It's assumed that the mechanical power loss of the driveline is fixed, the shift of the AMT can be completed instantly without delay;
(6) In the braking system, the relationship between braking pressure and braking torque is simplified to fixed braking gain, and the braking torque is realized without delay.

It can be seen that based on the above basic assumptions of system modeling, the longitudinal dynamics system of the i-HEV has the following characteristics: the AMT contains five discrete varying gears; the working modes of the hybrid power system with integrated engine, motor and hydraulic braking include engine-only driving, engine driving + power generation, battery-only driving, engine/drive motor co-driving, motor braking energy regeneration and motor/hydraulic co-braking; there is a nonlinear element of air resistance in vehicle longitudinal dynamics. The longitudinal vehicle dynamics system of i-HEV is a typical hybrid nonlinear system, and its dynamics behavior has nonlinear and discrete characteristics.

The longitudinal dynamics model of i-HEV is obtained as shown in Eq. (7.2) based on the characteristics of working mode switching of the i-HEV ACC multi-energy system and the drive/brake force balance relationship during longitudinal driving.

$$a_{\mathrm{f}} = \frac{\mathrm{d}v}{\mathrm{d}t} = \frac{K_{\mathrm{ehj}} T_{\mathrm{eh}} + K_{\mathrm{mj}} T_{\mathrm{m}} - mgf_{\mathrm{r}} \cos\alpha - \frac{C_D A \rho}{2} v^2 - mg \sin\alpha}{\delta m} \tag{7.2}$$

where C_{D} is air resistance coefficient, A is the equivalent upwind area, ρ is air density, m is vehicle mass, δ is the conversion factor of vehicle rotating mass, f_{r} is rolling resistance coefficient, α is road slope, T_{eh} is equivalent conventional power torque integrating engine torque (T_{e}, when the i-HEV system drives) and hydraulic braking torque (T_{h}, when the system brakes), j ($j = 1,2,\ldots 7$) is the driving/braking mode of the i-HEV system based on which mode the i-HEV is in and the transmission ratio of the engine, the motor and the AMT, and K_{ehj} and K_{mj} are respectively the coefficients of equivalent conventional power torque and equivalent motor torque in mode j.

The effective control set of system control input $\boldsymbol{u}$ is divided according to the i-HEV system's mode division and driving/braking state, as shown in Table 7.1.

The driving/braking working mode j of the i-HEV system is calculated based on Table 7.1, as shown in Eq. (7.3).

$$\begin{aligned}
&\boldsymbol{\Omega} = \bigcup_{j=1}^{7} \boldsymbol{C}^{\mathrm{j}} \\
&\boldsymbol{C}^1 = \{\boldsymbol{u} = [T_{\mathrm{eh}}, T_{\mathrm{m}}, i_{\mathrm{g}}]^{\mathrm{T}} | T_{\mathrm{ehmin}} \le T_{\mathrm{eh}} \le 0, T_{\mathrm{mbmin}} \le T_{\mathrm{m}} \le 0\} \\
&\boldsymbol{C}^2 = \{\boldsymbol{u} = [T_{\mathrm{eh}}, T_{\mathrm{m}}, i_{\mathrm{g}}]^{\mathrm{T}} | T_{\mathrm{eh}} = 0, 0 < T_{\mathrm{m}} \le T_{\mathrm{PMmax}}\} \\
&\boldsymbol{C}^{(\mathrm{i}+2)} = \{\boldsymbol{u} = [T_{\mathrm{eh}}, T_{\mathrm{m}}, i_{\mathrm{g}}]^{\mathrm{T}} | 0 < T_{\mathrm{eh}} \le T_{\mathrm{ehmax}}, T_{\mathrm{BSGmin}} \le T_{\mathrm{m}} \le T_{\mathrm{PMmax}}, i_{\mathrm{g}} = \boldsymbol{I}_{\mathrm{g}}(i)\} \\
&i = 1, 2, \ldots 5
\end{aligned} \tag{7.3}$$

Table 7.1 I -HEV system working mode and corresponding set of effective control quantities

Code	i-HEV working mode	Set of effective control quantities
1	Engine driving	$0 < T_{\mathrm{eh}} \le T_{\mathrm{ehmax}}, T_{\mathrm{m}} = 0, i_{\mathrm{g}} \in \boldsymbol{I}_{\mathrm{g}}$
2	Engine driving with power generation	$0 < T_{\mathrm{eh}} \le T_{\mathrm{ehmax}}, T_{\mathrm{mgmin}} \le T_{\mathrm{m}} < 0, i_{\mathrm{g}} \in \boldsymbol{I}_{\mathrm{g}}$
3	Engine/drive motor co-driving	$0 < T_{\mathrm{eh}} \le T_{\mathrm{ehmax}}, 0 < T_{\mathrm{m}} \le T_{\mathrm{mdmax}}, i_{\mathrm{g}} \in \boldsymbol{I}_{\mathrm{g}}$
4	Battery driving	$T_{\mathrm{eh}} = 0, 0 < T_{\mathrm{m}} \le T_{\mathrm{mmax}}, i_{\mathrm{g}} = 0$
5	Motor braking energy regeneration	$T_{\mathrm{eh}} = 0, T_{\mathrm{mbmin}} \le T_{\mathrm{m}} < 0, i_{\mathrm{g}} = 0$
6	Motor/hydraulic co-braking	$T_{\mathrm{ehmin}} \le T_{\mathrm{eh}} < 0, T_{\mathrm{mbmin}} \le T_{\mathrm{m}} < 0, i_{\mathrm{g}} = 0$

Where T_{ehmin}, T_{ehmax} respectively correspond to the maximum hydraulic braking torque (negative) and the maximum driving torque of the engine, T_{mbmin}, T_{mgmin}, T_{mdmax} respectively correspond to the maximum braking energy regeneration torque (negative), maximum power generation torque (negative) and maximum drive torque that can be provided by the equivalent motor, $\boldsymbol{I}_{\mathrm{g}}$ is the set of transmission ratios of the 5-gear AMT.

where $\boldsymbol{\Omega}$ is the set of effective control inputs of i-HEV ACC system based on Table 7.1, $\boldsymbol{C}^{\mathrm{j}}$ is the set of control inputs of driving/braking mode j of i-HEV which corresponds to 7 working modes (braking, battery driving, gear-1 driving, gear-2 driving, gear-3 driving, gear-4 driving, and gear-5 driving) and the Equation of j Eq. (7.4) is shown as follows.

$$j = k \;\; if \;\; \boldsymbol{u} \in \boldsymbol{C}^{\mathrm{k}}, \quad k = 1, 2, \ldots 7 \tag{7.4}$$

Corresponding coefficients K_{ehj}, K_{mj} engine torque instruction T_{e}, and EVB hydraulic torque T_{h} in each i-HEV driving/braking mode can be expressed as:

$$\begin{aligned} & T_{\mathrm{e}} = 0, \; T_{\mathrm{h}} = T_{\mathrm{eh}}, \; K_{\mathrm{eh1}} = \frac{1}{R_{\mathrm{w}}}, \; K_{\mathrm{m1}} = \frac{i_{\mathrm{m}} i_0}{R_{\mathrm{w}} \eta_{\mathrm{m}} \eta_0} \;\; if \;\; j = 1 \\ & T_{\mathrm{e}} = 0, \; T_{\mathrm{h}} = 0, \; K_{\mathrm{eh2}} = 0, \; K_{\mathrm{m2}} = \frac{i_{\mathrm{m}} i_0 \eta_{\mathrm{m}} \eta_0}{R_{\mathrm{w}}} \;\; if \;\; j = 2 \\ & T_{\mathrm{e}} = T_{\mathrm{eh}}, \; T_{\mathrm{h}} = 0, \; K_{\mathrm{ehk}} = \frac{\boldsymbol{I}_{\mathrm{g}}(k-2) i_0 \eta_{\mathrm{g}} \eta_0}{R_{\mathrm{w}}}, \\ & \qquad K_{\mathrm{m1}} = \frac{i_{\mathrm{m}} i_0 \eta_{\mathrm{m}} \eta_0}{R_{\mathrm{w}}} \;\; if \;\; j = k, \; k = 3, 4 \ldots 7 \end{aligned} \tag{7.5}$$

where i_0 is final drive ratio, η_{m}, η_{g}, η_0 are the average mechanical transmission efficiencies of the equivalent motor, transmission and main reducing gear, and R_{w} is the wheel radius assuming that the mechanical efficiency of all components remains unchanged during driving and braking.

As indicated by Eqs. (7.2) and (7.5), the acceleration of the ego vehicle a_{f} is determined by the engine torque, hydraulic braking torque, and motor torque and has a quadratic nonlinear relationship with vehicle velocity, which discretely changes with the current AMT gear, driving/braking mode and the working mode of the hybrid power system, showing a typical non-linear and mixing characteristic.

(I) Integrating the generalized longitudinal dynamics model of the i-HEV ACC car-following system

The generalized longitudinal dynamics model of i-HEV ACC couples the vehicle dynamics and inter-vehicle kinematics which are correlated with each other through the velocity and acceleration of the vehicles. Based on the established longitudinal dynamics model of i-HEV, the longitudinal dynamics of i-HEV and inter-vehicle kinematics are integrated to establish a unified generalized longitudinal dynamics model of the i-HEV ACC car-following system.

In the inter-vehicle kinematics relationship, Δd is the difference between the relative distance between vehicles and the safe distance expected by the driver, and relative velocity Δv is the difference between the velocity of the preceding vehicle and that of the ego vehicle, as shown in Eq. (7.6):

$$\begin{aligned} \Delta d &= d - d_{\mathrm{des}} \\ \Delta v &= v_{\mathrm{p}} - v \end{aligned} \tag{7.6}$$

where v_{p} is the velocity of the preceding vehicle, d is the actual distance measured by the radar, and d_{des} is the safe distance expected by the driver. A linear expected distance model is adopted, as shown in Eq. (7.7).

$$d_{\mathrm{des}} = d_0 + \tau_{\mathrm{h}} v \tag{7.7}$$

where τ_{h} is the safe time interval, and d_0 is the fixed minimum distance.

If system status $\boldsymbol{x} = [\Delta d,\ \Delta v,\ v]^{\mathrm{T}}$ and the effect of road slope α on vehicle movement is ignored ($\alpha = 0$), the generalized longitudinal dynamics model of the i-HEV ACC car-following system is as shown in Eq. (7.8):

$$\begin{aligned} \frac{\mathrm{d}(\Delta d)}{\mathrm{d}t} &= \Delta v - \tau_{\mathrm{h}} \frac{\mathrm{d}v}{\mathrm{d}t} \\ \frac{\mathrm{d}(\Delta v)}{\mathrm{d}t} &= a_{\mathrm{p}} - \frac{\mathrm{d}v}{\mathrm{d}t} \\ \frac{\mathrm{d}v}{\mathrm{d}t} &= \frac{K_{\mathrm{ehj}} T_{\mathrm{eh}} + K_{\mathrm{mj}} T_{\mathrm{m}} - mgf_{\mathrm{r}} - \frac{C_D A \rho}{2} v^2}{\delta m} \end{aligned} \tag{7.8}$$

where j is the driving/braking mode of the i-HEV system, and a_{p} is the acceleration of the preceding vehicle.

It can be seen from Eq. (7.8) that the relative inter-vehicle movement Δd and Δv are related to the system control input $\boldsymbol{u}$ through the acceleration of the ego vehicle a_{f}. The change rate of relative distance error Δd is related to the relative velocity Δv and the acceleration of the ego vehicle a_{f} while the change rate of the relative velocity Δv is related to the acceleration of the preceding vehicle a_{p} in addition to the acceleration of the ego vehicle a_{f}. The acceleration of the preceding vehicle is regarded as a disturbance in the i-HEV ACC car-following process. The acceleration of the preceding vehicle a_{p} is set as 0 in the design of the controller. In fact, in the process of vehicle car-following, assuming a uniform velocity of the preceding vehicle can reduce fluctuation in the control of car-following, which helps avoid the situation where the ego vehicle frequently follows the acceleration and deceleration of the preceding vehicle. Although it lowers the car-following performance of the system to some extent, an eased car-following process can reduce vehicle driving/braking energy loss, which can improve extra fuel economy, while the vehicle's car-following performance can be compensated via the real-time feedback control of distance error and relative velocity.

When the i-HEV system driving/braking mode is j, Eq. (7.8) is written as:

$$\frac{\mathrm{d}\boldsymbol{x}}{\mathrm{d}t} = \boldsymbol{p}^{\mathrm{j}}(\boldsymbol{x}, \boldsymbol{u}) \tag{7.9}$$

In the time domain, the generalized longitudinal dynamics model of i-HEV ACC car-following system can be described as follows:

$$\frac{\mathrm{d}\boldsymbol{x}}{\mathrm{d}t} = \boldsymbol{p}^{\mathrm{j}}(\boldsymbol{x}, \boldsymbol{u}) \quad if\ \boldsymbol{u} \in \boldsymbol{C}^{\mathrm{j}} \tag{7.10}$$

It can be seen from Eqs. (7.8) and (7.10) that the generalized longitudinal dynamics model of i-HEV ACC is composed of the inter-vehicle kinematics relationship described with relative distance error and relative velocity as well as the longitudinal dynamics characteristics of i-HEV. The inter-vehicle kinematics and the vehicle longitudinal dynamics are related through vehicle acceleration. The established system model has characteristics such as multiple inputs/outputs, non-linearity and hybridity. System input variables include relative distance error, relative velocity and the ego vehicle's velocity. System control quantities include engine, motor, hydraulic brake and AMT transmission ratio, etc. The system is non-linear because the vehicle's acceleration is directly proportional to the square of vehicle velocity. Hybridity is manifested in the discrete change of AMT and driving/braking system model. Due to the above characteristics of the established i-HEV ACC generalized longitudinal dynamics system, it is more difficult to design a real-time i-HEV ACC control algorithm optimized for multiple objectives including safety, economy, and comfort.

Because i-HEV ACC requires coordinated optimal control for multiple objectives including safety, economy, and comfort, from the perspective of the function of the i-HEV ACC system, the goal that conforms to the driver's expectation is the coordinated and optimized control of safety, fuel economy and driving comfort of system car-following under the condition that a certain driving distance is accomplished. The generalized longitudinal dynamics model of i-HEV ACC as shown in Eq. (7.10) is defined in the time domain. In order to more accurately measure fuel economy in the car-following process of i-HEV ACC, Eq. (7.10) needs to be transformed into a mathematical description in the distance space domain. According to the system state equation in the time domain shown in Eq. (7.10), the equation of system state in the space domain of driving distance can be further expressed as:

$$\frac{\mathrm{d}\boldsymbol{x}}{\mathrm{d}s} = \frac{\mathrm{d}\boldsymbol{x}}{\mathrm{d}t}\cdot\frac{\mathrm{d}t}{\mathrm{d}s} = \frac{\boldsymbol{p}^{\mathrm{j}}(\boldsymbol{x}, \boldsymbol{u})}{v}, \quad if\ \boldsymbol{u} \in \boldsymbol{C}^{\mathrm{j}} \tag{7.11}$$

Furthermore, a discrete i-HEV ACC generalized longitudinal dynamics model is established for the design of MPC system. The step size per unit distance is set as a constant h, and by assuming that the control quantity $\boldsymbol{u}$ remains constant within the step size, the Equation is set as follows:

$$\boldsymbol{x}_{\mathrm{k}} = \boldsymbol{x}(kh), \quad \boldsymbol{u}_{\mathrm{k}} = \boldsymbol{u}(kh) \tag{7.12}$$

By adopting Euler's method to discretize the system, the following equation can be obtained:

$$\begin{aligned} \boldsymbol{x}_{\mathrm{k}+1} &= \boldsymbol{x}_{\mathrm{k}} + h\frac{\mathrm{d}x}{\mathrm{d}s}|_{x=x_{\mathrm{k}}} \\ &= \boldsymbol{x}_{\mathrm{k}} + \frac{h}{v(kh)}\boldsymbol{p}^{\mathrm{j}}(\boldsymbol{x}_{\mathrm{k}}, \boldsymbol{u}_{\mathrm{k}}) \ , v(kh) \neq 0 \\ if \ & \boldsymbol{u}_{\mathrm{k}} \in \boldsymbol{C}^{\mathrm{j}} \end{aligned} \tag{7.13}$$

If $\boldsymbol{A} = [0, 0, 1]$, then:

$$\boldsymbol{x}_{\mathrm{k}+1} = \boldsymbol{x}_{\mathrm{k}} + \frac{h}{\boldsymbol{A}\boldsymbol{x}_{\mathrm{k}}}\boldsymbol{p}^{\mathrm{j}}(\boldsymbol{x}_{\mathrm{k}}, \boldsymbol{u}_{\mathrm{k}}) \quad if \ \boldsymbol{u}_{\mathrm{k}} \in \boldsymbol{C}^{\mathrm{j}} \tag{7.14}$$

Equation (7.14) is written as $\boldsymbol{x}_{\mathrm{k}+1} = \boldsymbol{f}(\boldsymbol{x}_{\mathrm{k}}, \boldsymbol{u}_{\mathrm{k}})$. The above equation shows that the hybridity of the established i-HEV ACC generalized longitudinal dynamics model mainly comes from the change of the system control quantities. The vehicle system dynamics characteristics vary in different system control modes, which indicate that for i-HEV with a hybrid power system, it's very difficult to comprehensively optimize the vehicle control because its dynamics control involves the switching and coordinated optimization of multiple energy systems' working modes.

7.3.1.2 I-HEV ACC Multi-performance Objective Design for Safety, Economy, and Comfort

In the process of car-following, the i-HEV ACC system can operate with different driving/braking powers, AMT gears, and engine and motor energy allocation strategies. Different driving/braking powers represent different levels of safety and comfort, while different AMT gears, engine and motor energy allocation strategies represent different control strategies of the hybrid power system that affect the economy of the vehicle. The above differences in the car-following process actually provide an infinite number of i-HEV ACC car-following methods, as well as the potential to coordinate and optimize safety, comfort and fuel economy during the car-following process.

In the design of the i-HEV ACC system, the longitudinal driving comfort of the driver is taken as the constraint, and the focus is put on the coordination and optimization of system safety and economy. The multi-objective function of i-HEV ACC is designed by superposing weight coefficients of various performance indicators, as shown in Eq. (7.15).

$$\begin{gathered} \boldsymbol{g}(\boldsymbol{x}, \boldsymbol{u}) = q\boldsymbol{L}_{\mathrm{s}}(\boldsymbol{x}, \boldsymbol{u}) + r\boldsymbol{L}_{\mathrm{m}}(\boldsymbol{x}, \boldsymbol{u}) \\ \text{Subj. to}: \quad \boldsymbol{u} \in \boldsymbol{\Omega}_{\mathrm{c}} \end{gathered} \tag{7.15}$$

where $\boldsymbol{g}(\boldsymbol{x}, \boldsymbol{u})$ is the i-HEV ACC multi-performance objective function that coordinates safety and economy, q and r are the coefficients for safety indicator $\boldsymbol{L}_{\mathrm{s}}$ and economy indicator L_m respectively, $\boldsymbol{\Omega}_\mathrm{c}$ is the constraint set of system control quantities that meet the driver's comfort requirement. It can be seen from the above equation that the preference of the system's car-following performance and fuel economy performance can be adjusted by designing different weight coefficients of safety and economy, so as to achieve the coordination and optimization of safety and economy of i-HEV ACC.

In the following part, performance objectives including longitudinal driving safety, fuel economy, and longitudinal driving comfort will be designed.

(1) Longitudinal driving safety

The objective of longitudinal driving safety is achieved by considering the inter-vehicle distance and the relative velocity. On the one hand, a certain safe distance should be maintained between the ego vehicle and the preceding vehicle to enable sufficient braking response time and braking distance. On the other hand, in order to increase the traffic flow as much as possible and, from the perspective of safety, to reduce situations where the preceding vehicle cutting into and being overtaken by the preceding vehicle, the ego vehicle and the preceding vehicle should keep the same velocity as much as possible. Safety control objective $\boldsymbol{L}_\mathrm{s}$ of the car-following system is established according to Ref. [2].

$$\boldsymbol{L}_\mathrm{s}(\boldsymbol{x}, \boldsymbol{u}) = \omega_{\Delta d} \Delta d^2 + \omega_{\Delta v} \Delta v^2 \tag{7.16}$$

where $\omega_{\Delta d}$ and $\omega_{\Delta v}$ are the weight coefficients of relative distance error Δd and relative velocity Δv respectively. As shown in Eq. (7.16), safety indicator $\boldsymbol{L}_\mathrm{s}$ is expressed with the 2-norm of Δd and Δv, which tends to respond to and control larger car-following errors, while for smaller car-following errors, the system does not respond or controls with smaller acceleration/deceleration, which to a certain extent reduces the car-following performance of rapid acceleration/deceleration to improve the fuel economy of the system.

(2) System fuel economy

Because the i-HEV ACC system has multiple working modes including engine-only driving, battery-only driving, engine driving + power generation, engine/motor co-driving, motor braking energy regeneration and motor/EVB co-braking, etc., its energy consumption includes fuel consumption and power consumption. Except for the part from the braking energy regeneration, the power consumed by the battery mainly comes from the generator which turns the mechanical energy produced by the internal combustion engine into electric energy. Therefore, the fuel economy of the i-HEV system should be evaluated by comprehensively considering the fuel consumption of the engine and the fuel consumption corresponding to the battery's power consumption.

Since the fuel economy indicator in the time domain cannot reflect the fuel consumption characteristics expected by the driver within a driving distance, the fuel economy indicator of i-HEV in the space domain of driving distance is established based on the vehicle's driving distance, as shown in Eq. (7.17).

$$
\begin{aligned}
\boldsymbol{L}_{\mathrm{m}}(\boldsymbol{x},\boldsymbol{u}) &= \frac{\mathrm{d}m_{\mathrm{eh}}(\boldsymbol{x},\boldsymbol{u})}{\mathrm{d}s} + \frac{\mathrm{d}m_{\mathrm{b}}(\boldsymbol{x},\boldsymbol{u})}{\mathrm{d}s} \\
&= \frac{\mathrm{d}m_{\mathrm{eh}}(\boldsymbol{x},\boldsymbol{u})}{\mathrm{d}t}\cdot\frac{\mathrm{d}t}{\mathrm{d}s} + \frac{\mathrm{d}m_{\mathrm{b}}(\boldsymbol{x},\boldsymbol{u})}{\mathrm{d}t}\cdot\frac{\mathrm{d}t}{\mathrm{d}s} \\
&= \frac{1}{v}(\dot{m}_{\mathrm{eh}}(\boldsymbol{x},\boldsymbol{u}) + \dot{m}_{\mathrm{b}}(\boldsymbol{x},\boldsymbol{u}))
\end{aligned}
\tag{7.17}
$$

where $\boldsymbol{L}_{\mathrm{m}}$ is the economic indicator of the i-HEV system in the space domain of driving distance, $\dot{m}_{\mathrm{b}}(\boldsymbol{x},\boldsymbol{u})$ is the instantaneous equivalent fuel consumption of the battery in the time domain, and $\dot{m}_{\mathrm{eh}}(\boldsymbol{x},\boldsymbol{u})$ is the instantaneous fuel consumption corresponding to the equivalent conventional power torque T_{eh} in the time domain.

The insignificant energy consumed in the EVB hydraulic braking process is ignored here because it mainly relies on the solenoid valve to push the brake pedal, while the fuel consumption of the engine in the driving process is the main consideration. Based on the driving/braking mode j of the i-HEV system, the auxiliary variable $\boldsymbol{\Gamma}^{\mathrm{j}}$ is introduced to calculate the instantaneous fuel consumption corresponding to the equivalent conventional power torque T_{eh} in the time domain:

$$
\begin{aligned}
\dot{m}_{\mathrm{eh}}(\boldsymbol{x},\boldsymbol{u}) &= \boldsymbol{\Gamma}^{\mathrm{j}}\dot{m}_{\mathrm{e}}(\boldsymbol{x},\boldsymbol{u}) \\
\boldsymbol{\Gamma}^{\mathrm{j}} &= \begin{cases} 0 & j = 1, 2 \\ 1 & j = 3, 4 \ldots 7 \end{cases}
\end{aligned}
\tag{7.18}
$$

where $\dot{m}_{\mathrm{e}}(\boldsymbol{x},\boldsymbol{u})$ is the instantaneous fuel consumption of the engine in the time domain.

It can be seen from Eqs. (7.17) and (7.18) that the economic indicator of the i-HEV system in the distance space domain is related to the driving/braking working mode j of the i-HEV system, the instantaneous fuel consumption of the engine and the instantaneous equivalent fuel consumption of the battery in the time domain, and is also related to the current vehicle velocity to some extent. The instantaneous fuel consumption of the engine and the instantaneous equivalent fuel consumption of the battery are respectively quantitatively calculated as follows.

In this layer of multi-objective steady-state optimization, the steady-state efficiency characteristics of each system are mainly considered, to calculate the optimal energy allocation strategy of the engine and motor under the steady state. Because the fuel consumption characteristics of the engine under the steady state are highly nonlinear, it's difficult to accurately describe those characteristics with a simple and explicit mathematical equation. The instantaneous fuel consumption of the engine is calculated by looking up the table based on the steady-state fuel consumption MAP graph, as shown in Eq. (7.19).

$$
\dot{m}_{\mathrm{e}}(\boldsymbol{x},\boldsymbol{u}) = M_{\mathrm{e}}(T_{\mathrm{e}},\omega_{\mathrm{e}}) \tag{7.19}
$$

where M_e is the MAP chart of the steady-state fuel consumption of the engine; ω_e is the engine speed; T_e is the engine torque. The instantaneous fuel consumption of the engine is a function of engine torque and speed.

In the working process of the hybrid power system, the battery replenishes and consumes electric energy based on the charging and discharging requirements of the system. According to the principle of energy conservation, the electric energy consumed by or supplemented to the battery comes from the charging and discharging of the engine and the braking energy regeneration of the motor. Based on this basic idea, the energy consumed by the battery is described as the equivalent fuel consumption, and the equivalent fuel consumption factor of the battery is introduced to define the instantaneous equivalent fuel consumption of the battery $\dot{m}_b$ as:

$$\dot{m}_b(\boldsymbol{x}, \boldsymbol{u}) = \lambda P_b(T_m, \omega_m) \tag{7.20}$$

where $P_b(T_m, \omega_m)$ is the battery power consumed (or stored) under the torque T_m and rotational speed ω_m of the equivalent motor EM, λ is the battery's equivalent fuel consumption factor under the current working scenario. The factor measures the average conversion relationship between fuel and electricity of the hybrid power system under the overall working scenario, so its estimate is associated with the vehicle's current driving condition and the energy (fuel and electricity) allocation strategy. The estimate will be carried out in the last chapter of this paper. The equivalent fuel consumption of the battery $\dot{m}_b$ is determined both by the power of the battery and the equivalent fuel consumption factor.

The battery's electric power is calculated with the equivalent motor power and the charging and discharging efficiency of the motor and battery. For the nickel-metal hydride batteries used in the i-HEV system, the charging and discharging efficiency can always remain at a high level when the normal operating range of battery SOC is 40–60% [3]. The charging and discharging efficiency of the battery is set as a constant η_0 based on the data provided by the battery manufacturer.

Since the working efficiency of the motor is highly non-linear, it's difficult to describe its characteristics accurately by using simple dominant mathematical equations, so the equivalent charge–discharge efficiency of the motor under the given motor torque T_m and motor speed ω_m is obtained by looking up the table. In terms of the function of the equivalent motor EM, the driving mode dominates when the motor discharges and the generation mode dominates when the braking energy is recovered. In these two cases, the EM efficiency $\eta_m(T_m, \omega_m)$ is expressed as:

$$\eta_m(T_m, \omega_m) = M_{PM}(T_m, \omega_m) \tag{7.21}$$

where M_{PM} is the charge–discharge efficiency MAP of PM motor.

When the equivalent motor generates electricity, the BSG generation mode dominates and the chart of BSG generation efficiency should be applied. The equivalent motor EM efficiency $\eta_m(T_m, \omega_m)$ is expressed as:

$$\eta_{\mathrm{m}}(T_{\mathrm{m}}, \omega_{\mathrm{m}}) = M_{\mathrm{BSG}}(T_{\mathrm{m}}K, \frac{\omega_{\mathrm{m}}}{K}) \tag{7.22}$$

where M_{BSG} is BSG's charge–discharge efficiency MPA chart.

Without wheel sliding and slipping taken into consideration, the engine speed ω_{e} and motor speed ω_{m} and vehicle velocity are converted with wheel radius, and transmission ratios of the main reducer and the transmission, as shown in Eq. (7.23).

$$\omega_{\mathrm{e}} = \frac{v}{R_{\mathrm{w}}} i_0 i_{\mathrm{g}}, \quad \omega_{\mathrm{m}} = \frac{v}{R_{\mathrm{w}}} i_0 i_{\mathrm{m}} \tag{7.23}$$

(3) Longitudinal driving comfort

The driver's longitudinal driving comfort is manifested in two aspects: on the one hand, the relative distance error and relative velocity is maintained when the ego vehicle is following the preceding vehicle, which has been reflected in the indicator of vehicle longitudinal car-following safety; on the other hand, it is mainly measured by vehicle acceleration/deceleration. Excessive vehicle acceleration or excessive braking deceleration will increase the unsmooth feel for the driver and reduce his or her driving comfort. The acceleration/deceleration is limited based on Reference [4] to meet drivers' requirements for longitudinal driving comfort, as shown in the following equation:

$$\begin{aligned} &\text{drivingcondition:} \quad a_{\mathrm{f}} \le a_{\mathrm{dmax}} \\ &\text{braking condition:} \quad a_{\mathrm{bmin}} \le a_{\mathrm{f}} \end{aligned} \tag{7.24}$$

where a_{bmin} and a_{dmax} are respectively the lower limit of braking deceleration velocity and the upper limit of driving acceleration to meet the driver's longitudinal driving comfort requirement, which are set as − 2.5 and 2 m/s^2 respectively. Combined with Eq. (7.8), the constraint of Eq. (7.24) on the ego vehicle's acceleration is transformed into the constraint on the system's state variables and control inputs, as shown below:

$$\begin{aligned} &\frac{K_{\mathrm{ehj}} T_{\mathrm{eh}} + K_{\mathrm{mj}} T_{\mathrm{m}} - mgf_{\mathrm{r}} - \frac{C_D A \rho}{2} v^2}{\delta m} \ge a_{\mathrm{bmin}} \\ &\frac{K_{\mathrm{ehj}} T_{\mathrm{eh}} + K_{\mathrm{mj}} T_{\mathrm{m}} - mgf_{\mathrm{r}} - \frac{C_D A \rho}{2} v^2}{\delta m} \le a_{\mathrm{dmax}} \end{aligned} \tag{7.25}$$

It can be seen from Eqs. (7.24) and (7.25) that the longitudinal driving comfort can be achieved by limiting the acceleration and deceleration of the vehicle, which is eventually transformed, in the form of constraint, into limits on the system's control quantities including engine torque, motor torque and hydraulic brake torque and integrated into the calculation of the multi-objective coordinated optimization of i-HEV ACC. For convenience, the constraint space of the control quantities T_{eh} and T_{m} in Eq. (7.25) is marked as $\boldsymbol{\Omega}_{\mathrm{c}}$.

Furthermore, the multi-performance objective function of i-HEV ACC coordinated with safety and economy can be obtained, as shown in Eq. (8.12).

$$
\begin{aligned}
\boldsymbol{g}(\boldsymbol{x}, \boldsymbol{u}) &= q\boldsymbol{L}_{\mathrm{s}}(\boldsymbol{x}, \boldsymbol{u}) + r\boldsymbol{L}_{\mathrm{m}}(\boldsymbol{x}, \boldsymbol{u}) \\
&= q(w_{\Delta d}\Delta d^2 + w_{\Delta v}\Delta v^2) + r(\boldsymbol{\Gamma}^{\mathrm{j}}\dot{m}_{\mathrm{e}}(\boldsymbol{x}, \boldsymbol{u}) + \dot{m}_{\mathrm{b}}(\boldsymbol{x}, \boldsymbol{u})) \\
&= qw_{\Delta d}\Delta d^2 + qw_{\Delta v}\Delta v^2 + \frac{r}{v}\boldsymbol{\Gamma}^{\mathrm{j}}M_{\mathrm{e}}(T_{\mathrm{e}}, \omega_{\mathrm{e}}) + \frac{\lambda r}{v}P_{\mathrm{b}}(T_{\mathrm{m}}, \omega_{\mathrm{m}}) \\
&= qw_{\Delta d}\Delta d^2 + qw_{\Delta v}\Delta v^2 + \frac{r}{v}\boldsymbol{\Gamma}^{\mathrm{j}}M_{\mathrm{e}}(T_{\mathrm{e}}, \frac{v}{R_{\mathrm{w}}}i_0 i_{\mathrm{g}}) + \frac{\lambda r}{v}P_{\mathrm{b}}(T_{\mathrm{m}}, \frac{v}{R_{\mathrm{w}}}i_0 i_{\mathrm{m}})
\end{aligned}
\tag{7.26}
$$

It can be seen from Eq. (7.26) that the performance objective function of i-HEV ACC is highly non-linear and discrete, and its main fuel economy indicators of the engine and the battery are obtained by looking up the component efficiency table, which cannot be easily described with mathematical equations. Due to discrete system states such as engine on/off and braking, the objective function also has the characteristics of discrete change. Such nonlinear objective function presents great challenges to the establishment of the i-HEV ACC nonlinear model prediction problem and its optimal solution.

7.3.1.3 Designing the i-HEV ACC System Constraints

i-HEV ACC involves a typical system car-following control problem. On the one hand, the driver always requires the vehicle to follow steadily in the process of car-following, that is, the car-following error should be gradually reduced. On the other hand, the design of the nonlinear predictive control system requires ensuring the stability of the control system. Therefore, additional prediction terminal constraints should be added to the predictive control problem of the established i-HEV ACC nonlinear model to meet the driver's requirements for steady car-following and control system stability [5].

The prediction terminal constraint is set with the acceleration of the preceding vehicle a_p regarded as disturbance, as shown in Fig. 7.9. For the driver, the i-HEV ACC car-following process is a process of gradually converging the car-following error, in which the driver always expects the vehicle to catch up with the preceding vehicle within a certain period. The i-HEV ACC car-following problem can be further described as: under the assumption that the preceding vehicle is running at a uniform velocity, various components of the system are coordinated to optimize performance objectives including car-following safety, economy, and comfort, to achieve car-following in the prediction domain, and the car-following error is kept at 0 to ensure good following stability.

If the prediction distance domain is set as $k + N$, the prediction terminal meets the following constraints:

$$
\begin{aligned}
\Delta d_{\mathrm{k+N}} &= 0 \\
\Delta v_{\mathrm{k+N}} &= 0
\end{aligned}
\tag{7.27}
$$

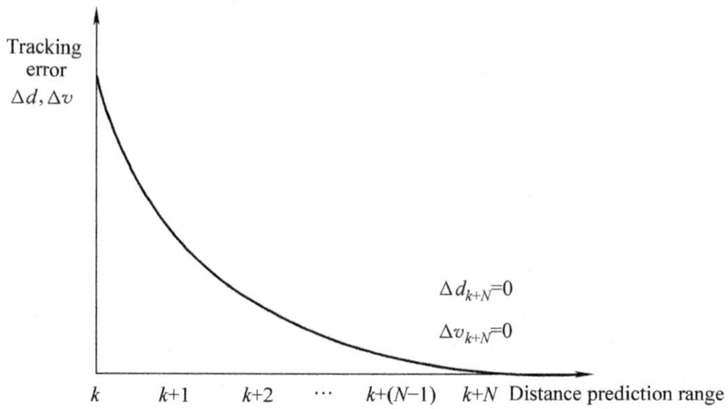

Fig. 7.9 Schematic diagram of i-HEV's gradually stabilized car-following process

In order to solve the prediction and optimization problems of the i-HEV ACC model while increasing the freedom level of optimization and optimizing the economy as much as possible on the condition that the car-following safety and comfort is achieved, the constraints shown in Eq. (7.27) are softened, that is, the vehicle is only required to be within a certain range of car-following error at the prediction terminal:

$$|\boldsymbol{x}_{\mathrm{k+N}}| \leq \boldsymbol{\varepsilon}_0 \tag{7.28}$$

where $\boldsymbol{\varepsilon}_0 = [\varepsilon^0_{\Delta d}, \varepsilon^0_{\Delta v}, \varepsilon^0_{\mathrm{v}}]^{\mathrm{T}}$, $\varepsilon^0_{\Delta d}, \varepsilon^0_{\Delta v}$ are respectively the constraint values of distance error and relative velocity at the terminal, and $\varepsilon^0_{\mathrm{v}} = v_{\max}$ is the maximum allowable velocity of i-HEV ACC.

In order to ensure car-following safety, the distance error Δd and relative velocity Δv of the system should be within the range of the car-following error expected by the driver, so as to avoid frequent cutting-in of preceding vehicles caused by excessive inter-vehicle distance and the danger caused by close car-following. The one-dimensional allocation model of driver's expected car-following error is applied to design the constraints of relative distance error and relative velocity based on Ref. [6].

$$\begin{aligned} \Delta d_{\min}.SDE^{-1} \leq \Delta d \leq \Delta d_{\max}.SDE^{-1} \\ \Delta v_{\min}.SVE^{-1} \leq \Delta v \leq \Delta v_{\max}.SVE^{-1} \end{aligned} \tag{7.29}$$

where $\Delta d_{\min}$ and $\Delta v_{\min}$ are the lower limits of distance error and relative velocity respectively; $\Delta d_{\max}$ and $\Delta v_{\max}$ are the upper limits of distance error and relative velocity respectively; *SDE* and *SVE* are the sensitivity of the driver to car-following errors and are functions of vehicle velocity, as shown in Eq. (7.30).

$$
\begin{aligned}
SDE^{-1} &= k_{\mathrm{SDE}} v_{\mathrm{x}} + d_{\mathrm{SDE}} \\
SVE^{-1} &= k_{\mathrm{SVE}} v_{\mathrm{x}} + d_{\mathrm{SVE}}
\end{aligned}
\tag{7.30}
$$

According to the results of the experiment of Chinese drivers' driving characteristics on urban roads and expressways in Ref. [6], the values of the above parameters are respectively set as $\Delta d_{\mathrm{min}} = -7$ m, $\Delta d_{\mathrm{max}} = 10$ m, $\Delta v_{\mathrm{min}} = -3$ m/s, $\Delta v_{\mathrm{max}} = 3$ m/s, $k_{\mathrm{SDE}} = 0.06$, $d_{\mathrm{SDE}} = 0.12$, $k_{\mathrm{SVE}} = 0.005$, $d_{\mathrm{SVE}} = 0.91$。

In addition, the velocity of i-HEV ACC should be maintained within a certain allowed range.

$$
v_{\mathrm{min}} \leq v \leq v_{\mathrm{max}} \tag{7.31}
$$

where v_{min} and v_{max} are respectively the minimum and maximum allowable velocities of i-HEV ACC, which are 20 and 120 km/h respectively according to the working characteristics of the i-HEV system.

By combining Eqs. (7.29), (7.30) and (7.31), the effective set of i-HEV ACC state variable $\boldsymbol{x}$ is defined as shown in Eq. (7.32):

$$
\boldsymbol{\chi}_{\mathrm{x}} = \left\{ \boldsymbol{x} = [\Delta d, \Delta v, v]^{\mathrm{T}} \middle| \begin{array}{l} \Delta d_{\mathrm{min}} . SDE^{-1} \leq \Delta d \leq \Delta d_{\mathrm{max}} . SDE^{-1} \\ \Delta v_{\mathrm{min}} . SVE^{-1} \leq \Delta v \leq \Delta v_{\mathrm{max}} . SVE^{-1} \\ v_{\mathrm{min}} \leq v \leq v_{\mathrm{max}} \end{array} \right\} \tag{7.32}
$$

By combining Eqs. (7.3) and (7.25), the effective set of i-HEV ACC control quantity $\boldsymbol{u}$ is defined as shown in Equation:

$$
\boldsymbol{\Omega}_{\mathrm{u}} = \left\{ \boldsymbol{u} = [T_{\mathrm{eh}}, T_{\mathrm{m}}, i_{\mathrm{g}}]^{\mathrm{T}} \middle| \boldsymbol{u} \in \boldsymbol{\Omega} \bigcap \boldsymbol{\Omega}_{\mathrm{c}} \right\} \tag{7.33}
$$

7.3.1.4 Derivation of i-HEV ACC Nonlinear Model Prediction Problem

The theoretical derivation of the multi-objective steady-state optimal control problem of i-HEV ACC is conducted in this section based on the nonlinear model prediction theory with the generalized longitudinal dynamics model of i-HEV ACC established in Eq. (7.14) and the i-HEV multi-objective function of the current position point in the space domain of driving distance designed with Eq. (7.26).

The essence of MPC is to solve a local rolling optimization problem online. Given the predicted distance domain D, the optimized objective function of i-HEV ACC can be expressed as an integral based on the performance indicator function of each position point, as shown in Eq. (7.34).

$$
\begin{aligned}
J(\boldsymbol{x}) &= \int\limits_{s_0}^{s_0+D} g(\boldsymbol{x},\boldsymbol{u})\mathrm{d}s \\
&= \int\limits_{s_0}^{s_0+D} (qw_{\Delta d}\Delta d^2 + qw_{\Delta v}\Delta v^2)\mathrm{d}s \\
&+ \int\limits_{s_0}^{s_0+D} \left[\frac{r}{v}\Gamma^{\mathrm{j}} M_{\mathrm{e}}\left(T_{\mathrm{e}}, \frac{v}{R_{\mathrm{w}}} i_0 i_{\mathrm{g}}\right) + \frac{\lambda r}{v} P_{\mathrm{b}}\left(T_{\mathrm{m}}, \frac{v}{R_{\mathrm{w}}} i_0 i_{\mathrm{m}}\right)\right]\mathrm{d}s \\
&= \sum_{i=0}^{N-1} \int\limits_{s_0+ih}^{s_0+(i+1)h} (qw_{\Delta d}\Delta d^2 + qw_{\Delta v}\Delta v^2)\mathrm{d}s \\
&+ \sum_{i=0}^{N-1} \int\limits_{s_0+ih}^{s_0+(i+1)h} \left[\frac{r}{v}\boldsymbol{\Gamma}^{\mathrm{j}} M_{\mathrm{e}}\left(T_{\mathrm{e}}, \frac{v}{R_{\mathrm{w}}} i_0 i_{\mathrm{g}}\right) + \frac{\lambda r}{v} P_{\mathrm{b}}\left(T_{\mathrm{m}}, \frac{v}{R_{\mathrm{w}}} i_0 i_{\mathrm{m}}\right)\right]\mathrm{d}s
\end{aligned} \tag{7.34}
$$

where s_0 is the current position point, N is the total number of predicted steps in the distance domain, and h is the step size of the predicted distance. The above equation is discretized to establish the multi-performance (longitudinal driving safety, fuel economy and longitudinal driving comfort) optimized objective function of i-HEV ACC in the discrete states, as shown in Eq. (7.35).

$$
\begin{aligned}
J(\boldsymbol{x}_{\mathrm{k}}) &= h\sum_{i=0}^{N-1} g(\boldsymbol{x}_{\mathrm{k+i|k}}, \boldsymbol{u}_{\mathrm{k+i|k}}) \\
&= h\sum_{i=0}^{N-1} (qw_{\Delta d}\Delta d^2_{\mathrm{k+i|k}} + qw_{\Delta v}\Delta v^2_{\mathrm{k+i|k}}) \\
+h\sum_{i=0}^{N-1} &\left[\frac{r}{v_{\mathrm{k+i|k}}}\boldsymbol{\Gamma}^{\mathrm{j}} M_{\mathrm{e}}\left(T_{\mathrm{ek+i|k}}, \frac{v_{\mathrm{k+i|k}}}{R_{\mathrm{w}}} i_0 i_{\mathrm{g}}\right) + \frac{\lambda r}{v_{\mathrm{k+i|k}}} P_{\mathrm{b}}\left(T_{\mathrm{mk+i|k}}, \frac{v_{\mathrm{k+i|k}}}{R_{\mathrm{w}}} i_0 i_{\mathrm{m}}\right)\right]
\end{aligned} \tag{7.35}
$$

where the system is in the current position k, $\boldsymbol{x}_{\mathrm{k}}$ is the current states of the system, $J(\boldsymbol{x}_{\mathrm{k}})$ is the multi-performance indicator function of i-HEV ACC's predicted distance domain, $g(\boldsymbol{x}_{\mathrm{k+i|k}}, \boldsymbol{u}_{\mathrm{k+i|k}})$ is the multi-performance indicator function of step i in the future, $\boldsymbol{x}_{\mathrm{k+i|k}}$ and $\boldsymbol{u}_{\mathrm{k+i|k}}$ are the predicted system state and the system control quantity of step i, $\Delta d_{\mathrm{k+i|k}}$, $\Delta v_{\mathrm{k+i|k}}$, $v_{\mathrm{k+i|k}}$ are respectively the predicted distance error, relative velocity and ego vehicle's velocity of step i, $T_{\mathrm{ek+i|k}}$ and $T_{\mathrm{mk+i|k}}$ are respectively the predicted engine and motor torque control quantity of step i.

On the other hand, due to the possible large disturbance and uncertainty in the system, as the system state changes, the fixed prediction terminal constraint as shown in Eq. (7.28) may make the controller not able to control the system within the constraint range, leading to no solution for this optimization problem. In addition, the terminal constraint can be adjusted as the system state changes, and there is no need to force the system to stay in a hard terminal state. Based on the above analysis, the relaxation variable $\boldsymbol{\varepsilon} = [\varepsilon_{\Delta d}, \varepsilon_{\Delta v}, \varepsilon_{\mathrm{v}}]^{\mathrm{T}}$ is introduced into the objective function as a square penalty term [7], and the objective function shown in Eq. (7.35) is further modified as:

$$\boldsymbol{J}(\boldsymbol{x}_{\mathrm{k}}) = h\sum_{i=0}^{N-1} \boldsymbol{g}(\boldsymbol{x}_{\mathrm{k+i|k}}, \boldsymbol{u}_{\mathrm{k+i|k}}) + \frac{1}{2}\boldsymbol{\varepsilon}^{\mathrm{T}}\boldsymbol{\rho}\boldsymbol{\varepsilon} \tag{7.36}$$

where $\boldsymbol{\rho} = \mathrm{diag}(\rho_1, \rho_2, 0)$ is the penalty coefficient of relaxation variable $\boldsymbol{\varepsilon}$, and ρ_1, ρ_2 change with the initial sate of the system $\boldsymbol{x}_{\mathrm{k}}$. When the system's car-following errors are small, larger values are set to ensure that the vehicle can follow the preceding vehicle steadily without error; when the system's car-following errors are large, smaller values are set to achieve the optimization of vehicle safety and economy within a larger optimization range.

Based on the multi-performance objective function of i-HEV ACC shown in Eq. (7.36) and the generalized longitudinal dynamics model of i-HEV ACC, the steady-state multi-objective optimization problem of i-HEV ACC is established based on the multi-objective nonlinear model predictive control, as shown in Eq. (7.37).

$$\begin{aligned}
&\min_{u_{\mathrm{i+k|k}}, i=0:N-1} \quad \boldsymbol{J}(\boldsymbol{x}_{\mathrm{k}}) = h\sum_{i=0}^{N-1} \boldsymbol{g}(\boldsymbol{x}_{\mathrm{k+i|k}}, \boldsymbol{u}_{\mathrm{k+i|k}}) + \frac{1}{2}\boldsymbol{\varepsilon}^{\mathrm{T}}\boldsymbol{\rho}\boldsymbol{\varepsilon} \\
&\text{Subj. to.:} \\
&\quad (1)\ \boldsymbol{x}_{\mathrm{1+i+k|k}} = \boldsymbol{f}(\boldsymbol{x}_{\mathrm{i+k|k}}, \boldsymbol{u}_{\mathrm{i+k|k}}) \\
&\quad (2)\ \boldsymbol{x}_{\mathrm{i+k|k}} \in \boldsymbol{\chi}_{\mathrm{x}},\ \boldsymbol{u}_{\mathrm{i+k|k}} \in \boldsymbol{\Omega}_{\mathrm{u}} \\
&\quad (3)\ |\boldsymbol{x}_{\mathrm{N+k|k}}| \le \boldsymbol{\varepsilon} \\
&\quad (4)\ i = 0, 1, 2, \ldots N-1
\end{aligned} \tag{7.37}$$

In the above nonlinear model prediction and optimization problem, the future state of the system is predicted based on the established i-HEV ACC generalized longitudinal dynamics model, and coordinated optimization is performed on the safety, economy, and comfort performance indicators of the car-following system within the predicted distance domain, so as to achieve local optimization of the car-following system within the predicted distance domain. In order to overcome the uncertainty of the system caused by external environment disturbance, the rolling optimization method is applied for real-time control, that is, the first value of the optimal control sequence is set as the actual control input of the vehicle at the position point k, and the local optimization process is repeated at the position point $k+1$ to realize the closed-loop feedback control of the system.

7.3.1.5 Optimal Solution of i-HEV ACC Nonlinear Model Prediction Problem

The predictive optimization control problem of the i-HEV ACC nonlinear model established by Eq. (7.37) has the following 3 characteristics: the optimization objective function is strongly nonlinear, in which only discrete data on the characteristics of engine fuel consumption and motor efficiency are available from test bench, so it is difficult to model these data with an analytical mathematical model; the established i-HEV ACC generalized longitudinal dynamics model is hybrid and nonlinear; there are strong constraints on the system states and control inputs in the designed predictive optimization problem. By far, there is no mature and general real-time optimization method for the optimal solution of this kind of hybrid and nonlinear model predictive control problem with constraints. The solution method based on offline optimization and online synthesis has been attracting the attention of scholars at home and abroad.

The essential difference between model predictive control and optimal control is that optimal control is an off-line computational control law in an infinite time domain, while predictive control is an online optimization solution of open-loop optimal control in a finite time domain [7]. Therefore, the basic theory of optimal control can also be used to solve this kind of predictive optimization problem. Further analysis of Eq. (7.37) shows that the model predictive optimization problem divides the optimal calculation of control quantity and the model's prediction process into N stages in the prediction domain, which is a typical multi-stage decision problem. For a given initial system state $\boldsymbol{x}_{\text{k}}$, Eq. (7.37) can be approximated as an online rolling open-loop optimal control problem. In this paper, the dynamic programming (DP) algorithm is applied to find the optimal solution for the hybrid nonlinear model predictive control problem with constraints as shown in (7.37). The offline optimization control law of the set of the system's initial states is obtained based on the idea of offline optimization and online synthesis. In the process of online control, the MAP chart is obtained by offline optimizing and calculating the initial state variable and control rate of the system, and the real-time control law of the system is obtained by looking up the table to solve the real-time problem of the control system.

In order to perform online table lookup, the feasible region of the state variables of the optimization problem needs to be determined. Firstly, the discretized value space of distance error Δd, relative velocity Δv and velocity v is defined:

$$\begin{aligned}
&\boldsymbol{X}_{\Delta d} = \{\Delta d | \Delta d = \Delta d_{\text{min}} + s_1 \delta_{\Delta d}\},\ s_1 = 1, 2, \ldots \frac{\Delta d_{\text{max}} - \Delta d_{\text{min}}}{\delta_{\Delta d}} \\
&\boldsymbol{X}_{\Delta v} = \{\Delta v | \Delta v = \Delta v_{\text{min}} + s_2 \delta_{\Delta v}\},\ s_2 = 1, 2, \ldots \frac{\Delta v_{\text{max}} - \Delta v_{\text{min}}}{\delta_{\Delta v}} \\
&\boldsymbol{X}_{\text{v}} = \{v | v = v_{\text{min}} + s_3 \delta_{\text{v}}\},\ s_3 = 1, 2, \ldots \frac{v_{\text{max}} - v_{\text{min}}}{\delta_{\text{v}}}
\end{aligned} \tag{7.38}$$

As shown in Eq. (7.38), the state variables of the control system are mainly determined by the ranges of distance error, relative velocity and velocity allowed in the cycle process. Further, based on the constraints of i-HEV ACC on state variables, the discrete feasible region of state variables for the optimization problem $\boldsymbol{X}_{\mathrm{D}}$ can be defined as:

$$\boldsymbol{X}_{\mathrm{D}} = \{\boldsymbol{x} = [\Delta d, \Delta v, v] | \Delta d \in \boldsymbol{X}_{\Delta d}, \Delta v \in \boldsymbol{X}_{\Delta v}, v \in \boldsymbol{X}_{\mathrm{v}}\} \bigcap \chi_{\mathrm{x}} \tag{7.39}$$

For a given vehicle velocity v_{k}, the engine speed ω_{e} is related to the vehicle velocity and current transmission gear, which can be expressed as:

$$\omega_{\mathrm{e}}(v_{\mathrm{k}}, i_{\mathrm{g}}) = \frac{v_{\mathrm{k}}}{R_{\mathrm{w}}} i_0 i_{\mathrm{g}} \tag{7.40}$$

If the range of engine speed is set as $[\omega_{\mathrm{emin}}, \omega_{\mathrm{emax}}]$ for the discrete AMT gear set $\boldsymbol{I}_{\mathrm{g}}$, then for a given vehicle velocity v_{k}, the effective gear set is:

$$\boldsymbol{G}_{\mathrm{vk}} = \{i_{\mathrm{g}} | \omega_{\mathrm{emin}} \le \omega_{\mathrm{e}}(v_{\mathrm{k}}, i_{\mathrm{g}}) \le \omega_{\mathrm{emax}}, i_{\mathrm{g}} \in \boldsymbol{I}_{\mathrm{g}}\} \tag{7.41}$$

Control quantity $\boldsymbol{u} = [T_{\mathrm{eh}}, T_{\mathrm{m}}, i_{\mathrm{g}}]^{\mathrm{T}}$ is discretized, and the set of effective control quantities $\boldsymbol{\Omega}_{\mathrm{D}}$ under a given state variable $\boldsymbol{x}_{\mathrm{k}}$ is further calculated:

$$\begin{aligned}
\boldsymbol{\Omega}_{\mathrm{eh}} &= \{T_{\mathrm{eh}} | T_{\mathrm{eh}} = T_{\mathrm{ehmin}} + n_1 \delta_{\mathrm{T_{eh}}}\},\ n_1 = 1, 2, \ldots \frac{T_{\mathrm{ehmin}} - T_{\mathrm{ehmin}}}{\delta_{\mathrm{T_{eh}}}} \\
\boldsymbol{\Omega}_{\mathrm{T_m}} &= \{T_{\mathrm{m}} | T_{\mathrm{m}} = T_{\mathrm{mmin}} + n_2 \delta_{\mathrm{T_m}}\},\ n_2 = 1, 2, \ldots \frac{T_{\mathrm{mmax}} - T_{\mathrm{mmin}}}{\delta_{\mathrm{T_m}}} \\
\boldsymbol{\Omega}_{\mathrm{D}} &= \{\boldsymbol{u} = [T_{\mathrm{d}}, T_{\mathrm{m}}, i_{\mathrm{g}}]^{\mathrm{T}} | T_{\mathrm{eh}} \in \boldsymbol{\Omega}_{\mathrm{eh}}, T_{\mathrm{m}} \in \boldsymbol{\Omega}_{\mathrm{T_m}}, i_{\mathrm{g}} \in \boldsymbol{G}_{\mathrm{vk}}\} \bigcap \boldsymbol{\Omega}_{\mathrm{u}}
\end{aligned} \tag{7.42}$$

Due to the introduction of relaxation factor $\boldsymbol{\varepsilon}$, the control quantities of the rolling optimization problem include $\boldsymbol{u}_{\mathrm{k+i|k}}, i = 0, 1, \ldots N - 1$ and the relaxation factor, that is, for any given initial state $\boldsymbol{x} \in \boldsymbol{X}_{\mathrm{D}}$, there is a sequence of control quantities $\boldsymbol{u}_{\mathrm{k+i|k}}$ and relaxation factor $\boldsymbol{\varepsilon}$, so that the minimum value is set for objective function $\boldsymbol{J}(\boldsymbol{x}_{\mathrm{k}})$. In this section, the step-by-step DP method is adopted to solve Eq. (7.37). The solution process is shown as follows:

(1) Define the discretization space of relaxation factor $\boldsymbol{\varepsilon}$:

$$\boldsymbol{E} = \{\boldsymbol{\varepsilon} | \boldsymbol{\varepsilon} = \boldsymbol{\varepsilon}_{\min} + n\Delta\boldsymbol{\varepsilon}, \boldsymbol{\varepsilon}_{\min} \le \boldsymbol{\varepsilon} \le \boldsymbol{\varepsilon}_{\max}\},\ n = 1, 2 \ldots \frac{\boldsymbol{\varepsilon}_{\max} - \boldsymbol{\varepsilon}_{\min}}{\Delta\boldsymbol{\varepsilon}} \tag{7.43}$$

(2) Give relaxation factor $\boldsymbol{\varepsilon}_s$:

$$\boldsymbol{\varepsilon}_{\mathrm{s}} = \boldsymbol{\varepsilon}_{\min} + s\Delta\boldsymbol{\varepsilon},\ s \in [1, 2, \ldots \frac{\boldsymbol{\varepsilon}_{\max} - \boldsymbol{\varepsilon}_{\min}}{\Delta\boldsymbol{\varepsilon}}] \tag{7.44}$$

(3) For a given relaxation factor $\boldsymbol{\varepsilon}_s$, optimized solution is performed on the objective function:

$$\boldsymbol{U}_{\mathrm{k}}^{*}(\boldsymbol{x}_{\mathrm{k}}, \boldsymbol{\varepsilon}_{\mathrm{s}}) = \arg\ \min \boldsymbol{J}^{*}(\boldsymbol{x}_{\mathrm{k}}, \boldsymbol{\varepsilon}_{\mathrm{s}}) \tag{7.45}$$

Basic steps:

1. Let $i = N$,

$$\begin{aligned} & if\ |\boldsymbol{x}_{\mathrm{N+k|k}}| \le \boldsymbol{\varepsilon}_{\mathrm{s}} \\ & \quad \boldsymbol{J}_{\mathrm{N}}^{*}(\boldsymbol{x}_{\mathrm{N+k|k}}) = \frac{1}{2}\boldsymbol{\varepsilon}_{\mathrm{s}}^{T}\boldsymbol{\rho}\boldsymbol{\varepsilon}_{\mathrm{s}} \\ & else \\ & \quad \boldsymbol{J}_{\mathrm{N}}^{*}(\boldsymbol{x}_{\mathrm{N+k|k}}) = +\infty \end{aligned} \tag{7.46}$$

2. Let i = N-1,

$$\begin{aligned} \boldsymbol{J}_{\mathrm{N-1}}^{*}(\boldsymbol{x}_{\mathrm{N-1+k|k}}) = & \min_{\boldsymbol{u}_{\mathrm{N-1+k|k}} \in \boldsymbol{\Omega}_{\mathrm{D}}} [\boldsymbol{g}(\boldsymbol{x}_{\mathrm{N-1+k|k}}, \boldsymbol{u}_{\mathrm{N-1+k|k}}) \\ & + \boldsymbol{J}_{\mathrm{N}}^{*}(\boldsymbol{f}(\boldsymbol{x}_{\mathrm{N-1+k|k}}, \boldsymbol{u}_{\mathrm{N-1+k|k}})],\ \boldsymbol{x}_{\mathrm{N-1+k|k}} \in \boldsymbol{X}_{\mathrm{D}} \end{aligned} \tag{7.47}$$

3. Let i = N-2, N-3, … 0, repeat step 2), then:

$$\begin{aligned} \boldsymbol{J}_{\mathrm{i}}^{*}(\boldsymbol{x}_{\mathrm{i+k|k}}) = & \min_{\boldsymbol{u}_{\mathrm{i+k|k}} \in \boldsymbol{\Omega}_{\mathrm{D}}} [\boldsymbol{g}(\boldsymbol{x}_{\mathrm{i+k|k}}, \boldsymbol{u}_{\mathrm{i+k|k}}) \\ & + \boldsymbol{J}_{\mathrm{i+1}}^{*}(\boldsymbol{f}(\boldsymbol{x}_{\mathrm{i+k|k}}, \boldsymbol{u}_{\mathrm{i+k|k}})],\ \boldsymbol{x}_{\mathrm{i+k|k}} \in \boldsymbol{X}_{\mathrm{D}}, \boldsymbol{x}_{\mathrm{i+1+k|k}} \in \boldsymbol{\chi}_{\mathrm{x}} \end{aligned} \tag{7.48}$$

4. When i = 0, it's obtained:

$$\boldsymbol{J}^{*}(\boldsymbol{x}_{\mathrm{k}}, \boldsymbol{\varepsilon}_{\mathrm{s}}) = \min_{\boldsymbol{u}_{\mathrm{k}} \in \boldsymbol{\Omega}_{\mathrm{D}}} [\boldsymbol{g}(\boldsymbol{x}_{\mathrm{k}}, \boldsymbol{u}_{\mathrm{k}}) + \boldsymbol{J}_{1}^{*}(\boldsymbol{f}(\boldsymbol{x}_{\mathrm{k}}, \boldsymbol{u}_{\mathrm{k}})],\ \boldsymbol{x}_{\mathrm{k}} \in \boldsymbol{X}_{\mathrm{D}}, \boldsymbol{x}_{\mathrm{k+1|k}} \in \boldsymbol{\chi}_{\mathrm{x}} \tag{7.49}$$

(4) By this point, the optimal objective function $\boldsymbol{J}^{*}(\boldsymbol{x}_{\mathrm{k}}, \boldsymbol{\varepsilon}_{\mathrm{s}})$ and the corresponding optimal control law $\boldsymbol{U}_{\mathrm{k}}^{*}(\boldsymbol{x}_{\mathrm{k}}, \boldsymbol{\varepsilon}_{\mathrm{s}})$ can be obtained under the given $\boldsymbol{\varepsilon}_{\mathrm{s}}$. For all $\boldsymbol{\varepsilon}_{\mathrm{s}} \in \boldsymbol{E}$, solve:

$$[\boldsymbol{U}_{\mathrm{k}}^{*}, \boldsymbol{\varepsilon}^{*}] = \arg\ \min_{\boldsymbol{\varepsilon}_{\mathrm{s}} \in \boldsymbol{E}}(\boldsymbol{J}^{*}(\boldsymbol{x}_{\mathrm{k}}, \boldsymbol{\varepsilon}_{\mathrm{s}})) \tag{7.50}$$

The optimal relaxation factor $\boldsymbol{\varepsilon}^{*}$ and optimal control law $\boldsymbol{U}_{\mathrm{k}}^{*}$ can be obtained under the given system state $\boldsymbol{x}_{\mathrm{k}}$:

$$\boldsymbol{U}_{\mathrm{k}}^{*} = [\boldsymbol{u}_{\mathrm{k|k}}^{*}, \boldsymbol{u}_{\mathrm{k+1|k}}^{*}, \ldots, \boldsymbol{u}_{\mathrm{k+N-1|k}}^{*}]^{\mathrm{T}} \tag{7.51}$$

(5) Take the optimal control quantity of the current position point k as the optimal control law under the system state variable $\boldsymbol{x}_{\mathrm{k}}$, that is:

$$\boldsymbol{u}_{\mathrm{k}}^{*} = [\boldsymbol{I}\ 0\ \ldots\ 0]\boldsymbol{U}_{\mathrm{k}}^{*} \tag{7.52}$$

where $\boldsymbol{I}$ is a third-order identity matrix.

Through the above optimization solution, for any initial state $\boldsymbol{x}_{\mathrm{k}} \in \boldsymbol{X}_{\mathrm{D}}$, there is an optimal control quantity $\boldsymbol{u}_{\mathrm{k}}^{*}$ and the initial state $\boldsymbol{x}_{\mathrm{k}}$ and the corresponding optimal control quantities form one-to-one corresponding arrays. Store these arrays in memory in proper order. When the online table lookup is performed, for any state $\boldsymbol{x}$, if $\boldsymbol{x} \in \boldsymbol{X}_{\mathrm{D}}$, then the corresponding control quantity $\boldsymbol{u} = \boldsymbol{u}_{\mathrm{k}}^{*}$; if $\boldsymbol{x} \in \boldsymbol{X}_{\mathrm{D}}$, the cubic spline interpolation is performed to obtain the optimal control quantity.

7.3.2 Multi-system Dynamic Coordination

Using the multi-objective steady-state optimization control algorithm based on nonlinear model prediction theory, i-HEV ACC acheives the coordinated optimization of multi-performance objectives (vehicle safety, economy, and comfort) considering the steady-state efficiency characteristics of the components, and the optimal control torques of the engine, motor and hydraulic braking are calculated. As the i-HEV system has a variety of driving and braking modes, including switch co-operation between engine, electric motor and hydraulic brake and other different system working states. From the perspective of driver's driving comfort and fast response characteristics, it's required to coordinatedly control multiple parts according to the characteristics of switch and co-operation between sub-system modes and the difference in their control methods and response characteristics to realize the smooth switch between system modes and guarantee the quick and the continuous dynamic response of the vehicle. This is the problem that the multi-system dynamic coordination layer of i-HEV ACC needs to solve.

The difference in the response velocity, accuracy, and control difficulty of various components of i-HEV's driving and braking systems will directly influence the design and control effect of the multi-system dynamic coordinated control algorithm. In this chapter, the control method and the response characteristics of engine, motor and hydraulic brake in i-HEV's driving/braking systems are analyzed and modeled. On this basis, the dynamic coordination control algorithm is respectively designed for the driving system and braking system combining open loop and closed loop.

7.3.2.1 Analysis and Modeling of i-HEVs' Critical Powertrain Response Characteristics

Since the i-HEV system is transformed from traditional HEV and AMT systems and the gear-shifting process control are completed by the component, only the mode switching and co-operation between engine and motor in the driving system and between the motor and the hydraulic in the braking system are under dynamically coordinated control.

The control of the engine, motor, hydraulic brake and other components involved in the dynamic coordination of the i-HEV system is realized through the bus, that is, the vehicle controller sends the expected engine torque, motor torque and hydraulic braking pressure to the CAN bus, and each component receives corresponding instructions and performs corresponding operations. From the perspective of the i-HEV ACC vehicle controller, only the transfer characteristics between the expected control instructions and the actual output of each component are considered and the corresponding control is carried out. The dynamic response characteristics of the engine, motor and hydraulic braking are modeled as follows.

The engine controls the throttle position through the electronic throttle, and the engine ECU controls the fuel injection and ignition timing based on real-time engine intake. Because the engine torque output involves many aspects such as throttle position control, intake, fuel injection and gas combustion in the cylinder, it's slow in the output torque response and difficult to accurately control the torque. Internal working processes such as engine intake, fuel injection, ignition and combustion are ignored. It is assumed that various delay characteristics of engine work are converted into the dynamic characteristics of engine output torque, which is approximate by using the first-order inertial link, as shown in Eq. (7.53) [8].

$$T_{\mathrm{e\_out}} = \frac{1}{1+\tau_{\mathrm{e}} s} T_{\mathrm{e\_des}} \tag{7.53}$$

where $T_{\mathrm{e\_out}}$ is the engine's output torque, $T_{\mathrm{e\_des}}$ is the expected engine torque and τ_{e} is the first-order delay time constant. Based on the experimental results of engine torque response and Ref. [9], the least square method is used to identify the characteristic constant of engine torque dynamic response, and $\tau_{\mathrm{e}} = 0.2$ is obtained.

Figure 7.10 shows the correlation curves of the actual engine torque and the identified engine torque under the expected engine torque instruction, in which the solid line is the expected engine torque, the dotted line is the engine torque collected in the experiment, and the solid-dotted line is the engine torque result identified by the model as shown in Eq. (7.53). It can be seen from the result that there is a certain delay in the torque identified by the model when the engine starts, and the first-order delay model can better approximate the actual engine torque after the engine starts.

The rotating magnetic field and the corresponding electromagnetic force are generated by the motor through the current, which can quickly make the motor rotate and output power. It has the characteristics of fast response as well as simple control and high precision because it is directly controlled by the current. For motor

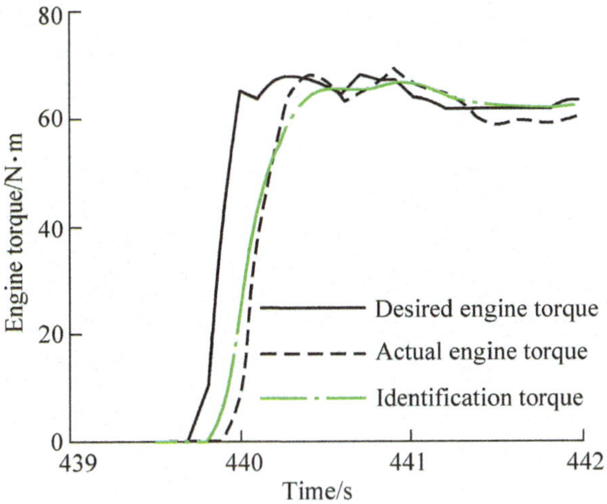

Fig. 7.10 Correlation curves of the engine's dynamic torque response

modeling, based on Ref. [10], first-order delay is adopted to approximate the torque response of the motor. Let T_{m_out} be the actual output torque of the motor and T_{m_des} be the expected torque of the motor, then the following equation can be obtained:

$$T_{m\_out} = \frac{1}{1 + \tau_m s} T_{m\_des} \quad (7.54)$$

Based on the result of the motor torque response experiment, the least square method is also used to identify the dynamic characteristic constant of the motor torque response, and $\tau_m = 0.01$ is obtained.

Figure 7.11 shows the correlation curves of the actual motor torque and the identified motor torque under the expected motor torque instruction, in which the solid line is the expected motor torque, the dotted line is the motor torque collected in the experiment, and the solid-dotted line is the motor torque result calculated by the model as shown in Eq. (5.54). It can be seen from the result that the actual motor torque can rapidly track the expected motor torque, and the response of the established motor model is almost consistent with that of the actual motor torque.

The i-HEV ACC system requires the vehicle to achieve active braking force control, and the braking torque generated by the braking energy recovered by the motor is not enough to achieve that purpose, so additional hydraulic braking is also needed in strong braking. A parallel braking system consisting of the driver and the electronic vacuum brake (EVB) which pushes the pushrod through the solenoid valve requires little change to be performed on the original braking system and can be used as a mature and reliable active hydraulic braking system for good compatibility, simple structure and low cost. The mechanism of EVB is as follows: the EVB cavity is divided into two parts by the spring diaphragm containing the pushrod; one part is connected with the engine intake pipe, and the other is connected with the atmosphere.

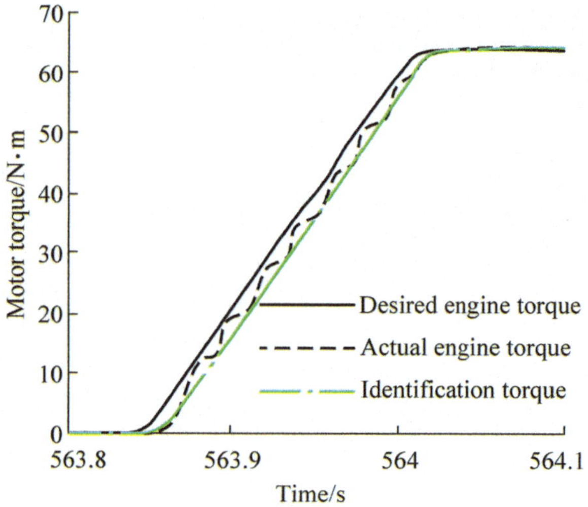

Fig. 7.11 Correlation curve of the motor's dynamic torque response

The pushrod is pushed through the control solenoid valve to push the pedal to move under the vacuum power of the engine and generate the braking force.

Because the control instruction obtained by the upper controller is the braking torque, and EVB controls the solenoid valve based on the braking pressure collected by the pressure sensor to carry out the feedback control of the braking pressure, the braking torque instruction calculated by the upper layer should be converted to the expected hydraulic braking pressure of EVB.

Under the condition that the maximum braking force of the road surface is not exceeded, the hydraulic braking torque is generated by the friction between the brake disc and the brake caliper, which can be expressed as [11]:

$$p = \frac{T_{\mathrm{hj}}}{A_{\mathrm{p}} f_{\mathrm{b}} r_{\mathrm{b}}} \tag{7.55}$$

where p is the wheel cylinder pressure; T_{hj} is the braking torque generated at the corresponding wheel; A_{p} is the piston area of the wheel cylinder; f_{b} is the friction coefficient between the brake disc and the brake caliper; r_{b} is the radius of friction on the brake disc. It can be seen from the above equation that, without considering stability, EVB hydraulic braking pressure shows a linear function relationship with vehicle's braking torque, as shown in the following equation [12]:

$$p_{\mathrm{EVB}} = \frac{T_{\mathrm{h}}}{k_{\mathrm{pf}}} \tag{7.56}$$

where T_{h} is the vehicle braking torque corresponding to EVB hydraulic braking pressure, and k_{pf} is the proportional coefficient.

EVB hydraulic braking system includes EVB controller and EVB actuator. EVB controller and EVB actuator are regarded as a whole, and the transfer function between the expected braking pressure and the actual braking pressure is obtained by using the model identification method. Emergency braking rarely occurs for i-HEV ACC, and the braking pressure of EVB is basically under 2 MPa. In this section, EVB pressure's step response experiment results under 1.6 and 2 MPa are selected for EVB model identification.

As shown in Fig. 7.12, it can be seen from some results that EVB has a certain delay due to the inertia of the system and the need to overcome the static friction force at the early stage of startup, while it has a certain overshoot in the rising process. Based on the model identification result of EVB in Ref. [13], the second-order-plus-dead-time (SOPDT) transfer function structure is used to represent the braking pressure transfer function of EVB $\boldsymbol{G}_{\text{EVB}}(s)$, as shown in the following equation:

$$\boldsymbol{G}_{\text{EVB}}(s) = \frac{K\omega_{\text{n}}^2}{s^2 + 2\xi\omega_{\text{n}}s + \omega_{\text{n}}^2}e^{-Ls} \tag{7.57}$$

where L is delay time, and $K > 0$, $\xi > 0$, $\omega_{\text{n}} > 0$ are constants.

Each parameter is identified based on the method used in Ref. [14], to obtain the pressure response transfer function of EVB:

$$\boldsymbol{G}_{\text{EVB}}(s) = \frac{1}{0.005s^2 + 0.08s + 1}e^{-0.1s} \tag{7.58}$$

With Eq. (7.57) substituted into Eq. (7.58), the relationship between EVB hydraulic braking pressure and expected braking torque can be obtained as follows:

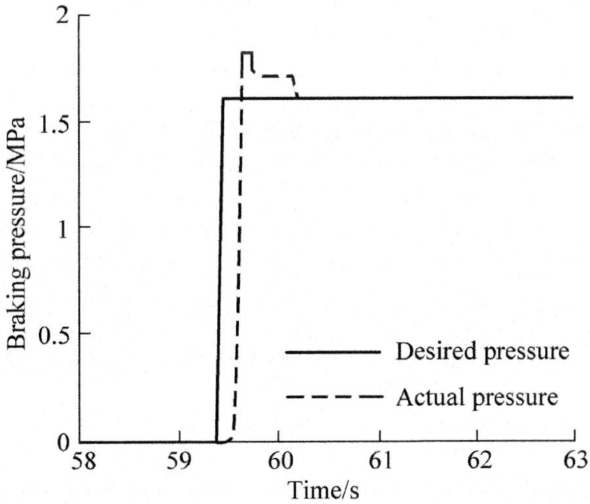

Fig. 7.12 EVB braking pressure response curve

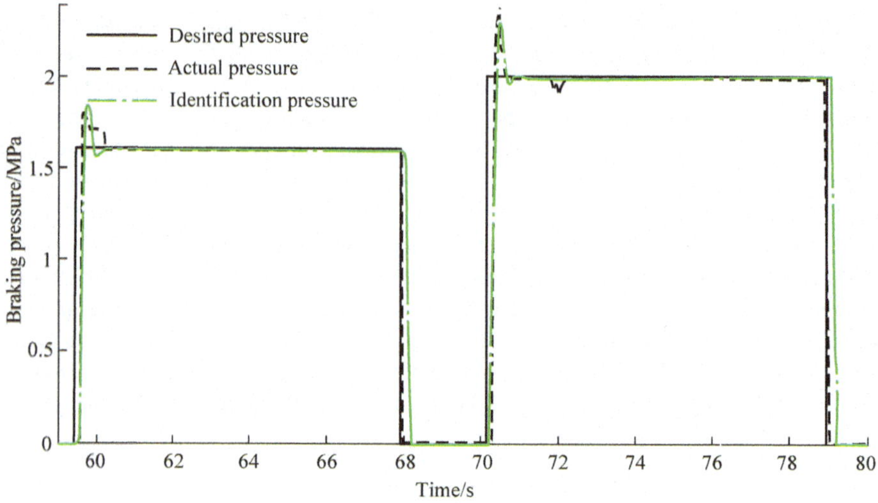

Fig. 7.13 Correlation curves of EVB braking pressure

$$G_{\mathrm{EVB}}(s) = \frac{1}{0.005s^2 + 0.08s + 1} e^{-0.1s} \frac{T_{\mathrm{h}}^*}{k_{\mathrm{pf}}} \tag{7.59}$$

Figure 7.13 shows the response curves of EVB hydraulic braking pressure under step response, in which the solid line is the expected braking pressure of EVB, the dotted line is the EVB braking pressure collected in the experiment, and the solid-dotted line is the braking pressure identified by the model. It can be seen from the figure that a good consistency is maintained between the identified EVB pressure response curve and the actual pressure curve, and the error between the braking pressure identified by the model and the actual braking pressure is within 0.2 MPa. In addition, EVB brake pressure has a certain overshoot and delay as shown in the figure. This is because the movement of the braking pedal mainly comes from the vacuum in the cavity despite that EVB uses the electromagnetic valve to push the brake pedal, which makes the system slower in response and vulnerable to the influence of vacuum pressure fluctuation, and the hydraulic itself also has a certain delay and pressure fluctuation. Therefore, EVB hydraulic braking has characteristics including slow response velocity and difficulty to control.

7.3.2.2 Dynamic Coordination of i-HEV Driving System

As shown by the engine torque and motor torque response characteristics in Figs. 7.10 and 7.11, the motor's torque response time is about 0.01 s and the engine's torque response time is about 0.2 s. The motor's torque can track the expected torque without error because it's easy to control the motor, while it's hard to accurately control the

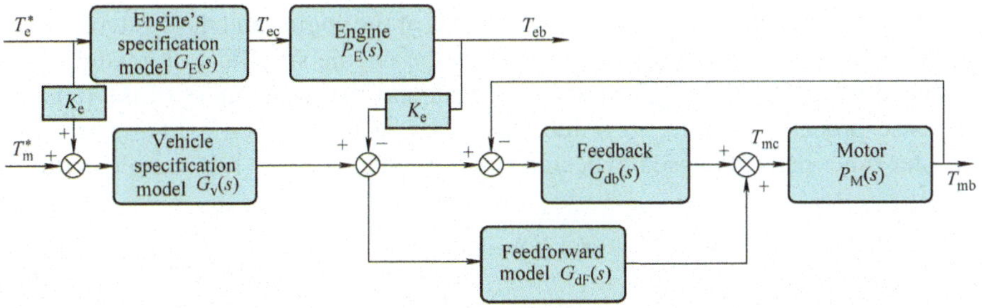

Fig. 7.14 Structure diagram of the control system for the dynamic coordination of the i-HEV driving system

engine torque due to the complexity in engine control. Based on the above characteristics of engine and motor (torque response velocity, precision, and control difficulty), the dynamic and coordinated control algorithm of the i-HEV driving system is designed utilizing engine open-loop and motor closed-loop feedback control. The basic structure is shown in Fig. 7.14.

As shown in Fig. 7.14, the optimal engine torque and motor torque obtained with optimal allocation calculation in the upper layer are respectively $T_e{}^*$, $T_m{}^*$, the engine's specification model is $\boldsymbol{G}_E(s)$, the engine's component model is $\boldsymbol{P}_E(s)$, the vehicle's expected specification model is $\boldsymbol{G}_v(s)$, the motor's feedback compensator is $\boldsymbol{G}_{db}(s)$, the motor's component model is $\boldsymbol{P}_M(s)$, and the motor's feedforward compensator is $\boldsymbol{G}_{dF}(s)$, K_e is the conversion coefficient between the engine torque and the motor torque.

The input of the controller includes the expected engine torque $T_e{}^*$ and the expected motor torque $T_m{}^*$. The expected engine torque obtains the actual controllable expected torque instruction through the engine's specification model $\boldsymbol{G}_E(s)$, and directly sends the control instruction to the engine. At the same time, to meet the power demand of the vehicle, the motor torque is used to feedback compensate the difference between the total power demanded by the vehicle and the actual driving power output. The motor's feedback controller uses the two-degree-of-freedom model matching control structure with feedforward and feedback.

The torque response characteristics expected by the driver are the main consideration in the design of vehicle specification model $\boldsymbol{G}_v(s)$. Because the response characteristics of the specification model will affect the design of the feedforward compensator, in order to decrease the order of the feedforward compensator, a first-order delay transfer function is used to approximate the specification model based on the characteristics of driver's characteristics and the model matching, as shown in the following equation:

$$\boldsymbol{G}_v(s) = \frac{1}{\tau_d s + 1} \tag{7.60}$$

where s is Laplace operator; τ_d is time constant.

The actual dynamic response characteristics of the engine and the system comfort are the main considerations in the design of the engine specification model $\boldsymbol{G}_{\mathrm{E}}(s)$, balanced with the fast response of the system and the smoothness of the dynamic response process. In order to reduce the order of the system, a first-order delay transfer function is also used to design $\boldsymbol{G}_{\mathrm{E}}(s)$, as shown in the following equation:

$$\boldsymbol{G}_{\mathrm{E}}(s) = \frac{1}{\tau_{\mathrm{E}} s + 1} \tag{7.61}$$

where τ_{E} is the time constant of the engine specification model.

As shown in Fig. 7.14, on the engine's open-loop control loop, the optimized engine torque T_{e}^{*} performs open-loop control with the engine specification model $G_{\mathrm{E}}(s)$ and the engine model $P_{\mathrm{E}}(s)$, to obtain the engine output torque as shown in the following equation:

$$T_{\mathrm{eb}} = T_{\mathrm{e}}^{*} \boldsymbol{G}_{\mathrm{E}}(s) \boldsymbol{P}_{\mathrm{E}}(s) \tag{7.62}$$

On the motor's closed-loop control loop, the motor control torque T_{mc} is composed of the superposition of the motor's feedforward loop and the feedback compensation loop, as shown in the following equation.

$$\begin{aligned} T_{\mathrm{mc}} = {} & [(T_{\mathrm{e}}^{*} K_{\mathrm{e}} + T_{\mathrm{m}}^{*}) \boldsymbol{G}_{\mathrm{v}}(s) - T_{\mathrm{eb}} K_{\mathrm{e}}] \boldsymbol{G}_{\mathrm{dF}}(s) \\ & + \{[(T_{\mathrm{e}}^{*} K_{\mathrm{e}} + T_{\mathrm{m}}^{*}) \boldsymbol{G}_{\mathrm{v}}(s) - T_{\mathrm{eb}} K_{\mathrm{e}}] - T_{\mathrm{mb}}\} \boldsymbol{G}_{\mathrm{db}}(s) \end{aligned} \tag{7.63}$$

On the other hand, the motor's output torque T_{mb} can be expressed as:

$$T_{\mathrm{mb}} = T_{\mathrm{mc}} \boldsymbol{P}_{\mathrm{M}}(s) \tag{7.64}$$

From Eqs. (7.63) and (7.64), it can be obtained that:

$$(T_{\mathrm{e}}^{*} K_{\mathrm{e}} + T_{\mathrm{m}}^{*}) \boldsymbol{G}_{\mathrm{v}}(s) - T_{\mathrm{eb}} K_{\mathrm{e}} = T_{\mathrm{mb}} \frac{\boldsymbol{G}_{\mathrm{db}}(s) \boldsymbol{P}_{\mathrm{M}}(s) + 1}{\boldsymbol{P}_{\mathrm{M}}(s) [\boldsymbol{G}_{\mathrm{dF}}(s) + \boldsymbol{G}_{\mathrm{db}}(s)]} \tag{7.65}$$

On the other hand, to make the system's output track the system's expected input, there is:

$$T_{\mathrm{eb}} K_{\mathrm{e}} + T_{\mathrm{mb}} = (T_{\mathrm{e}}^{*} K_{\mathrm{e}} + T_{\mathrm{m}}^{*}) \boldsymbol{G}_{\mathrm{v}}(s) \tag{7.66}$$

Therefore, the motor's feedforward transfer function $\boldsymbol{G}_{\mathrm{F}}(s)$ is obtained from Eqs. (7.65) and (7.66):

$$\boldsymbol{G}_{\mathrm{dF}}(s) = \frac{1}{\boldsymbol{P}_{\mathrm{M}}(s)} \tag{7.67}$$

Because the PID controller has better robustness and stability, PID control is used in the design of the algorithm for the feedback compensator $\boldsymbol{G}_{\text{db}}(s)$, as shown in the following Eq. (7.68).

$$\boldsymbol{G}_{\text{db}}(s) = k_{\text{dp}} + k_{\text{di}}\frac{1}{s} + k_{\text{dd}}s \tag{7.68}$$

where k_{dp} is the proportional term coefficient, k_{di} is the integral term coefficient, and k_{dd} is the differential term coefficient.

7.3.2.3 Dynamic Coordination of the i-HEV Braking System

Based on the characteristics of EVB hydraulic braking's and motor torque's response velocity, accuracy and control difficulty, and similar to the dynamic coordination of the driving system, the control system for the dynamic coordination of the i-HEV braking system is designed by using EVB hydraulic braking open-loop control and motor closed-loop feedback control. Its basic structure is shown in Fig. 7.15.

As shown in Fig. 7.15, the optimal hydraulic braking torque and motor torque obtained from the upper optimization allocation are T_{h}^{*} and T_{m}^{*} respectively, the hydraulic braking specification model is $\boldsymbol{G}_{\text{h}}(s)$, the hydraulic braking system model is $\boldsymbol{P}_{\text{h}}(s)$, the motor feedback compensator is $\boldsymbol{G}_{\text{bb}}(s)$, the motor component model is $\boldsymbol{P}_{\text{M}}(s)$, and the motor feedforward compensator is $\boldsymbol{G}_{\text{bF}}(s)$.

According to Eqs. (7.55) and (7.58), the braking torque response of EVB hydraulic braking has the same transfer characteristics as the braking pressure response, so the transfer function of EVB's expected torque and output torque can be obtained as follows:

$$\boldsymbol{P}_{\text{h}}(s) = \boldsymbol{G}_{\text{EVB}}(s) \tag{7.69}$$

The controller input includes the expected torque of hydraulic braking and the expected motor torque. The expected hydraulic braking torque obtains the actual

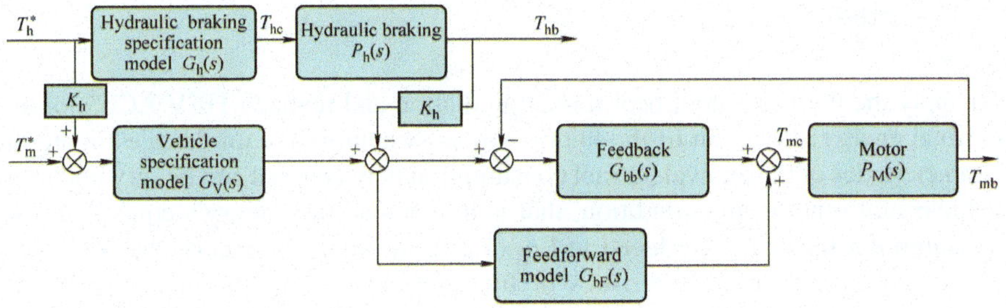

Fig. 7.15 Structure diagram of the control system for the dynamic coordination of the i-HEV braking system

controllable expected torque instruction through the hydraulic braking specification model, and the control instruction is directly sent to the EVB controller. At the same time, the motor conducts feedback compensation on the difference between the actual braking torque demanded by the vehicle and the total braking torque. The motor feedback controller also uses a two-degree-of-freedom model matching control structure with feedforward and feedback.

The dynamic response characteristics of the actual hydraulic braking system and system comfort are the main considerations in the design of hydraulic braking specification model $\boldsymbol{G}_{\mathrm{h}}(s)$. Similar to the specification model of engine, a first-order delay transfer function is also used to approximate the specification model of hydraulic braking, as shown in the following equation:

$$\boldsymbol{G}_{\mathrm{h}}(s) = \frac{1}{\tau_{\mathrm{hyd}} s + 1} \tag{7.70}$$

where τ_{hyd} is the time constant of the hydraulic brake specification model.

Similinar to the dynamic coordination of the driving system, the part of motor feedforward compensation can be calculated as follows:

$$\boldsymbol{G}_{\mathrm{bF}}(s) = \frac{1}{\boldsymbol{P}_{\mathrm{M}}(s)} \tag{7.71}$$

Algorithm design is performed on the feedback compensation controller $\boldsymbol{G}_{\mathrm{bb}}(s)$ by using PID control, as shown in the Eq. (7.72)

$$\boldsymbol{G}_{\mathrm{bb}}(s) = k_{\mathrm{bp}} + k_{\mathrm{bi}} \frac{1}{s} + k_{\mathrm{bd}} s \tag{7.72}$$

where k_{bp} is the proportional term coefficient, k_{bi} is the integral term coefficient, and k_{bd} is the differential term coefficient.

7.3.3 Estimation of Battery's Equivalent Fuel Consumption Factor

To apply the theory of nonlinear model prediction and realize i-HEV ACC's online optimal energy allocation of the hybrid power system, it's required to estimate the characteristics of the equivalent fuel consumption between fuel and battery under the vehicle's current driving condition, that is to estimate the battery's equivalent fuel consumption factor λ introduced in Eq. (7.20), in order to estimate the equivalent energy consumption characteristics of fuel and electric energy under the current driving condition.

The i-HEV platform designed in this section integrates sensor and communication systems such as GPS, radar, camera and wireless V2I communication to obtain the vehicle's current driving condition characteristics. Based on the variation curve of velocity and acceleration under this driving condition, the battery's equivalent fuel consumption factor is estimated in this section based on Refs. [15, 16]. In this section, the engine fuel consumption characteristics are first modeled for analysis and simplification, and then the globally optimized power allocation and control problem of the hybrid power allocation system is established based on the velocity and acceleration variation curves under the working scenario. Lagrange operator is also introduced to solve this global optimization problem and estimate the battery's equivalent fuel consumption factor under this working scenario.

7.3.3.1 Simplification of Engine Fuel Consumption Model

From the perspective of energy conversion, the optimal energy allocation strategy of the hybrid power system is mainly aimed at the optimal allocation of engine drive energy and battery discharge/charge energy, so as to achieve the optimal economy of the vehicle system. Therefore, from the perspective of system energy optimization, a simplified model of engine fuel consumption and engine output power is established in this section to describe the mathematical relationship between them.

Based on engine torque T_e and rotational speed w_e, the engine power can be described as:

$$P_e = T_e \omega_e \tag{7.73}$$

The engine's instantaneous fuel consumption $\dot{m}_e$ (g/s) can be expressed as:

$$\dot{m}_e = \frac{\omega_e T_e \boldsymbol{M}_e(\omega_e, T_e)}{3.6 \times 10^6} \tag{7.74}$$

where $\boldsymbol{M}_e(\omega_e, T_e)$ is the instantaneous fuel consumption of the engine obtained by looking up the table based on the steady-state fuel consumption MAP of the engine, and its unit is (g/kWh)

Based on Eqs. (7.73) and (7.74) and the steady-state engine fuel consumption MAP, the relationship between engine fuel consumption and power and the corresponding fitting curve can be obtained, as shown in Fig. 7.16.

As shown in Fig. 7.16, point O is the engine power and their corresponding fuel consumption corresponding to the engine's all operating points, and the solid black line is the fitting curve. As indicated by point O in the figure, the fuel consumption increases with the increase of engine power, and the engine power shows a quadratic mathematical relationship with its corresponding fuel consumption. The engine power and fuel consumption characteristics are fitted based on different power points of the engine and corresponding fuel consumption, and the fitting result is shown in Eq. (7.75).

$$\dot{m}_e(P_e) = aP_e^2 + bP_e + c \tag{7.75}$$

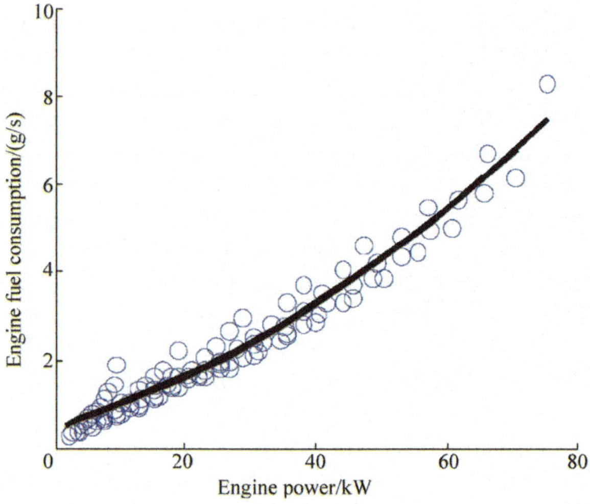

Fig. 7.16 Correlation curve of the relationship between engine fuel consumption and power

where a, b, and c are fitting coefficients, and $a = 0.0006$, $b = 0.0478$, $c = 0.4433$.

As shown in Fig. 7.16, the fitted engine power-fuel consumption characteristics curve can reflect the quadratic change relationship between the actual engine power and the corresponding fuel consumption, which meets the requirements for estimating the battery's equivalent fuel consumption factor.

7.3.3.2 Estimation of the battery's Equivalent Fuel Consumption Factor Based on Working Scenario Characteristics

The vehicle's driving condition characteristics CYC are identified by the i-HEV intelligent information interaction system and obtained through information fusion and processing. Its characteristics can be described by vehicle velocity sequence $v_{\mathrm{cyc}}(k)$ and acceleration sequence $a_{\mathrm{cyc}}(k)$. According to the vehicle dynamics equation, the vehicle's demanded power sequence $P_{\mathrm{d}}(k)$ in the driving cycle can be calculated as:

$$P_{\mathrm{d}}(k) = (ma_{\mathrm{cyc}}(k) + mgf_{\mathrm{r}} + \frac{C_{\mathrm{d}}A\rho v_{\mathrm{cyc}}(k)^2}{2})v_{\mathrm{cyc}}(k) \tag{7.76}$$

If the driving power of the engine at moment k is set as $P_{\mathrm{e}}(k)$ and the power consumed by the battery is $P_{\mathrm{b}}(k)$, then the required power of the vehicle is equal to the superposition of the power of the engine and the battery, as shown in the equation below:

$$P_{\mathrm{d}}(k) = P_{\mathrm{e}}(k)\eta_{\mathrm{e}} + P_{\mathrm{b}}(k)\eta_{\mathrm{b}} \tag{7.77}$$

where η_e and η_b are respectively the average mechanical efficiency of engine power transfer, and the total average efficiency of the charge and discharge of electric energy and power transfer.

In order to achieve the lowest fuel consumption of the hybrid power system under this working scenario, the corresponding optimization objective of the hybrid power system can be expressed as the engine fuel consumption under this working scenario:

$$\boldsymbol{L}_{\text{cyc}}(P_b, P_e) = \sum_{k=1}^{P} \dot{m}_e(P_e(k))\Delta t \tag{7.78}$$

where P is the length of the driving cycle and Δt is the discrete time step.

On the other hand, for a given working scenario CYC, the energy balance of battery charge and discharge at the beginning and end of the working scenario is required, as shown in Eq. (7.79).

$$\sum_{k=1}^{P} P_b(k) = 0 \tag{7.79}$$

In this section, the equivalent fuel consumption factor of the battery that can reflect the equivalent relationship between fuel and electric energy under the working scenario is obtained by calculating the characteristics of the optimal allocation of fuel and electric energy under the working scenario, without considering the constraints of the engine, motor, battery and other components. The operator λ is introduced and the Lagrange function is defined:

$$\boldsymbol{L}_{\text{cyc}}(P_b, P_e, \lambda) = \sum_{k=1}^{P} \dot{m}_e(P_e(k))\Delta t - \lambda \sum_{k=1}^{P} P_b(k)\Delta t \tag{7.80}$$

The problem of optimal energy allocation of the hybrid power system is further transformed into the global optimization problem as shown in Eq. (6.80) above. Substitute Eq. (7.75) into Eq. (7.80), and set $P_b(k)$ as the optimization variable, then:

$$\begin{aligned}
\boldsymbol{L}_{\text{cyc}}\left(P_b, \lambda_{\text{cyc}}\right) &= \sum_{k=1}^{P} \dot{m}_e\left(\frac{P_d(k) - P_b(k)\eta_b}{\eta_e}\right)\Delta t - \lambda_{\text{cyc}} \sum_{k=1}^{P} P_b(k)\Delta t \\
&= \sum_{k=1}^{P} \left[a\left(\frac{P_d(k) - P_b(k)\eta_b}{\eta_e}\right)^2 + b\left(\frac{P_d(k) - P_b(k)\eta_b}{\eta_e}\right) + c\right]\Delta t - \lambda_{\text{cyc}} \sum_{k=1}^{P} P_b(k)\Delta t \\
&= s_0 + \sum_{k=1}^{P} [s_1(k)P_b(k)^2 + s_2(k)P_b(k)]\Delta t - \lambda_{\text{cyc}} \sum_{k=1}^{P} P_b(k)\Delta t
\end{aligned} \tag{7.81}$$

where,

$$
\begin{aligned}
s_0 &= \sum_{k=1}^{P}\left[a\frac{P_d(k)^2}{\eta_e^2} + bP_d(k) + c\right] \\
s_1(k) &= a\left(\frac{\eta_b}{\eta_e}\right)^2 \\
s_2(k) &= -(2aP_d(k) + b)\frac{\eta_b}{\eta_e}
\end{aligned} \tag{7.82}
$$

Kuhn-Tucker conditions need to be satisfied for Eq. (7.82) to be optimal, as shown below:

$$
\begin{aligned}
\frac{\partial \boldsymbol{L}_{\mathrm{cyc}}(P_b, \lambda_{\mathrm{cyc}})}{\partial P_b} &= 0 \\
\frac{\partial \boldsymbol{L}_{\mathrm{cyc}}(P_b, \lambda_{\mathrm{cyc}})}{\partial \lambda_{\mathrm{cyc}}} &= 0
\end{aligned} \tag{7.83}
$$

It can be obtained from Eqs. (7.82) and (7.83):

$$
\begin{aligned}
2s_1(k)P_b^*(k) + s_2(k) - \lambda_{\mathrm{cyc}} &= 0 \\
P_b^*(k) &= \frac{\lambda - s_2(k)}{2s_1(k)}
\end{aligned} \tag{7.84}
$$

By summing Eq. (7.84) from $k = 1$ to P, and combining with Eq. (7.79), the Lagrange operator λ can be obtained as:

$$
\lambda_{\mathrm{cyc}} = \frac{\sum_{k=1}^{P}\frac{s_2(k)}{2s_1(k)}}{\sum_{k=1}^{P}\frac{1}{2s_1(k)}} \tag{7.85}
$$

In the calculation process of the above optimal energy control strategy of the hybrid power system, Lagrange operator is introduced to convert the optimization problem with the energy balance constraints of battery charge and discharge into an unconstrained problem to be solved. For a given driving cycle CYC, the fixed λ can be obtained through the above calculation, and the global optimization problems as shown in Eq. (7.78) can be rewritten into P instantaneous optimization problems as shown in the following equation and carried out respectively.

$$
\begin{aligned}
&\min_{P_e(k), P_b(k)} \boldsymbol{L}_{\mathrm{cyc}}(P_b(k), P_e(k), k) = \dot{m}_e(P_e(k)) - \lambda_{\mathrm{cyc}} P_b(k) \\
&\textit{subj.to:} \\
&P_d(k) = P_e(k)\eta_e + P_b(k)\eta_b \\
&k = 1, 2, \ldots P
\end{aligned} \tag{7.86}
$$

From the perspective of optimal energy allocation, with the introduced factor λ_{cyc}, Eq. (7.86) translates the problem of global optimal energy allocation of the hybrid power system into the problem of optimally allocating fuel and electric energy at every moment. This also indicates that for the hybrid system, by learning the characteristics of the global working scenario and introducing them to the current energy allocation of the hybrid power system, we can convert the global optimization with complicated calculation and poor realizability to the optimal instantaneous allocation of energy with small amount of calculation and strong operability. An analysis of Eq. (8.72) from the dimensional perspective shows that the unit of the first item on the right side of the equation is the engine's instantaneous fuel consumption. For this optimization problem, from the perspective of uniform dimension, the second item can be regarded as battery power $P_b(k)$ converted into the engine's instantaneous fuel consumption with factor λ_{cyc} which reflects the equivalent conversion relationship between fuel and electricity under the current working scenario and is thus the battery's equivalent fuel consumption factor. Although the velocity curve of i-HEV ACC in the process of car-following is not exactly the same as the working scenario of CYC, its operation velocity, acceleration and deceleration characteristics are the same as the identified working scenario CYC. Therefore the factor λ_{cyc} can reflect the characteristics of fuel and electricity allocation of i-HEV ACC in driving and can be used in Eq. (7.20) for the equivalent calculation of the vehicle fuel economy and the subsequent optimization control.

7.4 Simulation Analysis of i-HEV ACC

7.4.1 The Forward Simulation Platform Structure Design

The i-HEV ACC simulation platform is mainly used in the development, debugging and simulation analysis of the control strategy of the vehicle's longitudinal assistant driving system. As a basic platform for studying the comprehensive performance of i-HEV, it should meet the following basic requirements: (1) the simulation model has good consistency with the real-life vehicle and can well reflect the real-life vehicle's dynamics characteristics; (2) the simulation platform is modular, flexible and convenient for adding or removing system components according to needs, and it can be quickly transplanted to other vehicle simulation models; (3) the basic structure and information flow of the simulation system are consistent with those of the real-life vehicle, which can be used for the rapid development, performance evaluation and strategy optimization of the control system.

Based on the structural characteristics of i-HEV of information sharing between multi-energy driving/braking system and environmental awareness system, a hierarchically modular and flexible forward simulation platform of i-HEV ACC is designed

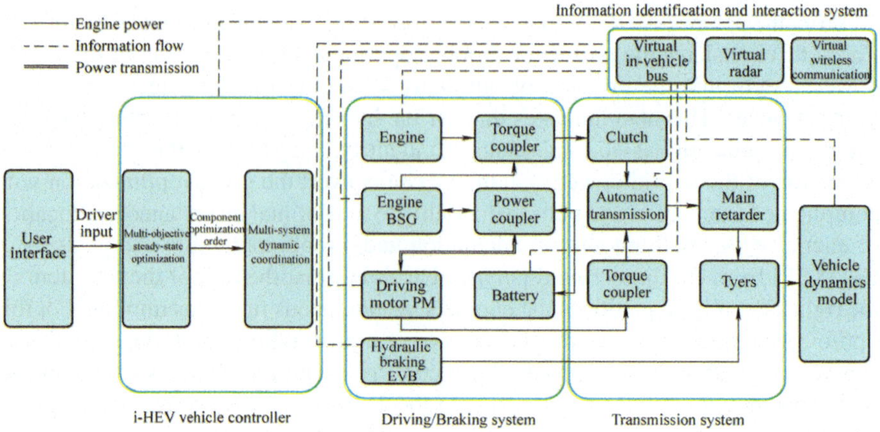

Fig. 7.17 Structure of i-HEV ACC forward simulation platform

with real-life vehicle components as unit modules and the information flow of real-life vehicle control system and components as the information flow direction in the model, the overall structure of which is as shown in Fig. 7.17.

As shown in Fig. 7.17, the information and energy flow of the designed i-HEV ACC simulation platform is consistent with that of the real-life vehicle. Each power component receives the vehicle controller's instructions and outputs the corresponding power, while the vehicle controller receives the information inputs such as the feedback of vehicle states, environmental identification, and driver operation, formulates the corresponding control strategy and controls the vehicle movement. The simulation platform has modules including user interface, i-HEV controller, driving/braking system, driveline, vehicle dynamics model, and information identification and interaction system.

The user interface module is used to simulate the operation of the driver and formulate signals such as accelerator pedal, brake pedal, cruise switch and cruise velocity setting, which are inputted into the vehicle controller of i-HEV which includes sub-modules such as multi-objective steady-state optimization controller and multi-system dynamic coordination controller. The driving/braking system includes models of components such as engine, BSG, PM drive motor, battery and EVB hydraulic brake; the transmission system mainly includes models of components such as clutch, AMT automatic transmission and main reducer; with longitudinal dynamics characteristics as the main consideration in its design, the vehicle dynamics model simulates the relationship between the longitudinal driving force and the drag force of the real-life vehicle, and calculates its velocity and acceleration; the information identification and interaction system simulates the vehicle CAN bus communication network, radar and wireless communication systems to obtain the information of vehicle status, traffic environment and inter-vehicle movement, and realizes the information interaction between various components of the vehicle and the vehicle controller.

As indicated by Fig. 7.17, the i-HEV ACC forward simulation platform involves complex mechanical transmission, power transmission and information flow. Starting from the idea of modular modeling, the mechanical transfer between power components is simulated by the torque coupler, the transfer of electric energy between the electric driving system and the battery is simulated by the power coupler, and the information flow is simulated by the virtual in-vehicle bus.

7.4.2 *Forward Simulation Platform Model*

To balance the accuracy and simplicity of the model, the first-order and second-order transfer functions such as Eqs. (7.53), (7.54) and (7.59) are used to model the drive/braking systems such as the engine, motor and EVB models. The i-HEV high-voltage battery pack is modeled based on Ref. [17] with the Rint model.

In the information identification and interaction system, the virtual intra-vehicle bus receives and sends the state information of each component in a fixed period without considering the delay; the virtual wireless communication is used to simulate the V2I communication to obtain the real-time traffic information; the virtual radar contains the simulation of the relative distance and relative velocity between the ego vehicle and the preceding vehicle. Reference [18] is used to model the dynamics models of the transmission system and the vehicle. The final and complete i-HEV ACC forward simulation platform established is as shown in Fig. 7.18. The platform consists of six subsystems: user interface subsystem, information identification and interaction subsystem, vehicle controller, driving/braking subsystem, transmission subsystem and vehicle dynamics model.

The established i-HEV ACC forward simulation platform uses the original HEV parameters, in which the vehicle's longitudinal dynamics parameters and parameters of engine, electric generator, drive motor, power battery, transmission and final drive are as shown in Tables 7.2, 7.3, 7.4, 7.5, 7.6 and 7.7, and the MAP charts of motor drive, brake energy regeneration efficiency, and BSG's power generation efficiency are as shown in Figs. 7.19, 7.20, and 7.21.

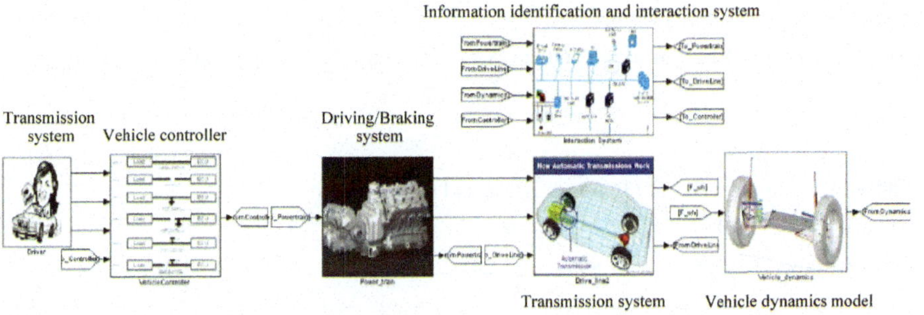

Fig. 7.18 Modules of the i-HEV ACC model

Table 7.2 i-HEV's longitudinal dynamics parameters

Parameter	Value
Vehicle mass m (kg)	1544
Wheel base l (m)	2.675
Tyre's rolling radius R_w (m)	0.307
Drag coefficient C_D	0.316
First-degree term of vehicle speed with rolling resistance coefficient speed f_2	0.0001
Mechanical transmission efficiency of transmission η_g	0.92
Mechanical transmission efficiency of drive motor η_{md}	0.95
Mechanical transmission efficiency of electric generator η_{mg}	0.9
Conversion factor of vehicle rotating mass δ	1.08
Engine/generator speed ratio i_{e_BSG}	0.625
Drive motor/transmission output shaft speed ratio i_{PM_AMT}	1.19

Table 7.3 Engine parameters

Displacement (L)	Rated power (kW)	Max. torque (Nm)	Max. RPM (r/min)
1.3	63 (6000r/min)	120	7000

Table 7.4 Generator parameters

Rated power (kW)	Rated torque (Nm)	Max. power (kW)	Max. torque (Nm)
4	20	8	40

Table 7.5 Drive motor parameters

Rated power (kW)	Rated torque (Nm)	Max. power (kW)	Max. torque (Nm)
15	64	25	135

Table 7.6 Power battery parameters

Battery capacity (A.h)	Working voltage (V)	Unit cell voltage (V)	Unit cell capacity (A.h)
6	288	1.2	6

Table 7.7 Parameters of 5T136 transmission and main reducer

Gear	1	2	3	4	5	R	i0
Velocity ratio	3.273	1.895	1.241	0.919	0.756	3.214	4.333

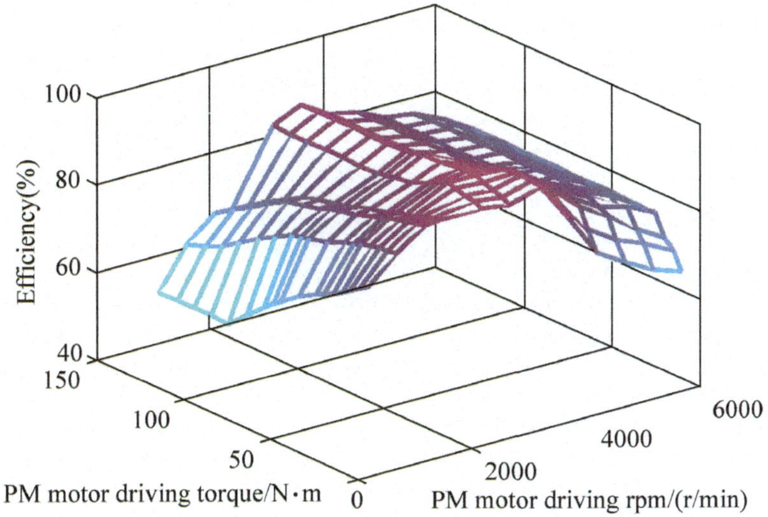

Fig. 7.19 MAP chart of PM motor's driving efficiency

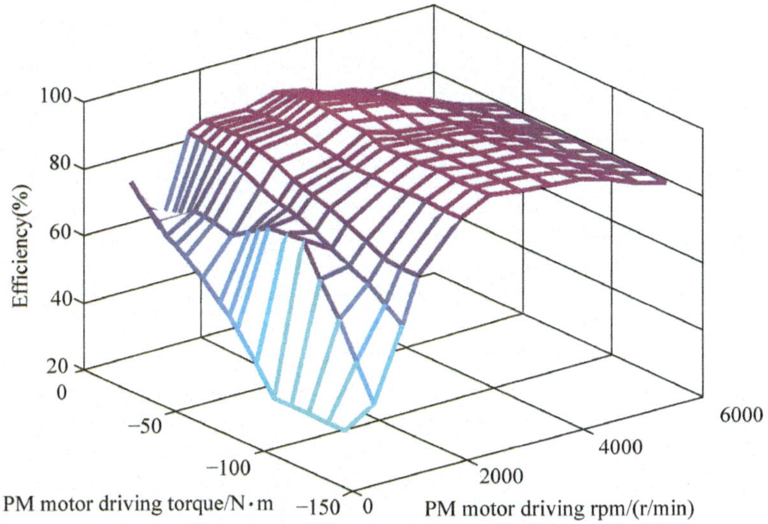

Fig. 7.20 MAP chart of PM motor's brake energy regeneration efficiency

7.4.3 Verification of Forward Simulation Platform

To verify the accuracy of the established i-HEV forward simulation platform, a real-life vehicle test is performed on the i-HEV in which the control torques of components including engine, motor, EVB hydraulic brake generated by the actual control system

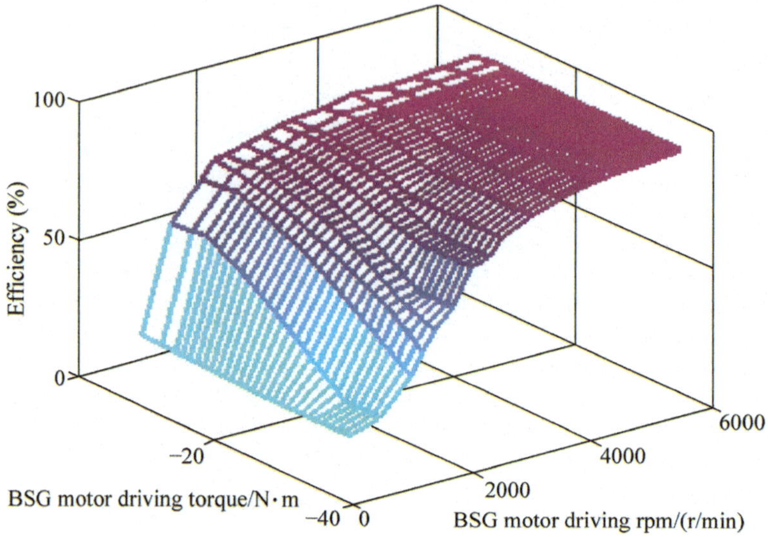

Fig. 7.21 MAP chart of BSG's generation efficiency

of i-HEV are sent to each component's model; the simulated vehicle velocity is then calculated through the driving/braking system model, the transmission system model and the vehicle longitudinal dynamics model and compared with the actual vehicle velocity. The contrast result is shown in Fig. 7.22.

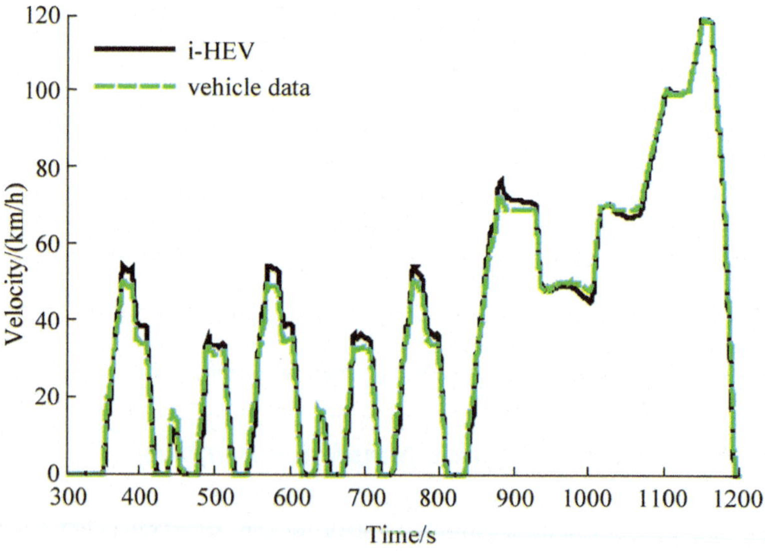

Fig. 7.22 Verification of i-HEV longitudinal dynamics model

As shown in Fig. 7.22, the solid line represents the simulated vehicle velocity and the dashed line the actual vehicle velocity. The figure shows that the simulated velocity's trend is consistent with that of the actual vehicle velocity. There's a certain error between the simulated and the actual vehicle velocities in the low-speed stage with the maximum error within 10% and the average error within 5%. Considering it's hard to get the accurate parameters of road surface due to its complexity, the established i-HEV forward simulation model can basically reflect the real-life vehicle's dynamics performance, and thus can be used in the development of the i-HEV ACC control system, the simulation of algorithms and the contrast evaluation.

7.4.4 Simulation Contrast of Working Scenarios

(1) Simulation 1: i-HEV ACC versus IV ACC

The scenario of sharp acceleration and deceleration of the preceding vehicle is designed in this section to reflect the car-following safety performance of each system in their response characteristics in case of rapid change in the preceding vehicle's velocity.

1. Sharp deceleration of the preceding vehicle

The scenario of sharp deceleration is mainly used to simulate and analyze the works of various vehicle components and the change of vehicle velocity in case of a strong braking during the cruise. The velocity of the preceding vehicle is set to begin with 70 km/h and decelerate at a constant braking deceleration speed of – 2 m/s^2, as shown in Fig. 7.23a.

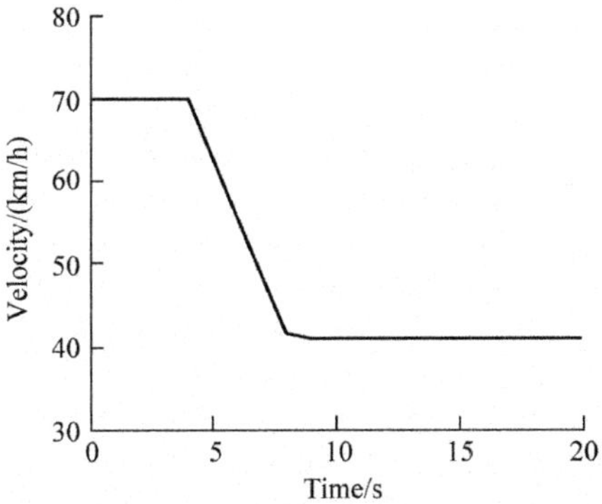

a) Sharp deceleration of the preceding vehicle

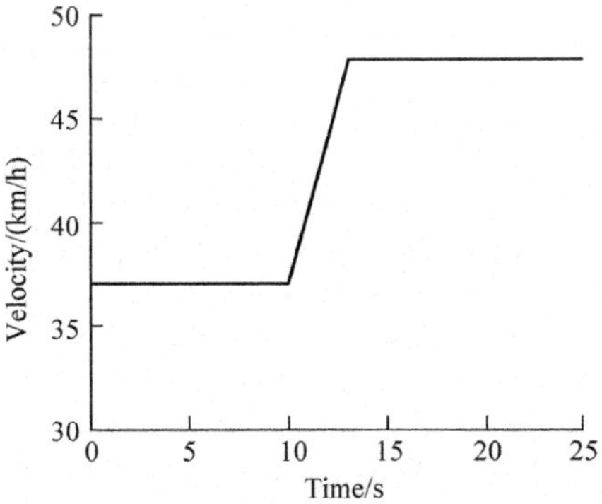

b) Sharp acceleration ofthe preceding vehicle

Fig. 7.23 Deceleration of the preceding vehicle

2. Sharp acceleration of the preceding vehicle

The scenario of sharp acceleration is mainly used to simulate and analyze the works of various vehicle components and the change of vehicle velocity in case of a strong acceleration of the preceding vehicle. The velocity of the preceding vehicle is set to begin with 37 km/h and accelerate at a constant acceleration of 1 m/s^2 for 3 s and then the speed is maintained, as shown in Fig. 7.23b.

(2) Simulation 2: i-HEV ACC versus driver HEV

In order to comprehensively compare the car-following safety and fuel economy between the i-HEV ACC and the driver HEV, the ISUZU ACC driving cycle test is adopted in this section to simulate and evaluate the performance of each system, which includes the urban driving cycle and the high-speed driving cycle. Because the original ISUZU urban driving cycle contains the vehicle's start-stop cruise process, at which the vehicle velocity is zero, while ACC generally requires the vehicle velocity to be higher than 20 km/h. Based on the characteristics of the original driving cycle, the minimum vehicle velocity is increased and the maximum velocity is limited under 50 km/h so that the contrast scenario of urban driving cycle is redesigned as shown in Fig. 7.24a, b is the high-speed contrast driving cycle.

(3) Simulation 3: i-HEV ACC versus simple superposition ACC; ISUZU driving cycle in Simulation 2 is used

The summary of contrast between simulated working scenarios of i-HEV ACC and IV ACC, HEV and simple-superposition HEV ACC is shown in Table 7.8.

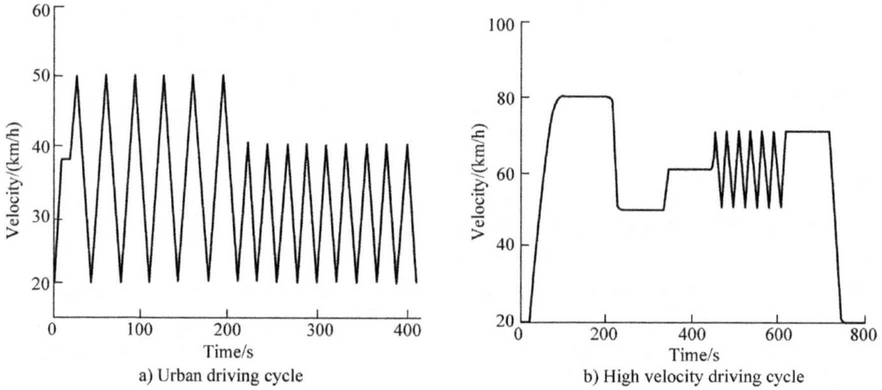

Fig. 7.24 ISUZU ACC driving cycle

Table 7.8 Contrast of i-HEV ACC's simulated working scenarios

No.	System	Working scenario	Main contrast indicators
1	IV ACC	Sharp deceleration	Car-following safety
		Sharp acceleration	Car-following safety
2	HEV	ISUZU ACC driving cycle	Car-following safety, fuel economy
3	Simple-superposition HEV ACC	ISUZU ACC driving cycle	Fuel economy

7.4.5 Evaluation Indicators for Simulation and Contrast

In order to comprehensively analyze and compare the simulation results, it is required to define quantitative indicators of car-following safety, driving comfort and fuel economy.

(1) Car-following safety indicator

The car-following safety in the longitudinal movement of a vehicle mainly refers to the relative velocity and safe distance maintained between the ego vehicle and the preceding vehicle in the process of car-following. Based on the definition of distance error and relative velocity in the process of car-following, the car-following safety indicator S is designed as shown in the following equation.

$$S = \frac{1}{N}\sum_{k=1}^{N}\left(\left|\frac{\Delta d(k)}{K_{\mathrm{DV}}}\right| + |\Delta v(k)|\right) \tag{7.87}$$

where N is all the sampling data points in the driving cycle, K_{DV} is the weight coefficient of relative distance error, $\Delta d(k)$ is the error between the actual inter-vehicle distance and the expected safe distance at moment k, and $\Delta v(k)$is the relative velocity between vehicles at moment k.

(2) Driving comfort indicator

The acceleration in the process of car-following is used to represent the driving comfort indicator. The smaller the acceleration amplitude, the better the driving comfort, and vice versa. At the same time, from the perspective of driver comfort constraint, if the acceleration meets $a_{\mathrm{b\,min}} \leq a_{\mathrm{f}} \leq a_{\mathrm{d\,max}}$, it indicates that the system meets the basic requirements of driver comfort.

(3) Fuel economy indicator

Fuel economy indicator is expressed as equivalent fuel consumption under the driving cycle of the hybrid power system, that is, the superposition of engine fuel consumption m_{e} and battery equivalent fuel consumption m_{b} under the driving cycle, where the equivalent fuel consumption of the battery is determined by the change of its

SOC. With γ set as the conversion coefficient of the equivalent fuel consumption of the battery SOC, the fuel economy of the system can be expressed as:

$$EQF = m_e + \gamma \Delta_{SOC} \tag{7.88}$$

Different conversion coefficients for the equivalent fuel consumption of the battery γ have a certain influence on the final fuel consumption result of the system. In order to make the fuel consumption of each system comparible to some extent and ensure the objectivity of the results, it should be expressed based on the results of engine fuel consumption and battery's power consumption in the simulation as much as possible. Therefore, although the battery equivalent fuel consumption factor γ has been estimated based on the current vehicle's driving cycle, it's only suitable for the optimal design of the controller. To evaluate the simulation fuel economy results, it is necessary to re-calculate the battery equivalent fuel consumption factor based on the engine fuel consumption and power consumption.

For the same hybrid power system, the conversion coefficient of battery SOC's equivalent fuel consumption is mainly affected by two factors: (1) the difference of working scenarios of the hybrid power system; (2) the difference in the energy allocation strategies of the hybrid power system. Therefore, to evaluate the fuel economy of the hybrid power system under different control strategies, the conversion coefficient of battery SOC's equivalent fuel consumption under different working scenarios and different strategies should be calculated respectively.

For the purpose of fuel economy contrast in this chapter, ISUZU driving cycle is used in all working scenarios. The hybrid power allocation strategies of all simulated systems for contrast are mainly divided into two categories: the energy allocation strategy based on the coordinated optimization of car-following safety and fuel economy corresponding to i-HEV and the real-life vehicle's energy allocation strategy based on rules adopted by the contrast system. Therefore, to calculate the conversion coefficient of battery SOC's equivalent fuel consumption, it is necessary to calculate the conversion coefficient of the equivalent fuel consumption of i-HEV battery SOC under the ISUZU driving cycle γ_{iHEV} and the conversion coefficient of battery SOC's equivalent fuel consumption based on the regular allocation strategy γ_{Rule}.

In order to avoid the influence of a single controller parameter on the conversion coefficient of battery SOC's equivalent fuel consumption, the engine fuel consumption obtained under different controller parameters is linearly fitted with the change of SOC under the driving cycle to obtain a unified conversion coefficient of battery SOC's equivalent fuel consumption. The corresponding system equivalent fuel consumption can be expressed as:

$$\begin{aligned} EQF_{iHEV} &= \gamma_{iHEV} * \Delta_{SOC} + f_{Eng} \\ EQF_{Rule} &= \gamma_{Rule} * \Delta_{SOC} + f_{Eng} \end{aligned} \tag{7.89}$$

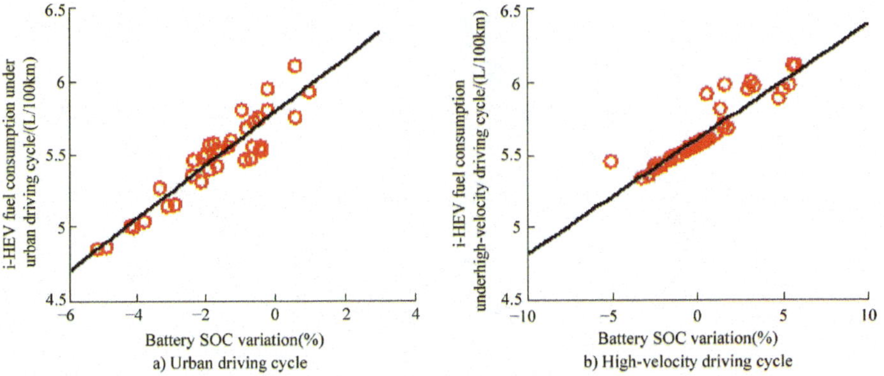

Fig. 7.25 Variation curves of i-HEV fuel consumption and battery SOC under different driving cycles

where EQF_{iHEV} is the equivalent fuel consumption representing the fuel economy of the i-HEV system, EQF_{Rule} is the equivalent fuel consumption representing the fuel economy of the HEV system based on the regular energy allocation strategy, f_{Eng} is the engine's fuel consumption, and Δ_{SOC} is the change of the battery SOC under the driving cycle.

By changing the weight coefficients of the controller's car-following and fuel economy, we can get the simulation results of engine fuel consumption and battery SOC variation under different controller parameters, where the variation of battery SOC is the difference between the SOC at the end of the driving cycle and the SOC at the beginning of the driving cycle. The variation curves of fuel consumption and battery SOC under the i-HEV, HEV urban cycle and high-speed cycle conditions are shown in Figs. 7.25 and 7.26, respectively. Points O are the simulation results of engine fuel consumption and battery SOC variation under different controller parameters, and the black straight line is the result of engine fuel consumption under fitted battery SOC variation. The slope obtained through linear fitting is the conversion coefficient of battery SOC's equivalent fuel consumption.

According to the fitted straight lines in Figs. 7.25 and 7.26, the conversion coefficients of i-HEV battery SOC's equivalent fuel consumption γ_{iHEV} under the urban driving cycle and the high-speed driving cycle are 18.01 and 7.91 respectively, and those of HEV battery SOC γ_{HEV} are 18.89 and 5.26 respectively.

7.4.6 Simulation Contrast Strategy and Contrast Method

The simulation contrast scheme designed involves the contrast and evaluation of system simulation with different system structures and different control modes. In this

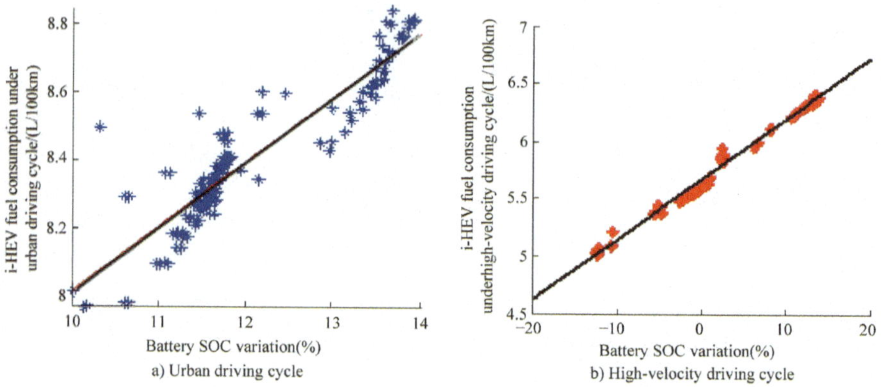

Fig. 7.26 Variation curves of HEV fuel consumption and battery SOC under different driving cycles

section, appropriate contrast strategies are designed according to the characteristics of each contrast system.

(1) Simulation 1: i-HEV ACC versus IV ACC

In order to reduce the influence of control systems and parameter design differences on the contrast of i-HEV ACC and IV ACC's car-following safety performance, the same vehicle power demand is used as the input of i-HEV ACC and IV ACC, as shown in Eq. (7.90), where $T_{\text{iHEV\_w}}$ is the demand torque at the wheel end of i-HEV and $T_{\text{IV\_w}}$ is the demand torque at the wheel end of IV. Under this assumption, the system response characteristics and corresponding car-following safety of i-HEV and IV in the sharp deceleration and acceleration scenarios are compared.

$$T_{\text{iHEV\_w}} = T_{\text{IV\_w}} \tag{7.90}$$

(2) Simulation 2: i-HEV ACC versus HEV

It's required to address two issues to achieve the driver mode HEV: developing an HEV energy allocation strategy for contrast and designing a longitudinal car-following model for the driver.

To address the issue of HEV energy allocation strategy for contrast, we adopt an existing rule-based hybrid power allocation strategy which has been tested repeatedly and matched with parameters. It is proved to be an energy allocation strategy with good comprehensive effect of fuel economy, and thus can be used as the contrast control strategy for i-HEV ACC.

In terms of the modeling of driver's longitudinal car-following, the Gipps longitudinal car-following model [19] based on the safe distance model proposed by Peter G. Gipps in the 1980s is widely used at present. The improved Gipps longitudinal driver model is used based on Ref. [20] as expressed in the following equation.

$$v_n(t+\tau) = \min\left\{\begin{array}{l} v_n(t) + 2.5a_n\tau\left(1 - \frac{v_n(t)}{V_n}\right)\sqrt{0.025 + \frac{v_n(t)}{V_n}} \\ b_n\tau + \sqrt{b_n^2\tau^2 - b_n\left[2\left[x_{n-1}(t) - s_{n-1} - x_n(t) - v_n(t)\tau\right] - \frac{v_{n-1}(t)^2}{\hat{b}}\right]} \end{array}\right. \tag{7.91}$$

where a_n is the maximum acceleration allowed by the driver, b_n is the maximum deceleration allowed by the driver, b_n is the maximum deceleration allowed by the driver of the preceding vehicle, s_n is the effective length of the preceding vehicle, V_n is the driver's expected vehicle velocity in the absence of any vehicle, $v_n(t)$ is the vehicle velocity at moment t, $v_{n-1}(t)$ is the velocity of the preceding vehicle at moment t, $x_n(t)$ is the position of the vehicle at moment t, $x_{n-1}(t)$ is the position of the preceding vehicle at moment t, and τ is the driver's reaction time.

Some parameters are simplified for the simplification of model design, as shown in the following equations.

$$\begin{aligned} b_n &= -2.0a_n \\ s_{n-1} &= 4(m) \\ \hat{b} &= \min(-3.0, (b_n - 3.0)/2) \end{aligned} \tag{7.92}$$

As shown in Eq. (7.92), the first equation on the right side of the Gipps model represents the process in which the driver follows the set vehicle velocity without any preceding vehicle, and the second equation represents the process in which the driver follows the movement of the preceding vehicle. Since the simulation contrast involved in this chapter mainly considers the working scenario where the preceding vehicle is present, the first car-following equation on the right side of Eq. (7.91) without the preceding vehicle is not considered, and the driver's longitudinal car-following model is modified as follows:

$$v_n(t+\tau) = b_n\tau + \sqrt{b_n^2\tau^2 - b_n\left[2\left[x_{n-1}(t) - s_{n-1} - x_n(t) - v_n(t)\tau\right] - \frac{v_{n-1}(t)^2}{\hat{b}}\right]} \tag{7.93}$$

As shown in Eqs. (7.92) and (7.93), the parameters adjustable in the simplified driver's longitudinal car-following model include the maximum acceleration allowed by the driver a_n and the driver's reaction time τ. a_n represents the driver's demand for vehicle acceleration, and τ represents the driver's speed of reaction to the environment change. With these two parameters modified, the car-following characteristics of different drivers can be comprehensively reflected.

A single controller parameter is not sufficient to analyze and evaluate the integrated performance of the i-HEV ACC system and the HEV in driver mode. Therefore, the values of a_n and τ are modified based on the Gipps model to change the longitudinal car-following characteristics of the driver; on the other hand, the weight coefficients of safety and fuel economy q and r for the optimal objective function of i-HEV ACC

Table 7.9 Designed simulation and contrast parameters

Category	Parameter	Value
Parameters of the driver's mode	a_n	0.6–2.5 m/s^2
	T	0.2–2 s
	V_n	120 km/h
i-HEV ACC	Q	0.01–100
	R	1

as shown in Eq. (7.94) are modified. Thus, different driver car-following results under different driver car-following characteristics and different safety and fuel economy weights of i-HEV ACC can be obtained, based on which the comprehensive performance of safety, economy, and comfort of the i-HEV ACC system and the HEV system are further analyzed.

$$g(x, u) = qL_s(x, u) + rL_m(x, u) \tag{7.94}$$

The values of the maximum acceleration allowed by the driver a_n, the driver's reaction time τ, and weight coefficients q and r are shown in Table 7.9.

(3) Simulation 3: i-HEV ACC versus simple-superposition HEV ACC

The simple-superposition HEV ACC hierarchical control algorithm is taken as the contrast control algorithm [21], and the structure of the control system is shown in Fig. 7.27. The expected acceleration is determined in the upper layer with the linear regulation method of car-following errors. The control in the lower layer is consistent with the energy allocation strategy used in the driver mode HEV, that is, the vehicle energy allocation control strategy that has been applied in the real-life vehicle is used for the energy allocation of the hybrid system.

The equation of the expected acceleration based on the linear adjustment of car-following error is shown as follows.

$$a_d = K_1 \Delta d + K_2 \Delta v \tag{7.95}$$

where a_d is the expected acceleration, and K_1 and K_2 are adjustable parameters of the controller.

In order to avoid acceleration impact caused by excessive car-following error, the calculated expected acceleration is saturated as shown in Eq. (7.96).

$$a_d = \begin{cases} 2, & a_d > 2 \\ a_d, & -2.5 < a_d < 2 \\ -2.5, & a_d < -2.5 \end{cases} \tag{7.96}$$

Similar to the HEV contrast, the controller parameters of the two control systems are modified to avoid the influence of the differences in the designed controller

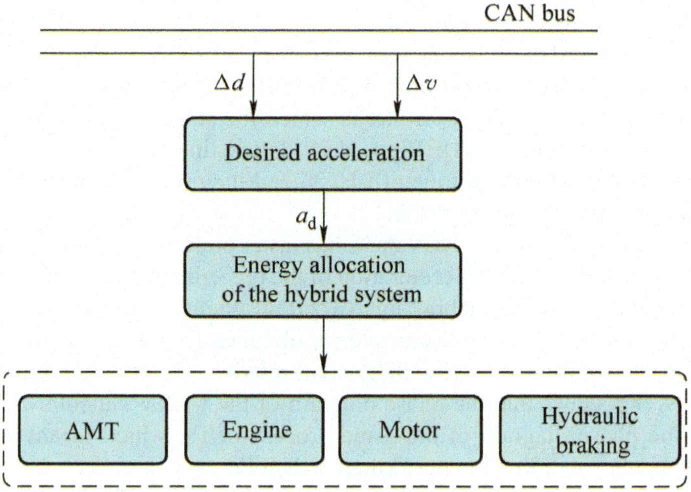

Fig. 7.27 Structure of the simple-superposition HEV ACC control system

Table 7.10 Designed simulation and contrast parameters

Category	Parameter	Value
Simple-superposition ACC	K_1	0–10
	K_2	0–10
i-HEV ACC	Q	0.01–100
	R	1

parameters on the characteristics of the two systems. A comprehensive simulation is conducted on the i-HEV ACC system and the simple-superposition ACC system. Systematic analysis and contrast are conducted based on a series of simulation results.

For the simple-superposition ACC, the coefficients of car-following error linear regulation K_1 and K_2 are modified; for i-HEV ACC, the weight coefficients of safety and fuel economy q and r for the optimal objective function of i-HEV ACC are modified. The value ranges of all parameters are shown in Table 7.10.

7.4.7 Analysis of i-HEV ACC Versus IV ACC Simulation Contrast Results

(1) Scenario of sharp deceleration of the preceding vehicle

Figure 7.28 shows the simulation contrast results of i-HEV ACC and IV ACC in the scenario of sharp deceleration, where (a) is the variation curve of the velocity of the preceding vehicle, i-HEV and IV, (b) is the variation curve of the relative velocity, (c) is the variation curve of the distance error, (d) is the phase diagram of car-following error, and (e) is the braking torque at the wheel end, in which the solid line represents the state of i-HEV and the dash line represents the state of IV. Figure 7.28f shows the variation curve of i-HEV's motor braking and EVB hydraulic braking torque, as well as IV's EVB hydraulic braking torque, in which the solid line represents the EVB hydraulic braking torque of i-HEV, the dash-dotted line represents the PM motor energy regenerative braking torque of i-HEV, and the gray dash line represents the EVB hydraulic braking torque of IV.

As shown in Fig. 7.28a, b, i-HEV ACC has faster braking response characteristics than IV ACC during the sharp deceleration of the preceding vehicle, which improves the car-following safety during braking to a certain extent. This can also be seen from the difference in the following distance error obtained from the velocity integration in Fig. 7.28c. In addition, the phase diagram of relative velocity and relative distance error Fig. 7.28d shows that the phase diagram of the i-HEV car-following process is within the phase diagram of the same process of IV, which means i-HEV can maintain a better car-following performance than IV.

The braking process of the two systems are analyzed in Fig. 7.28e, f. As Fig. 7.28e shows, in the braking process of i-HEV, EVB hydraulic braking and motor regenerative braking are dynamically coordinated in real time and can quickly track the system's expected braking torque. It can respond to the braking demand of the system faster than IV which only relies on EVB hydraulic braking. As shown in Fig. 7.28f, the regenerative braking torque of the motor in the braking process is dynamically adjusted around its allowable maximum braking torque (100 Nm) value so that the expected total braking torque is dynamically coordinated on the one hand, and the energy recovery can also be improved by maintaining a large braking torque on the other hand. Although IV has a higher hydraulic braking torque than i-HEV, its braking effect is worse than that of i-HEV due to its slower system response.

(2) Scenario of sharp acceleration of the preceding vehicle

Figures 7.29 and 7.30 show the simulation comparison results of i-HEV ACC and IV ACC under the condition of sharp acceleration, in which Fig. 7.29a is the variation curve of the velocity of the preceding vehicle, i-HEV and IV, and Fig. 7.29b is the variation curve of relative velocity. Figure 7.30a is the variation curve of range error, Fig. 7.30b is the phase diagram of following error, Fig. 7.30c is the variation curve of the torque of i-HEV PM motor, engine and BSG, Fig. 7.30d is the variation of i-HEV gear, and Fig. 7.30e is the output torque at the transmission's rear end converted between i-HEV and IV systems.

As shown in Figs. 7.29 and 7.30a, b, during the acceleration of the preceding vehicle, i-HEV always accelerates faster with the preceding vehicle than IV with better following safety under the same expected driving torque. During the car-following process, the i-HEV's car-following error phase diagram composed of the relative velocity and relative distance error is smaller than that of IV. Under the current driving power demand, i-HEV can follow the preceding vehicle faster with basically zero distance error.

Further analysis of the mode switching and co-operation process between engine and motor shows that i-HEV can follow the vehicle's expected drive torque faster

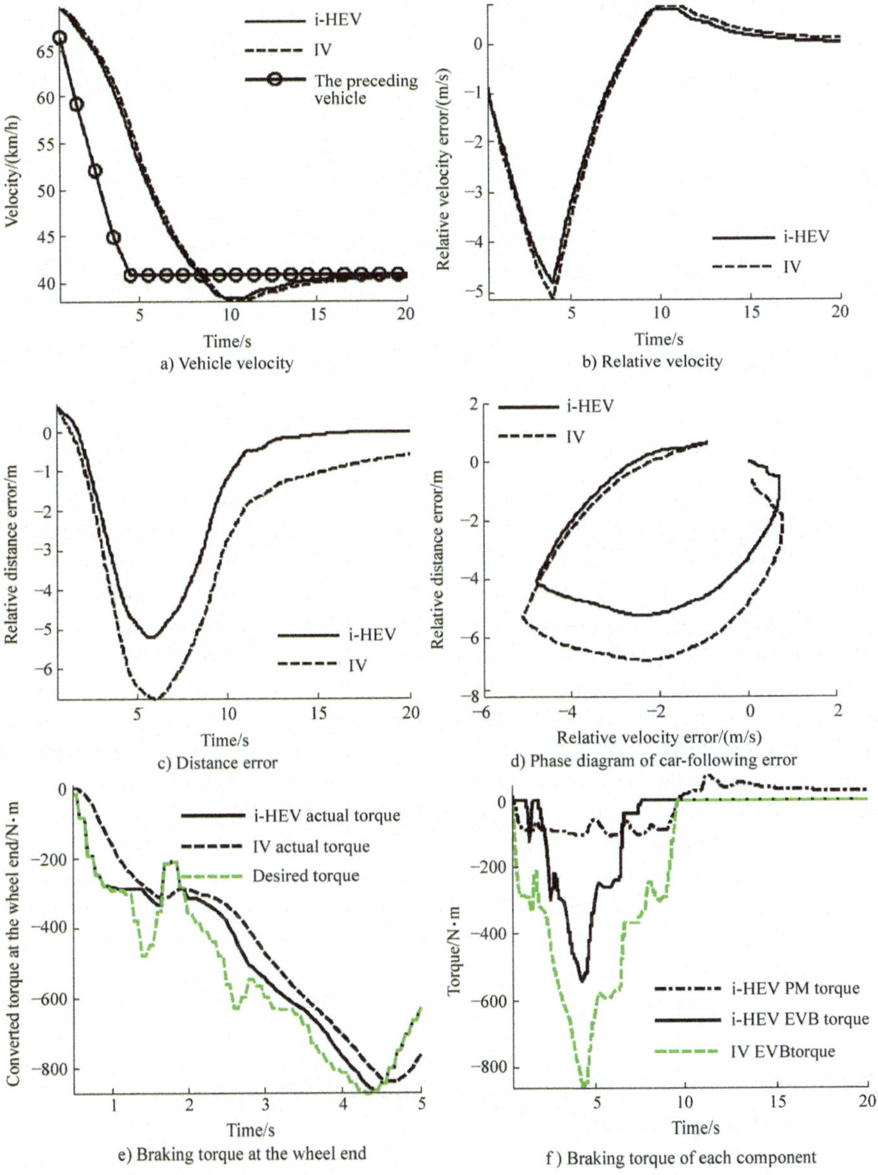

Fig. 7.28 Scenario of sharp deceleration of the preceding vehicle

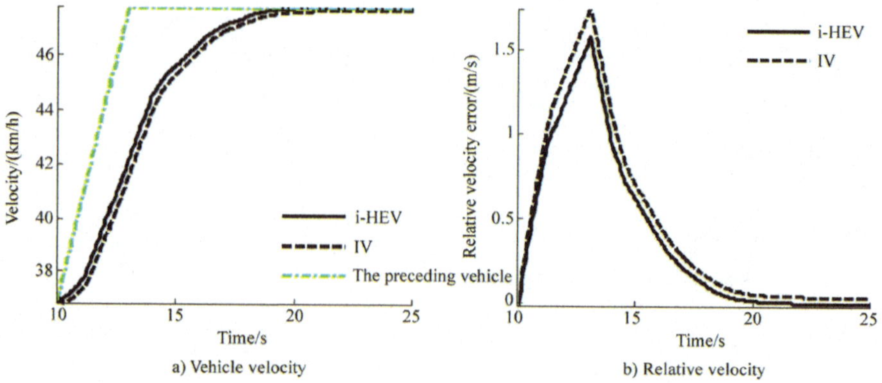

Fig. 7.29 Scenario of sharp acceleration of the preceding vehicle (Part 1)

than IV with the dynamic coordination control of the engine and motor. In Fig. 7.30e, the solid line is the actual torque at the rear end of the i-HEV transmission, the dotted black line is the actual torque at the rear end of the IV transmission, and the dotted green line is the expected torque at the rear end of the vehicle transmission. Due to the slower engine response, the response speed of the actual output torque of IV containing only the power source from engine is slower than that of i-HEV based on the coordination control of motor and engine during the rapid increase of expected torque. Further analysis of Fig. 7.30c shows that the motor's torque is not immediately unloaded in the early stage of the engine starting, but compensate the delay in engine response based on the vehicle's driving demand to ensure the vehicle's actual driving power rapidly follow the demanded driving power, which leads to the faster system response characteristics of i-HEV than those of IV in the process of acceleration.

The fast response characteristics of the system can further satisfy the driver's characteristics, reduce the driver's frequent intervention, and reduce the probability of preceding vehicle cutting-in, and improve the stability of the traffic system and the system's car-following safety.

7.4.8 Analysis of i-HEV ACC Versus HEV Simulation Contrast Results

(1) General contrast of the simulation results of car-following safety and fuel economy

Figure 7.31 shows a series of simulation results of car-following safety and fuel economy of i-HEV ACC and driver HEV with varied control parameters in the ISUZU driving cycle, in which Fig. 7.31a is the simulation result of urban driving cycle, Fig. 7.31b is the simulation result of high-speed driving cycle, point O is the

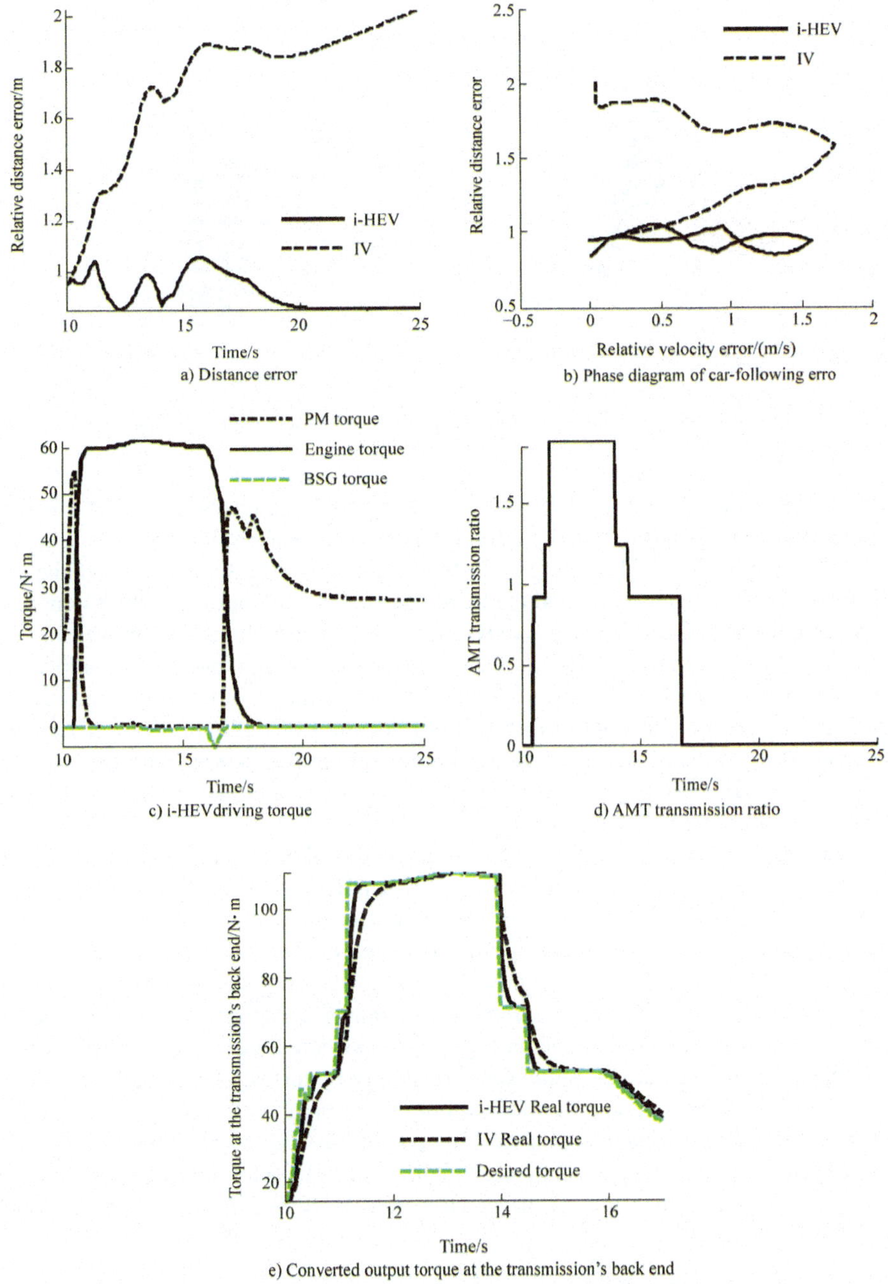

Fig. 7.30 Scenario of sharp acceleration of the preceding vehicle (Part 2)

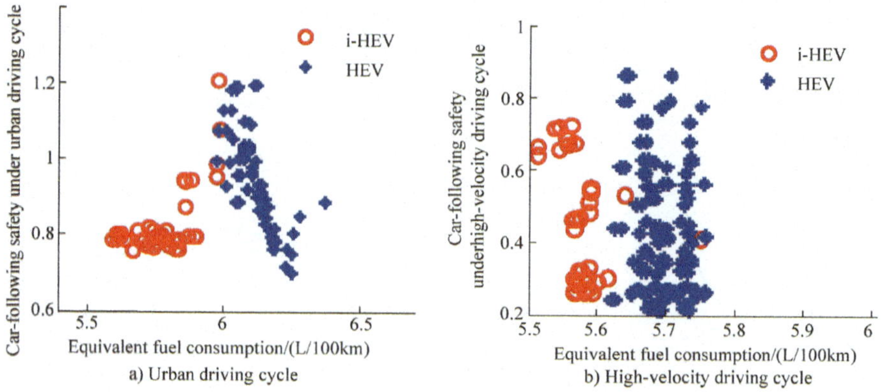

Fig. 7.31 General contrast result of car-following safety and economy of i-HEV and HEV

simulation result of i-HEV ACC with varied control parameters, and point * is the simulation result of the driver HEV following the preceding vehicle.

As shown in Fig. 7.31, the closer to the origin point, the smaller the evaluation factor of car-following safety and the lower the fuel consumption, so the system has a more integrated car-following safety and fuel economy. It can be seen from the simulation results that the i-HEV ACC is closer to the origin point compared with the driver HEV and has better comprehensive performance of car-following safety and fuel economy. With optimized energy allocation of the hybrid power system based on the on-board environment awareness system and power optimization and allocation, i-HEV ACC provides higher fuel economy than driver HEV under the same car-following safety conditions.

The simulation result of safety and economy in the urban driving cycle in Fig. 7.31a shows that the car-following safety indicator of i-HEV ACC is about 0.8. Because vehicle velocity varies frequently in the urban driving cycle, the economy of the hybrid power system has a great optimization potential—about 10% improvement from that of driver HEV at the same level of car-following safety with coordinated optimization of safety, economy and various power components. The simulation result of high-speed driving cycle in Fig. 7.31b shows that the driver can follow the movement of the preceding vehicle in a relatively better way because there is less volatility in acceleration and deceleration in this driving cycle; the system is in the engine-only driving mode in most working conditions and the potential to optimize the hybrid power system is relatively smaller. The fuel economy of i-HEV ACC is about 3% higher than that of driver HEV, and both systems can maintain good car-following safety.

(2) Analysis of simulation cases

With the car-following safety indicators of both HEV and i-HEV set as about 0.8, the simulation results of ISUZU urban driving cycle are analyzed with weight coefficients of i-HEV and HEV $q = 10, r = 1, \tau = 0.3, a_n = 0.8$.

The simulation results are shown in Figs. 7.32, 7.33 and 7.34. In Fig. 7.32, (a) is the variation curve of the preceding vehicle, i-HEV and HEV's velocity, (b) is the variation curve of relative velocity, (c) is the variation curve of distance error, and (d) is the phase diagram of car-following error.

In Fig. 7.33, (a) is the variation curve of engine torque, (b) is the variation curve of PM motor's torque, (c) is the variation curve of BSG's torque, (d) is the variation curve of transmission ratio, (e) is the variation curve of vehicle acceleration, and (f) is the variation curve of battery SOC in the driving cycle.

Figure 7.34 shows the engine's working point in the driving cycle. In all figures, the solid line represents the variation curve of i-HEV, while the dashed line the variation curve of HEV.

Figure 7.32a–d show that both i-HEV ACC and HEV can follow the preceding vehicle well, but i-HEV ACC has a better comprehensive performance of coordinating distance error and relative velocity. i-HEV ACC can more smoothly follow the velocity of the preceding vehicle without overshoot and always maintain a small following error compared with HEV—its relative following velocity is kept within

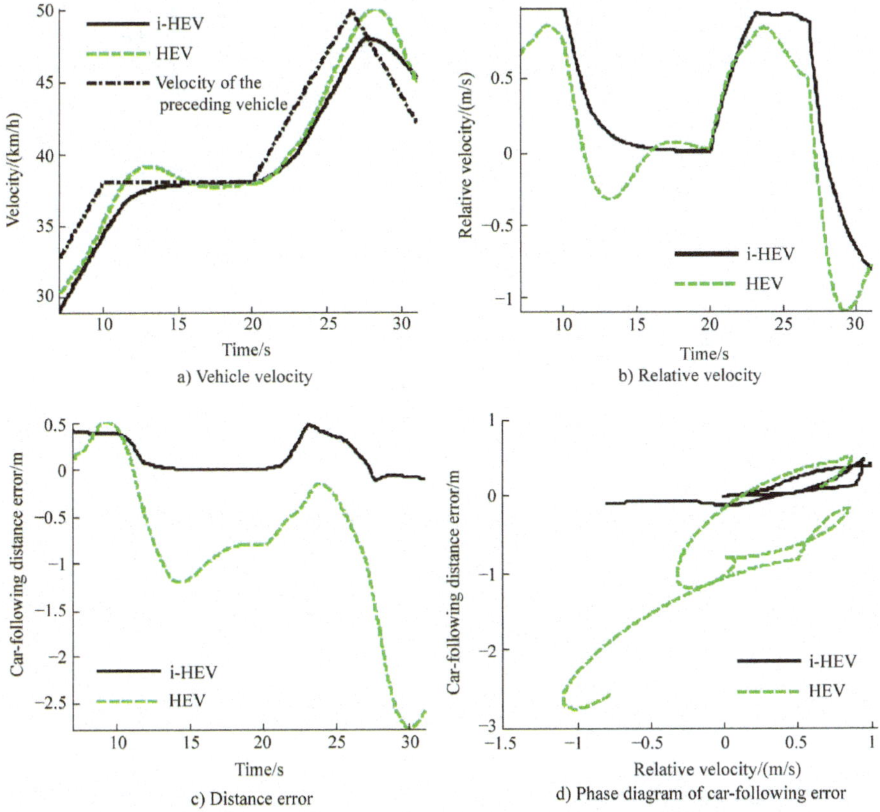

Fig. 7.32 Simulation results of ISUZU driving cycle (Part 1)

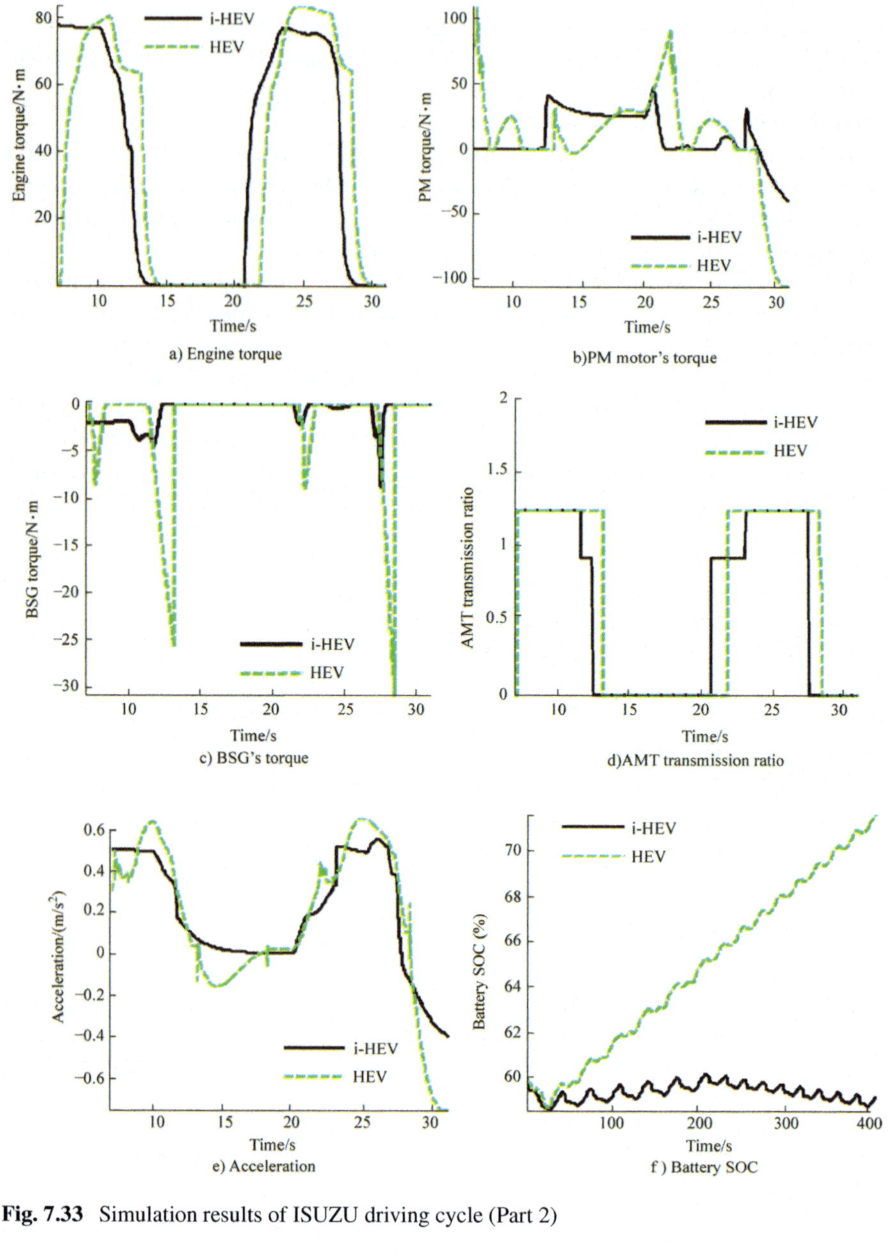

Fig. 7.33 Simulation results of ISUZU driving cycle (Part 2)

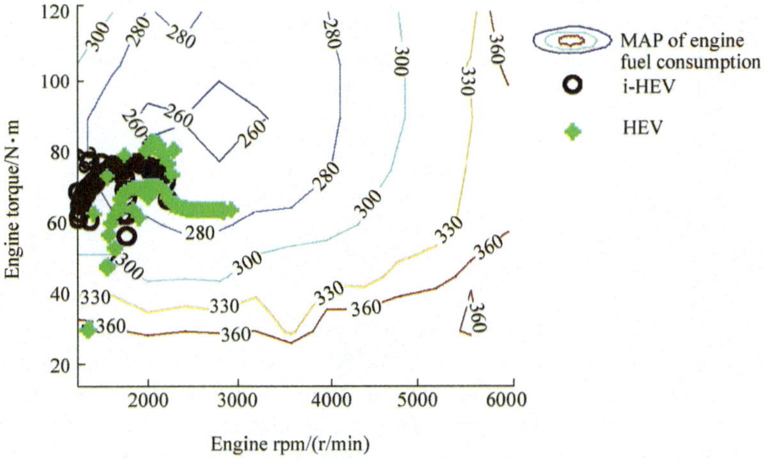

Fig. 7.34 Simulation results of ISUZU driving cycle (Part 3)

1 m/s and its relative distance error within 0.5 m. This can be further seen from the phase diagram of relative velocity and relative distance error in Fig. 7.32 d. As shown in this figure, the car-following error phase diagram of i-HEV is located near 0 and smaller than that of HEV. Although HEV has a slightly faster velocity response, it has a certain overshoot and does not have the problem of coordinating distance following.

The vehicle acceleration curve in Fig. 7.33e shows that both i-HEV ACC and HEV follow the movement of the preceding vehicle with a small acceleration to maintain good driving comfort. i-HEV ACC has smoother acceleration variation and smaller variation amplitude, resulting in better driving comfort compared with HEV. As the acceleration and deceleration during the movement of i-HEV ACC is more gentle, the additional energy loss caused by rapid acceleration and deceleration is also reduced, which can be translated into some fuel economy gain. From this point of view, there is a consistency between driving comfort and fuel economy.

The variation curves of various power components of the i-HEV and the HEV in Fig. 7.33a–d show that the i-HEV ACC's engine, motor, and AMT work together in an optimized manner to keep a good fuel economy, which is better than that of HEV, by identifying traffic information and optimizing the coordination of safety and economy in the car-following process. As can be seen from the variation curve of engine in Fig. 7.33a, the preceding vehicle accelerates at 20 s. Due to the large power demand at this moment, the i-HEV starts the engine directly and enters the mode of engine driving with power generation to meet the acceleration demand of the ego vehicle. Meanwhile, in order to further optimize the engine's working point, the AMT is kept in the 3rd gear, higher than that of the HEV. When the preceding vehicle begins to decelerate at 27 s, the i-HEV shuts down the engine earlier than the HEV and changes to the battery-only mode. Both the i-HEV and the HEV enter the braking energy recovery mode when the preceding vehicle further decelerates.

Table 7.11 Result of fuel consumption/100 km in the ISUZU driving cycle

Item	Engine fuel consumption (L/100 km)	SOC change (%)	Equivalent fuel consumption (L/100 km)
Driver HEV	8.4	11.7	6.19
i-HEV ACC	5.46	-0.9	5.62

Figure 7.34 shows the variation curve of engine working point, where O represents the working point of i-HEV and * that of HEV. As can be seen from the figure, both i-HEV and HEV can ensure that the engine is in an optimal working range with optimized engine working point. However, compared with the rule-based HEV energy allocation strategy, the engine operating point of the i-HEV is generally in a better engine fuel consumption MAP, which also improves the fuel economy of the vehicle to a certain extent. Figure 8.31f is the variation curve of battery SOC. It can be seen from the Figure that the battery SOC of i-HEV is in balance at the beginning and the end of the driving cycle, while the battery SOC of HEV is significantly improved under the rule-based energy allocation strategy.

The simulation result of driving cycle is as shown in Table 7.11. The engine fuel consumption of the HEV is 8.4 L/100 km in the ISUZU driving cycle, and its SOC increases by 11.7% at the end of the cycle; through a coefficient-based conversion, the final vehicle equivalent fuel consumption of HEV is 6.19 L/100 km. The engine fuel consumption of i-HEV is 5.46 L/100 km in the ISUZU driving cycle, and its SOC decreases by 0.9.7% at the end of the cycle; through a coefficient-based conversion, the final vehicle equivalent fuel consumption of i-HEV is 5.62 L/100 km. This shows a 9.2% improvement in fuel economy of the i-HEV from the driver HEV mode with basically the same car-following effect.

7.4.9 Analysis of i-HEV ACC Versus Simple-Superposition HEV ACC Simulation Contrast Results

(1) General contrast of the simulation results of car-following safety and fuel economy

Figure 7.35 shows a series of simulation results of car-following safety and fuel economy of i-HEV ACC and simple-superposition HEV ACC with varied control parameters in the ISUZU driving cycle, in which Fig. 7.35a is the simulation result of urban driving cycle, Fig. 7.35b is the simulation result of high-speed driving cycle, O is the simulation result of i-HEV ACC with varied control parameters, and *point is the simulation result of the simple-superposition HEV ACC following the preceding vehicle.

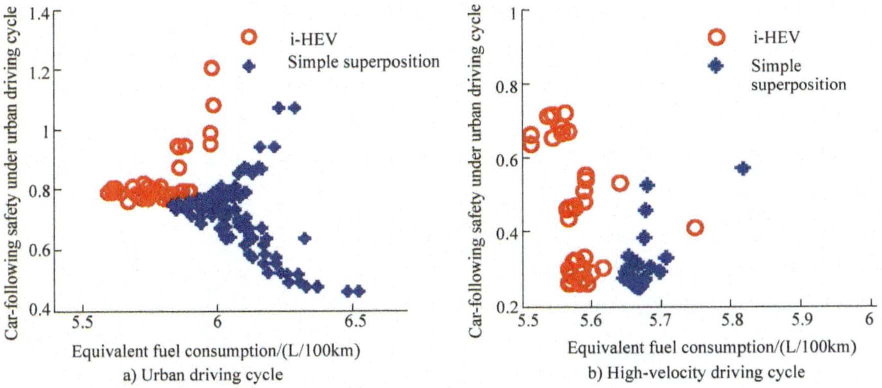

Fig. 7.35 General contrast result of car-following safety and economy in the ISUZU driving cycle

It can be seen from the simulation results that both i-HEV ACC and simple-superposition HEV ACC can achieve good car-following safety. Compared with simple-superposition HEV ACC, i-HEV ACC takes vehicle driving condition information into a comprehensive consideration, coordinates following safety and fuel economy to optimize energy allocation, and has comprehensively optimized following safety and fuel economy, resulting in a better fuel economy with the same following safety.

Figure 7.35 shows that when the car-following safety indicator is about 0.8 in the urban driving cycle, the fuel consumption of i-HEV ACC is 5.6 L/100 km, 5% higher than that of the simple-superposition ACC which is 5.9 L/100 km; when the car-following safety indicator is about 0.3 in the high-speed driving cycle, the optimal fuel economies of the i-HEV ACC and the simple-superposition ACC are 5.55 L/100 km and 5.65 L/100 km respectively, showing a 2% improvement from the latter. The main reason for the smaller improvement of fuel economy in the high-speed driving cycle than that in the urban driving cycle is that the engine is in a relatively optimal working range, with a smaller potential for fuel economy optimization.

(2) Analysis of simulation cases

With the car-following safety indicators of both i-HEV and simple-superposition HEV ACC set as about 0.8, the simulation results of ISUZU urban driving cycle are analyzed with weight coefficients of i-HEV and simple-superposition HEV ACC $q = 10, r = 1, K_1 = 4, K_2 = 0.1$.

Figures 7.36, 7.37 and 7.38 show partial simulation results of i-HEV and simple-superposition HEV ACC in the ISUZU driving cycle. In Fig. 7.36, (a) is the variation curve of the preceding vehicle, i-HEV and simple-superposition HEV ACC's velocity, (b) is the variation curve of relative velocity, (c) is the variation curve of distance error, and (d) is the phase diagram of car-following error. In Fig. 7.37, (a) is the variation curve of engine torque, (b) is the variation curve of PM motor's torque, (c) is the variation curve of BSG's torque, (d) is the variation curve of AMT transmission ratio, (e) is the variation curve of vehicle acceleration, and (f) is the variation

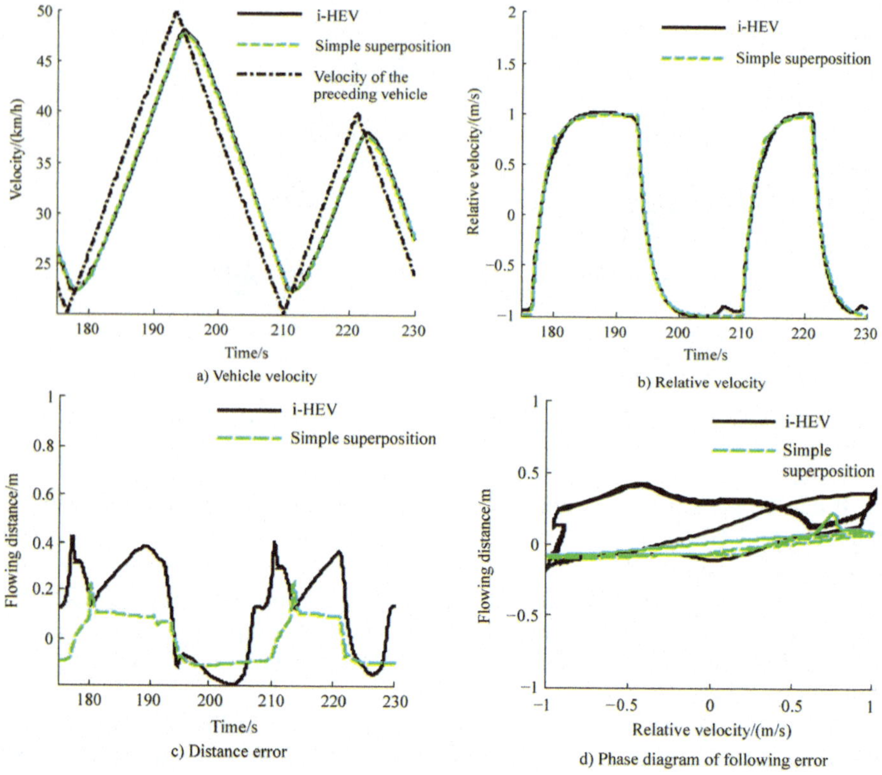

Fig. 7.36 Simulation results of ISUZU driving cycle (Part 1)

curve of battery SOC in the driving cycle. Figure 7.38 shows the engine's working point in the driving cycle. In all figures, the solid line represents the variation curve of i-HEV, while the dashed line the variation curve of HEV.

It's shown in Fig. 7.36a–d that both i-HEV and simple-superposition algorithm can follow the preceding vehicle and maintain a good following safety. As shown in the variation curve of relative velocity in Fig. 7.36b that both algorithms can maintain the relative velocity within 1 m/s and are basically the same. As shown by the distance error in Fig. 7.36c, both algorithms can follow the expected distance quite well with error kept within 0.4 m. I-HEV ACC's following distance error is slightly larger, but the difference between it and the simple-superposition algorithm is only about 0.2 m, which can be ignored for longitudinal following.

As shown in Fig. 7.37e, both i-HEV and the simple-superposition algorithm can keep the vehicle acceleration in the car-following process within 0.5 m/s^2, which meets the requirement of longitudinal driving comfort. It can be seen from the contrast result that the acceleration trajectories of the two algorithms arc basically consistent, with only minor differences at a few moments.

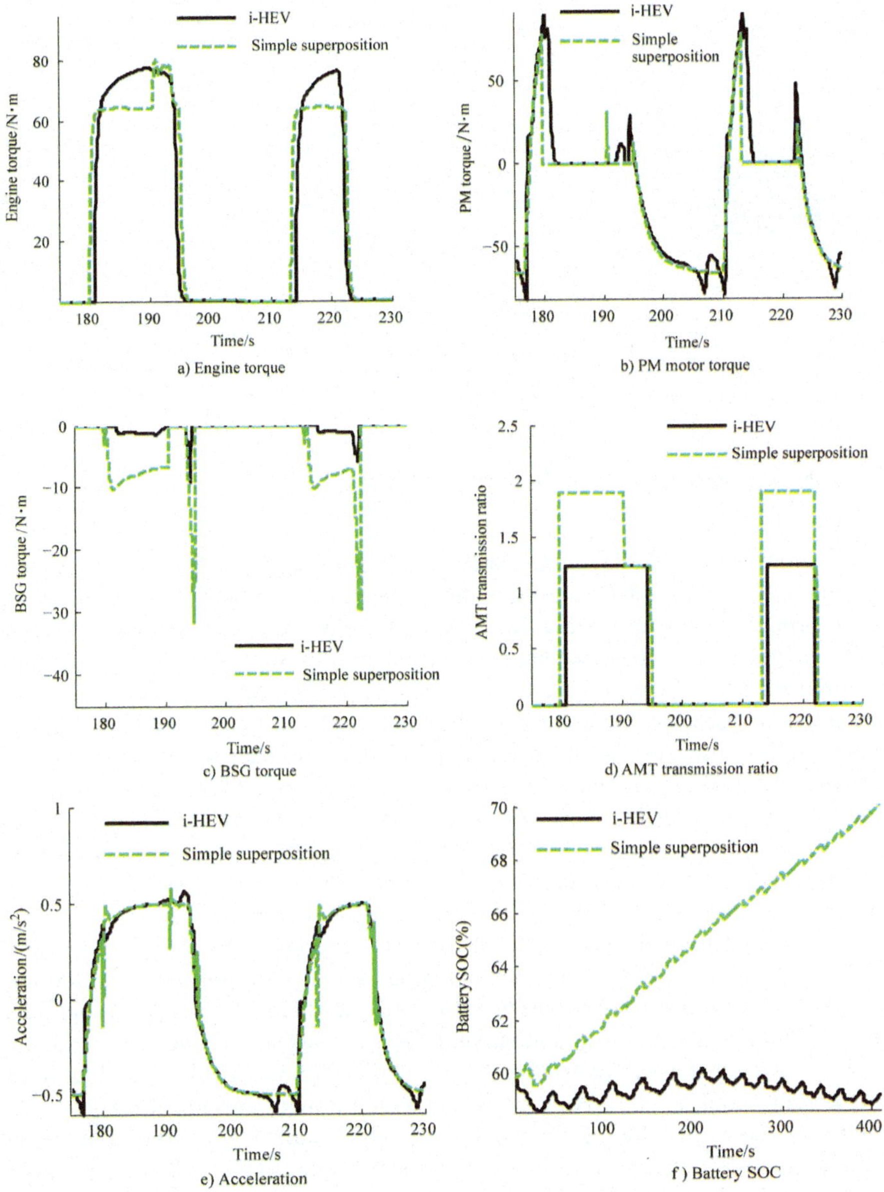

Fig. 7.37 Simulation results of ISUZU driving cycle (Part 2)

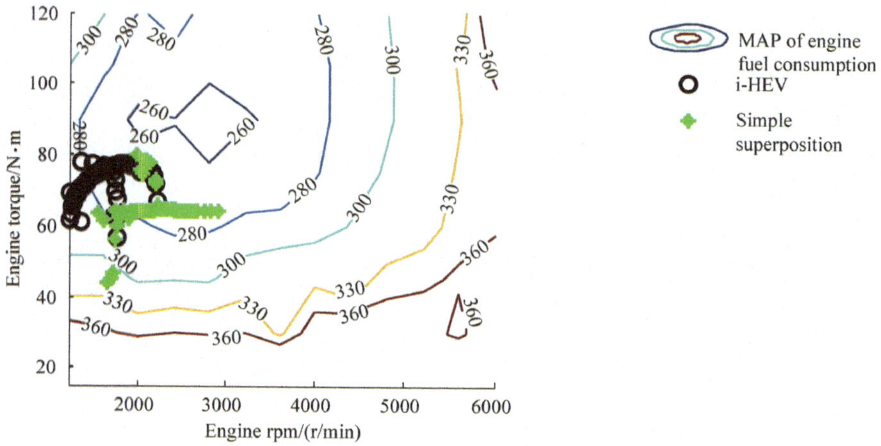

Fig. 7.38 Simulation results of ISUZU driving cycle (Part 3)

As shown in Fig. 7.38a–d, the variation curves of each power component show that both the i-HEV ACC and the simple-superposition algorithm can make the vehicle run in a relatively optimal working mode of hybrid power system, and i-HEV ACC has better fuel economy with coordinated optimization of engine starting time and transmission. With the acceleration and deceleration of the preceding vehicle, the system experiences different working modes of hybrid power systems, such as battery-only driving, engine driving with power generation, engine-only driving, and braking energy regeneration. And as shown in Fig. 7.38a, i-HEV engine starts later and shuts down earlier than the simple-superposition solution, and the engine working point is in a more economical range, which improves the vehicle fuel economy to a certain extent. It also can be seen from Fig. 7.38d of transmission gear that i-HEV always run at the 3rd gear in the process of the engine driving, while the simple-superposition algorithm runs at the 2nd gear sometimes. It can be seen that i-HEV can optimize the working point of torque with the same driving power through coordinated optimization of engine working range and transmission gear, to optimize the efficiency of the engine and improve the fuel economy of the vehicle.

As shown in the variation curve of the engine's working point in Fig. 7.38, the working point of the i-HEV engine is within a more optimized range. In the figure, O represents the working point of i-HEV, and * represents that of the simple-superposition algorithm. As shown in the figure, both i-HEV and simple-superposition algorithm can ensure the engine is in an optimal working range by optimizing its working point. However, compared with the energy allocation strategy of rule-based simple-superposition algorithm, the engine's working point of i-HEV is generally in a better engine fuel consumption MAP, which also improves the fuel economy of the vehicle to a certain extent.

As shown in the variation curve of battery SOC in Fig. 7.37f, there is a more balanced SOC variation in the i-HEV driving cycle, which can improve the comprehensive fuel economy of the system. It can be seen from the figure that i-HEV achieves appropriate energy allocation between the engine and the battery by designing appropriate equivalent fuel consumption factor of battery and basically maintains the balance of battery SOC at the beginning and end of the driving cycle, while the simple-superposition algorithm has higher SOC at the end of the driving cycle. In fact, as long as the driving cycle requirement is met, trying to maintain the balance of battery SOC is the ideal energy allocation solution which can avoid efficiency loss caused by charging and discharging the battery twice, so as to improve the fuel economy of the system. This also shows from another side that the i-HEV ACC control system based on model prediction optimization can improve the comprehensive fuel economy of the system by estimating the equivalent fuel consumption factor of the battery under working conditions and by calculating and optimizing the energy demand under local prediction.

As shown in Fig. 7.37a, b, the system switches smoothly from the battery driving mode to the engine driving mode through the coordination of motor and engine torque. Due to the slower response of the engine, the torque of the driving motor does not disappear completely once the engine is started, but is gradually reduced to make the system smoothly switch from the battery driving mode to the engine driving mode, so as to keep the system's driving power basically unchanged.

The simulation result of the ISUZU driving cycle is as shown in Table 7.12, in which the simple-superposition HEV ACC algorithm achieves 7.64 L/100 km engine fuel consumption in the ISUZU driving cycle and SOC increases by 9.4% at the end of the driving cycle, and the final vehicle equivalent fuel consumption obtained with coefficient-based conversion is 5.86 L/100 km, while the vehicle equivalent fuel consumption of i-HEV ACC in the ISUZU driving cycle is 5.62 L/100 km, indicating a 4.1% improvement in the fuel economy of i-HEV from that of the simple-superposition algorithm.

Table 7.12 Fuel consumption results in the ISUZU driving cycle

Contrast item	Engine's fuel consumption (L/100 km)	SOC variation (%)	Equivalent fuel consumption (L/100 km)
HEV ACC	7.64	9.4	5.86
i-HEV ACC	5.46	– 0.9	5.62

7.5 Experimental Study of i-HEV ACC

7.5.1 *Experiment Platform Design of i-HEV*

In order to realize the functions of i-HEV ACC, the experiment platform for real-life i-HEV is designed in this section by modifying the braking system, sensing system and communication system of an existing HEV system, on which basis, RCP, the i-HEV ACC control system, is developed based on DSPACE.

The original structure of HEV is shown in Fig. 7.39. The vehicle controller communicates with each power component through CAN network. The vehicle controller monitors in real time through CAN bus and controls the coordination of the engine, the motor, the battery, AMT and other components. The engine controller ECU controls the engine, the motor controller MCU controls the PM drive motor and the BSG at the same time, the battery controller BCU monitors the battery in real time and manages the charge and discharge, and the AMT controller TCU controls the AMT gear shift and clutch operation. From the perspective of designing i-HEV, the original HEV system provides a complete power system and transmission system, but it still requires modification to the braking system, sensing system, communication system and the vehicle control system to realize the longitudinal auxiliary control of i-HEV.

In order to ensure that the modified i-HEV platform has the same functions as the original system and ensure the safety of the vehicle, the modification of the HEV system to achieve the i-HEV platform shall be based on the following basic principles: (1) the original system's wiring harness shall not be changed in the i-HEV system to ensure the normal operation of the original system in the event of new i-HEV functions failing; (2) the modified i-HEV shall have good scalability, which facilitates the addition of new functions; (3) new control functions can be quickly developed and verified in the i-HEV control system.

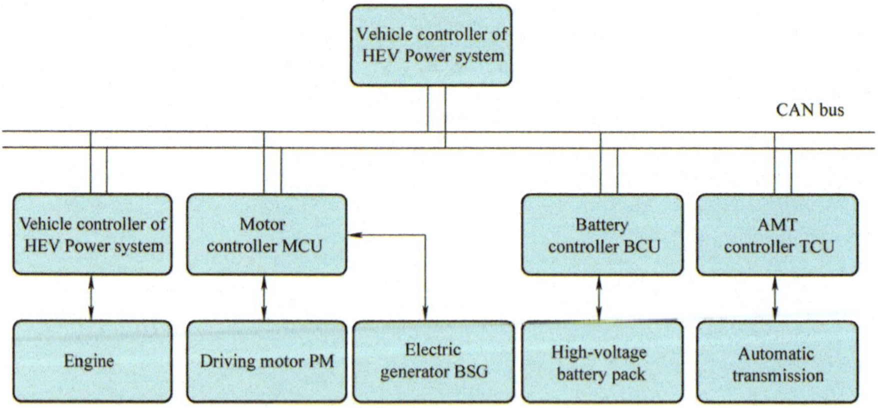

Fig. 7.39 Original structure of the HEV system

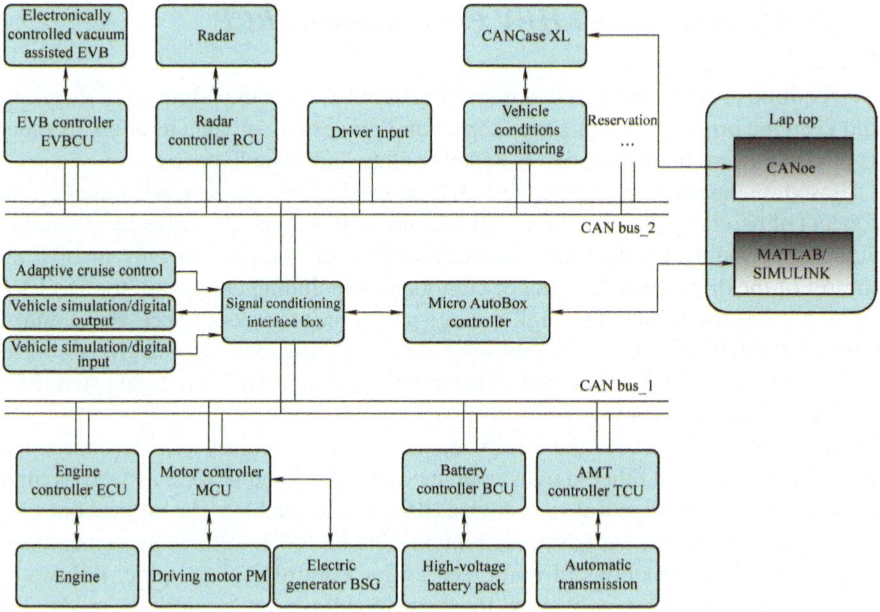

Fig. 7.40 Overall structure of the i-HEV system

Based on the above analysis of the original HEV system and the basic principles of system modification, the overall structure of the real-life vehicle experiment platform for i-HEV ACC is designed as shown in Fig. 7.40.

As shown in Fig. 7.40, the vehicle controller of the original HEV system is replaced by a signal conditioning interface box and a Micro Autobox controller from DSPACE, Germany, in order to achieve rapid controller prototyping and facilitate rapid development and verification of new functions of the i-HEV. The controller is connected with a laptop computer, and the controller algorithm is developed based on MATLAB/ SIMULINK.

The designed i-HEV uses double CAN network communication. The original system's CAN bus is reserved without any change to ensure the normal operation of the network communication of the original system. The intelligent system control CAN bus for i-HEV is added separately to control the components related to intelligent control and part of environment identification sensor communication, mainly used for radar communication, EVB control, acceptance of the driver's cruise information input, and vehicle condition monitoring, etc. The monitoring of vehicle condition is connected with a laptop computer through CANCase XL, and real-time monitoring of information is carried out on CAN bus through software CANOE.

In addition, the i-HEV ACC switch is introduced to ensure that i-HEV ACC meets the driver's needs. The ACC function can only be started when the switch is pressed by the driver. The signal is directly input into the Micro Autobox controller through the signal conditioning interface box.

7.5.2 *Developing the i-HEV Control System RCP*

It is required to develop the i-HEV control system based on the original HEV system, and keep the original system's monitoring and processing of vehicle safety as much as possible from the perspective of system safety and reliable operation.

Based on the above analysis, the i-HEV control system is designed based on the concept of parallel program control to make the i-HEV ACC algorithm have separate modules including input, control strategy and control output and only involve the control of the driving and braking of components including engine, motor and EVB under normal conditions. The high-voltage power on/off process, accessory control, fault control functions such as monitoring and processing of the original HEV control in the original system are reserved. The structure of the i-HEV control system RCP is designed as shown in Fig. 7.41.

As shown in Fig. 7.41, the i-HEV control system receives vehicle component state, radar, EVB and other signals contained in CAN1 and CAN2, as well as input signals from the intelligent system such as driver ACC enabler. The online control of i-HEV ACC is carried out based on vehicle state and inter-vehicle movement state to calculate the instructions of optimized torque control of the engine, PM motor, BSG and EVB braking, as well as the corresponding system working mode. The original HEV control system receives the i-HEV control instructions and outputs the corresponding instructions to control the engine, PM motor, BSG and other power systems as well as other accessory systems according to the current working state of the system components. The EVB hydraulic brake is directly controlled by the i-HEV ACC.

In order to realize the online control of i-HEV ACC, the optimization control law obtained by off-line optimization calculation of step-by-step dynamic programming

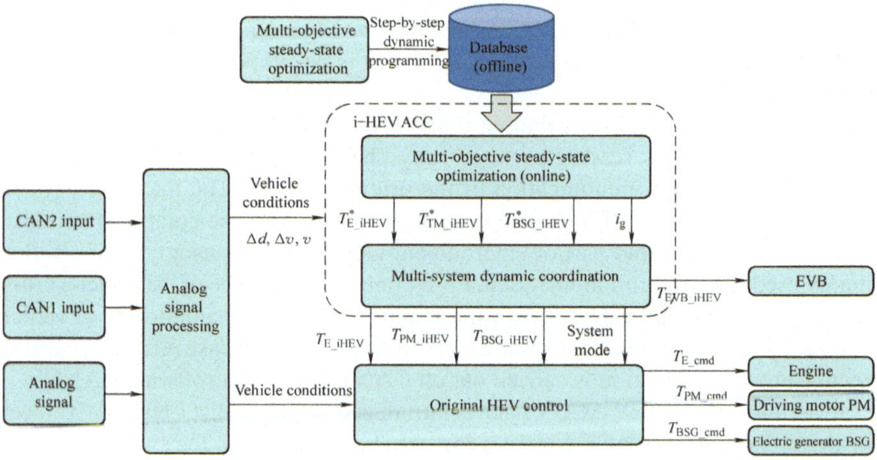

Fig. 7.41 Structure of the i-HEV control system RCP

is stored in the memory of dSPACE, and the control instructions for the engine, PM motor, BSG and hydraulic braking are obtained through online table look-up.

On this basis, the i-HEV control system RCP is designed. The i-HEV control system mainly includes signal input, signal conditioning, i-HEV control algorithm, I-HEV control output, HEV vehicle control algorithm and HEV vehicle output modules. i-HEV ACC RCP module mainly includes ACC optimization control algorithm, calculation of the i-HEV system's working mode and output processing modules.

7.5.3 *Designing the Hardware Platform of i-HEV*

In order to realize the ACC functions of i-HEV, it is required to transform the control system, braking system, environment awareness system and CAN communication system of the traditional HEV system structure. The wiring harness connection structure of the i-HEV is designed as shown in Fig. 7.42.

As shown in Fig. 7.42, the dSPACE Micro Autobox controller is connected to the signal conditioning interface box to realize all control functions of the i-HEV, including IO, ADC/DAC, CAN communication and other signal connections. The signal conditioning interface box is connected to the wiring harness of the original HEV system and is used to control the power components and accessories of the original HEV system as well as to collect the input of relevant vehicle status signals, including various sensor signals of the original HEV system, CAN bus 1 communication, and control of various relays of the system. The radar and EVB mainly communicate with the vehicle control system through CAN bus 2 to realize the corresponding signal interaction and vehicle component control.

Based on the existing experimental equipment, MicroAutobox 1401/1501 motherboard is selected for the controller, which has an 18-bit ADC and an 8-bit DAC, a large amount of digital I/O, a dual-wire CAN bus and a RS232 bus to meet the most requirements of the vehicle controller. The signal conditioining interface box is mainly used for MicroAutoBox controller to achieve input signal filtering, power driving

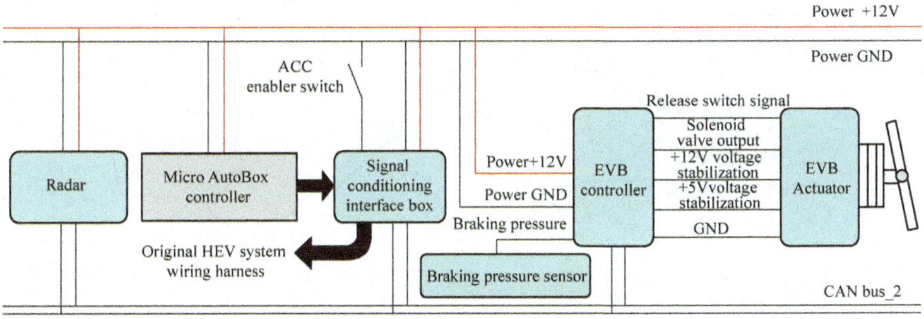

Fig. 7.42 Wiring harness connection structure of the i-HEV

of output signals, input and output signals over-current and over-voltage protection, etc., to ensure the safe operation of the AutoBox controller while achieving the compatibility of Micro AutoBox with I/O signals and methods of all kinds of voltage and current ranges by preprocessing the vehicle's input signals. In order to speed up the development of the i-HEV ACC control system, a mature millimeter-wave radar system produced by ADC, Germany, is adopted. The radar uses 77 GHz microwave frequency and transmits data with an on-board CAN bus.

Considering the active braking control demand of i-HEV ACC, the electronic vacuum booster (EVB), which is widely applied in foreign ADAS, is used to achieve braking assistance with the vacuum at the engine throttle, and to realize the active control of vehicle braking pressure by actively controlling the opening and closing of the vacuum tube solenoid valve. In the battery-only driving mode of the i-HEV, the engine does not work and cannot provide vacuum assistance. Therefore, in order to ensure that the EVB braking system work normally when the engine is shut down, an EVB is added on the basis of the traditional EVB, the basic structure of which is shown in Fig. 7.43. The vacuum tank is used for vacuum storage. When the engine is started, the vacuum pump is shut down and the engine provides vacuum pressure to the vacuum tank. When the engine is shut down and there is not sufficient vacuum pressure in the vacuum tank, the vacuum pump is started to provide vacuum pressure to the vacuum tank. The EVB braking system relies on the vacuum pressure stored in the vacuum tank to actively control the hydraulic braking pressure through the solenoid valve.

The EVB controller uses PI algorithm to control the braking pressure. the expected EVB braking pressure is set as P_{des}, and the actual pressure is P_{act}, then the duty ratio of solenoid valve control PWM is:

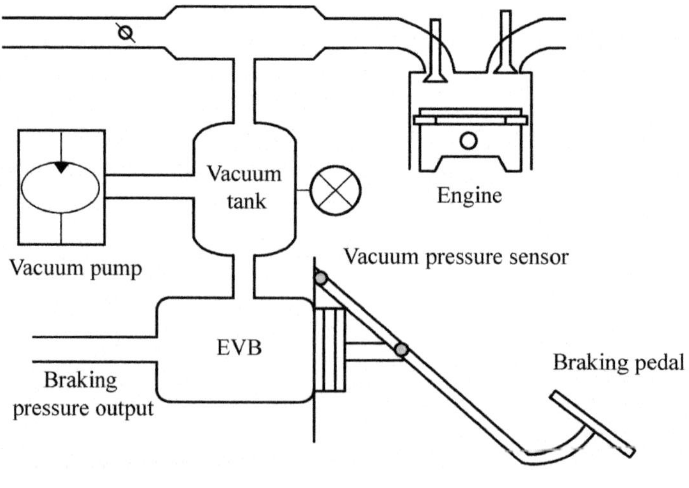

Fig. 7.43 Structure of the active braking system EVB

$$\begin{aligned} e &= P_{\text{des}} - P_{\text{act}} \\ PWM &= K_{\text{PWM}}(k_{\text{p}}e + k_{\text{I}} \int edt + C) \end{aligned} \quad (7.97)$$

where kp is the proportional control gain of the pressure tracking error e, k_{I} is the integral control gain, and K_{PWM} is the conversion factor of the duty ratio of solenoid valve control. To ensure that EVB has a faster startup response, the initial control quantity C is set.

The corresponding wiring harness interfaces are defined based on the selection and design of each system, as shown in Table 7.13.

Because the i-HEV experimental platform is designed by modifying a traditional HEV vehicle, the logic of AMT gear shift is mainly based on the accelerator pedal and automatic speed control, and does not accept the instructions of the vehicle

Table 7.13 List of wiring harness modified solenoid valve in the i-HEV

Component	Wire harness	Function
Radar	12 V	Power supply
	GND	
	CAN 2 high	CAN communication
	CAN 2 low	
Micro AutoBox	12 V	Power supply
	GND	
	Wire harness connected to the interface box	Vehicle control
Signal conditioning interface box	Wire harness of the original HEV system	Vehicle control
	12 V	Power supply
	GND	
	CAN 2 high	CAN communication
	CAN 2 low	
	ACC enabler switch input	Driver ACC enabler
EVB controller	12 V	Power supply
	GND	
	Input of brake pressure signal	Brake pressure input
	Input of switch release signal	Driver brake pedal signal
	Solenoid valve output	EVB solenoid valve control signal
	12 V constant-voltage output	Power supply to EVB solenoid valve
	5 V constant-voltage	Power supply to release control
	GND	EVB ground

control system. Therefore, the experimental research of i-HEV ACC carried out in this chapter mainly verifies the comprehensive performance of the i-HEV ACC algorithm through the experimental analysis of the car-following effect of the i-HEV ACC and the engine's working point. Furthermore, the driving cycle is designed based on the characteristics of ISUZU driving cycle adopted in the simulation within a narrow speed range and on the premise that AMT does not shift gears, which is studied in comparison with HEV and simple-superposition HEV ACC modes in experiments.

7.5.4 *Experimental Verification Scheme for the Comprehensive Performance of i-HEV ACC*

The working scenarios of ACC involve different movement modes of the preceding vehicle such as constant velocity, acceleration and deceleration, and the components of the hybrid power system such as engine, hydraulic brake, motor and AMT have different working characteristics. By combining the demands for verifying the performance of the hybrid power system and complex vehicle movement, the experiment scenarios of i-HEV ACC are designed to include constant velocity, mild, moderate and sharp acceleration/deceleration under low, medium and high velocity. Considering the experimental sites and the safety of the running system, the experiment velocity of i-HEV ACC is limited to be equal or below 60 km/h.

Combining the working characteristics and the main working conditions of systems including engine and motor, 5 experiment scenarios (mild acceleration/deceleration, moderate acceleration/deceleration, sharp acceleration/deceleration, low-velocity sine, high-velocity sine) are designed to verify the basic functions and effectiveness of the i-HEV ACC system.

(1) Scenario of the preceding vehicle acceleration/deceleration

This includes 3 scenarios: mild, moderate and sharp acceleration/deceleration. As shown in Fig. 7.44, the black solid line, the gray dashed line and the black solid-dot line represent respectively the velocity variation curves of the mild, moderate and sharp acceleration/deceleration of the preceding vehicle.

Mild acceleration/deceleration scenario is designed to verify the car-following performance of i-HEV ACC when the preceding vehicle is running at lower velocity or mild acceleration. The velocity of the preceding vehicle is set to start at 20 km/h, then accelerate before 10 s at 0.2 m/s^2 for 15 s, be kept for 15 s, and then decelerate at $-$ 0.2 m/s^2 for 15 s to 20 km/h. Sharp acceleration/deceleration scenario is used to verify the car-following performance of i-HEV ACC, the working characteristics of power systems and the coordination between components during the switch between modes when the preceding vehicle sharply accelerates. The velocity of the preceding

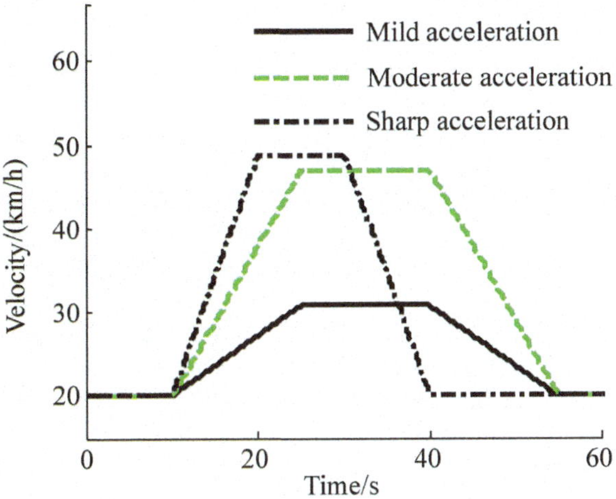

Fig. 7.44 Scenario of the preceding vehicle running at constant acceleration and deceleration

vehicle is set to start at 20 km/h, then accelerate before 10 s at 0.8 m/s^2 for 10 s, be kept for 10 s, and then decelerate at – 0.8 m/s^2 for 10 s to 20 km/h.

(2) Sinusoidal scenario of the preceding vehicle

The sinusoidal scenario of the preceding vehicle is usually used to simulate the comprehensive performance of the control system in the dynamic car-following mode. To reflect the working characteristics of the system under different vehicle velocities, we design the low-velocity and high-velocity sinusoidal scenarios in this section to test and evaluate the control system of i-HEV ACC. The working condition of the preceding vehicle is set as shown in Fig. 7.45, in which the black solid line and the grey dashed line represent the velocity variation curves in the low-velocity and high-velocity sinusoidal scenarios of the preceding vehicle, in which the frequency of the sinusoidal scenario is 0.1 rad/s and the acceleration amplitude is 0.3 m/s^2. The initial velocity of the low-velocity sine is 20 km/h and that of the high-velocity sine is 40 km/h.

7.5.5 Contrast Experiment Schemes Based on i-HEV ACC Driving Cycles

Referring to the characteristics of the ISUZU driving cycle used in the simulation, constant acceleration/deceleration movements are used as the driving cycles in the contrast experiment. Based on the gear-shift pattern of AMT, the vehicle velocity range is set to be 35–45 km/h. To fully verify the functions and characteristics of different types of algorithms, the movement of the preceding vehicle is divided into 3 acceleration/deceleration periods, i.e. 24, 18 and 12 s, and each period is consecutively run for 4 cycles.

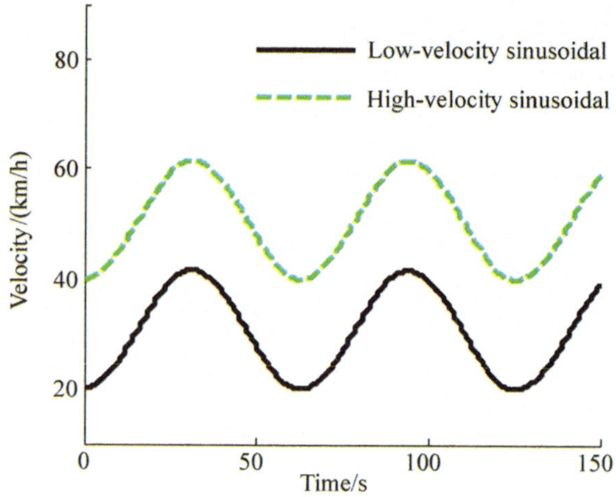

Fig. 7.45 Sinusoidal scenario

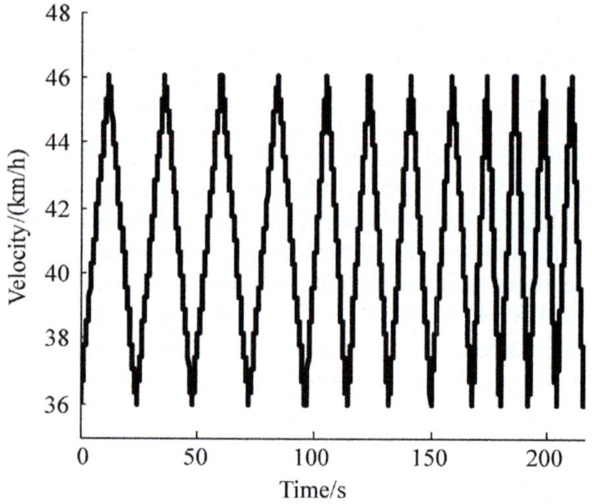

Fig. 7.46 i-HEV ACC cycle test condition

To verify the comprehensive performance of i-HEV ACC, the worst car-following situations during the cruise are considered to include car-following errors such as maximum/minimum relative velocity and maximum/minimum distance errors to represent the safety of the system in the car-following process, the maximum acceleration/deceleration to represent the driving comfort for the driver, and the working point of the engine to represent the fuel economy of the system. The experiment results of working scenarios (preceding vehicle's mild, moderate and sharp acceleration/deceleration, low-velocity and high-velocity sinusoidal scenario) are summarized to obtain the car-following safety and driving comfort results of i-HEV ACC as shown in Table 7.14. The engine's working points in all scenarios are as shown in Figs. 7.47, 7.48, 7.49, 7.50 and 7.51. Because the system is in the battery-only mode in the mild acceleration/deceleration scenario, Fig. 7.47 shows the torque value of the PM motor.

As shown in Table 7.14, i-HEV ACC can maintain good car-following safety in all kinds of constant acceleration/deceleration and sinusoidal scenarios. In mild acceleration/deceleration, low-velocity and high-velocity sinusoidal scenarios, the maximum following distance error is 1.8 m, the minimum distance error is – 0.2 m, the maximum relative velocity is 0.7 m/s, and the minimum relative velocity is –

Table 7.14 Result of i-HEV ACC car-following safety and driving comfort experiment

Scenario	Indicator	Experiment result
Mild acceleration/deceleration of the preceding vehicle	Max. relative velocity (m/s)	0.40
	Min. relative velocity (m/s)	– 0.40
	Max. distance error (m)	1.1
	Min. distance error (m)	0.1
	Max. acceleration (m/s^2)	0.25
	Max. deceleration (m/s^2)	– 0.27
Moderate acceleration/deceleration of the preceding vehicle	Max. relative velocity (m/s)	1.50
	Min. relative velocity (m/s)	– 1.00
	Max. distance error (m)	2.0
	Min. distance error (m)	– 0.5
	Max. acceleration (m/s^2)	0.97
	Max. deceleration (m/s^2)	– 0.66
Sharp acceleration/deceleration of the preceding vehicle	Max. relative velocity (m/s)	1.75
	Min. relative velocity (m/s)	– 1.70
	Max. distance error (m)	2.6
	Min. distance error (m)	– 0.6
	Max. acceleration (m/s^2)	0.90
	Max. deceleration (m/s^2)	– 1.12
Low-velocity sine	Max. relative velocity (m/s)	0.70
	Min. relative velocity (m/s)	– 0.55
	Max. distance error (m)	1.8
	Min. distance error (m)	– 0.2
	Max. acceleration (m/s^2)	0.40
	Max. deceleration (m/s^2)	– 0.35
High-velocity sine	Max. relative velocity (m/s)	0.60
	Min. relative velocity (m/s)	– 0.75
	Max. distance error (m)	1.1
	Min. distance error (m)	0
	Max. acceleration (m/s^2)	0.37
	Max. deceleration (m/s^2)	– 0.37

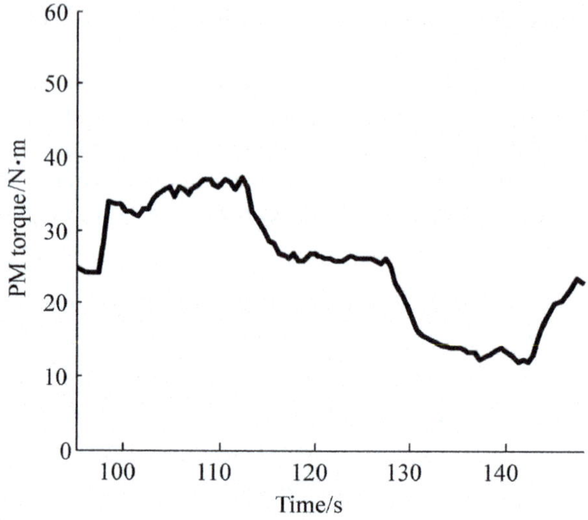

Fig. 7.47 Mild acceleration/deceleration of the preceding vehicle

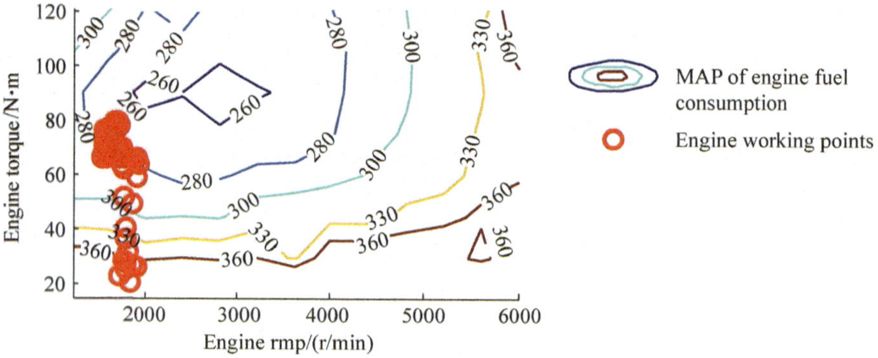

Fig. 7.48 Moderate acceleration/deceleration of the preceding vehicle

0.75 m/s, showing good car-following performance. In moderate and sharp acceleration/deceleration scenarios, the rapid change of the preceding vehicle's velocity results in increased relative velocity errors: the maximum relative velocity is 1.75 m/s, and the minimum relative velocity is − 1.7 m/s. However, the maximum distance error 2.6 m and the minimum distance error − 0.6 m show that the system doesn't immediately follow the preceding vehicle as the latter sharply accelerates/decelerates; instead, it follows the preceding vehicle at a smaller acceleration by taking into account the fact that the safe distance increases with velocity, so that a smaller distance error is kept during the car-following process. This helps reduce vehicle acceleration, improve driving comfort, reduce fuel consumption caused by rapid acceleration, and improve the vehicle's fuel economy.

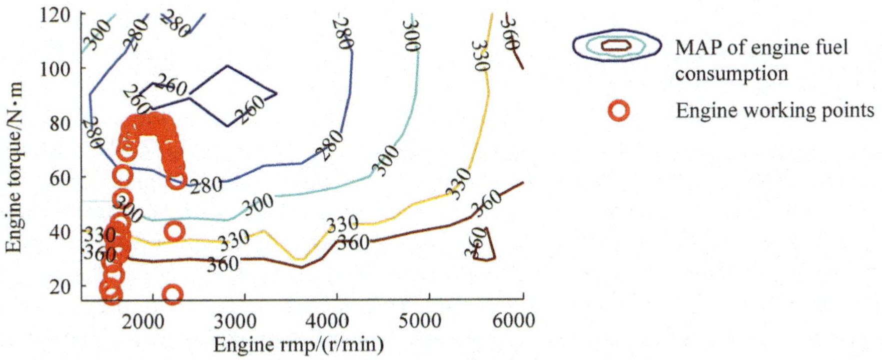

Fig. 7.49 Sharp acceleration/deceleration of the preceding vehicle

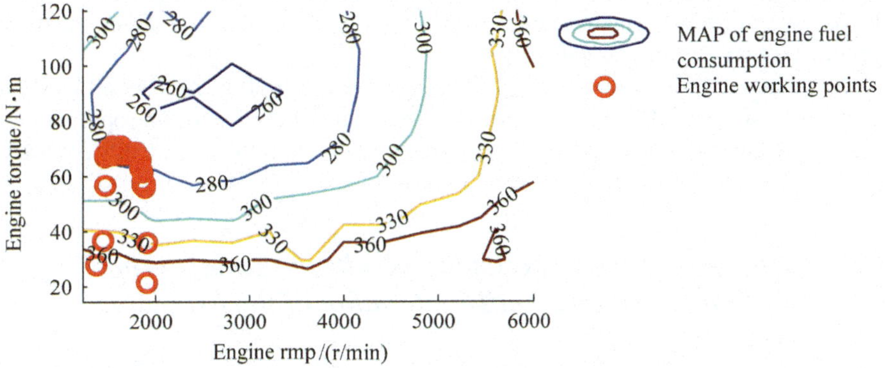

Fig. 7.50 Low-velocity sine of the preceding vehicle

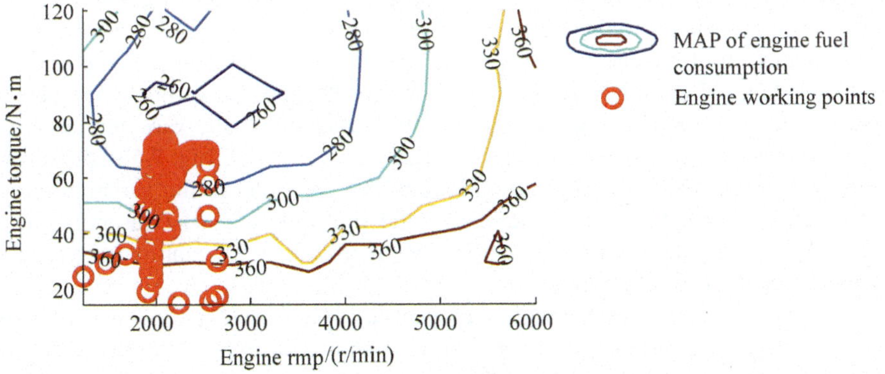

Fig. 7.51 High-velocity sine of the preceding vehicle

Table 7.14 also shows that the average acceleration amplitudes of the ego vehicle in all working scenarios are kept within 1.2 m/s^2, a range that can achieve the driving comfort required by the driver.

It can be seen from Figs. 7.47, 7.48, 7.49, 7.50 and 7.51 that the i-HEV ACC system has optimized fuel economy in all scenarios. As shown in Fig. 7.47 of the mild acceleration scenario, the ego vehicle is running at a lower velocity and demands less power due to the smaller acceleration of the preceding vehicle 0.2 m/s^2, so the system is always in the battery-only mode to keep the engine away from a poor working range. According to the basic principle of traditional HEV's energy allocation, the system has optimized fuel economy at this moment. In moderate and sharp acceleration/deceleration scenarios and low-velocity and high-velocity sinusoidal scenarios, as shown in Figs. 7.48, 7.49, 7.50 and 7.51, i-HEV can make sure that the engine is always in an optimal working range by optimizing its working points the engine operating point optimization of the i-HEV can ensure that the engine. According to the basic principle of traditional HEV's energy allocation, the system also has optimized fuel economy at this moment.

It can also be seen from the car-following effects of i-HEV ACC in all working scenarios that the i-HEV ACC control system designed based on offline optimization and online table lookup can meet the requirement for controlling the real-life vehicle.

7.5.6 Analysis of the Experiment Results in the Scenario of Sharp Acceleration/Deceleration of the Preceding Vehicle

Figures 7.52, 7.53 and 7.54 show partial results in the scenario of sharp acceleration/deceleration of the preceding vehicle. Figure 7.52a shows the variation curve of the velocity of the preceding vehicle and i-HEV ego vehicle, Fig. 7.52b shows the variation curve of battery SOC. Figure 7.53a shows the variation curve of relative velocity, Fig. 7.53b shows the variation curve of distance error, Fig. 7.53c shows the variation curve of PM motor torque, Fig. 7.53d shows the variation curve of engine torque, Fig. 7.53e shows the variation curve of BSG, Fig. 7.53f shows the variation curve AMT transmission ratio, and Fig. 7.54 shows the variation curve of acceleration. In Fig. 7.52a, the black solid line represents the variation curve of i-HEV's velocity and the grey dashed line represents the variation curve of the preceding vehicle's velocity.

It can be seen from the variation curves of the system velocity, relative velocity and car-following distance error of i-HEV as shown in Figs. 7.52a and 7.53a, b that the relative velocity of i-HEV ACC is kept within 1.75 m/s as the preceding vehicle accelerates at 0.8 m/s^2 and the ego vehicle can follow the preceding vehicle with a relative distance error close to zero. The purpose of following the preceding vehicle with smaller acceleration is achieved to improve the fuel economy and driving comfort.

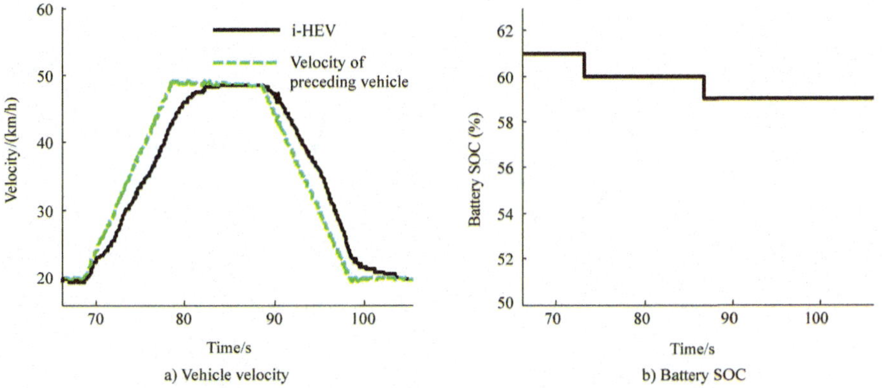

Fig. 7.52 Scenario of sharp acceleration/deceleration of the preceding vehicle (Part 1)

It can be seen from the curves of PM drive motor, engine and BSG in Fig. 7.53c–e that the i-HEV ACC achieves optimal fuel economy through appropriate energy allocation. The driving power is small when the preceding vehicle starts to accelerate, and the vehicle is in the battery-only driving mode. Due to the large acceleration of the preceding vehicle, the system switches to the mode of engine driving with power generation in response to the power demand generated from the vehicle's large acceleration. When the preceding vehicle is running at a constant velocity, the system switches from engine driving to battery driving despite the velocity as high as almost 50 km/h, due to the low power demand of the system in this mode. When the preceding vehicle decelerates at $-$ 0.8 m/s^2, the driving torque of PM motor decreases rapidly due to the sharp decrease in the velocity of the preceding vehicle, and the PM motor enters the braking energy regeneration mode to meet the vehicle's demand for sharp deceleration.

It can also be seen from the variation curves of PM drive motor torque and engine torque in Fig. 7.53c, d that they maintain good dynamic coordinated control performance when the system is switching between battery driving and engine driving. As the velocity of the preceding vehicle increases and the car-following error begins to increase, BSG starts the engine when t is about 72 s with a torque of 40 Nm to enter the process of switching and coordination between the motor and the engine driving. The variation of PM motor torque in Fig. 7.53c shows that PM motor's drive torque quickly rises to 95 Nm to compensate the delay in power response caused by slower torque response in the gear shift of transmission and the start of engine, and with the access of engine and increase of engine torque, the PM motor's drive torque begins to decline, and the system finally switches to the mode of engine driving with power generation from BSG.

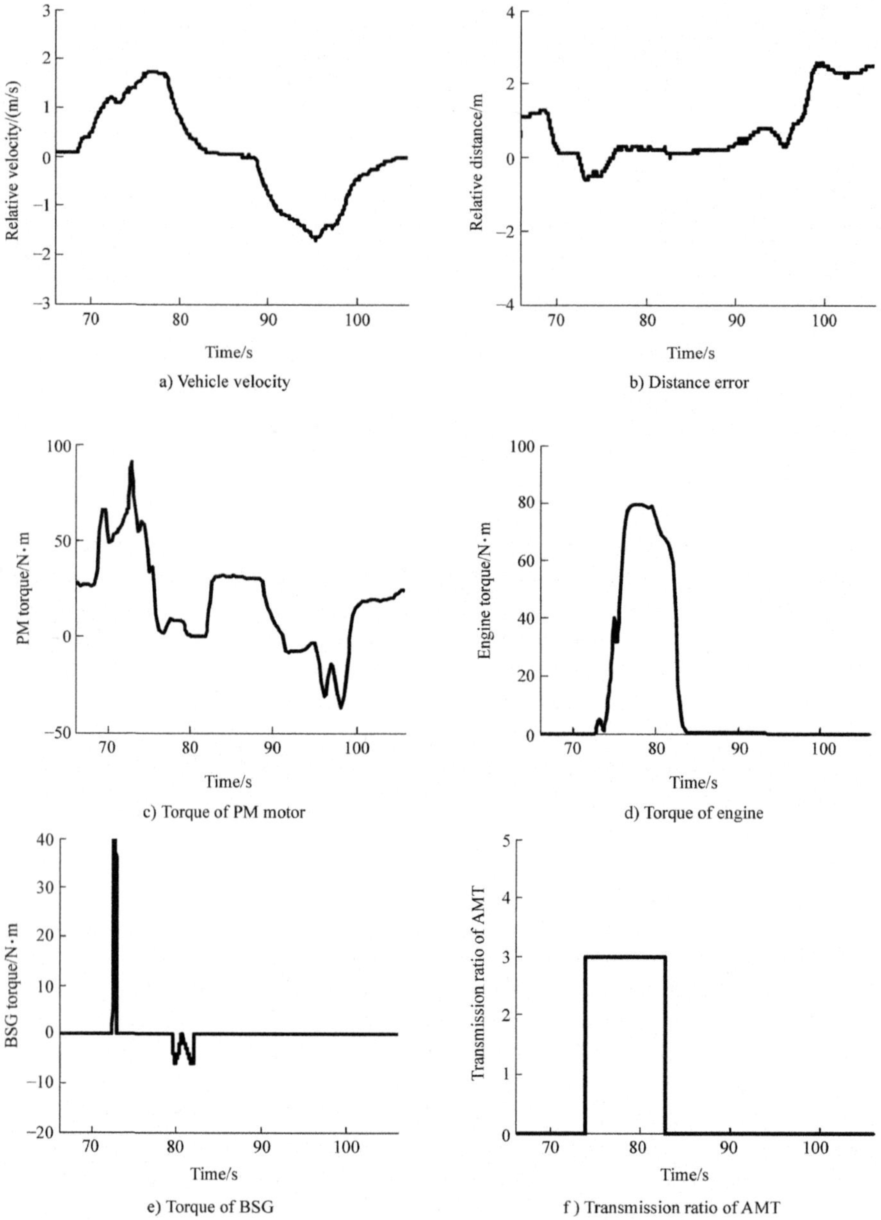

Fig. 7.53 Scenario of sharp acceleration/deceleration of the preceding vehicle (Part 2)

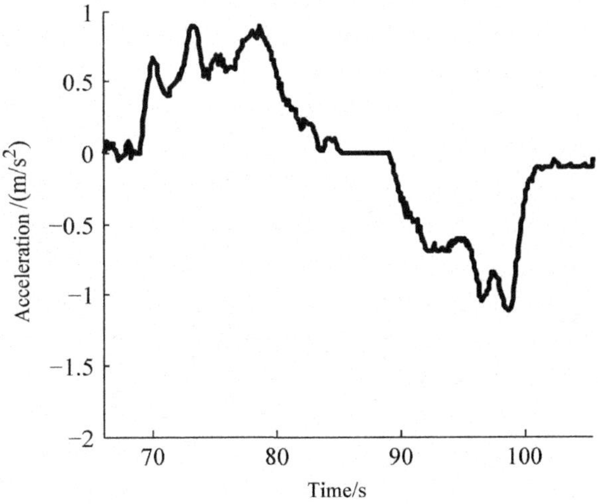

Fig. 7.54 Scenario of sharp acceleration/deceleration of the preceding vehicle (Part 3)

7.5.7 Analysis of Driving Cycle Contrast Experiment Results

Based on the verification of the comprehensive performance of the i-HEV ACC system, this section carries out contrast experiments on the comprehensive performance of i-HEV ACC, the driver HEV mode and the simple-superposition HEV ACC, including fuel economy, car-following safety and driving comfort. Limited by experimental conditions, only a set of controller parameters are compared, and the same controller parameters as those used in the simulation are used to design the driver HEV and the simple-superposition HEV ACC system controller. To address the HEV energy distribution problem, the control strategy that has been applied to real-life vehicles is used for power distribution.

(1) General experimental results of fuel economy and car-following safety

The fuel economy of each contrast system is measured by the equivalent fuel consumption indicator of the system. Due to the limited experimental data, the method of fitting a large number of data used in the simulation contrast cannot be used to calculate the fuel economy results. Instead, the SOC zero-passing method is adopted to obtain the corresponding equivalent fuel consumption of the system by fitting the engine fuel consumption under different variations of SOC. The results of engine fuel consumption under different SOC variations of each contrast system in the driving cycle are shown in Fig. 7.55.

As shown in Fig. 7.55, point O is the result of battery SOC varation and engine fuel consumption of the i-HEV algorithm in the driving cycle, * is the result of battery SOC varation and engine fuel consumption of the HEV algorithm, • is the result of battery SOC varation and engine fuel consumption of the simple superposition algorithm, the black solid line is the fitting straight line of the equivalent fuel consumption of the i-HEV algorithm, the solid-dotted line is the fitting straight line of the equivalent fuel consumption of the HEV algorithm, and the dashed line is the fitting straight

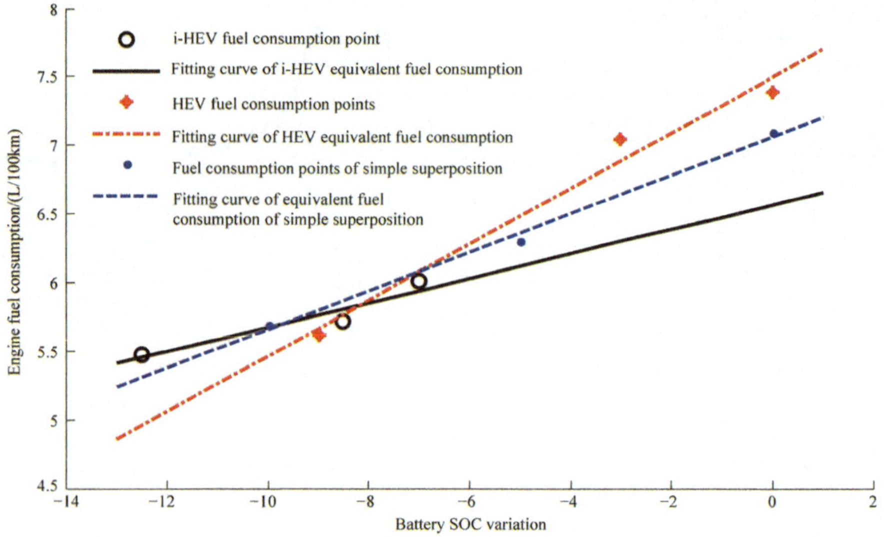

Fig. 7.55 Equivalent fuel consumption result in the driving cycle

line of the equivalent fuel consumption of simple superposition algorithm. It can be seen from the figure that when the variation of battery SOC is 0, it corresponds to the equivalent fuel consumption value of each system by comprehensively considering the engine fuel consumption and the battery power consumption. I-HEV algorithm has the minimum equivalent fuel consumption in SOC equilibrium, followed by the simple-superposition algorithm, while the driver HEV mode has the highest fuel consumption. The following safety indicator of each system is measured by the linear superposition of relative velocity and car-following distance error.

Experimental results of fuel economy and car-following safety of each system in the driving cycle are shown in Table 7.15.

As shown in Table 7.15, the i-HEV algorithm has the best fuel economy in the driving cycle, which is 6.57 L/100 km, 7.07% higher than the simple superposition algorithm and 12.4% higher than the driver HEV algorithm. All systems can maintain good following safety in general, and their safety indicators are all below 0.5. The

Table 7.15 Experimental results of fuel economy and car-following safety of each contrast system

Contrast system	Fuel economy indicator (L/100 km)	Car-following safety indicator
i-HEV algorithm	6.57	0.42
Simple-superposition algorithm	7.07	0.38
HEV algorithm	7.50	0.28

safety indicator of i-HEV is 0.42, which is basically equal to that of the simple superposition algorithm 0.38, while the safety indicator of HEV algorithm is 0.28, better than that of i-HEV algorithm, which is related to the coordination and optimization of economy and safety performance of the i-HEV algorithm, as well as the design of controller parameters. In summary, on the premise that the i-HEV system basically meets the following safety requirement, priority is given to the fuel economy of the system, which will compromise the car-following performance to some extent.

(2) Contrast analysis of experimental results

As shown in Figs. 7.56, 7.57, 7.58, 7.59, 7.60 and 7.61, the variation curves of component and vehicle states of the i-HEV algorithm, HEV algorithm and the simple superposition algorithm in different periods are drawn according to 3 types of periods and acceleration as well as the constant acceleration/deceleration of the preceding vehicle included in the designed driving cycle. In Figs. 7.56 and 7.57 the movement period of the preceding vehicle is 24 s, and the amplitude of the constant acceleration/deceleration of the preceding vehicle is 0.23 m/s^2. In Figs. 7.58 and 7.59 the movement period of the preceding vehicle is 18 s, and the amplitude of the constant acceleration/deceleration of the preceding vehicle is 0.31 m/s^2. In Figs. 7.60 and 7.61 the movement period of the preceding vehicle is 12 s, and the amplitude of the constant acceleration/deceleration of the preceding vehicle is 0.46 m/s^2.

In Figs. 7.56, 7.58, and 7.60, (a) is the velocity variation curve of the preceding vehicle, the i-HEV algorithm, the HEV algorithm and the simple-superposition algorithm, in which the black dotted line is the velocity curve of the preceding vehicle, (b) is the variation curve of relative velocity, (c) is the variation curve of distance error, (d) is the variation curve of battery SOC, (e) is the variation curve of PM motor drive torque. In Figs. 7.57, 7.59, and 7.61, (a) is the variation curve of the engine drive torque, (b) is the variation curve of the BSG torque, (c) is the variation curve of AMT transmission ratio, (d) is the variation curve of acceleration, and (e) is the working point of the engine. In all figures, the solid black line represents the vehicle state curve of the i-HEV algorithm, the gray dashed line is the vehicle state curve of the HEV algorithm, and the solid-dotted line is the vehicle state curve of the simple-superposition algorithm.

As it can be seen from the variation curves of vehicle velocity, relative velocity and car-following distance error in different movement periods in Figs. 7.56, 7.58 and 7.60a–c, the i-HEV algorithm, HEV algorithm and simple superposition algorithm are all able to follow the movement of the preceding vehicle and maintain good following safety. It's observed that the velocity following errors of all systems are kept within 1 m/s^2 and the distance error is kept within 2 m, which meets the basic requirement of car-following safety.

As it's observed from Figs. 7.57, 7.59 and 7.61d, both the i-HEV algorithm and the simple-superposition algorithm can follow the movement of the preceding vehicle and the acceleration amplitude is kept within 2 m/s^2, thus maintaining good driving comfort. For the driver HEV mode, however, as shown in the fast period car-following process in Fig. 7.45d, the driver is not able to smoothly follow the movement of the preceding vehicle that rapidly changed, resulting in large following acceleration and poor driving comfort. This also explains the role of ACC in improving the driving comfort to a certain extent.

As observed from the working states of the engine, PM motor and BSG of each system in Figs. 7.56, 7.57, 7.58, 7.59, 7.60e, 7.61a, b, the energy allocation strategy of the i-HEV algorithm is more reasonable than the rule-based energy allocation

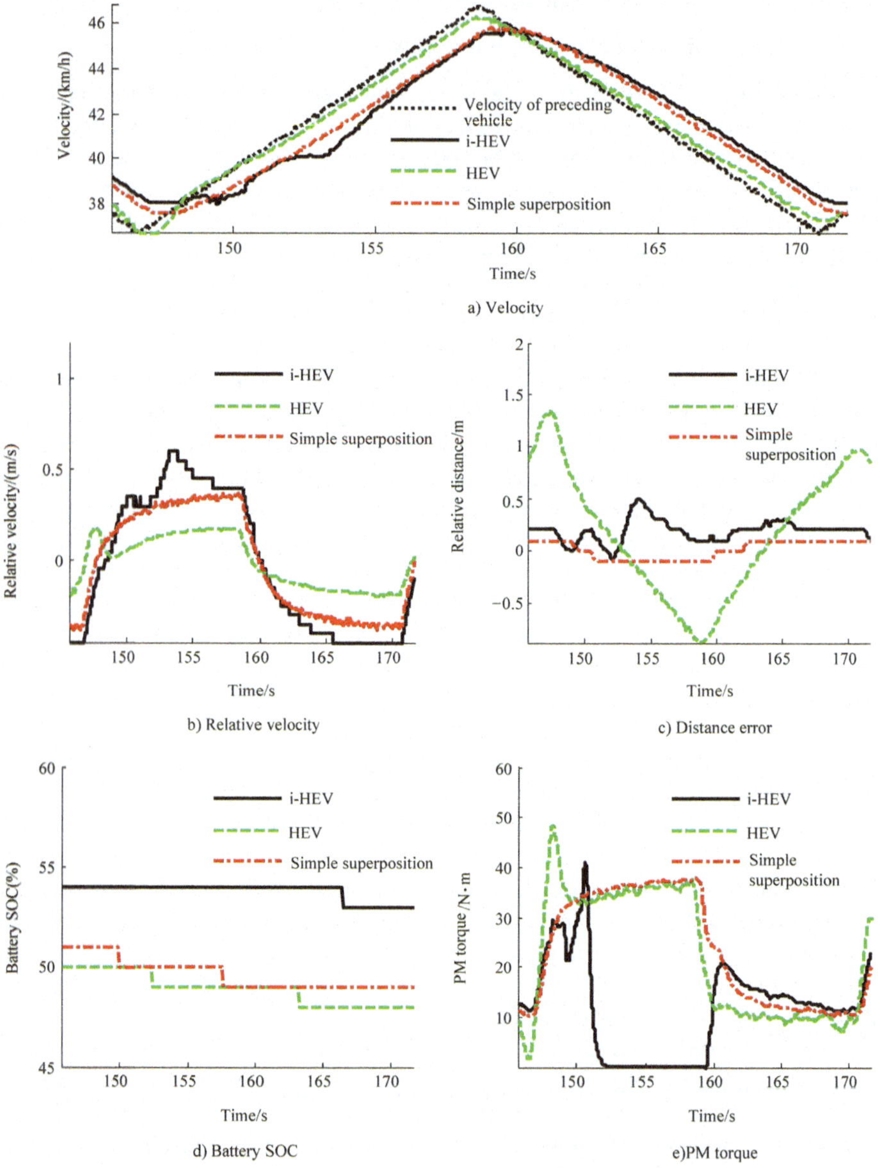

Fig. 7.56 Scenario of acceleration/deceleration in the slow period (Part 1)

strategy of the HEV algorithm and the simple superposition algorithm, so that the system has better fuel economy. Due to the overall small power demand of the system in the slow period, the rule-based energy allocation strategy puts the system in the battery working mode. However, when it moves into the medium period, due to insufficient battery SOC, the engine is forced to drive and generate electricity at the same time. In the process of deceleration, the engine has a low working point and higher fuel consumption. As for the energy allocation strategy of i-HEV based

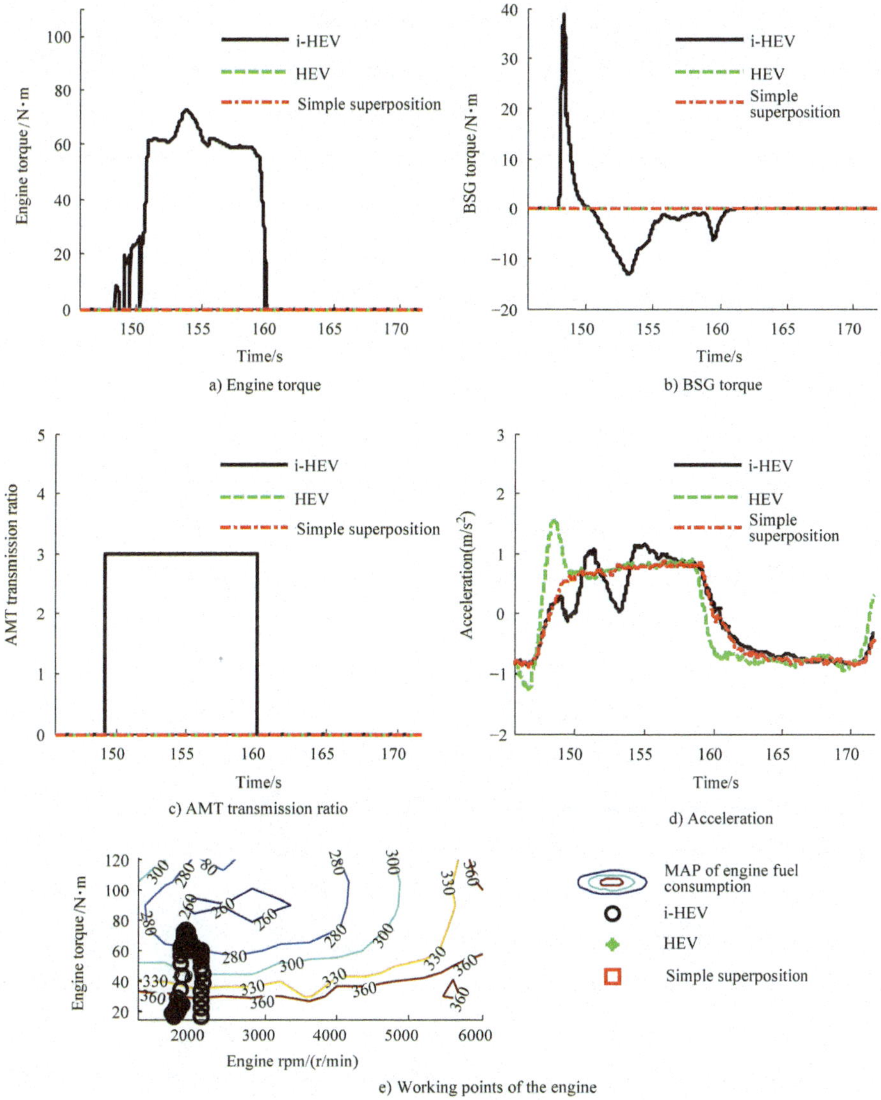

Fig. 7.57 Scenario of acceleration/deceleration in the slow period (Part 2)

on coordination optimization algorithm, the optimization is based on the characteristics of the entire driving cycle, so the system switches to the mode of engine driving + power generation in each acceleration period of the preceding vehicle. For this kind of driving cycle that requires large power consumption, the algorithm

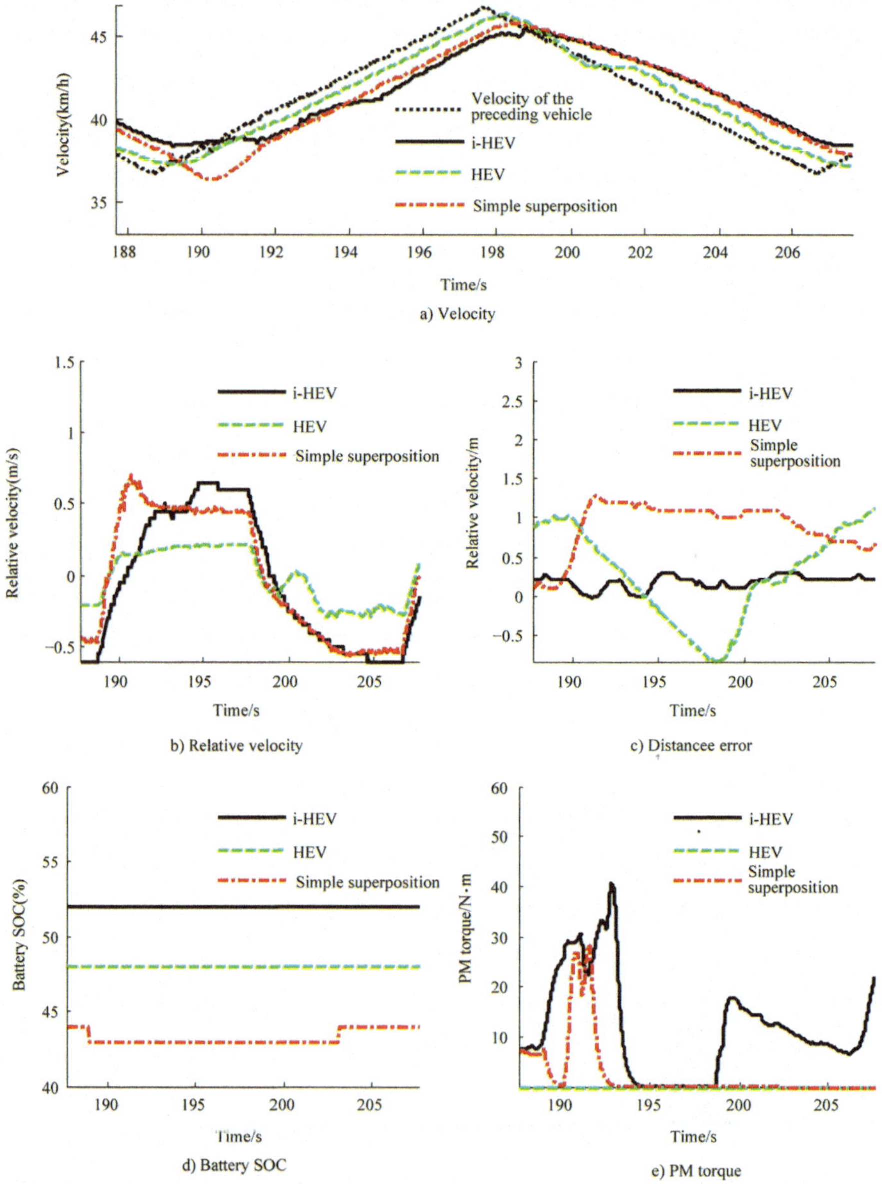

Fig. 7.58 Scenario of acceleration/deceleration in the medium period (Part 1)

can use the advantage that the engine can both drive and generate power under high power demand, to optimize the engine's working points, reduce the battery power consumption and improve fuel economy.

As it can be seen from the distance and velocity following errors in Figs. 7.56, 7.58 and 7.60c, the i-HEV algorithm has a more reasonable comprehensive effect

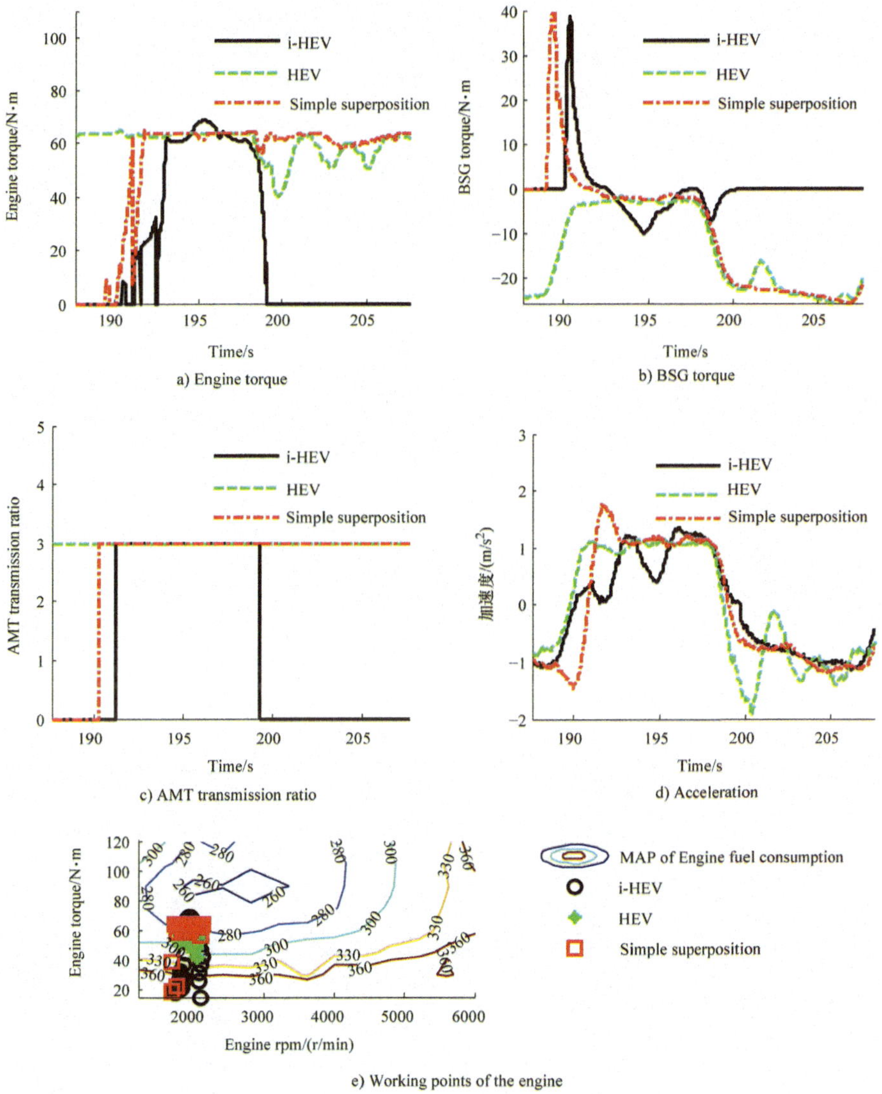

Fig. 7.59 Scenario of acceleration/deceleration in the medium period (Part 2)

of coordinating the following distance and velocity compared with the HEV algorithm, which can further reduce the energy loss in the process of vehicle acceleration and deceleration and improve the fuel economy. Although it's a little bit slower in following the velocity of the preceding vehicle, it can keep a good balance

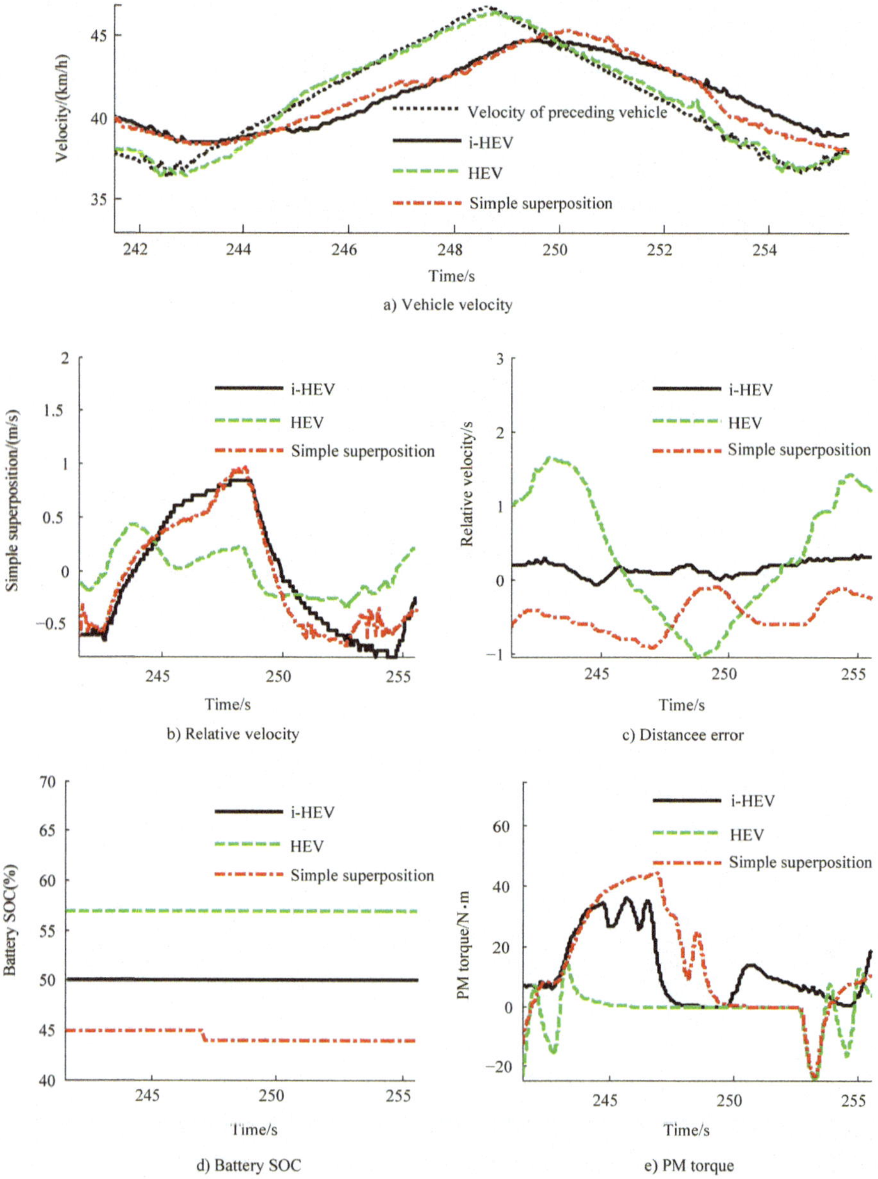

Fig. 7.60 Scenario of acceleration/deceleration in the fast period (Part 1)

between speed and distance and maintain a good following distance, thus reducing the vehicle's driving power and improving the fuel economy to a certain extent. It also shows the advantage of ACC following the preceding vehicle in terms of both velocity and distance over the driver.

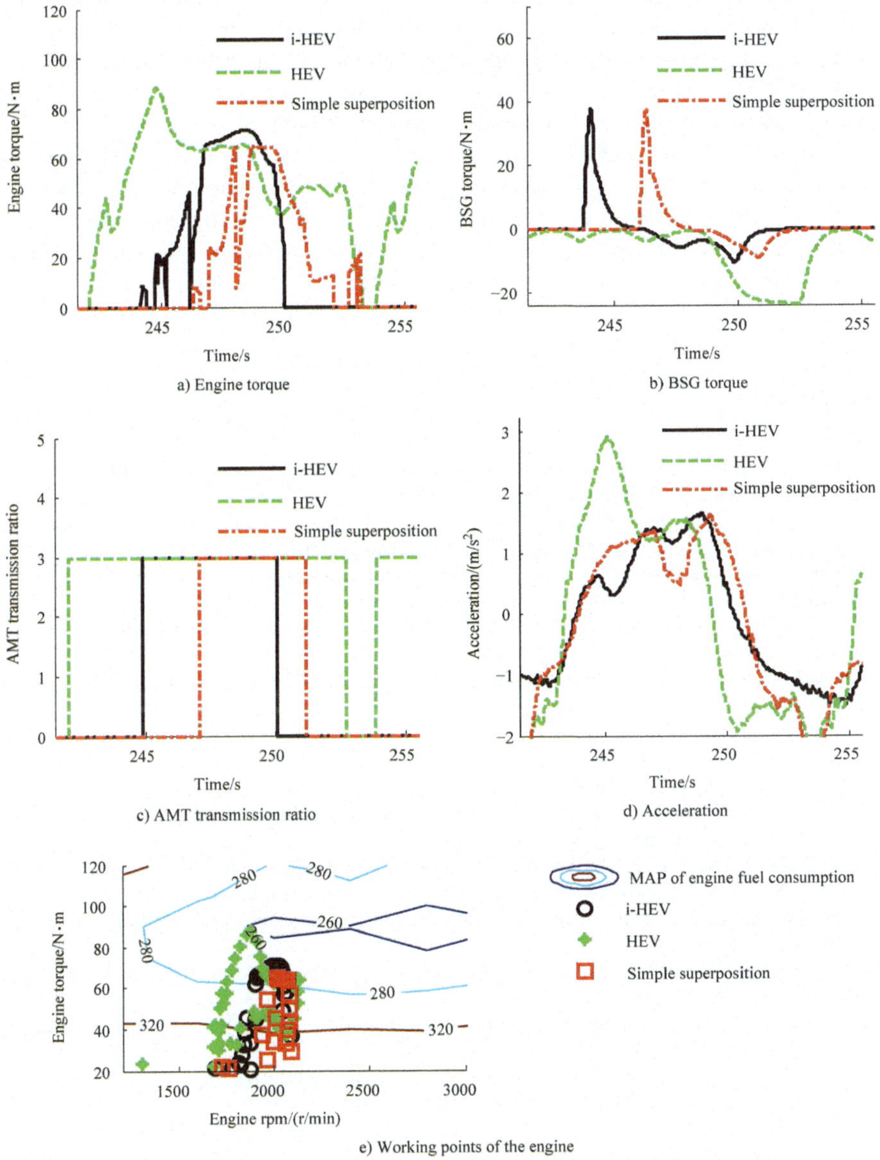

Fig. 7.61 Scenario of acceleration/deceleration in the fast period (Part 2)

It can also be seen from the working process of the engine, PM motor and BSG in Figs. 7.60e and 7.61a, b that the coordinated and optimized i-HEV ACC system can optimize the control of various components and improve the fuel economy better than the simple superposition HEV ACC algorithm. When there are bigger errors in the velocity and distance from those of the preceding vehicle, the i-HEV algorithm can start the engine and put it into the mode of driving with power generation earlier than the simple-superposition algorithm. It can also shut down the engine in time in deceleration to ensure the optimized engine operation to some extent while reducing the consumption of electricity, thus reducing the fuel consumption caused by electricity compensation required by mandatory charging.

Figures 7.57, 7.59 and 7.61e are the variation curves of the engine's working points, where O represents the working point of the i-HEV engine, * represents the working point of the HEV engine, and □ represents the working point of engine of the simple superposition algorithm. As observed from the figures, the engine is in an optimal range under all algorithms since it's optimized by the two systems.

References

1. Ferrara A, Pisu P (2004) Minimum sensor second-order sliding mode longitudinal control of passenger vehicles. IEEE Trans Intell Transp Syst 5(1):20–32
2. Bageshwar L, Garrard L, Rajamani R (2004) Model predictive control of transitional maneuvers for adaptive cruise control vehicles. IEEE Trans Veh Technol 53(5):1573–1585
3. Ouyang M, Li J, Yang F et al (2008) A new automotive powertrain system: configuration, modeling and control. Tsinghua University Press, Beijing
4. Yia K, Do Kwon Y (2001) Vehicle-to-vehicle distance and velocity control using an electronic-vacuum booster. JSAE Rev 22(4):403–412
5. Qian J, Zhao J, Xu Z (2007) Prediction control. National Defense Industry Press, Beijing
6. Li S (2009) Automotive adaptive cruise control with multi-objective coordination. Tsinghua University, Beijing
7. Koot MWT (2006) Energy management for vehicular electric power systems. Technische Universiteit Eindhoven, Eindhoven
8. Ying C (2005) Research on the control method of the multi-energy power system of hybrid electric vehicles. Tsinghua University, Beijing
9. Li P, Hu D, Zhang J (2010) System recognition. China Water & Power Press, Beijing
10. Sakai SI, Hori Y (2000) Advantage of electric motor for anti skid control of electric vehicle. In: Proceedings of the 9th international conference on power electronics and motion control. Budapest, Hungary
11. Lei Z (2009) Coordinated control of braking energy regeneration and braking stability of electric vehicle. Tsinghua University, Beijing
12. Dezao H (2004) Research on automotive longitudinal active collision avoidance system. Tsinghua University, Beijing
13. Wang J, Zhang D, Li K et al (2011) Design of an automobile integrated electronic vacuum booster system. China J Highway Transp 24(1):115–121
14. Quan Y, Zhang W, Xu X (2002) Step response identification method for second-order plus delay model. Control Theory Appl 19(6):954–956
15. Kessels JTBA, Koot M, de Jager B et al (2007) Energy management for the electric powernet in vehicles with a conventional drivetrain. IEEE Trans Control Syst Technol 15(3):494–505

16. Koot M, Kessels JTBA (2005) Energy management strategies for vehicular electric power systems. IEEE Trans Veh Technol 54(3):771–782
17. He XL (2001) Battery modeling for HEV simulation model development. SAE 2001-01-0960
18. Wipke KB, Cuddy MR, Burch SD (1999) ADVISOR 2.1: a user friendly advanced powertrain simulation using a combined backward/forward approach. IEEE Trans Veh Technol 48(6):1751–1761
19. Gipps PG (1981) A Behavioral car-following model for computer simulation. Transp Res Board (15-B):105–111
20. Lee K, Peng H (2004) Identification and verification of a longitudinal human driving model for collision warning and avoidance systems. Int J Veh Auton Syst 2(1–2):3–17
21. Luo Y, Chen T, Li K (2010) Control system of intelligent hybrid electric vehicle. J Tsinghua Univ (Sci Technol) 50(8):29–39

Chapter 8
Implementation of Intelligent EV Charging/ Battery-Swapping Scheduling Based on Multi-source Information Fusion

Different from i-HEVs, the optimal control of intelligent electric vehicles requires guaranteeing the efficient and safe operation of the traffic system and the power grid system due to the existence of charging process. Therefore, a charging/battery-swapping scheduling solution is needed for intelligent electric vehicles. In this chapter, a charging/battery-swapping scheduling strategy is designed according to the different charging/battery-swapping characteristics of fast-charging, slow-charging and battery-swapping electric vehicles, as well as the different driving characteristics of electric private cars, electric taxis and electric buses, to realize the comprehensive optimization of traffic and power grid.

8.1 Studies on the Charging/Battery-Swapping Schedule of EVs

With the surge of global population, rapid economic growth and increasing consumption of fossil fuels, the problem of environmental pollution is becoming more and more serious. As the most ideal pollution-free transportation tool that arises at the right moment, electric vehicle technologies are being vigorously developed by major automobile manufacturers and universities around the world [1]. However, there are many problems that need to be solved before electric vehicles completely replace traditional ICE vehicles. First of all, the driving ranges per charge are short due to the limitation of battery technology, which affects the reliability and convenience of electric vehicles. Secondly, the large-scale application of electric vehicles may cause local traffic congestion and make travels less easy. Thirdly, due to the uncertainty of charging time, place and power of electric vehicles, problems such as excessive voltage drop, excessive load, excessive power loss and decreased power quality of local power grids will occur when there is a massive access of electric vehicles to the power grid for unordered charging [2].

K. Li et al., *The Intelligent Environment Friendly Vehicle*,
Key Technologies on New Energy Vehicles,
https://doi.org/10.1007/978-981-19-4851-0_8

Scheduling the charging/battery-swapping behavior of intelligent electric vehicles based on information including traffic, power grid, vehicle location, and energy can not only alleviate the worries of EV drivers for range, but also reduce the impact of large-scale application of electric vehicles on traffic and power grid. This chapter aims to study how to schedule the charging/battery-swapping of electric vehicles by combining the information of road traffic network and power grid, while optimizing the performance of different types of electric vehicles, road traffic networks and power grids.

There have been a lot of studies on large-scale EV charging scheduling around the world, which improve the convenience of electric vehicles and reducing the impact of large-scale EV charging behavior on traffic and power grid to varying degrees, but there are still the following shortcomings:

(1) The EV charging strategy only considers a single optimization goal, which cannot guarantee the optimization of the whole system. Most of the researches on EV charging scheduling strategy so far only consider the optimization of power grid or the improvement of driving convenience. There is a lack of researches on charging scheduling that take into account the large-scale operation of electric vehicles, road traffic network, power grid and charging station, which cannot guarantee the optimization of the overall performance of the system.
(2) The actual characteristics of large-scale EV operation have not been considered, and there is a lack of diversification in the types of EV. Most of the existing researches focus on fast-charging private cars, which cannot fully reflect the actual large-scale application of electric vehicles in the future.

To address the above shortcomings, this chapter will study different types of electric vehicles and put forward a large-scale EV charging/battery-swapping scheduling strategy based on a thorough consideration of their driving and charging characteristics, as well as the traffic and power grid information. A "EV–grid–network" simulation platform is established to verify the feasibility of the proposed strategy. Our specific works include:

(1) Proposing charging/battery-swapping scheduling strategy for electric vehicles based on the comprehensive optimization of traffic and power grid. Based on the charging/battery-swapping characteristics of fast/slow charging and battery-swapping EVs as well as the driving characteristics of electric private cars, electric taxis and electric buses, by considering the operation performances of traffic efficiency, power grid load, power loss, and voltage deviation and factors including the efficiency of charging stations, an EV charging/battery-swapping scheduling strategy that fits vehicle types is formulated to achieve the optimal overall performance of "EV-grid-traffic network".
(2) Establishing a simulation platform for electric vehicle charging/battery-swapping scheduling. Based on the "vehicle-network-road" intelligent interactive system established in the early-stage study by the research group, models of different EV types and their corresponding charging/battery-swapping station models are established according to the characteristics of different types of EV,

characteristics of charging/battery-swapping stations, the coverage area of the power grid, and the size of EVs. Also with the expansion of the power grid model, the simulation platform of "different types of EV–traffic network–power grid" is established.

(3) Verifying the charging/battery-swapping scheduling strategy of electric vehicles by simulation. The simulation platform is used to verify the proposed charging/battery-swapping scheduling strategy of electric vehicles for the comprehensive optimization of traffic and power grid, according to the type of electric vehicles as well as their reasonable application environment. The advantages and disadvantages of the proposed strategy in the optimization of traffic, power grid and charging station are also evaluated.

8.1.1 Designing the Process of EV Charging/Battery-Swapping

For a complex group such as electric vehicles, the scheduling of charging/battery-swapping should not only consider the current position and energy information of vehicles, the operation conditions of the road traffic network, the power grid and charging stations, but also the needs of drivers. In order to meet the requirement of reasonableness, the charging/battery-swapping scheduling process of electric vehicles is designed in this chapter as shown in Fig. 8.1, which includes three parts: judgement of charging/battery-swapping demand, driver's decision-making, and charging/battery-swapping scheduling.

Judgement of charging/battery-swapping demand is a prerequisite for charging/battery-swapping of electric vehicles. The scheduler determines whether the electric vehicle needs charging/battery-swapping according to the remaining electric capacity, the current position of the vehicle, the destination and the map information of the region where the vehicle is located. The driver's decision-making includes two aspects. First, the driver decides whether to charge or swap battery according to his actual situation and will. Second, after the charging/battery-swapping is scheduled, the driver chooses the driving path from several paths given by the system. The whole process reflects the leading role of the driver and ensures the rationality of the charging/battery-swapping process. In the stage of charging/battery-swapping scheduling, the result of the first decision made by the driver is based on to determine whether the driver needs charging/battery-swapping. If the driver decides to charge/swap the battery, the traffic and power grid information will be used to provide several charging/battery-swapping paths to the driver. The power grid and traffic network information will then be updated based on the driver's second decision to ensure the real-timeliness and accuracy of such information. In the process of charging/battery-swapping of electric vehicles, it depends, on the one hand, on the power information of the vehicle, the vehicle position, the destination and map information to determine whether there's a need for charging/battery-swapping; on the other hand, it is also necessary to recommend appropriate charging/battery-swapping stations and plan

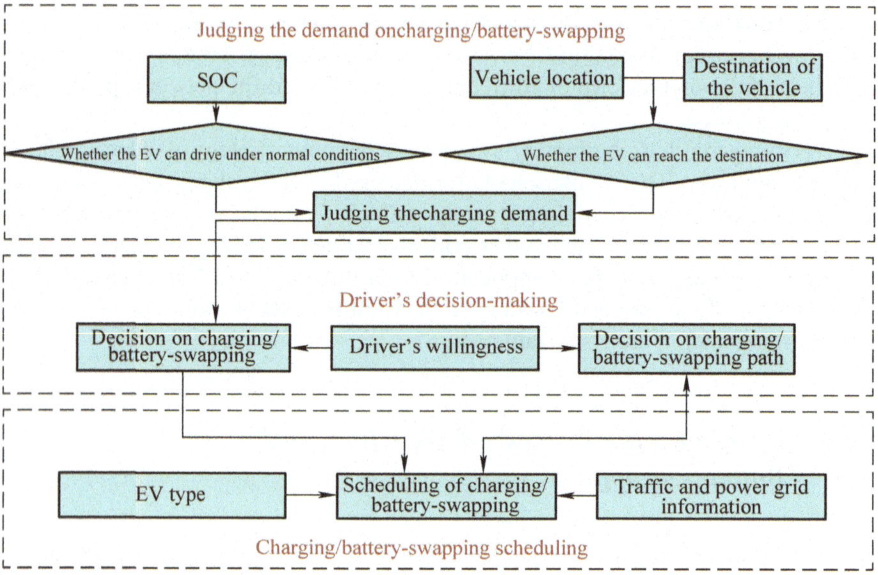

Fig. 8.1 Charging/battery-swapping scheduling process

reasonable charging/battery-swapping paths for the vehicle according to the real-time operation information of the road traffic network and the power grid. The whole process fully considers the interaction between EVs and the traffic and power grid information, so as to ensure the overall optimal operation of EVs, the road traffic network and the power grid.

Electric vehicles are divided into different types based on the application scenarios in the future. They are divided into the categories of charging and battery-swapping by the way they obtain power; private cars, taxis and buses by their usage. Private cars get power by charging and battery-swapping, taxis by charging and buses by battery-swapping. Since the charging/battery-swapping scheduling processes of charging/ battery-swapping vehicles with different uses are similar, this chapter only studies the scheduling of charging/battery-swapping behaviors of electric vehicles respectively, rather than differentiating the scheduling of electric vehicles with different uses.

This section addresses how to dispatch electric vehicles to charging and battery-swapping stations that meet the requirement of the convenience of the vehicle and ensure the orderly operation of the traffic network and power grid in the entire charging/battery-swapping scheduling process of electric vehicles. Therefore, this section will explain the following three aspects: judgement of charging/ battery-swapping demand, driver's decision-making, EV charging scheduling and EV battery-swapping scheduling.

8.1.2 Judgement of Charging/Battery-Swapping Demand and Driver's Decision-Making

The designed charging/battery-swapping scheduling process of electric vehicles requires judgment of charging/battery-swapping demand and decision-making of the driver before scheduling. This section introduces the judgment of charging/battery-swapping demand and the decision-making process of the driver.

(a) Judgement of charging/battery-swapping demand

During the driving process of an electric vehicle, it judges whether the electric vehicle has charging/battery-swapping demand at regular intervals. As shown in Fig. 8.2, the flow chart of the charging/battery-swapping demand judgment includes the following steps.

(1) Initialization of the map information. In the initial stage of charging demand judgment, the vehicle judges the current position of the vehicle and expected destination by itself and updates the local map information, including the current condition of the traffic network. Based on the above information, the path search algorithm is used to calculate the driving distance and time required to reach the destination, and then estimate the power required ($Power\_req$). Estimating the power required to reach the destination is the basis for judging whether the electric vehicle has charging/battery-swapping demand.
(2) Initialization of the electric vehicle's remaining power. In the process of judging charging/battery-swapping demand, it is necessary to make multiple judgments according to the state of electric vehicle's remaining power to determine whether the vehicle has charging/battery-swapping demand. In the process of driving, the electric vehicle uses a device to monitor the remaining power in real time.
(3) Determining whether the remaining power of the electric vehicle can meet its basic driving demand. When the vehicle detects that its remaining power (SOC_rem) is $N\%$ less than its battery capacity ($Battery\_Capacity$), i.e.:

$$SOC\_rem < N\% * Battery\_Capacity \tag{8.1}$$

the value range of N is generally 25–35; at this point, the system determines that the electric vehicle's remaining power cannot meet its basic driving demand, that is, if the condition of Eq. (8.1) is satisfied, then go to step (5), otherwise go to step (4).
(4) Determining if the electric car will make it to its destination. When the system determines the remaining power of the electric vehicle (SOC_rem) can meet its basic driving demand, that is, Eq. (8.1) is not valid, the system continues to determine whether the remaining power can support the electric vehicle to arrive the destination by judging whether the remaining power is greater than the power required by the vehicle to reach the destination in step (1) ($Power\_req$). If the electric vehicle's remaining power (SOC_rem) is greater than the power required by the vehicle to reach the destination ($Power\_req$), i.e.:

Fig. 8.2 Judgement of charging/battery-swapping demand

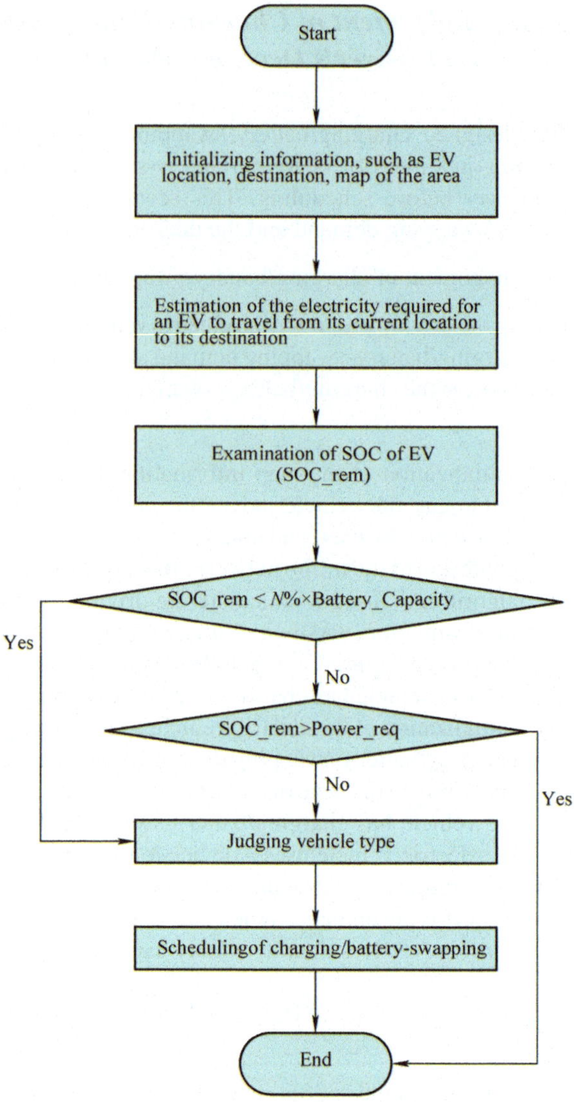

$$SOC\_rem > Power\_req \tag{8.2}$$

the system quits the scheduling process; if the electric vehicle's remaining power (SOC_rem) is less than the power required by the vehicle to reach the destination ($Power\_req$), that is Eq. (8.2) is not valid, then the system determines that the remaining power cannot support the electric vehicle to arrive the destination. In that case, go to step (5).

(5) Scheduling charging/battery-swapping. Based on the judgement in step (3) and (4), when the electric vehicle's remaining power cannot meet its basic driving demand or it can meet the basic driving demand but cannot support the vehicle to reach the destination, the system determines that the vehicle has the charging/battery-swapping demand, and thus scheduling the charging/battery-swapping according to the type of the electric vehicle; on the other hand, if the system determines that the electric vehicle's remaining power can meet its basic driving demand and can support the vehicle to reach the destination, then the system determines that the vehicle doesn't have the charging/battery-swapping demand and quit the scheduling process, and it continues to determine the vehicle's charging/battery-swapping demand in real-time.

(b) Driver's decision-making

Figure 8.3 is the flow chart of decision-making of the EV driver. When the system determines that the vehicle has the charging/battery-swapping demand, it enters the driver decision-making mode to allow the driver to make decision on whether to charge/swap the battery and their paths.

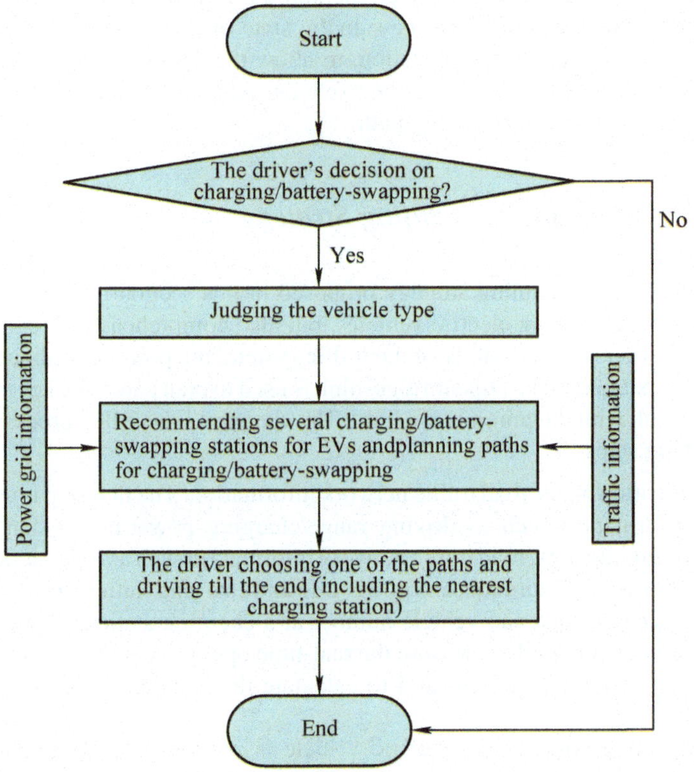

Fig. 8.3 Flow chart of driver's decision-making

The process of decision making includes the following steps:

(1) The driver decides whether to charge/swap the battery. When the vehicle has the charging/battery-swapping demand, the system enters the driver decision-making mode by taking into account the driver's will. The driver decides whether to charge/swap the battery according to his/her situation. If the driver refuses, the system quits the process of scheduling.
(2) Planning the charging/battery-swapping paths for the electric vehicle. If the driver agrees to charge/swap the battery, then the system recommends several charging/battery-swapping stations that can both ensure the driving convenience of the vehicle and the operational efficiency and economy of the traffic network and the power grid and plan corresponding paths by using the Dijkstra algorithm based on the charging/battery-swapping characteristics of the vehicle's charging solution (fast-charging, slow-charging, battery-swapping) and the driving characteristics of the vehicle's usage (private car, taxi, bus), as well as the comprehensive traffic, power grid and charging station information.
(3) The driver chooses a charging/battery-swapping path. The system obtains via calculation the charging/battery-swapping station options that meet the optimal traffic and power grid purpose and their corresponding paths, and then recommends those stations to the driver, including the nearest charging station to the current vehicle position and the standby state of the station when the vehicle reaches each recommended station, to allow the driver to choose a path from them. It's at the driver's discretion to choose whether to charge the vehicle and the charging/battery-swapping path.

8.1.3 EV Charging Scheduling Strategy

The EV charging scheduling strategy proposed in this section not only guarantees the convenience of using electric vehicles, but also comprehensively considers the operational efficiency and safety of the traffic system, the power distribution system and charging stations. The Dijkstra algorithm is used to plan paths, the key of which is to find the path with the minimum weight. The charging scheduling process includes the following steps:

(1) Initialization of EV road traffic network information. The first step is to initialize EV information such as driving range/charging power/power consumption/ remaining capacity, road traffic network information such as road section length/ congestion rate/saturation, and power grid and charging station information such as power grid load and vehicle number at a charging station. These pieces of information are used to measure the real-time operation of the traffic, the power grid, and charging stations and to calculate the weights during the search of paths.
(2) Determining whether the electric vehicle is near a charging station. For an electric vehicle that has a charging demand, the first thing to do is to judge whether it's near a charging station.

(3) Planning charging paths

(3.1) If the electric vehicle is near a charging station, the charging path from the current position of the vehicle to the charging station is directly planned;
(3.2) If the electric vehicle is not near a charging station, the comprehensive weight value considering the operation performance of the traffic system, the power grid system and charging stations is calculated. Then the Dijkstra algorithm is used to search the charging path with the minimum comprehensive weight value.

(4) Determining whether the charging load of the charging station exceeds its load threshold. For an electric vehicle with a planned charging path, it needs to determine whether the charging load of each charging station will exceed its load limit when the vehicle arrives based on the expected arrival time. If the charging load is expected to exceed the limit, then this charging station is excluded and the charging station with the next optimal comprehensive weight is searched.
(5) Determining whether the charging station is within the remaining range of the electric vehicle. If the charging station does not exceed its load threshold, it judges whether the recommended charging station is within the remaining range of the vehicle. If not, it will find the charging path with the shortest driving distance. If so, the vehicle goes directly to the charging station.
(6) Counting the charging parameters of the electric vehicle and the traffic flow of each road section. Then, the charging time is calculated, the charging load of the charging station and the number of vehicles in the charging station are counted, the traffic flow of each road section along the path from the current vehicle position towards the charging station is counted, and the traffic speed of each road section is calculated. The scheduling is quitted when all the vehicle calculation is done.

The flow chart of charging scheduling is as shown in Fig. 8.4.

(a) Determination of the objective function

It is a typical multi-objective optimization problem to conduct large-scale EV charging scheduling on the premise of ensuring the safe and efficient operation of the traffic network, the power grid, and charging stations. Determining a reasonable objective function on the premise of the comprehensive optimization of the system is the key to solve the multi-objective optimization problem.

The commonly used method to solve the multi-objective optimization problem is the weight coefficient method [3], that is, to assign a weight coefficient to each target, and then multiply all the targets by their corresponding weight coefficients, and then add them linearly into a new objective function. This method reflects the extent that the charging scheduling is biased to different optimization objectives, and the optimal solution of different aspects can be achieved by changing the weight coefficient.

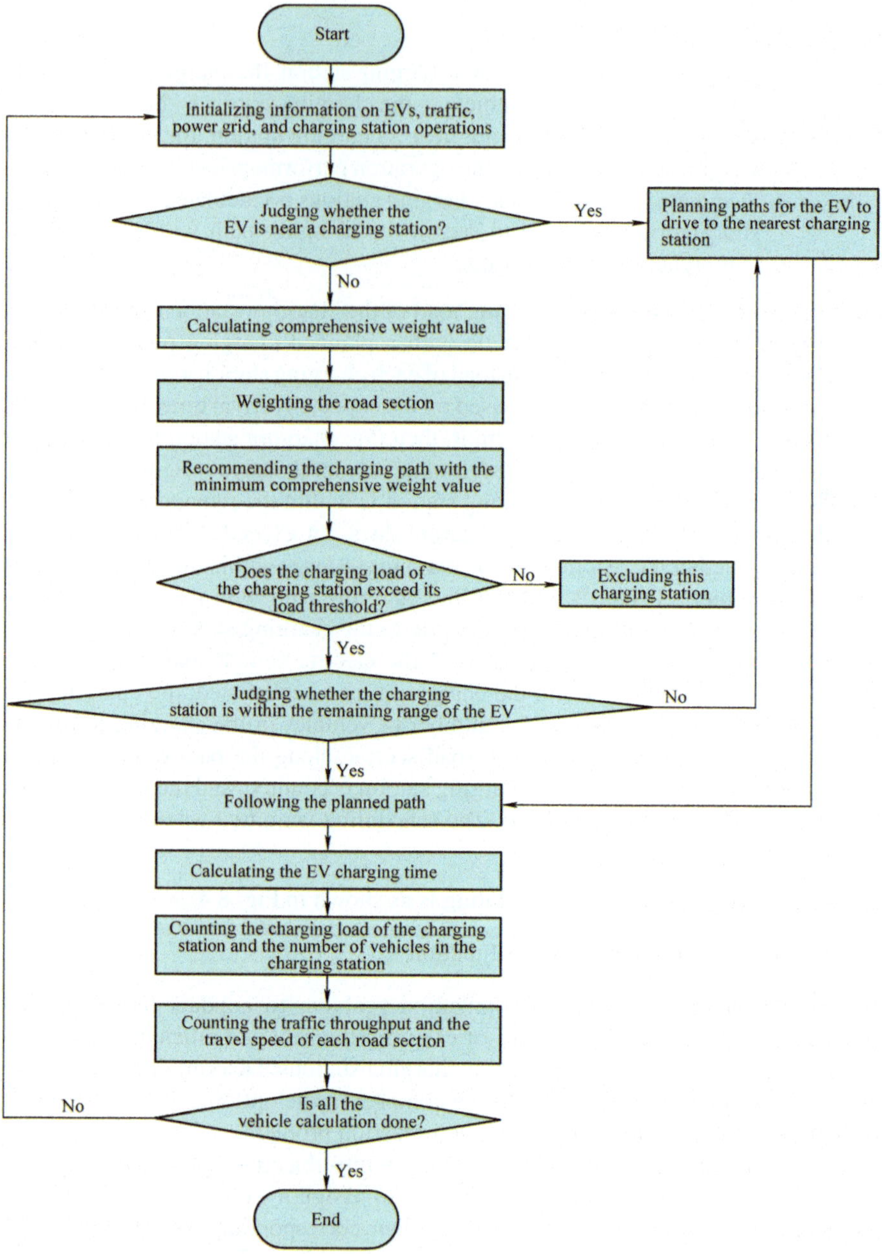

Fig. 8.4 Flow chart of charging scheduling

The three optimization objectives proposed in this section have different dimensions, so each objective needs to be turned dimensionless [4] first, and then the weighted summation of each objective is carried out. The optimization objective function established is shown in Eq. (8.3).

$$F = w_1 \frac{f_1}{\min f_1} + w_2 \frac{f_2}{\min f_2} + w_3 \frac{f_3}{\min f_3} \tag{8.3}$$

where f_1, f_2, and f_3 are respectively the reciprocals of the traffic speed of the road section near the charging station, the charging load, and the number of vehicles in the charging station; min f_1, min f_2, and min f_3 are the 1/the optimal traffic speed of the road section, the optimal power distribution network load, and the optimal number of vehicles in the charging station obtained by independently optimizing the traffic speed of the road section, the charging load, and the number of vehicles in the charging station without considering the other two optimization functions. w_1, w_2 and w_3 are the weight coefficients corresponding to the three optimization objectives respectively.

On the traffic side, the speed of the passable road section at each moment is calculated. For a passable section in the road traffic network $(p, q) \in A$, the traffic speed $V_{road}\{p, q\}(j)$ can be obtained from the traffic flow of the section at the end of the last moment $q\{p, q\}(j-1)$ and the "classical speed-flow model" [5]:

$$V_{road}\{p, q\}(j) = \frac{V_m\{p, q\}}{1 + q\{p, q\}(j-1)/C\{p, q\}} \tag{8.4}$$

where the traffic capacity of the road section $C\{p, q\}$ and the zero-flow speed of the road section $V_m\{p, q\}$ are the inherent properties of the road section and are known quantities; at the initial moment ($j = 0$), the traffic flow and traffic speed of the road section are respectively zero and the maximum speed limit of the section. Therefore, the traffic speed $V_{road}\{p, q\}(j)$ of the section $(p, q) \in A$ can be determined by the traffic flow at the end of the last moment $q\{p, q\}(j-1)$. The traffic flow of each road section can be measured by counting the road section the vehicle passes through at each moment.

Therefore, in order to ensure the smooth operation of the traffic system, the greater the speed of the road section, the better, that is, to ensure that its reciprocal as shown in Eq. (8.5) is as small as possible.

$$f_1 = 1/V_{road}\{p, q\}(j) \tag{8.5}$$

On the power distribution network side, the power loss will increase, the voltage profile will change and the load of charging stations will increase after massive electric vehicles are connected to the power distribution network. The operation performance of the power distribution network is related to the power flow of the network nodes. Therefore, limiting the charging load of the charging station can improve the operation performance of the power distribution network and ensure the power quality of the distribution system. The grid charging load is expressed in Eq. (8.6).

$$f_2 = Load\_Charging(r, t_0) \tag{8.6}$$

where M is the number of charging stations in the power distribution network; t_0 is the predicted time of EV arriving at the charging station [6]; $Load\_Charging(r, t_0)$ is the predicted load value of the charging station r at the moment t_0 (obtained by measuring the charging power of the EVs being charged at the charging station r at the moment t_0).

On the charging station side, when the EVs are massively connected to the charging stations, it is inevitable that there will be waiting for charging. In order to ensure that the EV does not have to wait too long for charging, the number of vehicles in each charging station should be limited. The number of vehicles in a charging station at different moments is counted, as shown in Eq. (8.7).

$$f_3 = Num\_Charging(r, t_0) \tag{8.7}$$

where $Num\_Charging(r, t_0)$ is the estimated number of vehicles in the charging station r at the moment t_0 [6].

The large-scale EV charging scheduling strategy as formulated above allows the electric vehicle to go to charging stations with smaller charging load and smaller number of electric vehicles through road sections with higher traffic speed.

(b) Determination of the weight coefficient of objective function based on fuzzy logic

The determination of weight coefficient is the key to solve the multi-objective optimization problem. Because the correlation between the three optimization objectives in the optimized objective function is uncertain, and the driving condition of electric vehicles is random and uncertain, a method to determine the weight coefficient based on fuzzy logic is proposed.

The flow chart of determining the weight coefficient of objective function based on fuzzy logic is shown in Fig. 8.5. The system consists of five parts: input value, output value, fuzzification, inference calculation and clearness. In this paper, three influencing factors, i.e. the traffic speed of road Section ($1/f_1$), the charging load (f_2) and the number of vehicles in a charging station (f_3) are selected as the inputs of the fuzzy inference process; the outputs of the process are the corresponding weight coefficients of the traffic network w_1, the power grid w_2, and the charging station w_3. The core of fuzzy inference is to determine the fuzzy membership function and the fuzzy rule. This part will introduce the fuzzification and the process of determining the fuzzy rule in detail.

For the fuzzification process, the traffic speed of the road section is fuzzified into three fuzzy sets: "low speed (L)", "medium speed (M)", and "high speed (H)"; the charging load is fuzzified into four fuzzy sets: "low load (L)", "medium load (M)", "high load (H)" and "extra-high load (EH)"; the number of vehicles in a charging

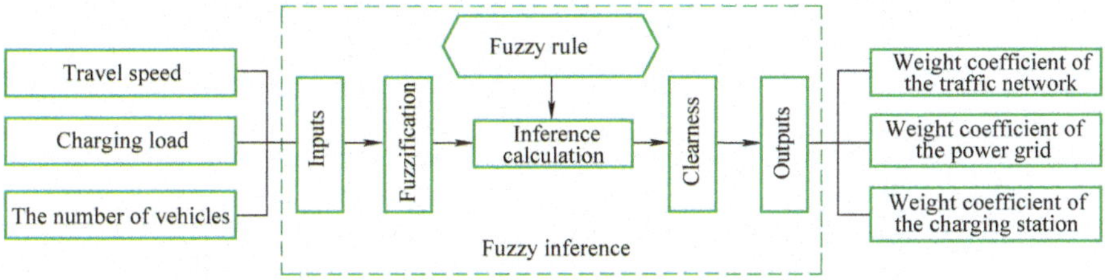

Fig. 8.5 Flow chart of determining the weight coefficient of objective function based on fuzzy logic

station is fuzzified into three fuzzy sets: "small (S)", "medium (M)" and "big (B)"; since the operation of the power grid becomes unsafe when the load is too high, the load on the side of the power grid when a safety problem occurs is set as "extra-high load (EH)" in addition to "high load (H)". When the load is "extra-high (EH)", special attention should be paid. Input and output membership functions are determined based on the existing research results [7]. In order to ensure the simplicity and practicability of fuzzy logic reference process, the triangular membership function and the trapezoidal membership function are used as the input and output membership function under the premise that the optimization effect of the proposed charging scheduling strategy is achieved. The process of determining the membership function of input and output variables is described as follows.

The first is the membership function of road section traffic speed. The eigenvalue of the membership function is determined by counting the average value of the road section traffic speed in each period under different weight coefficients. The average road section traffic speed in morning and evening peak hours is taken as the critical value of the "low speed (L)" membership function, which is also the starting value of the "medium speed (M)" membership function; the average road section traffic speed in off-peak hours is taken as the critical value of the "medium speed (M)" membership function, which is also the starting value of the "high speed (H)" membership function; the average road section traffic speed at night hours is taken as the critical value of the "high speed (H)" membership function, which is also the end point of the "medium speed (M)" membership function. The final membership function of traffic speed in the road section is shown in Fig. 8.6.

The simulation result of the membership function of charging load shows that the existing power distribution network is in danger when the charging load is 20 MW, and all the power grids are in danger when the charging load reaches 23 MW. Therefore, 20 MW and 23 MW are set as the starting and critical values of the "extra-high load (EH)" membership function respectively. The determination process of the "high load (H)", "medium load (M)", and "low load (L)" membership function is similar to that of the road section traffic speed membership function. The average charging loads in peak hours, off-peak hours and night hours are respectively set as the critical values of the "low load (L)", "medium load (M)", and "high load (H)" membership functions, and the starting and ending values of each set are determined on the same basis as the membership function of road section traffic speed. Charging vehicles are

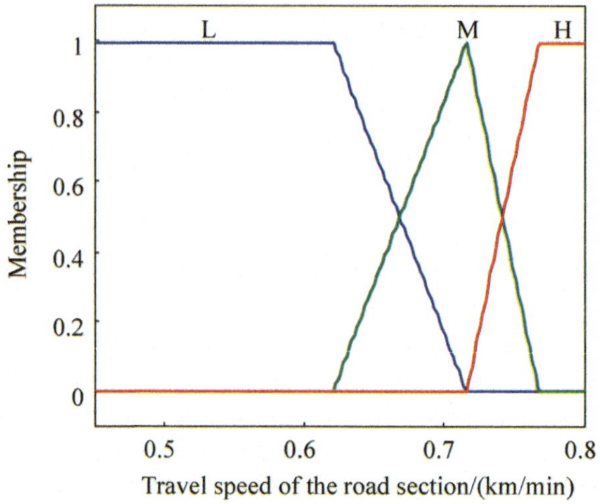

Fig. 8.6 Membership function of road section traffic speed

divided into private cars and taxis. These two types of electric vehicles are charged at different charging stations, so their membership functions of charging load need to be calculated respectively. The obtained membership functions of private cars and taxis are shown in Fig. 8.7a and b.

To determine the membership function of the number of vehicles in the charging station, the eigenvalue is set according to the capacity of the charging station. 0.3, 0.5 and 0.7 times of the station capacity are respectively taken as the critical values of the membership function of the "small (S)", "medium (M)" and "big (B)" number of vehicles. The membership functions of the number of vehicles in private car charging stations and taxi charging stations are respectively shown in Fig. 8.8a and b.

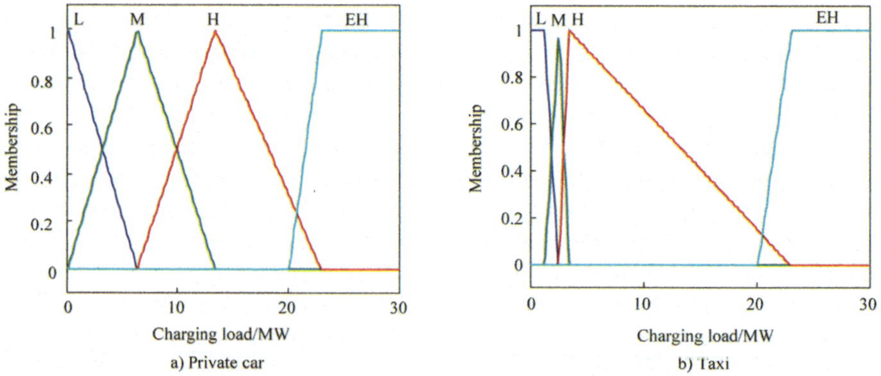

Fig. 8.7 Membership function of charging load

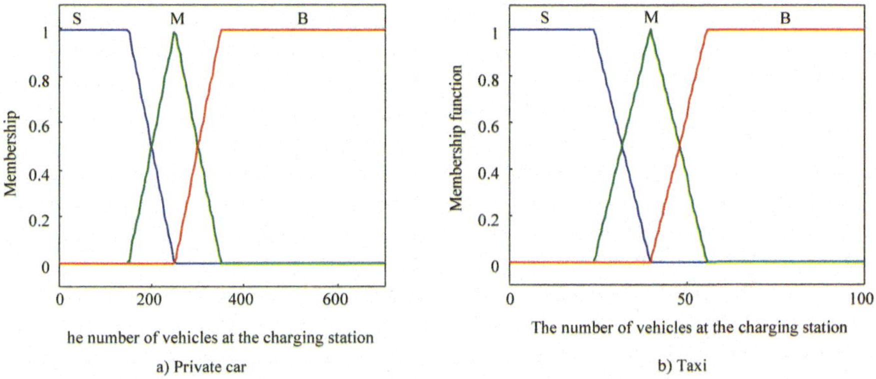

Fig. 8.8 Membership functions of the number of vehicles in charging stations

As for the membership function of the output variables, since the weights on the traffic side and the charging station side only have three levels, 0.3, 0.5 and 0.7 are set as their eigenvalues; the weight of the charging load has four levels, so 0.2, 0.4, 0.6 and 0.8 are set as their eigenvalues. Figure 8.9a–c are respectively the membership functions of the weights on the traffic side, power grid side and charging station side.

In order to ensure the orderly operation of the traffic network and the power grid, 36 fuzzy rules are formulated. The principles of formulating fuzzy rules are as follows:

(1) When the charging load is too high, the weight on the power grid side shall be the main weight, and the other two weights shall be smaller to ensure the safe operation of the power grid system.
(2) When the charging load is not too high and the vehicle velocity is low, the weight on the traffic side is set higher, and the other two weights shall be smaller to ensure the efficient operation of the traffic system.
(3) When the charging load is not too high and the vehicle velocity is not too low, the weight coefficients shall be set according to the normal rules.

(c) Constraints

Firstly, the constraint of EV accessibility is considered. In order to ensure the rationality of the algorithm, the charging station planned for the electric vehicle shall be within its remaining range. If the optimal charging path exceeds the remaining range of the electric vehicle, the nearest charging station shall be selected.

$$Mile\_rem(i) \geq Path\_Length(i) \tag{8.8}$$

where $Path\_Length(i)$ is the length of charging path planned for the electric vehicle i under the optimal strategy.

Then, a reasonable threshold value of charging station is set considering the charging needs of EV users in the area where the charging station is located.

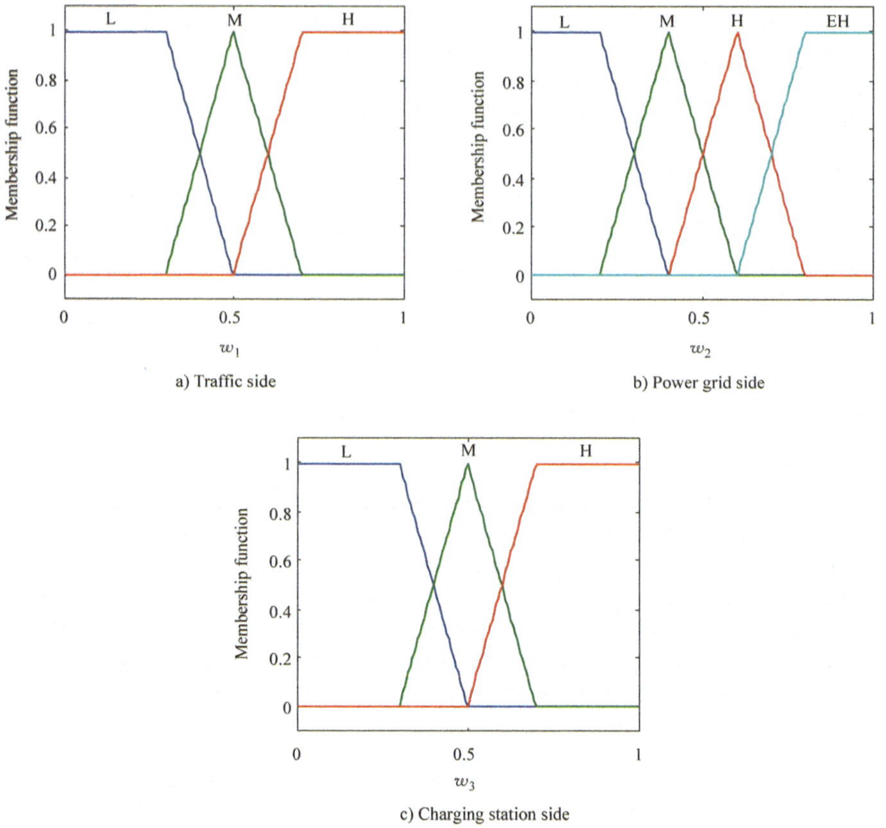

Fig. 8.9 Membership functions of the weights on the traffic side, power grid side and charging station side

$$Load\_Charging(r, t_0) < Power\_threshold(r) \tag{8.9}$$

where $Power\_threshold(r)$ is the load threshold value of the charging station r.

Considering the safe operation of the power grid, the power loss will increase, the voltage profile will change and the load of charging stations will increase after massive electric vehicles are connected to the power distribution network. In order to ensure the safe and reliable operation of the power distribution network, it is necessary to evaluate the economy and safety of the power distribution network of the charging station the EV may go to before charging scheduling, that is, to prejudge the power loss $P_{loss}(Load\_Charging(r, t_0), t_0)$ and voltage offset $V_{shift}(Load\_Charging(r, t_0), t_0)$ of the distribution network, to make sure they don't exceed the rational limit.

For the power loss of the power distribution network, the large-scale application of electric vehicles will cause changes in the load distribution of distribution network nodes connected by EV charging stations while increasing the power loss of the distribution network. Under the premise that the loads of the rest nodes, the access location of charging stations and the capacity of charging stations are definite, the power loss of the distribution network is mainly related to the load of access nodes of charging stations $Load\_Charging$, so the power loss of each power grid can be described as $P_{loss}(Load\_Charging(r, t_0), t_0)$. Based on existing studies [8], it is determined that the power loss meets the following conditions:

$$P_{loss}(Load\_Charging(r, t_0), t_0) < 7\% \tag{8.10}$$

where t_0 is the estimated time when the electric vehicle arrives at the charging station, $Load\_Charging(r, t_0)$ is the estimated load of the charging station r at the moment t_0, and $P_{loss}(Load\_Charging(r, t_0), t_0)$ is the estimated power loss of the distribution network where the charging station r is located at the moment t_0 (obtained by using the load flow calculation software MATPOWER with the estimated load of the charging station).

Similarly, the large-scale charging behavior of electric vehicles will also increase the voltage deviation of the distribution network. The voltage deviation is calculated with Eq. (8.11).

$$V_{shift}(Load\_Charging(r, t_0), t_0) = \sum_{i}^{N} \frac{|V_i - V_0|}{V_0 \cdot N} \tag{8.11}$$

where N is the number of nodes in the power distribution network; V_0 is the voltage of balance node; Vi is the voltage of node i (obtained by using the load flow calculation software MATPOWER with the estimated load of the charging station); $V_{shift}(Load\_Charging(r, t_0), t_0)$ is the estimated voltage deviation of the distribution network where the charging station r is located at the moment t_0.

According to the limit of supply voltage deviation stipulated in *GB/T 12,325-2008 Power quality—Deviation of Supply Voltage*, "the three-phase supply voltage deviation of 20 kV and below shall be ± 7% of the nominal voltage". Therefore, in order to ensure that the voltage offset of the power grid is within a reasonable range, the voltage offset of each node of the power grid is constrained:

$$V_{shift}(Load\_Charging(r, t_0), t_0) < 7\% \tag{8.12}$$

Finally, considering the constraints of the charging process, the charging strategy proposed in this chapter is only for charging electric vehicles whose remaining range is less than N% of the full range or insufficient to reach the destination:

$$Mile\_rem(i) \le N\% \cdot Mile\_ful(i) \tag{8.13}$$

$$Mile\_rem(i) \le Path\_Length(i) \tag{8.14}$$

where $Mile\_rem(i)$ is the remaining range of the electric vehicle i, and $Path\_Length(i)$ is the path length of the electric vehicle i.

(d) Solution of the objective function

Based on the proposed EV charging scheduling strategy, the road traffic network is abstracted into a weighted directed graph $M = (V, S, W)$. Intersections in the road traffic network are node V in the directed graph, the passable road sections in the road traffic network are arc S in the graph, and the weights of each road section are the arc length W in the graph. Each road section $S(p, q)$ corresponds to a road section weight $W(p, q)$, and the charging path is obtained with Dijkstra algorithm. In the optimization process, the comprehensive weight is assigned to the corresponding road section to plan the charging path from the starting point to each charging station with the minimum directed comprehensive weight F. The process of searching the path with the Dijkstra algorithm is shown in Fig. 8.10.

The solution of charging path with the Dijkstra algorithm includes the following steps:

(1) Determining the weight coefficient of each objective. At the beginning of each moment, the traffic speed of all sections of the road traffic network, the load of each charging station in the power grid and the number of vehicles in the charging station are counted, and the three parameters are input into the weight coefficient reference process of the objective function based on fuzzy logic to obtain the weight coefficient of each optimization objective.
(2) Introducing traffic information. The weight of the road section in the road traffic network is the length of the road section at the initial moment, as shown in Eq. 8.15.

$$W(p, q) = a(p, q) \tag{8.15}$$

where $a(p, q)$ is the length of the road section (p, q).

After the introduction of traffic information, the weight of each road section is the time needed to pass the section, and the weight of this moment is turned dimensionless, and the weight coefficient on the traffic side is given to this weight value:

$$\begin{aligned} W(p, q) &= w_1 * a(p, q)/V_{road}\{p, q\}(j) \Big/ a(p, q)/\min f_1 \\ &= w_1 * \min f_1 \Big/ V_{road}\{p, q\}(j) \end{aligned} \tag{8.16}$$

(3) Introducing the power grid and charging station information. The information of the power grid load and charging station are given to the road section near the charging station. At the same time, it's turned dimensionless and multiplied by its corresponding weight coefficient. Then, it's linearly weighted with the weight value on the traffic side:

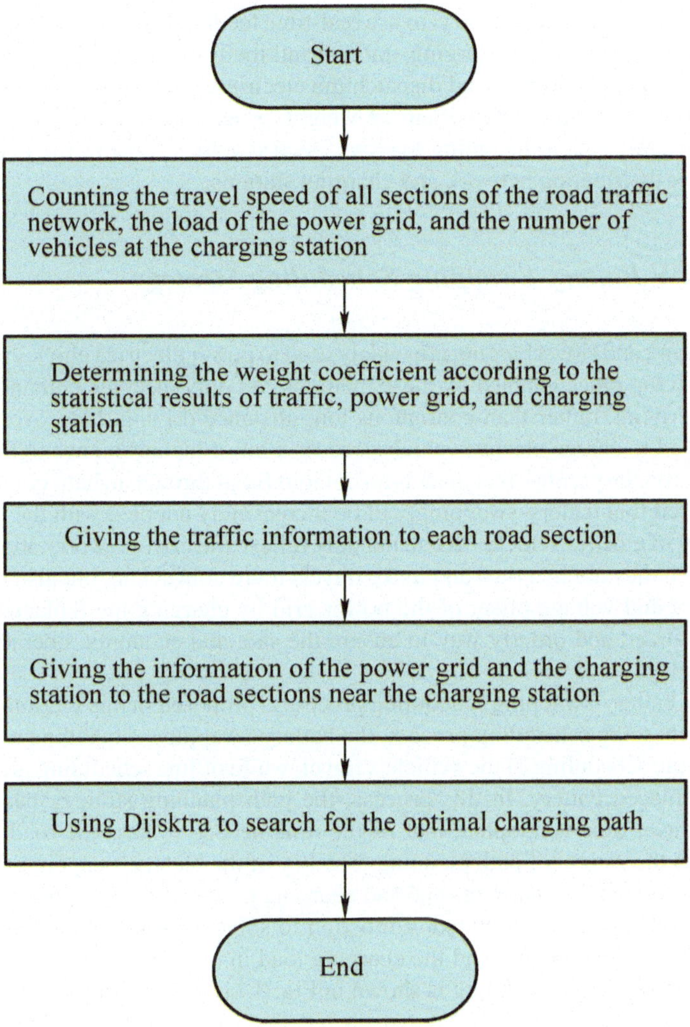

Fig. 8.10 Flow chart of charging path solution with the Dijkstra algorithm

$$W(p,q) = W(p,q) + w_2 * Load\_Charging(r,j)/\min f_2 + w_3 * Num\_Charging(r,j)/\min f_3 \tag{8.17}$$

where j is the current moment, (p,q) is the road section near the charging station r.

(4) Searching the path with the minimum weight with the Dijkstra algorithm. Dijkstra algorithm is used to search the charging path from the current position of the electric vehicle to each charging station with the minimum weight.

The core of this method is to use real-time feedback information of the power distribution network, charging stations and the traffic network to calculate the comprehensive weight and dispatch the electric vehicle to the charging station with the minimum comprehensive weight, so as to realize the optimization of the overall system including massive electric vehicles, the traffic network, the power distribution network and charging stations.

8.1.4 EV Battery-Swapping Scheduling Strategy

Fast charging and slow charging are widely used to power electric vehicles nowadays, but the driving range enabled by these methods can only meet the demand of short-distance driving, rather than continuous long-distance driving. Moreover, the long time needed to charge electric vehicles causes a lot of inconvenience to the drivers. Therefore, on the premise that there is no obvious breakthrough in battery technology, it's expected that battery-swapping will be increasingly adopted with the increasing popularity of electric vehicles. It can not only relieve the drivers' worry about driving range when they are in a rush and make travels more convenient, but also reduce the power loss and voltage offset of the power grid by charging the replaced batteries in a centralized and orderly way to ensure the safe and economic operation of the power grid.

An EV battery-swapping scheduling process is proposed in this section. Different from the charging scheduling process, the battery-swapping scheduling only covers the real-time scheduling of the vehicle's travel, while offline scheduling is conducted for the replaced battery. In this process, the path planning strategy based on the shortest travel time is adopted, that is, the time needed to pass the road section is selected as the weight of path planning. The key of the battery-swapping scheduling strategy is to plan the charging time and charging power of the swapped battery and design the appropriate optimization function to solve appropriate charging periods with the above information and the charging load in the power grid. The flow chart of battery-swapping scheduling is shown in Fig. 8.11, which includes the following steps.

(1) Information initialization. The operation information of the electric vehicle such as driving range, the road traffic network operation information such as the congestion rate of road section, and the power grid operation information such as power grid load are initialized. The battery-swapping path is then recommended for the electric vehicle and the charging of the swapped battery is scheduled based on the above information.
(2) Recommending battery-swapping stations and plan battery-swapping paths for the battery-swapping vehicle. Based on the congestion rate of each road section, Eq. (8.4) is used to calculate the real-time traffic speed of each road section and then to calculate the time needed to pass the road section. With the road passage time taken as the weight and based on the Dijkstra algorithm, the path from the current position of the vehicle to the battery-swapping station with the minimum

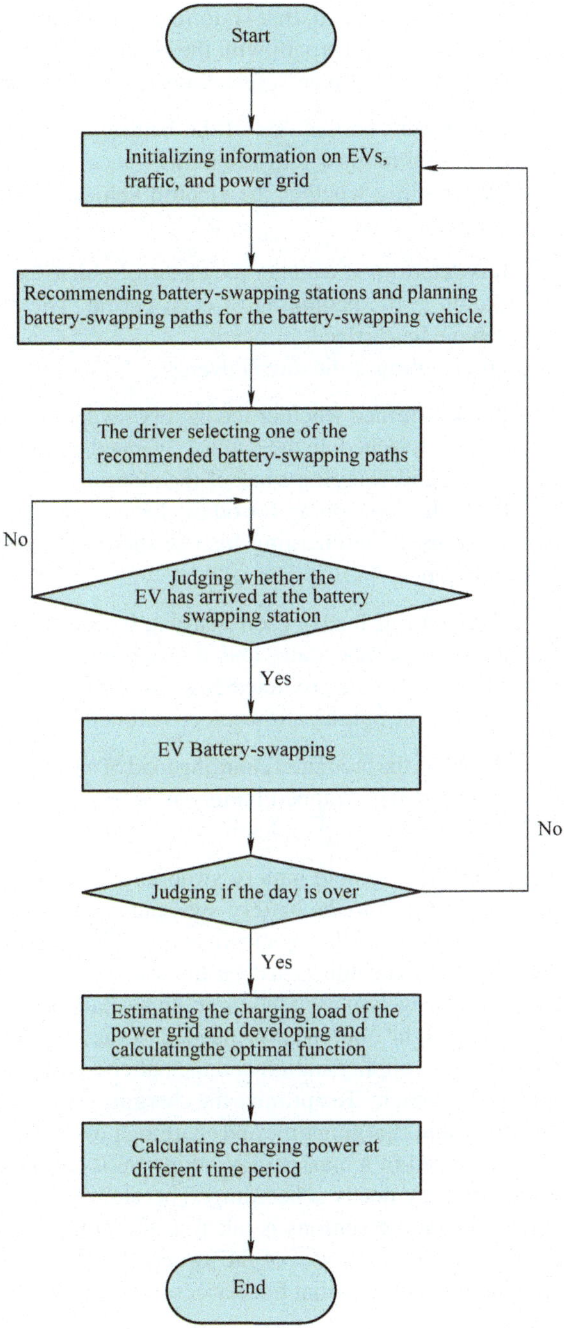

Fig. 8.11 Flow chart of EV battery-swapping scheduling

weight is searched, that is, to recommend the battery-swapping station and the battery-swapping path with the shortest driving time to the vehicle.

(3) Allowing the driver to select a battery-swapping path

The driver chooses one of the battery-swapping paths planned by the system according to his actual situation.

(4) Determining whether the electric vehicle has arrived at the battery-swapping station

It is determined whether the electric vehicle has arrived at the battery-swapping station at the beginning of each period. The judgment is completed once the vehicle has arrived.

(5) Determining if the day is over.

It's determined whether the battery-swapping station is closed or not. If closed, prepare to schedule the battery swapped within the day and arrange the charging power and charging time. If the station is not closed, it continues to schedule the battery-swapping for other electric vehicles from the initialization stage.

(6) Estimating the charging load of the power grid and developing the objective function

The charging load of each charging station on the next day is predicted according to the historical data, and the optimization objective function is developed according to the predicted charging load.

(7) Solving the optimization objective function of battery-swapping load scheduling

Based on the predicted charging load of each grid, the charging time and charging power of the swapped battery is solved by using the optimization function of the battery-swapping load.

The electric vehicle battery-swapping scheduling mainly includes two aspects: planning the path to the battery-swapping station for the vehicle and optimizing the charging process of the swapped battery. The battery-swapping path is planned with the Dijkstra algorithm based on the same principle as the charging path planning. However, the weight used in searching the path is no longer the "comprehensive optimization weight", but the time needed to pass the road section, that is, to recommend the battery-swapping station and the battery-swapping path with the shortest driving time to the vehicle. To optimize the charging process of the swapped battery is to optimize its initial charging time and charging power. Since a huge amount of batteries are to be swapped in a massive EV adoption, it requires very high computing capacity if each battery needs scheduling. Therefore the power grid accessed by charging/ battery-swapping stations is taken as the object of scheduling study. The charging time and charging power of the swapped battery is optimized based on the load of the charging stations that have access to the power grid.

(a) Determination of the objective funcion

It is the purpose of implementing the ordered battery-swapping scheduling to smooth the overall load of the power grid. The objective function of optimizing the power grid load is formulated, and the objective function based on the minimum standard deviation of the load and the minimum Peak-to-Valley (PV) of the load are respectively formulated [6].

The objective function based on the minimum standard deviation of the load means the optimization objective is the standard deviation of charging load, as shown in Eq. (8.18):

$$F = \min \sqrt{\frac{1}{n}\sum_{j=1}^{n}\left(P_{\mathrm{j}}'(\mathrm{r}) - \overline{P}(\mathrm{r})\right)^2} \tag{8.18}$$

where n is the number of time periods; $\overline{P}(\mathrm{r})$ is the daily average load of the power grid r; $P_{\mathrm{j}}'(\mathrm{r})$ is the total load of the power grid r in the period j after the scheduling.

The objective function based on the minimum PV of the load means the optimization objective is the PV of the charging load, as shown in Eq. (8.19):

$$F = \min[\max(P_{\mathrm{j}}'(\mathrm{r})) - \min(P_{\mathrm{j}}'(\mathrm{r}))] \tag{8.19}$$

where $\max(P_{\mathrm{j}}'(\mathrm{r}))$ and $\min(P_{\mathrm{j}}'(\mathrm{r}))$ are the peak load and valley load of the power grid r after the scheduling.

(b) Constraints

In terms of traveling, the battery-swapping station planned for the electric vehicle should be within its remaining range as shown in Eq. (8.8). At the same time, the condition for battery-swapping is set as "the remaining range is less than $N\%$ of the maximum range or the destination of the electric vehicle is not within its remaining range, as shown in Eqs. (8.13) and (8.14).

Since the battery-swapping station can charge the swapped battery during the valley hours, the load threshold of the battery-swapping station is not set. However, to ensure the safe and economic operation of the power grid, the voltage offset and power loss of the power grid are also constrained, as shown in Eqs. (8.10) and (8.12).

(c) Solution of the objective function

In this section, the particle swarm optimization algorithm is used to solve the objective function of battery-swapping scheduling. The steps of solution are as follows.

(1) Initializing the range, number, and acceleration of particles, that is, to initialize the value range of the battery-swapping power at each moment, the number of moments that needs the solution of the battery-swapping power, and the variation speed of the battery-swapping power. Since there is no constraint on the battery-swapping power above, a large range is set.
(2) Initializing the position of particles, that is, to initialize the charging power at each moment according to the predicted charging load. The normal particle swarm optimization algorithm initializes each particle in a random way. However, in order to speed up the convergence rate of the algorithm, in the initial

stage, the battery-swapping load at each moment is preliminarily set according to the predicted grid load at each moment and the principle of smoothing load curve.

(3) Evaluating the fitness function value of each particle. First, the initial charging power at each moment is substituted into the objective functions (8.18) and (8.19) to respectively calculate the standard deviation and PV of the load at this moment. Then, the updating speed and position of each particle are calculated according to the acceleration of the particle, that is, to update the load at each moment and calculate the fitness function value at this moment.

(4) Determining whether the result converges. Determine whether the result of this calculation converges compared with the last one. If it converges, the calculation is stopped. If it does not converge, continue to update the velocity and position of the particles.

(5) Determining whether the number of iterations has reached the upper limit N. In order to prevent an excessive amount of calculation, the maximum number of calculation is usually set when the particle swarm optimization algorithm is used to solve the problem. If the result does not converge after the number, the calculation will be stopped and the optimal value of the current position will be selected as the final optimization result. The solution process is shown in Fig. 8.12.

8.2 Establishing the Simulation Platform for EV Charging/ Battery-Swapping Scheduling

In order to verify the effectiveness of the large-scale EV charging/battery-swapping scheduling strategy for the comprehensive optimization of traffic and power grid proposed in this chapter, a simulation system for large-scale EV charging/battery-swapping scheduling is built in this section based on the Matlab simulation platform. Under the premise of comprehensive optimization of traffic and power grid, the system simulates and analyzes the charging/battery-swapping scheduling of different types of electric vehicles. The simulation platform includes different types of electric vehicle model, road traffic network model, power grid model, charging/ battery-swapping station model.

8.2.1 General Design of the Simulation Platform

A simulation platform based on MATLAB software environment is built in this section. MATLAB is used to establish the overall system of large-scale EV charging/ battery-swapping scheduling. MATPOWER, a simulation toolbox in MATLAB, is

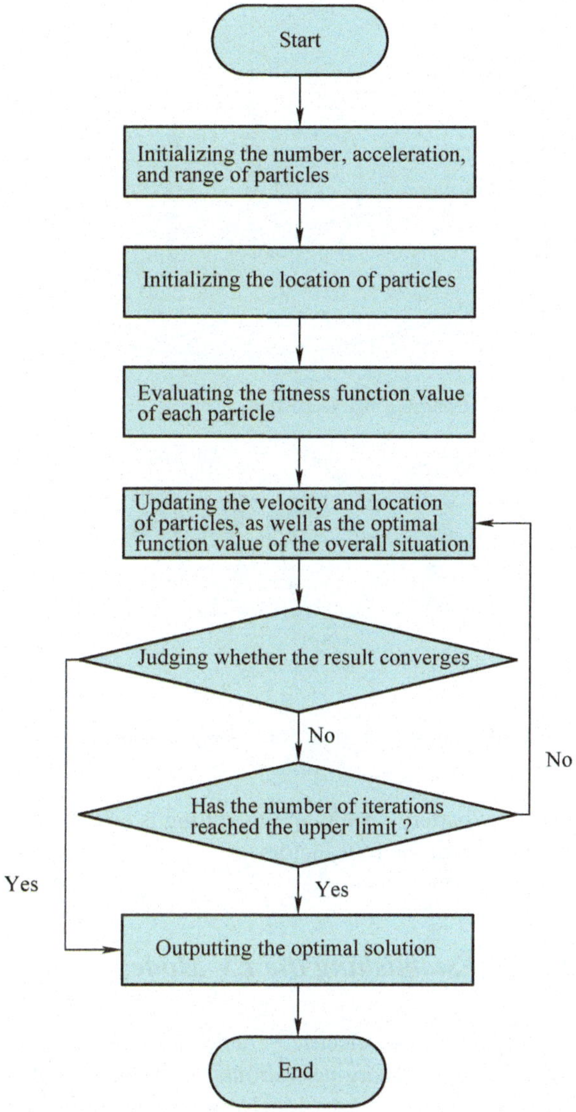

Fig. 8.12 Flow chart of the particle swarm optimization algorithm

used to conduct power flow analysis of the power grid to realize real-time comprehensive optimization of the traffic and the power grid. The established simulation system of EV charging/battery-swapping scheduling is shown in Fig. 8.13.

The attribute characteristics of electric vehicles, the traffic network, the power grid and charging stations are extracted respectively, and the interaction mechanism of the four factors is analyzed to establish the EV model, the traffic network model, the power grid model, and the charging station model. The simulation system of EV

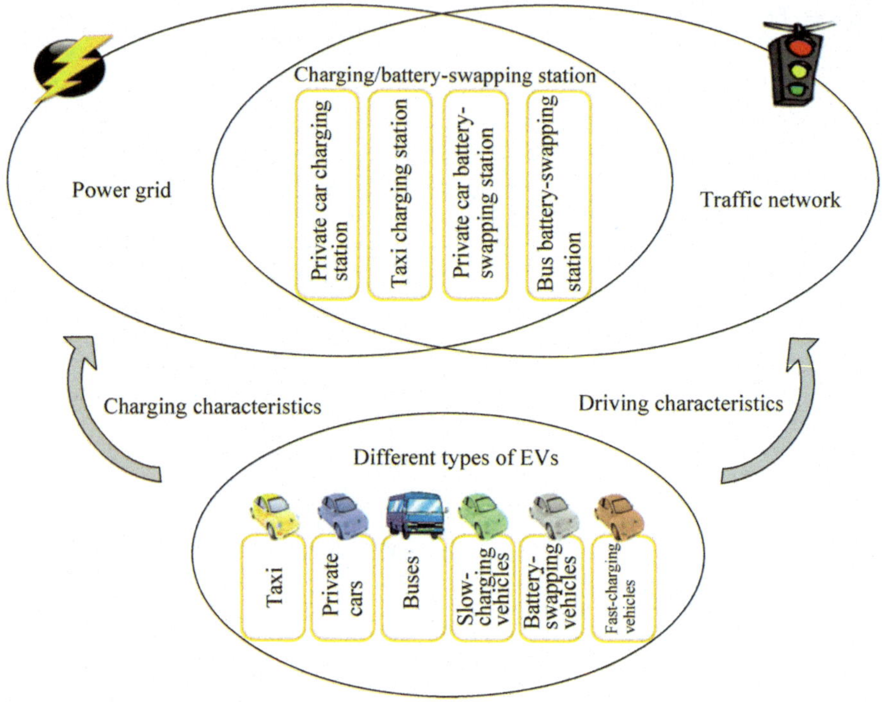

Fig. 8.13 Simulation system of EV charging/battery-swapping scheduling

charging/battery-swapping scheduling is established through the real-time dynamic interaction of the four factors.

8.2.2 Establishing the EV Model

As a typical load, electric vehicles have a direct impact on the traffic system, the power grid and charging stations. On the one hand, the driving behavior of massive electric vehicles causes local traffic congestion; on the other hand, the charging behavior of massive electric vehicles causes excessive load on the grid, obvious voltage offset, excessive power loss on the grid, and other grid safety and economic problems; different characteristics of different types of electric vehicles also have different impacts on the traffic and the power grid. Therefore, it is necessary to model the charging characteristics and driving characteristics of electric vehicles when discussing the impact of EV access on the traffic and the power grid. Considering that electric vehicles are divided into different categories and correspond to different charging and driving characteristics in practice, EVs are modeled from three dimensions (type of EV, charging characteristics, and driving characteristics)

Table 8.1 Characteristics of electric vehicles

Symbol	Meaning
EV_Type	EV type
Charge	Charging characteristics
Travel	Driving characteristics

Table 8.2 Type characteristics of electric vehicles

Symbol	Meaning
Function_Type	Function characteristics
Power_Type	Power supply characteristics

to obtain the characterization matrix of EVs, as shown in Eq. (8.20). Its value and significance are shown in Table 8.1.

$$EV = \{EV\_Type, Charge, Travel\} \tag{8.20}$$

In the simulation, electric vehicles are classified by function characteristics and power supply characteristics, and their values and significance are shown in Table 8.2.

In terms of function characteristics, electric vehicles can be divided into private cars, taxis and buses. Because buses of different functions have different driving characteristics such as driving time and routes, electric vehicles of different functions must be described in different categories. $Function\_Type$ represents the type of function, 1 represents private cars, 2 represents taxis, and 3 represents buses. Private cars have the most diverse functions, including commuting between work place and residence in the morning and evening rush hours on weekdays, as well as random travels with uncertain times and routes; taxis are large-scale commercial vehicles with long running times and relatively flexible driving routes; buses provide public transport services, and their driving routes and running times are relatively fixed. In this section, the bus routes are selected based on the actual bus routes inside the Third Ring Road of Beijing. 4 round-trip routes are finally selected as the research objects to ensure the uniform distribution of buses. In addition, the starting times of daily operation and release rule are also consistent with the actual arrangement of the bus company, for example, leaving in the morning and returning at night, and the release intervals of the busses are adjusted according to the change of passenger flow in different hours.

In terms of the type of power supply, electric vehicles can be divided into three categories: fast charging, slow charging and battery-swapping. $Power\_Type$ represents the type of power supply, 1 represents fast charging, 2 represents slow charging, and 3 represents battery-swapping. Fast-charging vehicles have the greatest impact on the safe and economic operation of the power distribution network and charging stations because of their high requirements on charging power, and fast-charging is the most widely adopted charging solution. The impact of slow-charging electric vehicles on the grid is relatively mild, and they can be charged at slow speeds using regular urban power without the need to go to a charging station. Therefore, the charging scheduling strategy proposed in this paper only covers fast-charging EVs, not slow-charging EVs. For battery-swapping electric vehicles, as there is no need to charge the battery immediately after the battery is swapped, they can be charged during low-peak hours to achieve "peak-shaving" of the power grid.

Based on the classification rules described above, the representation matrix of the type of electric vehicle can be obtained according to its function and power supply mode:

$$EV\_Type(i) = (Function\_Type, Power\_Type) \tag{8.21}$$

where i represents the ith electric vehicle.

Based on the classification by function and power supply mode, there are $3 \times 3 = 9$ types of electric vehicles. However, a further reasonable selection should be made according to the actual situation. In fact, although fast charging and slow charging are widely used in electric private cars, the driving ranges supported by those methods can only meet the demand of short distance driving, rather than continuous long-distance driving. Therefore, on the premise that there is no obvious breakthrough in battery technology, it's expected that battery-swapping will be increasingly adopted with the increasing popularity of electric vehicles. It can not only relieve the drivers' worry about driving range when they are in a rush and make travels more convenient, but also help reduce the power loss of the power grid by charging the replaced batteries in a centralized way to ensure the safe and economic operation of the power grid. Therefore, the power supply modes of private cars, as defined, include fast charging, slow charging and battery-swapping. As for taxis, most of the electric taxi demonstration projects that have been carried out in many cities in China adopt fast charging as the power supply mode, so fast charging is taken as the main power supply mode of taxis in this paper. As for buses, considering that its operation must strictly abide by the scheduling arrangement of the bus company, battery-swapping is adopted as the power supply method because it's most time-saving. In summary, 5 types of electric vehicles are finally chosen as the study objects, as shown in Table 8.3.

Table 8.3 Table of vehicle types

Power supply		Function		
		Function_Type		
		1-Private car	2-Taxi	3-Bus
Power_Type	1-Fast charging	(1, 1)	(2, 1)	–
	2-Slow charging	(1, 2)	–	–
	3-Battery-swapping	(1, 3)	–	(3, 3)

Table 8.4 Characterization matrix of electric vehicles' charging characteristic

Symbol	Meaning
CS	Charging station code
ArriveTime	Time to arrive at the charging station
WaitTime	Waiting time for charging
Power_Charge	Charging power
Power_Consume	Power consumption
ChargeTime	Charging time
EndTime	Ending time of charging

(a) Charging/battery-swapping characteristics of different EV types

In order to describe the charging/battery-swapping behaviors of the electric vehicles mentioned above, their characteristics are modeled in this paper.

For charging vehicles, their charging behavior mainly affects the operation of charging stations. Therefore, the characteristic quantities selected in this section include the charging station code, time to arrive at the charging station, waiting time for charging, charging time, charging power and power consumption to describe the charging behavior of electric vehicles. The characterization matrix of electric vehicles' charging characteristic can be obtained:

$$\begin{aligned} Charge(i) = (CS, ArriveTime, WaitTime, Power \\ \_Charge, Power\_Consume, ChargeTime, EndTime) \end{aligned} \tag{8.22}$$

The meanings of all symbols in the equation are as shown in Table 8.4. The detailed description of each characteristic is provided as follows.

Charging station code represents the code of the charging station recommended by the system for the electric vehicle. Charging station code is recommended to the driver through the system, and then generated after the driver selects the charging station.

$$CS(i) = \begin{cases} j, j \in [1, n]; & \text{vehicles need charging} \\ 0; & \text{vehicles donnot need charging} \end{cases} \tag{8.23}$$

i is the code of the electric vehicle; j is the code of the charging station; n is the number of charging stations; for vehicles that do not need to be charged, the charging station code is 0.

The time to arrive at the charging station refers to the time required for an electric vehicle to arrive at the charging station selected by the driver from the current position according to the planned path, which is calculated by the Dijkstra algorithm. The weight used in path search is the time needed to pass the road section.

Waiting time for charging refers to the difference between the end time of charging and the time of arriving of the last vehicle that arrives at the charging station, as shown in Eq. (8.24).

$$WaitTime(i) = EndTime(k) - ArriveTime(i) \tag{8.24}$$

k is the code of the last vehicle that finishes charging after the electric vehicle i arrives at the charging station.

Charging power is a fixed value for an electric vehicle, and its value is randomly determined within a certain range during initialization. The charging time is short for fast-charging electric vehicles, so their charging power is larger, usually 25–60 kW.

The power consumption is related to the function of electric vehicles. For electric private cars and electric taxis that get power by charging, the power consumption is smaller, usually 0.15–0.2 kW.

Charging time is calculated with the total battery capacity ($Battery\_Capacity$), the remaining battery capacity (SOC_rem) of the vehicle and the charging power of the charging station, as shown in Eq. (8.25).

$$ChargeTime = \frac{Battery\_Capacity - SOC\_rem}{Power\_Charge} \tag{8.25}$$

The charging end time is obtained by adding the time of the EV arriving at the charging station, the waiting time for charging and the charging time, as shown in Eq. (8.26).

$$EndTime = ArriveTime + WaitTime + ChargeTime \tag{8.26}$$

Battery-swapping vehicles are similar to charging vehicles. However, there is no waiting for battery-swapping vehicles because they can leave the battery-swapping station after the battery is swapped. Therefore, the characteristic quantities of battery-swapping vehicles only include the code of the battery-swapping station, the time of arrival at the battery-swapping station, the charging power, and the power consumption. Therefore, the characterization matrix of the battery-swapping characteristics can be obtained:

$$Battery\_switch(i) = (BSS, ArriveTime, Power\_Charge, Power\_Consume) \tag{8.27}$$

The meanings of all symbols in the equation are shown in Table 8.5.

Table 8.5 Characterization matrix of EV battery-swapping characteristics

Symbol	Meaning
BSS	Code of the battery-swapping station
ArriveTime	Time of arrival at the battery-swapping station
Power_Charge	Charging power
Power_Consume	Power consumption

The code of the battery-swapping station represents the code of the battery-swapping station recommended by the system for the electric vehicle. The code of the battery-swapping station is recommended to the driver through the system, and then generated after the driver selects the station.

$$BSS(i) = \begin{cases} j, j \in [1, n]; & \text{vehicles need charging} \\ 0; & \text{vehicles donnot need charging} \end{cases} \tag{8.28}$$

i is the code of the electric vehicle; j is the code of the battery-swapping station; n is the number of battery-swapping stations; for vehicles that do not need to be charged, the battery-swapping station code is 0.

The time to arrive at the battery-swapping station refers to the time required for an electric vehicle to arrive at the battery-swapping station from the current position according to the planned path, which is calculated by the Dijkstra algorithm. The weight used in path search is the time needed to pass the road section. The charging power in the battery-swapping characteristic refers to the charging power of the replaced battery at the battery-swapping station. Similarly, it is randomly determined within a certain range, and it is a constant value for an electric vehicle. Replaced batteries are charged in a centralized way with relatively gentle charging power, which is 2–4 kW. The power consumption is related to the function of the electric vehicle. For battery-swapping electric private cars, the power consumption is smaller and is 0.15–0.2 kW. For electric buses, the power consumption is higher due to their sizes, which is 0.8–1.2 kW.

(b) Driving characteristics of different types of electric vehicles

When an EV travels in the road traffic network, it's first determined whether the vehicle has charging demand according to its remaining capacity. For an EV with charging demand, the charging/battery-swapping should be planned according to its position in the road traffic network and driving speed; for an EV without charging demand, the system plans the path from the start position to the destination by using information including current vehicle position, vehicle velocity, and destination. In summary, it plans the charging/battery-swapping path for EVs with charging demand, and path towards the destination for other EVs. The information required in the process of EV travel scheduling is described in Fig. 8.14.

The vehicle's initial position, current position, destination, driving velocity, driving path, path length, range and power consumption per kilometer are selected as characteristic quantities to describe the driving characteristics of the EV:

$$\begin{aligned} Travel(i) = (&Loc\_Depart, Loc\_Now, Destination, Path, Path \\ &\_CS, Velocity, Path\_Length, Mile\_Rem, Mile \\ &\_Ful, Power\_Consume) \end{aligned} \tag{8.29}$$

The starting point of the vehicle is the initial position at which the electric vehicle enters the road traffic network. It is randomly determined from the nodes of the road

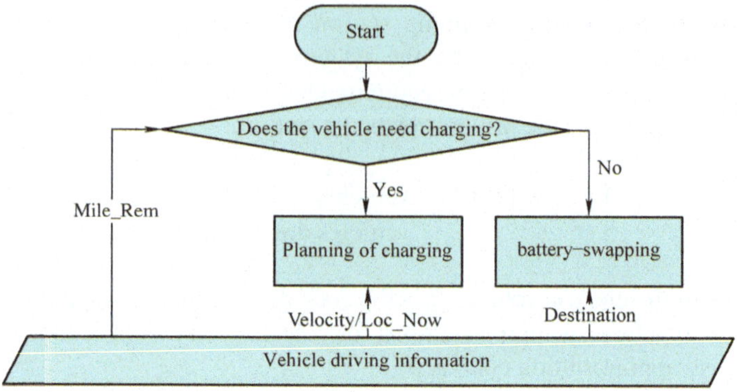

Fig. 8.14 EV travel scheduling

traffic network to be the starting point of the path that the vehicle has already traveled through.

$$Loc\_Depart(i) = j, j \in [1, n]; \tag{8.30}$$

i is the code of the electric vehicle; j is the starting point of the vehicle; n is the number of nodes in the road network;

The position of the vehicle at the current moment. When the length of the charging/ battery-swapping path or the length of the driving path is planned, the driving distance is calculated according to the vehicle's driving velocity and driving time, so as to obtain the current position of the vehicle. The destination of the vehicle is randomly generated in the road traffic network at the initial moment when the electric vehicle is put into operation.

$$Destination(i) = k, k \in [1, n]; \tag{8.31}$$

k is the destination of the vehicle.

In the definition of the driving path, for a vehicle that does not need to be charged, the driving path is the driving route planned by the system from its current position to the destination, including each node of the corresponding road. For a vehicle that needs to be charged, the driving path is the driving route planned by the system towards the charging station.

The velocity of the vehicle is the traffic speed of the road section $V_{road}\{p, q\}(j)$ where the vehicle is located, in which i is the code of the vehicle.

$$Velocity(i) = V_{road}\{p, q\}(j) \tag{8.32}$$

Path length refers to the total path length obtained by superposing the length of the section between each node in the driving or charging path planned for the vehicle. For example, the solution process for the length of the driving path $Path(i)$ of the vehicle i is shown in Eq. (8.33), in which n is the number of nodes in the driving path.

$$Path\_length(i) = \sum_{k=1}^{n} a(Path\{i\}(k), Path\{i\}(k-1)) \quad (8.33)$$

The mileage traveled refers to the distance traveled by the vehicle after the driving path or charging path is planned, as shown in Eq. (8.34), in which $t\_interval$ is the length of time per unit period.

$$Mile\_traveled(i, j) = Velocity(i) * t\_interval \quad (8.34)$$

The remaining mileage refers to the mileage supported by the current remaining battery capacity of the vehicle, as shown in Eq. (8.35). Measured by the difference between the remaining mileage of the vehicle at the last moment $Mile\_rem(i, j-1)$ and the distance traveled of the vehicle at the current moment $Mile\_traveled(i, j)$, the remaining mileage is used to estimate whether the electric vehicle can travel normally and can reach the destination.

$$Mile\_rem(i, j) = Mile\_rem(i, j-1) - Mile\_traveled(i, j) \quad (8.35)$$

The maximum mileage is the maximum distance that an electric vehicle can travel when the battery is fully charged. In the simulation process, the maximum mileage of the electric vehicle is generally set as 130–160 km.

The power consumption/km refers to the amount of battery power consumed by an electric vehicle to travel one kilometer. For the same electric vehicle, its power consumption/km is set as a constant. The power consumption/km of private cars and taxis is lower, about 0.15–0.2 kWh, while the power consumption/km of buses is higher, about 0.8–1.2 kWh. The symbols and meaning of the driving characteristics of electric vehicles is shown in Table 8.6.

8.2.3 Establishing the Charging/Battery-Swapping Station Model

As the link between electric vehicles and the power grid and the traffic network, the position and power supply characteristics of charging stations and battery-swapping stations can be modeled to establish the connection between electric vehicles and the traffic network and the grid. In this section, EV charging stations, electric taxi charging stations, electric private car battery-swapping stations and electric

Table 8.6 Driving characteristics of electric vehicles

Symbol	Meaning
Loc_Depart	Starting point of the vehicle
Loc_Now	Current position of the vehicle
Destination	Destination of the vehicle
Path	Driving path planned for the vehicle without charging demand
Path_CS	Charging path planned for the vehicle with charging demand
Velocity	Driving velocity
Path_Length	Length of road section
Mile_traveled	Mileage already traveled
Mile_Rem	Remaining mileage
Mile_Ful	Maximum mileage
Power_ Consume	Electric power consumed by an EV per kilometer

bus battery-swapping stations are modeled, including setting the geographic location of the charging/battery-swapping stations, the nodes in the charging/battery-swapping stations' access to the power grid, and the power supply characteristics of the charging/battery-swapping stations.

(a) Charging station model

The connection between the charging station and the traffic network and power grid can be established by setting the geographical location of the charging station in the road traffic network and its node location in the access to the power grid. Table 8.7 shows the symbols and meaning of the location information of the charging station.

It requires setting the geographical location of the charging station to establish the connection between the station and the traffic network. The node in the road traffic network is used to represent the geographical location of the charging station, as shown in Eq. (8.36).

$$Load\_CS(r) = [j] \tag{8.36}$$

r represents the code of the charging station, and j represents the node location of the charging station in the road traffic network.

In order to establish the connection between the charging station and the power grid, the access of the charging station to the power grid and the node of the power

Table 8.7 Characteristics of the location information of the charging station

Symbol	Meaning
Location_CS	Location in the road traffic network
Connection_CS	Location in the power grid

Table 8.8 Power supply characteristics of the charging station

Symbol	Meaning
Load_Charging	Charging load
Num_Charging	Number of vehicles being charged
Capacity_CS	Capacity of the charging station

grid for the access are set as shown in Eq. (8.37).

$$Connection\_CS(r) = [g, k] \tag{8.37}$$

g represents the code of the power grid that the charging station accesses to, and k represents the code of the grid of the power grid g at which the charging station r accesses to the power grid.

Because the clustered charging behavior of electric vehicles upon arriving at the charging station will increase the charging load, and the charging station is connected to the nodes of the power distribution network, the load of the charging station directly affects the operation of the power grid. Therefore, it is necessary to describe the power supply characteristics of the charging station. The characteristic quantities selected in this section include the charging load, the number of charging vehicles, the capacity of the charging station and the charging threshold:

$$\begin{aligned} &Charge\_Station \\ &\quad = \{Load\_Charging, Num\_Charging, Capacity\_CS, Power\_Threshold\} \end{aligned} \tag{8.38}$$

The power supply characteristics of the charging station and their corresponding symbols are shown in Table 8.8.

The charging load refers to the sum of the charging power of the vehicles being charged in the charging station at the current moment:

$$Load\_Charging(r, j) = \sum Power\_Charge(r, i) \tag{8.39}$$

where $Load\_Charging(r, j)$ is the charging power of the charging station r at the moment j; $Power\_Charge(r, i)$ is the charging power of the charging station r at the moment j.

The number of vehicles at the charging station refers to the number of electric vehicles currently parked at the charging station, including those currently being charged and those waiting to be charged. The number of vehicles at each charging station in each period is measured by calculating the period of each EV staying at the charging station. Equation (8.40) is used to judge whether a charging station is planned for the electric vehicle.

$$
\begin{aligned}
&EV\_CS(r, Arrive(i) : EndTime(i)) = 1; \quad \text{vehicles being planned for charging station}\\
&EV\_CS(r, Arrive(i) : EndTime(i)) = 0; \quad \text{vehicles not being planned for charging station}
\end{aligned}
\tag{8.40}
$$

$$
Num\_Charging(r, j) = \sum EV\_CS(r, j) \tag{8.41}
$$

$EV\_CS(r, Arrive(i){:}EndTime(i))$ represents the charging code of the electric vehicle i during its stay at the charging station r, which is used to count the number of vehicles in each charging station during in each period; $j \in \{Arrive(i){:}EndTime(i)\}$ is the time that the electric vehicle spends at the charging station.

The capacity of the charging station refers to the number of charging EVs that can be accommodated at the station at the same time, which is determined according to the amount of EVs put in the station. Charging threshold refers to the maximum charging power that can be provided by the charging station. The capacity of the charging station and the charging threshold are variable, which depend on the number of electric vehicles put in the station. In the simulation process, the charging power distribution of each charging station can be made more appropriate by adjusting the charging threshold.

(b) Battery-swapping station model

The connection between the battery-swapping station and the traffic network and the power grid can be established by setting the geographical location of the battery-swapping station in the road traffic network and its node location in the access to the power grid. Table 8.9 shows the symbols and meaning of the location information of the battery-swapping station.

Similarly, the nodes in the road traffic network are used to describe the geographical location of the battery-swapping station, as shown in Eq. (8.42).

$$
Location\_BBS(r) = [j] \tag{8.42}
$$

r represents the code of the battery-swapping station, and j represents the node location of the battery-swapping station in the road traffic network.

The access of the battery-swapping station to the power grid and the node of the power grid for the access are set as shown in Eq. (8.43).

$$
Connection\_BBS(r) = [g, k] \tag{8.43}
$$

Table 8.9 Power supply characteristics of the battery-swapping station

Symbol	Meaning
Location_BBS	Location in the road traffic network
Connection_BBS	Location in the power grid

Table 8.10 Power supply characteristics of the battery-swapping station

Symbol	Meaning
Load_Exchange	Battery-swapping load
Num_Exchange	Number of battery-swapping vehicles

g represents the code of the power grid that the battery-swapping station accesses to, and k represents the code of the node of the power grid g at which the charging station accesses to the power grid.

For a battery-swapping electric vehicle, the battery is swapped immediately after it arrives at the battery-swapping station, and the battery swapped from the vehicle is charged with other batteries in a concentrated way, so the characteristic quantities used to describe the power supply characteristics of the battery-swapping station in this paper include the batter-swapping load and the number of battery-swapping vehicles:

$$Battery\_Exchanging\_Station = \{Load\_Exchange, Num\_Exchange\} \tag{8.44}$$

See Table 8.10 for the power supply characteristics of the battery-swapping station and their corresponding symbols.

Battery-swapping load refers to the sum of the power being charged to swapped batteries at the station.

$$Load\_Charging(r, j) = \sum Power\_Charge(r, i) \tag{8.45}$$

where $Load\_Charging(r, j)$ is the charging power of the battery-swapping station r at the moment j; $Power\_Charge(r, i)$ is the power being charged to the swapped battery of the vehicle i at the battery-swapping station r at the moment j.

The number of battery-swapping vehicles refers to the number of electric vehicles currently swapping batteries at the battery-swapping station. The number of vehicles at each battery-swapping station in each period is measured by calculating the moment of each EV arriving at the charging station. Equation (8.46) is used to determine whether a battery-swapping station is planned for the electric vehicle.

$$\begin{array}{ll} EV\_BBS(r, Arrive(i)) = 1; & \text{vehicles being planned for charging station} \\ EV\_BBS(r, Arrive(i)) = 0; & \text{vehicles not being planned for charging station} \end{array} \tag{8.46}$$

$$Num\_Charging(r, j) = \sum EV\_BBS(r, j) \tag{8.47}$$

$EV\_BBS(r, Arrive(i) : EndTime(i))$ represents the battery-swapping code of the electric vehicle i during its stay at the battery-swapping station r; j is the period of the electric vehicle staying at the station. It's assumed in this research that the battery-swapping demand is constantly met by the full-capacity batteries stored in the station achieved by surplus storage and circulating charging, so there's no need for waiting and the time for battery-swapping is taken as zero.

8.2.4 Road Traffic Network Model and Power Grid Model

The traffic environment and power grid model have been studied. The traffic network and the power grid are modeled in the preliminary subject research, which are introduced in this section.

(a) Road traffic network model

The road traffic network model includes attributes such as road size, grade of the road section, road length, maximum speed limit of the road, and road capacity:

$$Map = \{Node, Connection, Road\_grade, Velocity, Capacity\} \quad (8.48)$$

as the representation matrix of the model; the corresponding relationship between each attribute and its symbol is shown in Table 8.11.

$Node$ represents each node in the road traffic network, i.e. each intersection in the actual road network; $Connection$ represents the connection relationship between nodes and the road length between nodes; $Road\_grade$ represents the grade of a road section; $Velocity$ represents the maximum speed limit of the road, which may vary with the grade of the road; $Capacity$ represents the capacity of a road. At the same time, in order to restore the road conditions of Beijing in a more authentic manner, the real-life road condition of the city on a certain day is based on to establish the basic congestion rate of each road section. The topology of the built road network is shown in Fig. 8.15.

(b) Power grid model

Used to verify the impact of large-scale electric vehicle application on the power grid, the power grid model is a key part of the simulation system for the electric vehicle charging/battery-swapping scheduling. Based on IEEE_33 standard power distribution network model, a power distribution network model including attributes such as topological structure, distribution capacity, and line impedance is established, and the power distribution network model is represented by Matrix (8.49).

Table 8.11 Characteristics of the road traffic network model

Symbol	Meaning
Node	Node
Connection	Connection relationship between nodes
Road_grade	Grade of the road section
Velocity	Maximum speed limit
Capacity	Road capacity

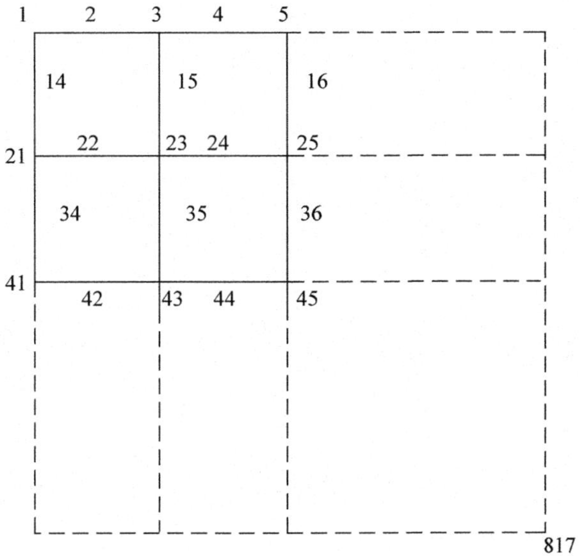

Fig. 8.15 Topology of the road traffic network

$$Grid = \{Node, Branch, Generator\} \tag{8.49}$$

where $Node$ is used to represent parameters of each node in the power grid such as type, power and voltage; $Branch$ is used to represent the parameters of inter-node branches in the power grid such as connection, resistance, and transmission power; $Generator$ represents the source node of the power grid, covering the information of the power supply node in the power grid such as location and capacity.

In this section, three power distribution network models of different scales are established by extending the above models, which use 25 nodes, 27 nodes and 33 nodes respectively, as shown in Fig. 8.16a–c.

8.3 Simulation Verification of the EV Charging/ Battery-Swapping Scheduling Strategy

In order to verify the effectiveness of the charging/battery-swapping scheduling strategy for different types of electric vehicles proposed above, a simulation verification is conducted in this section based on the simulation platform of the charging/ battery-swapping scheduling for electric vehicles. Simulation schemes are respectively designed for the charging scheduling strategy and the battery-swapping scheduling strategy, and the simulation result is analyzed from three perspectives (traffic, power grid and charging station). The simulation result shows that the large-scale application of electric vehicles causes problems including local traffic congestion, excessive grid load, declining grid performance, and long waiting time

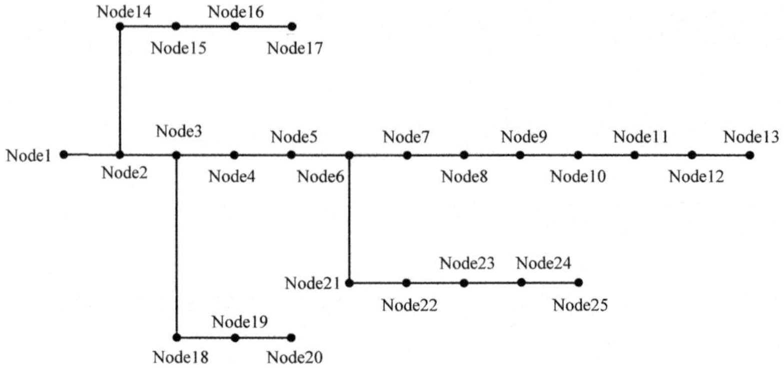

a) 25-node power distribution network

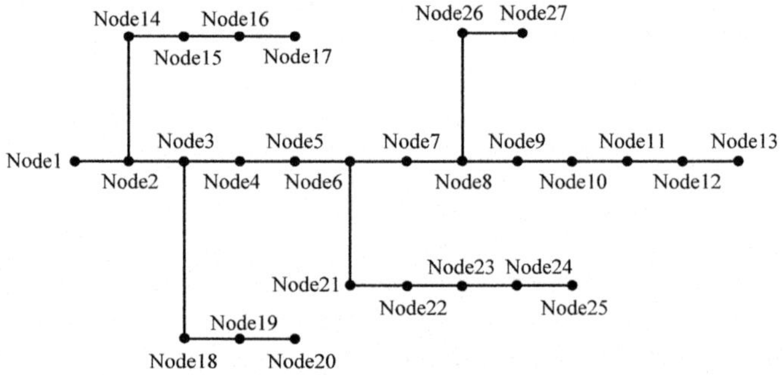

b) 27-node power distribution network

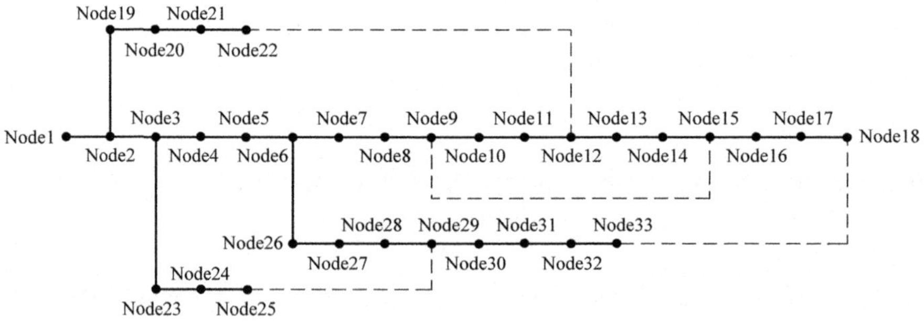

c) 33-node power distribution network

Fig. 8.16 Power distribution network model

for charging. The charging/battery-swapping scheduling strategy proposed in this section can reduce the influence of the above problems and optimize the operation of the whole system.

8.3.1 Designing the Simulation Scheme

In this section, simulation schemes are respectively designed for the charging scheduling strategy and the battery-swapping scheduling strategy, including designing the simulation scale, designing the simulation comparison scheme, and setting the simulation evaluation indicators.

(a) Designing the simulation scale

The actual urban road conditions within the Third Ring Road of Beijing are reasonably simplified to establish the road traffic network model covering the roads within the same area. Figure 8.17 shows the actual road conditions within the Third Ring Road of Beijing. The scale parameters of the model are shown in Table 8.12.

For the power grid model, nine 10 kV power distribution networks are built on the basis of the three distribution network models. Also considering the area and power load within the Third Ring Road of Beijing, a 110 kV transmission network with 118 nodes is introduced to cover the power consumption within the Third Ring Road, and nine distribution networks are connected to the nodes of the transmission network. The connection relationship between each distribution network and transmission network is shown in Table 8.13.

In consideration of the actual traffic congestion situation in Beijing, a 24-h simulation is carried out, and different vehicle numbers are introduced according to the driving habits of drivers in different periods. In the simulation, a certain number of electric vehicles are introduced into the system every 5 min. Table 8.14 shows the numbers of electric vehicles introduced in different periods, totaling 242,880 vehicles, including electric private cars (fast-charging, slow-charging and battery-swapping), electric taxis (fast-charging) and electric buses (battery-swapping). The numbers of vehicles introduced into the system by type are shown in Table 8.15.

The geographical distribution, the power grid connection, the capacity, and the load threshold of the charging/battery-swapping stations are set up based on the scale of the road traffic network, the scale of the power grid and the input of electric vehicles. The first step is to set the geographical location of charging/battery-swapping stations. In order to properly set the location of charging stations, the principle of combining practical research with reasonable assumption is adopted in this Section. 9 electric private car-oriented charging stations are determined according to the actual location of charging stations within the Third Ring Road of Beijing. Since no charging stations dedicated to electric taxis have been built within the Third Ring Road, 4 electric taxi charging stations are uniformly distributed within the Third Ring Road to provide charging service for electric taxis in the whole region, considering the characteristics of taxis including wide operating ranges and random routes. The proposed

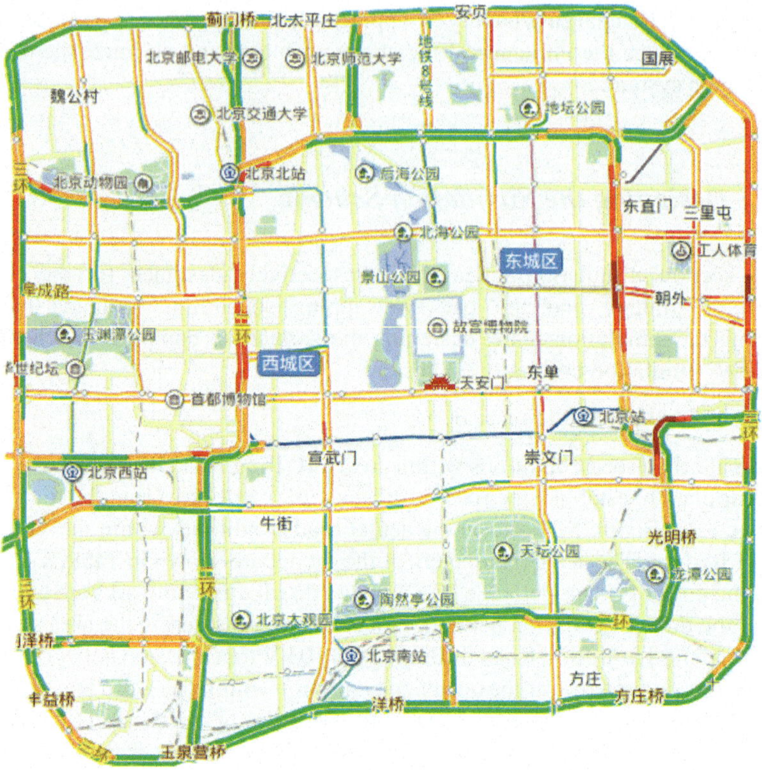

Fig. 8.17 Thumbnail view of roads within the Third Ring Road in Beijing

Table 8.12 Scale of the road traffic network

Area	Perimeter	Average road section length	Road density
165 km^2	45.5 km	1.08 km	1.63 km/km^2

location distribution of charging stations is shown in Fig. 8.18, where the diamonds represent the charging stations for electric private cars and the circles represent the charging stations for electric taxis. The node locations of each charging station in the road network are shown in Table 8.16.

The establishment of the battery-swapping station model also requires the definition of the location and power supply characteristics. Similar to charging stations, the principle of combining practical research with reasonable assumption is adopted to set the location of battery-swapping stations. 9 electric private car-oriented battery-swapping stations are determined according to the actual location of battery-swapping stations within the Third Ring Road of Beijing. Since no battery-swapping stations dedicated to electric buses have been built within the Third Ring Road of Beijing, 2 battery-swapping stations for electric buses are set at the northwest and

Table 8.13 Connection relationship between the power distribution network and the transmission network

Code of power distribution network	Size of power distribution network	Code of access node in the transmission network
1	33-node	14
2	25-node	22
3	25-node	102
4	33-node	44
5	33-node	50
6	33-node	58
7	27-node	97
8	27-node	84
9	27-node	93

Table 8.14 Number of electric vehicles introduced in different periods

Period	Number of vehicles introduced (per 5 min)
7:00–9:00	2300
9:00–17:30	900
17:30–19:30	2300
19:30–23:00	900
23:00–7:00	30

Table 8.15 Number of electric vehicles introduced by type

Vehicle type	Number of vehicles introduced (within 24 h)
Private car (fast-charging)	181,840
Private car (battery-swapping)	20,000
Private car (slow-charging)	10,000
Taxi	30,000
Bus	1040

southeast corners of the road traffic network considering buses have longer routes and fixed departure and destination locations. The location distribution of battery-swapping stations is shown in Fig. 8.19. The triangles represent the battery-swapping stations for electric private cars and the rectangles represent the battery-swapping stations for electric buses. The node locations of each battery-swapping station in the road network are shown in Table 8.17.

Considering the connection between charging/battery-swapping stations and the power grid, it is necessary to set the nodes at which the charging stations access to the distribution network. According to the calculation of power flow, the influence on the power grid is smaller when the node accessed by a charging station is close

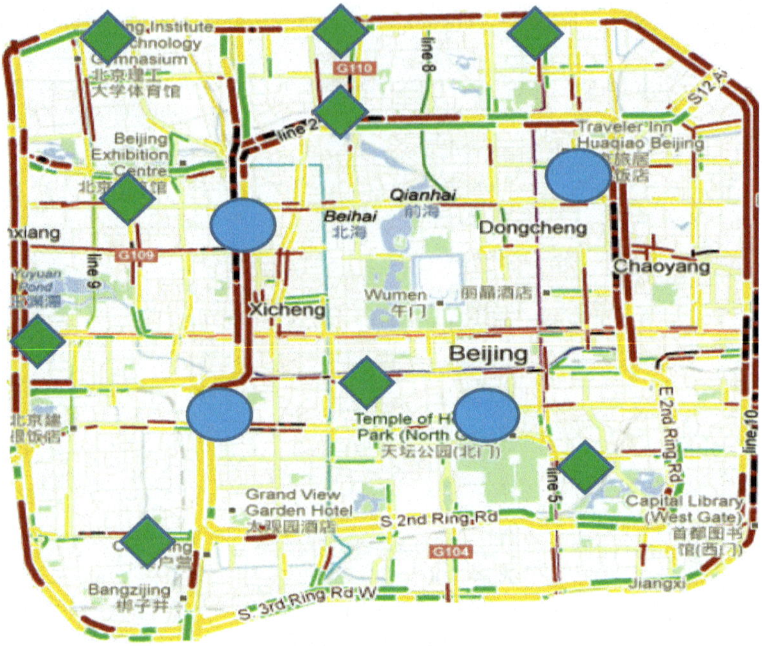

Fig. 8.18 Distribution of charging stations

Table 8.16 Location of charging stations in the road traffic network

Code of charging station for private cars	Code of charging station for taxis	Code of access node in the road traffic network	
		Charging station for private cars	Charging station for taxis
1	1	3	277
2	2	24	306
3	3	25	643
4	4	165	614
5	–	183	–
6	–	465	–
7	–	532	–
8	–	727	–
9	–	766	–

to the root node of the power grid. It's on this basis that the access nodes of the charging stations in the power grid are set, as shown in Tables 8.18 and 8.19, which respectively represent the access of charging stations for electric private cars and electric taxis to the power distribution network.

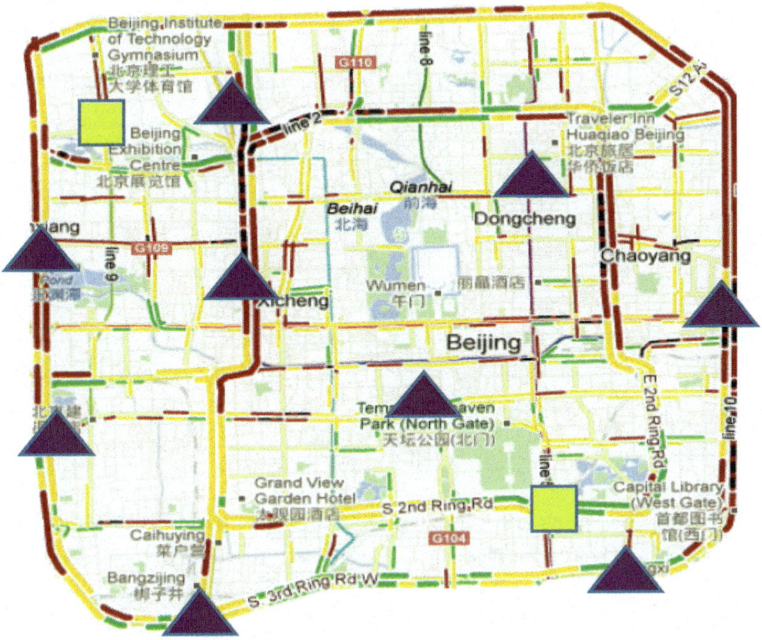

Fig. 8.19 Distribution of battery-swapping stations

Table 8.17 Location of battery-swapping stations in the road traffic network

Code of battery-swapping station for private cars	Code of battery-swapping station for buses	Code of access node in the road traffic network	
		Battery-swapping station for private cars	Battery-swapping station for buses
1	1	294	816
2	2	143	163
3	–	327	–
4	–	389	–
5	–	453	–
6	–	553	–
7	–	575	–
8	–	786	–
9	–	815	–

The access of battery-swapping stations to the distribution network is set. Tables 8.20 and 8.21 show the access of battery-swapping stations for electric private cars and buses to the power grid.

Table 8.18 Connection between each charging station for electric private cars and the distribution network

Code of charging station	Power distribution network accessed	Code of access node in the power distribution network
1	1	23
2	2	4
3	3	4
4	4	4
5	5	14
6	6	18
7	7	4
8	8	4
9	9	18

Table 8.19 Connection between each charging station for electric taxis and the distribution network

Code of charging station	Power distribution network accessed	Code of access node in the power distribution network
1	1	4
2	2	2
3	3	19
4	4	23

Table 8.20 Connection between each battery-swapping station for electric private cars and the power distribution network

Code of battery-swapping station	Power distribution network accessed	Code of access node in the power distribution network
1	1	19
2	2	19
3	3	14
4	4	14
5	5	18
6	6	14
7	7	18
8	8	14
9	9	4

The capacity and load threshold of each charging station are set according to the scale of electric vehicles put in. Tables 8.22 and 8.23 show the capacity and load threshold of charging stations for private cars and taxis.

Table 8.21 Connection between each battery-swapping station for electric buses and the power distribution network

Code of battery-swapping station	Power distribution network accessed	Code of access node in the power distribution network
1	1	4
2	2	4

Table 8.22 Capacity and load threshold of charging stations for private cars

Code of charging station	Capacity (*Capacity_Charging)*	Load threshold (*Power_Threshold*) (MW)		
		Valley hours	Off-peak hours	Peak hours
1	500	0.05	4.06	8.8
2	500	0.075	4.35	10.8
3	500	0.075	4.35	10.8
4	500	0.05	4.06	9.6
5	500	0.05	4.06	9.6
6	500	0.05	4.205	10.4
7	500	0.08	4.35	10.8
8	500	0.08	4.35	10.8
9	500	0.09	4.495	11.2

Table 8.23 Capacity and load threshold of charging stations for taxis

Code of charging station	Capacity (*Capacity_Charging)*	Load threshold (*Power_Threshold*) (MW)		
		Valley hours	Off-peak hours	Peak hours
1	80	1	2.2	2.5
2	80	1	2.2	2.5
3	80	2.3	3.2	3.8
4	80	2.3	3.2	3.2

(b) Simulation contrast scheme

(1) Analyzing the effect of the proposed charging scheduling strategy based on the comprehensive optimization of the traffic network and the power grid (hereinafter referred to as the "comprehensively optimized charging strategy") in improving traffic congestion, power grid operation, and charging station operation in comparison with the commonly used "traditional charging strategy" which refers to the strategy of recommending the charging station at the shortest distance. In this strategy, when the Dijkstra algorithm is used to search the shortest path, the section length in the road traffic network is adopted as the weight, that is, dispatching the electric vehicle to the nearest charging station for charging.

(2) Analyzing the role of the battery-swapping scheduling strategy (hereinafter referred to as the "ordered charging strategy") in balancing the load of the power grid compared with the "disordered charging strategy" which refers to the strategy of dispatching the electric vehicle to the nearest battery-swapping station to swap the battery; the battery is immediately charged with a certain power after the swapping without any scheduling.

(c) Evaluation indicators

Specific evaluation indicators need to be determined to conduct a quantitative analysis on the optimization effect of the proposed EV charging/battery-swapping scheduling strategy based on the comprehensive optimization of the traffic network and power grid. The traffic efficiency of the road traffic network, the optimization effect of the power grid, and the operation of charging stations are evaluated from the following three aspects.

On the traffic side, the traffic speed, road section saturation, and road congestion rate are selected to evaluate the optimization effect of the traffic system. Road section saturation is used to represent the ratio of the section flow to the actual traffic capacity; as shown in Eq. (8.50), the higher the saturation, the more congested the road.

$$Q\_flow\{p,q\}(j) = \frac{q\{p,q\}(j)}{C\{p,q\}} \tag{8.50}$$

$Q\_flow\{p,q\}(j)$ is the saturation of the road section (p,q) at the moment j.

The traffic speed of the road section represents the highest traffic speed of the road section at a certain moment. The traffic speed of the road section can be obtained by measuring the traffic flow of the road section at the current moment, as shown in Eq. (8.51). The lower the traffic speed of the road section, the lower the speed of the vehicles running on the road section and the traffic efficiency.

$$V_{road}\{p,q\}(j) = \frac{V_m\{p,q\}}{1 + q\{p,q\}(j)/C\{p,q\}} \tag{8.51}$$

Road congestion rate represents the ratio of the number of congested roads to the total number of roads at a certain moment; the congestion rates of roads near charging stations are counted, as shown in Eq. (8.52). The higher the congestion rate, the more congested the traffic.

$$R\_CS(j) = \frac{\sum Congestion\{p,q\}(\alpha/\beta/\chi)}{N\_CS} \tag{8.52}$$

$R\_CS(j)$ represents the congestion rate of roads near the charging station at the moment j; N_CS is the total number of roads near the charging station; $Congestion\{p,q\}(\alpha/\beta/\chi)$ represents the severity level (low, medium, high) of the road section (p,q); as shown in Eqs. (8.53), (8.54) and (8.55), we can know what kind of congestion state the road section is in or whether it is congested by comparing the traffic speed of the current road section and the maximum traffic speed of the section.

$$Congestion\{p,q\}(\alpha)=\begin{cases}1, & v_{road}\{p,q\}\in\{0.5v_m\{p,q\},0.7v_m\{p,q\}\}\\0, & v_{road}\{p,q\}\notin\{0.5v_m\{p,q\},0.7v_m\{p,q\}\}\end{cases} \tag{8.53}$$

$$Congestion\{p,q\}(\beta)=\begin{cases}1, & v_{road}\{p,q\}\in\{0.3v_m\{p,q\},0.5v_m\{p,q\}\}\\0, & v_{road}\{p,q\}\notin\{0.3v_m\{p,q\},0.5v_m\{p,q\}\}\end{cases} \tag{8.54}$$

$$Congestion\{p,q\}(\chi)=\begin{cases}1, & v_{road}\{p,q\}\in\{0,0.3v_m\{p,q\}\}\\0, & v_{road}\{p,q\}\notin\{0,0.3v_m\{p,q\}\}\end{cases} \tag{8.55}$$

On the power grid side, the grid load and the operating performance of the grid are evaluated. The load curve of the power grid directly represents the operating state of the power grid. The flatter the load curve and the smaller the PV, the better the operating performance of power grid. The indicators of voltage offset rate and power loss rate are selected to measure the safety and economy of the power grid's operation respectively. The lower the voltage offset rate and power loss rate, the better the safety and economy of the power grid's operation.

On the charging station side, the number of vehicles in the charging station and the waiting time are selected to measure the operation of the charging station. Firstly, by comparing the number of vehicles served by each charging station, we can judge whether the resource allocation of the charging stations is appropriate. Since the number of charging piles in each charging station is set to be the same, the more average the number of vehicles in each charging station, the more appropriate the resource allocation and the higher the charging efficiency. The waiting time for charging can reflect the convenience for EV drivers. The advantage of "comprehensive optimization charging strategy" can be intuitively felt by drivers. The shorter the waiting time for charging, and the smaller the number of vehicles waiting, the easier it is to drive an EV.

(d) Process of simulation verification

In this section, the simulation verification process is respectively designed for the charging scheduling strategy and the battery-swapping scheduling strategy. The first is the simulation verification process for the charging scheduling strategy, as shown in Fig. 8.20.

In the verification of the "comprehensively optimized charging strategy", only the driving and charging behavior of the fast-charging vehicles are considered, and the slow-charging vehicles and the battery-swapping vehicles are not considered. A certain number of EVs are put into the system every 5 min. After the EVs are put into the system, the driving and charging information of the vehicles previously put into the system are updated, the charging demand and charging decision of the EVs are judged, and then the charging scheduling strategy is adopted. The process is shown in Fig. 8.20.

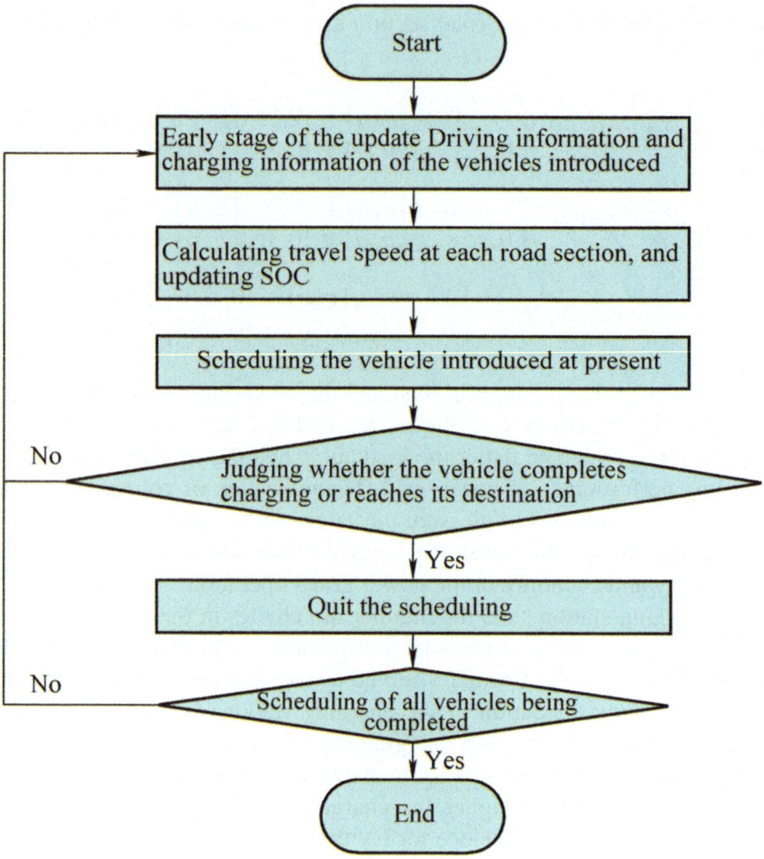

Fig. 8.20 Simulation verification process for the charging scheduling strategy

In the simulation verification of the battery-swapping scheduling strategy, considering the charging load of the charging station, the effect of the battery-swapping scheduling strategy on the distribution network is proved by changing the proportion of battery-swapping vehicles. The battery-swapping scheduling strategy includes two aspects: dispatching the electric vehicles to easily accessible battery-swapping stations and planning the charging of replaced batteries. The scheduling is conducted by power grids to smooth the load curve of the power grid. In order to avoid the excessive amount of data, the planning is carried out in hours. The process is shown in Fig. 8.21.

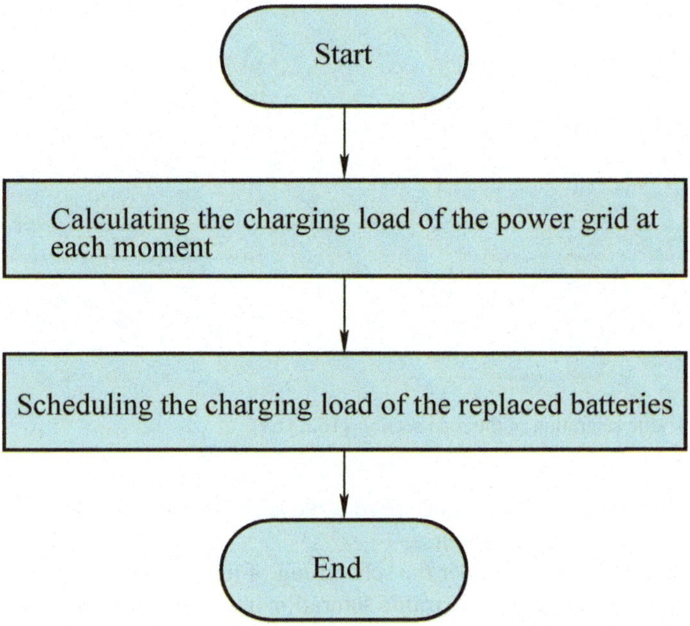

Fig. 8.21 Simulation verification process for the battery-swapping scheduling strategy

8.3.2 Analysis of the Simulation Result on the Traffic Side

With electric vehicles introduced based on the statistics of the actual traffic conditions on roads within the Third Ring Road of Beijing on a certain day, the traffic congestion around charging stations for electric private cars and electric taxis under the "traditional charging strategy" and the proposed "comprehensively optimized charging strategy" are compared. Considering the serious congestion in morning and evening rush hours in Beijing, the traffic saturation, traffic speed and congestion rate of road sections near charging stations during these hours are respectively measured as evaluation indicators.

(a) Traffic saturation of the road section

Figure 8.22a and b respectively shows the comparison results of the traffic saturation in the morning and evening peak hours of a certain road section under the "traditional charging strategy" and the proposed "comprehensively optimized charging strategy"; the road section {162, 163} near a charging station is selected as the research object, and the real-time traffic saturation and average traffic saturation of this road section are respectively measured. The real-time traffic saturation here is the traffic saturation of each moment in peak hours, and the average traffic saturation is the arithmetic mean of the real-time traffic saturation in peak hours; similarly, the real-time road section traffic speed is the road section traffic speed at each moment in peak hours,

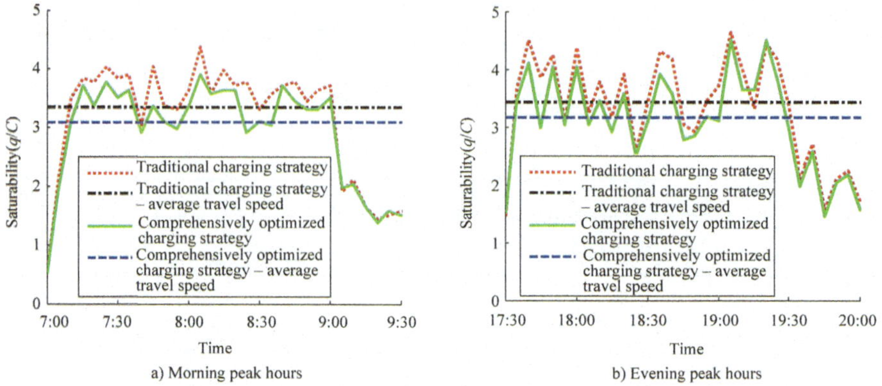

Fig. 8.22 Traffic saturation of the road section {162, 163}

and the average road section traffic speed is the arithmetic mean of the real-time road section traffic speed in peak hours.

As shown in the figure, under the scheduling of the "comprehensively optimized charging strategy", the real-time traffic saturation in morning and evening peak hours is generally lower than that under the scheduling of the "traditional charging strategy"; compared with the "traditional charging strategy", the average traffic saturation of the "comprehensively optimized charging strategy" in the morning peak hours decreased by 7.19% from 3.34 to 3.10, and the average traffic saturation of the evening peak hours decreased by 7.31% from 3.42 to 3.17.

(b) Traffic speed of the road section

Figure 8.23a and b shows the simulation results of the traffic speed of a road section in the road traffic network under the "traditional charging strategy" and the "comprehensively optimized charging strategy" respectively. The road section {162, 163} near the charging station is selected again as the research object to measure its real-time traffic speed and average traffic speed in the morning and evening peak hours under the two charging strategies. As shown in the figure, in the morning and evening peak hours, the real-time traffic speed of the road section {162, 163} under the scheduling of "traditional charging strategy" is generally higher than that under the scheduling of the "comprehensively optimized charging strategy"; compared with the "traditional charging strategy", the average traffic speed of the "comprehensively optimized charging strategy" in the road section in morning peak hours increases from 16.84 km/h to 17.59 km/h, and the average traffic speed in the evening peak hours increases from 16.57 km/h to 17.35 km/h.

(c) Road congestion rate

The road congestion rate is reflected by comparing different congestion states of roads near charging stations. In Figs. 8.24, 8.25, 8.26, 8.27, 8.28 and 8.29, different levels of congestion of roads near charging stations are compared. More specifically,

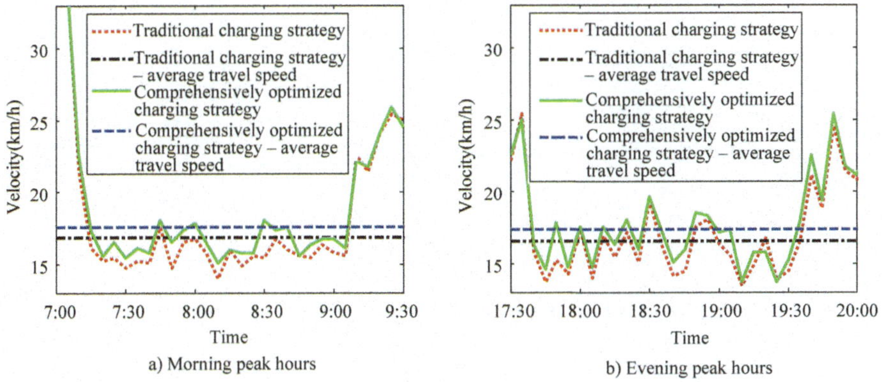

Fig. 8.23 Traffic speed of the road section {162, 163}

the simulation results of mild congestion, moderate congestion and severe congestion of roads near charging stations for electric private cars, charging stations for electric taxis and all charging stations in the morning and evening peak hours under the two charging strategies are obtained. The real-time congestion rate is the congestion rate at each moment in the peak hours, and the average congestion rate is the arithmetic mean of the real-time congestion rates in the peak hours.

(1) Comparison of mild congestion rate

Figures 8.24a–c and 8.25a–c are respectively the real-time mild congestion rate curve and the average mild congestion rate curve for roads near charging stations for electric private cars, charging stations for electric taxis, and all charging stations in the morning and evening peak hours under the two charging strategies.

As shown in Fig. 8.24, the mild congestion rate under the "traditional charging strategy" is higher than that under the "comprehensively optimized charging strategy" in most of the morning peak hours. For electric private car charging stations, the average mild congestion rate under the "traditional charging strategy" is 0.517, while the same rate under the "comprehensively optimized charging strategy" is 0.499; for electric taxi charging stations, the mild congestion rate under the "comprehensively optimized charging strategy" is 0.393, which is 15.302% higher than that under the "traditional charging strategy"; the average mild congestion rate near all charging stations decreases from 0.521 to 0.491 when the "comprehensively optimized charging strategy" is adopted for scheduling.

Figure 8.25 shows that in most of the evening and morning peak hours, the real-time mild congestion rates of electric private car charging stations, electric taxi charging stations and all charging stations under the "traditional charging strategy" are higher than those under the "comprehensively optimized charging strategy". By comparing the average congestion rate curves under the two strategies, it can be seen that for electric private car charging stations, the average mild congestion rate under the "traditional charging strategy" is 0.546, while the same rate under the

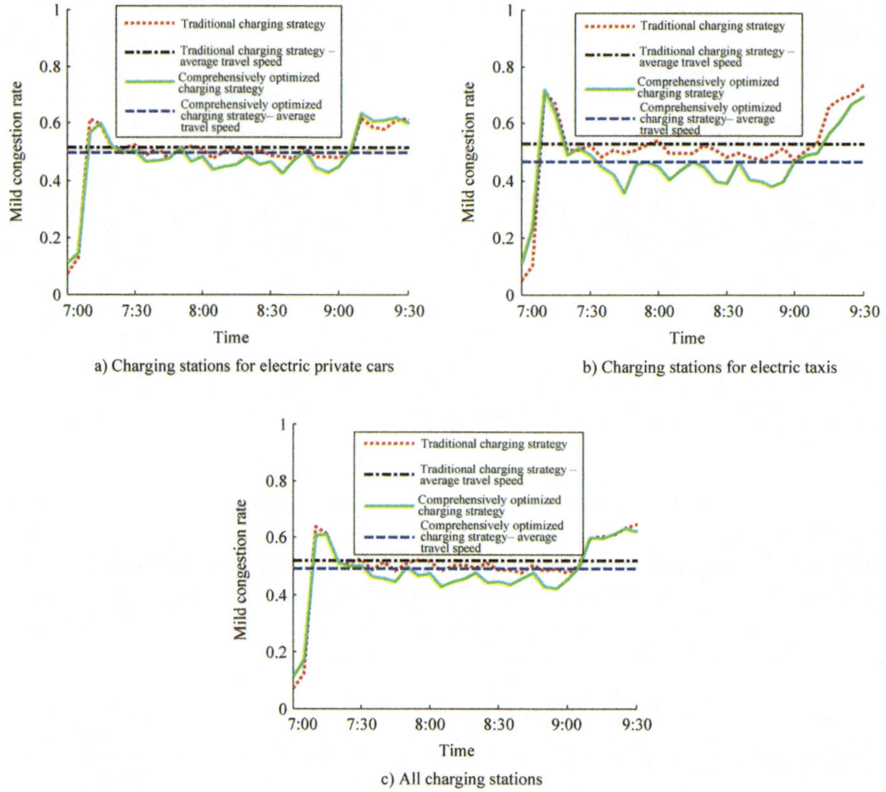

Fig. 8.24 Mild congestion rate curve for roads near charging stations in the morning peak hours

"comprehensively optimized charging strategy" is 0.502, representing a decrease of 8.059%; for electric taxi charging stations, the average mild congestion rate under the "traditional charging strategy" is 0.502, while the same rate under the "comprehensively optimized charging strategy" is 0.433, representing a decrease of 13.745%; for all charging stations, the average mild congestion rate is 0.535 under the "traditional charging strategy", while the same rate under the "comprehensively optimized charging strategy" is 0.485, representing a decrease of 9.346%.

The above comparison shows that the "comprehensively optimized charging strategy" can significantly reduce the mild congestion rate of road sections near charging stations during morning and evening peak hours compared with the "traditional charging strategy".

Figures 8.26a–c and 8.27a–c respectively show the comparison of the real-time moderate congestion rates and average moderate congestion rates of roads near electric private car charging stations, electric taxi charging stations and all charging stations in the morning and evening peak hours under the two charging strategies.

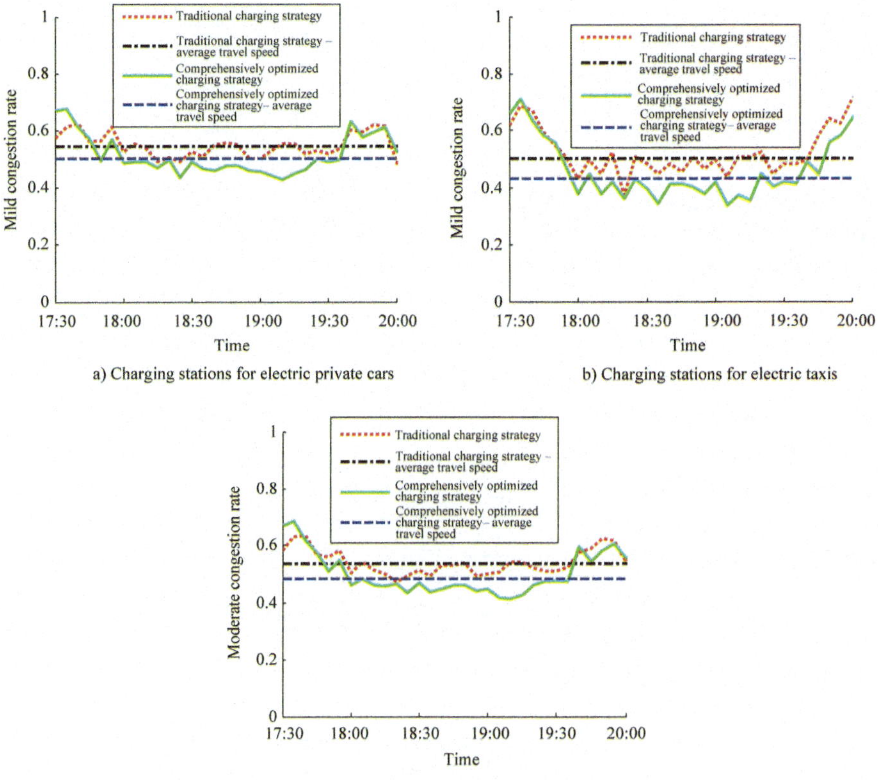

a) Charging stations for electric private cars

b) Charging stations for electric taxis

c) All charging stations

Fig. 8.25 Mild congestion rate curve for roads near charging stations in the evening peak hours

As shown in Fig. 8.26, the real-time moderate congestion rates of roads near electric private car charging stations, electric taxi charging stations and all charging stations under the "traditional charging strategy" are lower than those under the "comprehensively optimized charging strategy" in most of the morning peak hours. By comparing the average moderate congestion rates under the two strategies, we can see that for electric private car charging stations, the average moderate congestion rate under the "traditional charging strategy" is 0.329, while the same rate under the "comprehensively optimized charging strategy" is 0.375; for electric taxi charging stations, the average moderate congestion rate under the "traditional charging strategy" is 0.393, while the same rate under the "comprehensively optimized charging strategy" is 0.464; for all charging stations, the average moderate congestion rate under the "traditional charging strategy" is 0.345, while the same rate under the "comprehensively optimized charging strategy" is 0.397.

As shown in Fig. 8.27, the real-time moderate congestion rates of roads near electric private car charging stations, electric taxi charging stations and all charging stations under the "traditional charging strategy" are lower than those under the

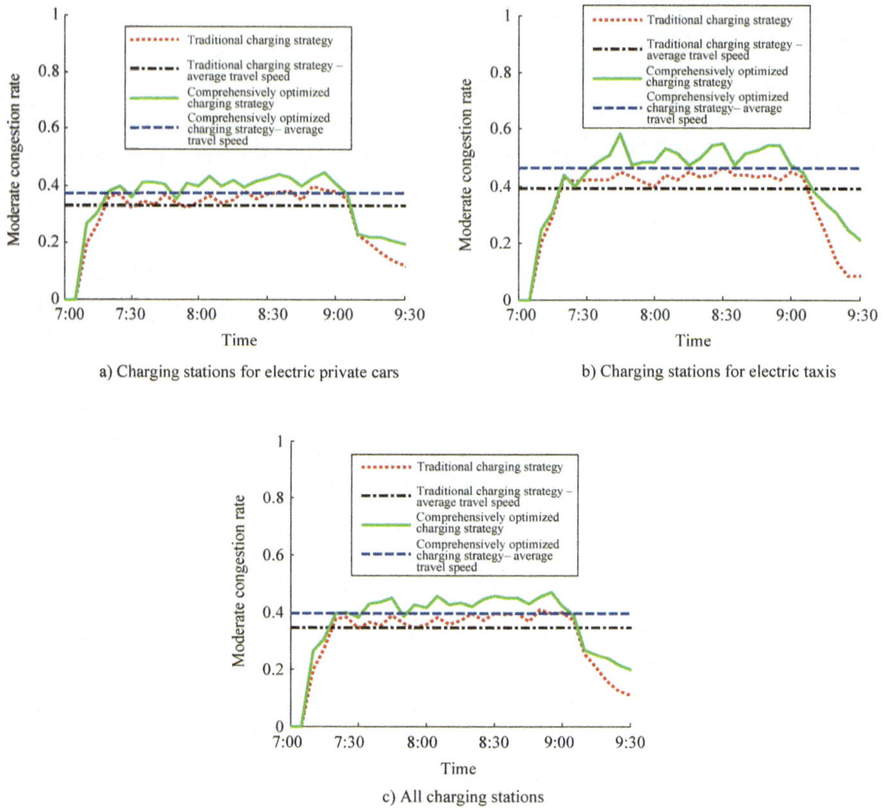

Fig. 8.26 Moderate congestion rate curve for roads near charging stations in the morning peak hours

"comprehensively optimized charging strategy" in most of the evening peak hours. By comparing the average moderate congestion rates under the two strategies, we can see that for electric private car charging stations, the average moderate congestion rate under the "traditional charging strategy" is 0.309, while the same rate under the "comprehensively optimized charging strategy" is 0.373; for electric taxi charging stations, the average moderate congestion rate under the "traditional charging strategy" is 0.393, while the same rate under the "comprehensively optimized charging strategy" is 0.464; for all charging stations, the average moderate congestion rate under the "traditional charging strategy" is 0.335, while the same rate under the "comprehensively optimized charging strategy" is 0.406.

The above comparison shows that the "traditional charging strategy" can reduce the moderate congestion rate of road sections near charging stations during morning and evening peak hours compared with the "comprehensively optimized charging strategy".

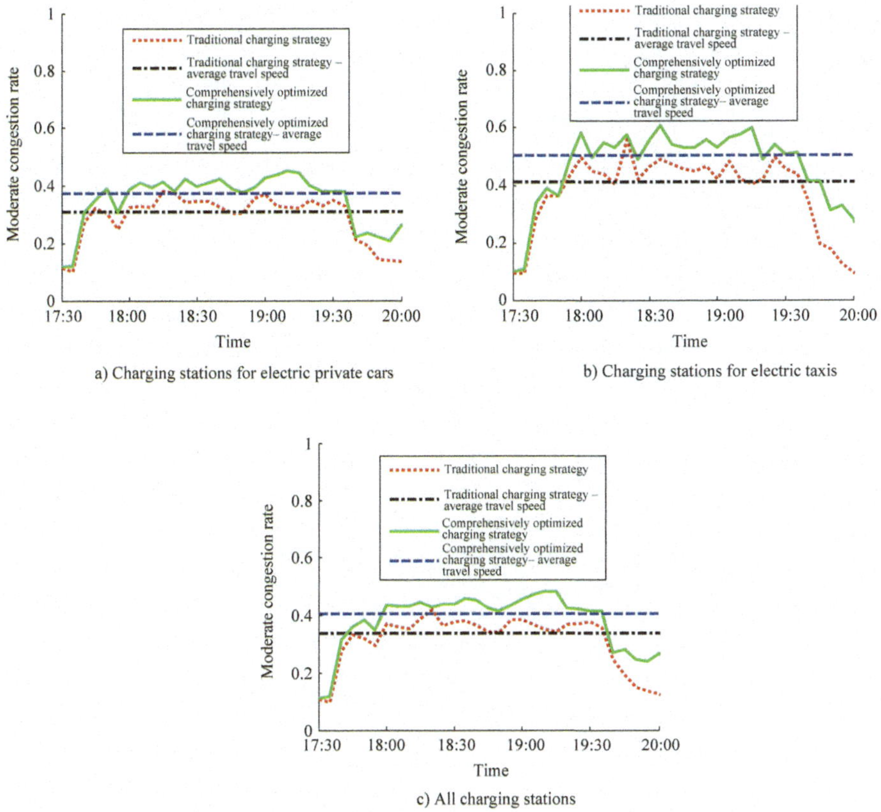

Fig. 8.27 Moderate congestion rate curve for roads near charging stations in the evening peak hours

Figures 8.28a–c and 8.29a–c respectively show the comparison of the real-time severe congestion rates and average severe congestion rates of roads near electric private car charging stations, electric taxi charging stations and all charging stations in the morning and evening peak hours under the two charging strategies.

As shown in Fig. 8.28, the real-time severe congestion rates of roads near electric private car charging stations, electric taxi charging stations and all charging stations under the "traditional charging strategy" didn't show obvious disadvantage or advantage over those under the "comprehensively optimized charging strategy" in the morning peak hours. By comparing the average severe congestion rates under the two strategies, we can see that for electric private car charging stations, the average severe congestion rate under the "traditional charging strategy" is 0.0648, while the same rate under the "comprehensively optimized charging strategy" is 0.0617, representing a decrease of 4.784%; for electric taxi charging stations, the average severe congestion rate under the "traditional charging strategy" is 0.0242, while the same rate under the "comprehensively optimized charging strategy" is 0.0273, representing

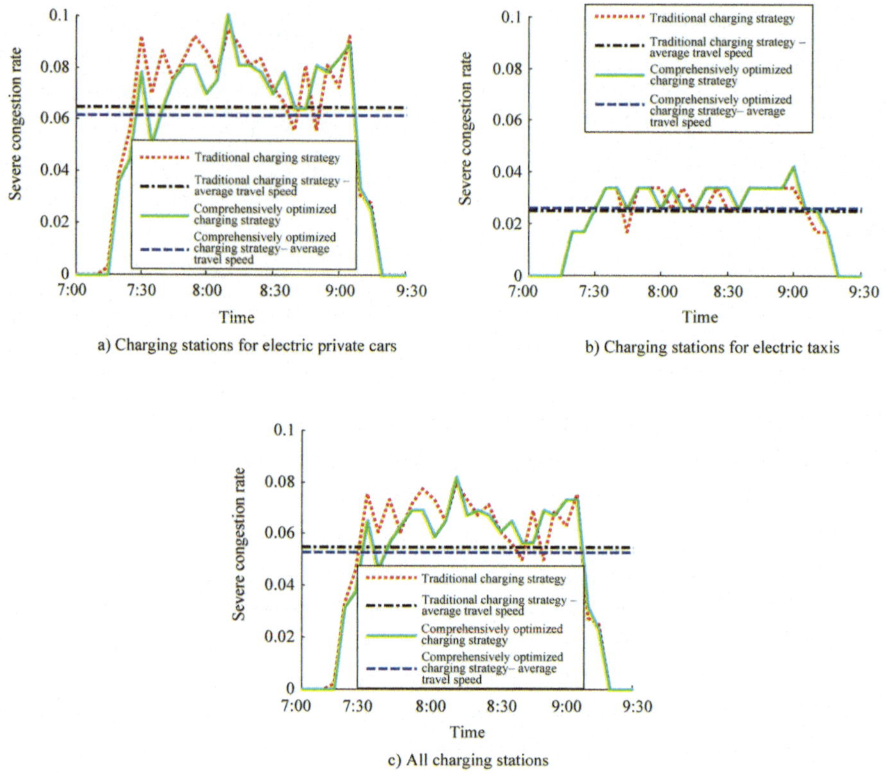

Fig. 8.28 Severe congestion rate curve for roads near charging stations in the morning peak hours

an increase of 1.104%; for all charging stations, the average severe congestion rate under the "traditional charging strategy" is 0.0549, while the same rate under the "comprehensively optimized charging strategy" is 0.0528, representing a decrease of 3.825%.

As shown in Fig. 8.29, the real-time severe congestion rates of roads near electric private car charging stations, electric taxi charging stations and all charging stations under the "traditional charging strategy" didn't show obvious disadvantage or advantage over those under the "comprehensively optimized charging strategy" in the evening peak hours either. By comparing the average severe congestion rates under the two strategies, we can see that for electric private car charging stations, the average severe congestion rate under the "traditional charging strategy" is 0.0627, while the same rate under the "comprehensively optimized charging strategy" is 0.0608; for electric taxi charging stations, the average severe congestion rate under the "traditional charging strategy" is 0.0242, while the same rate under the "comprehensively optimized charging strategy" is 0.0273; for all charging stations, the average severe congestion rate under the "traditional charging strategy" is 0.0531,

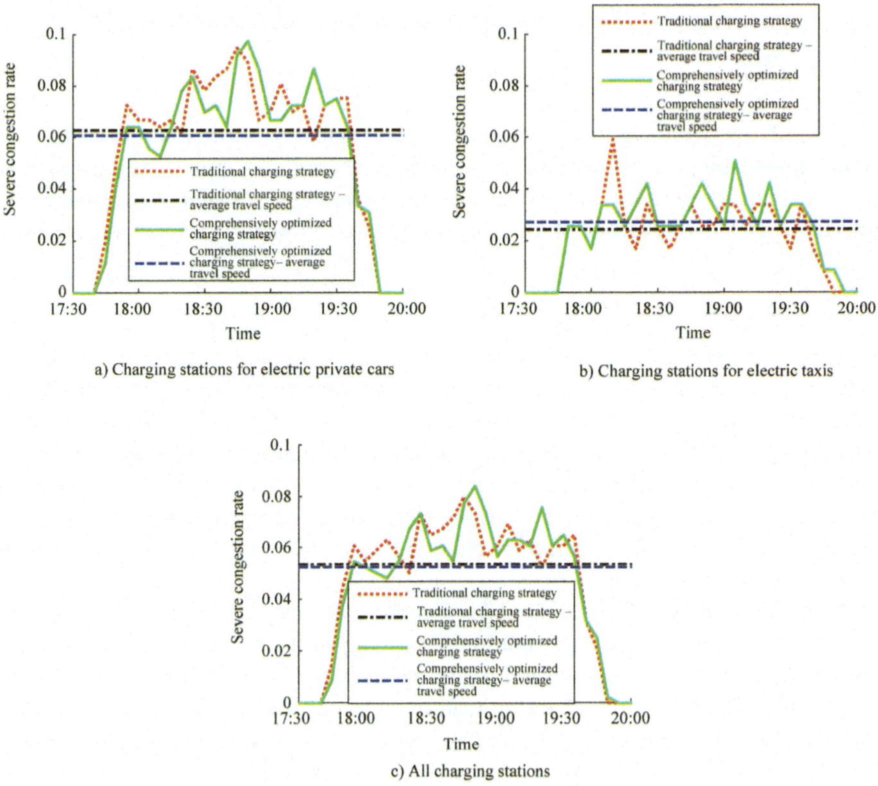

Fig. 8.29 Severe congestion rate curve for roads near charging stations in the evening peak hours

while the same rate under the "comprehensively optimized charging strategy" is 0.0525, representing a decrease of 1.130%.

According to the above analysis, the severe congestion of roads near electric private car charging stations in the morning and evening peak hours is improved under the "comprehensively optimized charging strategy" compared with the "traditional charging strategy" while the severe congestion of roads near electric taxi charging stations increases. However the congestion of roads near all charging stations is still improved due to the inherently lower congestion rate of roads near electric taxi charging stations. Therefore the "comprehensively optimized charging strategy" has a certain effect in improving the severe congestion of roads near charging stations.

In summary, the massive application of electric vehicles can cause different levels of congestion near the charging stations. In general, for charging stations for electric private cars and electric taxis, the "comprehensively optimized charging strategy" can reduce the mild and severe congestion rates of nearby roads to an extent. Although the improvement of severe congestion leads to the increasing moderate congestion rate, it eases the congestion near charging stations in general.

8.3.3 Analysis of the Simulation Result on the Power Grid Side

The simulation result on the power grid side is analyzed in this section. The load scheduling result of each charging station and each battery-swapping station under the proposed charging/battery-swapping scheduling strategy are analyzed respectively, and the load condition and the operation of each power grid are compared and analyzed.

(a) Load distribution of charging stations

Figures 8.30 and 8.31 show the variation in the charging load of each electric private car charging station and each electric taxi charging station in a day under different scheduling strategies. Figure 8.30a and b show the time–space distribution of charging load of each charging station for electric private cars under the "traditional charging strategy" and"comprehensively optimized charging strategy" respectively and Fig. 8.31a and b show the charging load curves of each charging station for electric taxis under the "traditional charging strategy" and the "comprehensively optimized charging strategy."

It can be seen from Fig. 8.30a and b that there is a large PV in the charging load of charging stations under the "traditional charging strategy". For example, charging station 7 has the highest load and charging station 2 has the lowest load during peak hours. The maximum charging load of charging station 7 is 24.46 MW, while the maximum charging load of charging station 2 is only 6.72 MW during peak hours, showing an obvious uneven distribution of load. Under the "comprehensively optimized charging strategy", the peak load of charging station 7 in peak hours decreases to 20.45 MW, while the peak load of charging station 2 rises to 14.25 MW, showing a reduced and more even PV of load among charging stations, which is beneficial to the operation of the power grid and charging stations.

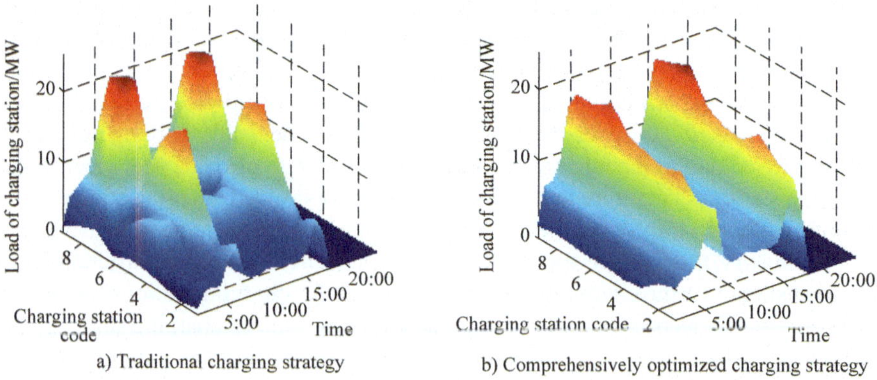

Fig. 8.30 Time–space distribution of charging load of each charging station for electric private cars

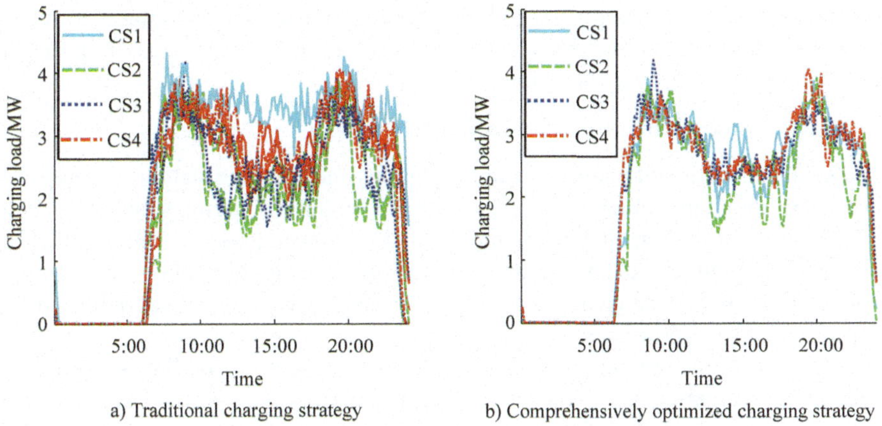

Fig. 8.31 Charging load curves of each charging station for electric taxis

As shown in Fig. 8.31a and b, the load distribution under the two strategies in morning and evening peak hours are neck to neck. However, the peak-valley values of the charging load under the "traditional charging strategy" varied greatly during off-peak hours in the daytime, so the load distribution between 9:00 and 17:30 is analyzed in detail.

Figure 8.32 shows the load distribution under the two charging strategies during off-peak hours in the daytime. It can be seen that the load distribution under the "comprehensively optimized charging strategy" is more even with a merely 1.753 MW PV at the same moment and a maximum PV at the same moment as high as 3.192 MW, an increase by 1.821 times.

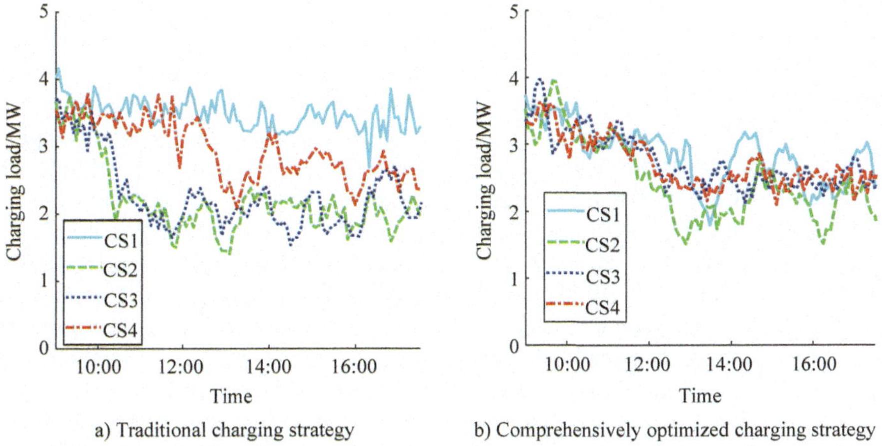

Fig. 8.32 Charging load curves of each charging station for electric taxis during off-peak hours in the daytime

The proposed EV charging scheduling strategy makes the charging load distribution of charging stations for electric private cars and electric taxis more even, achieving better resource allocation.

(b) Load distribution of battery-swapping stations

The load distribution of battery-swapping stations and charging stations is compared under the "unordered charging strategy" and the "ordered charging strategy" respectively. The distribution of power station load of power grid 7 is analyzed as an example. According to the layout of charging/battery-swapping stations in the power grid, electric private car charging station 7 and electric car battery-swapping station 7 are connected to power grid 7.

The scale of the simulation shows that there are about 10,000 battery-swapping vehicles, representing 4% of the entire fleet. If the proportion is too low, it won't be enough to show the effect of optimization of the proposed battery-swapping strategy ("ordered charging strategy").

Figures 8.33, 8.34, 8.35 and 8.36 show the respective results of different battery-swapping scheduling strategies when the proportion of battery-swapping vehicles is 4%, 30%, 50% and 70%. Figures 8.33a–c, 8.34a–c, 8.35a–c and 8.36a–c respectively show the charging and battery-swapping load distribution of power grid 7 under the "unordered charging strategy," the "ordered charging strategy" based on the minimum PV, and the "ordered charging strategy" based on the minimum standard deviation.

Figure 8.33a–c shows the load distribution of charging and battery-swapping stations in power grid 7 when battery-swapping vehicles account for 4% of total vehicles under the "unordered charging strategy," the "ordered charging strategy" based on the minimum PV, and the "ordered charging strategy" based on the minimum standard deviation. As shown in Fig. 8.33a, scheduling the battery-swapping stations to charge the swapped batteries under the "unordered charging strategy" results in the rise of load of power grid 7 during evening peak hours and exacerbates the peak-valley difference of the load, while under both the "ordered charging strategy" based on the minimum PV and the "ordered charging strategy" based on the minimum standard deviation, the battery-swapping stations charge the batteries in the valley hours of load, which does not increase the peak load of the power grid.

Figure 8.34a–c shows the load distribution of charging and battery-swapping stations in the power grid when battery-swapping vehicles account for 30% of total vehicles under different battery charging strategies. The peak load in the power grid is 19.45 MW and the valley load is 0.39 MW under the"unordered charging strategy"; under the "ordered charging strategy" based on the minimum PV, the peak load in the power grid declines to 15.49 MW by 20.36% and the valley load rises to 6.65 MW by 16.05 times; similarly, under the "ordered charging strategy" based on the minimum standard deviation, the peak load in the power grid declines by 20.3% and the valley load rises by 17.7 times as compared with the "unordered charging strategy."

Figure 8.35a–c shows the load distribution of charging and battery-swapping stations in the power grid when battery-swapping vehicles account for 50% of total vehicles under different battery charging strategies. The peak load in the power grid is 18.96 MW and the valley load is 0.47 MW under the "unordered charging strategy";

under the "ordered charging strategy" based on the minimum PV, the peak load in the power grid declines to 11.24 MW by 40.72% and the valley load rises to 8.68 MW by 17.47 times; similarly, under the "ordered charging strategy" based on the minimum

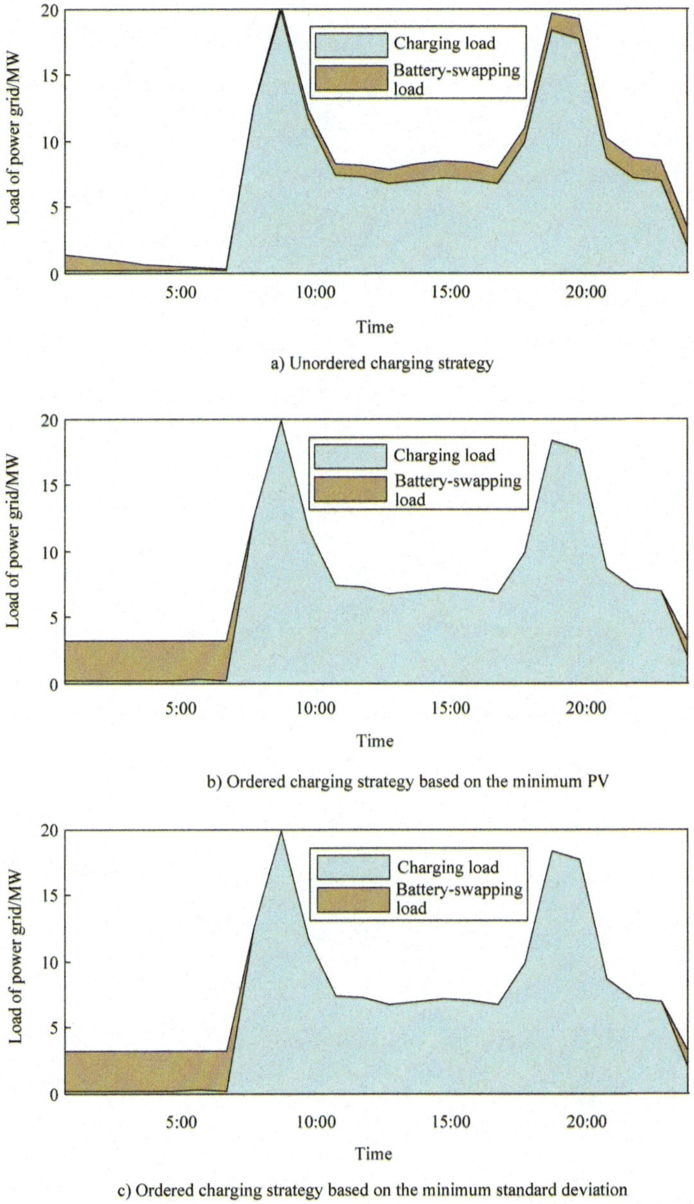

Fig. 8.33 Scheduling result when the proportion of battery-swapping vehicles is 4%

standard deviation, the peak load in the power grid declines by 40.72% and the valley load rises by 17.53 times as compared with the "unordered charging strategy."

Figure 8.36a–c shows the load distribution of charging and battery-swapping stations in the power grid when battery-swapping vehicles account for 70% of total

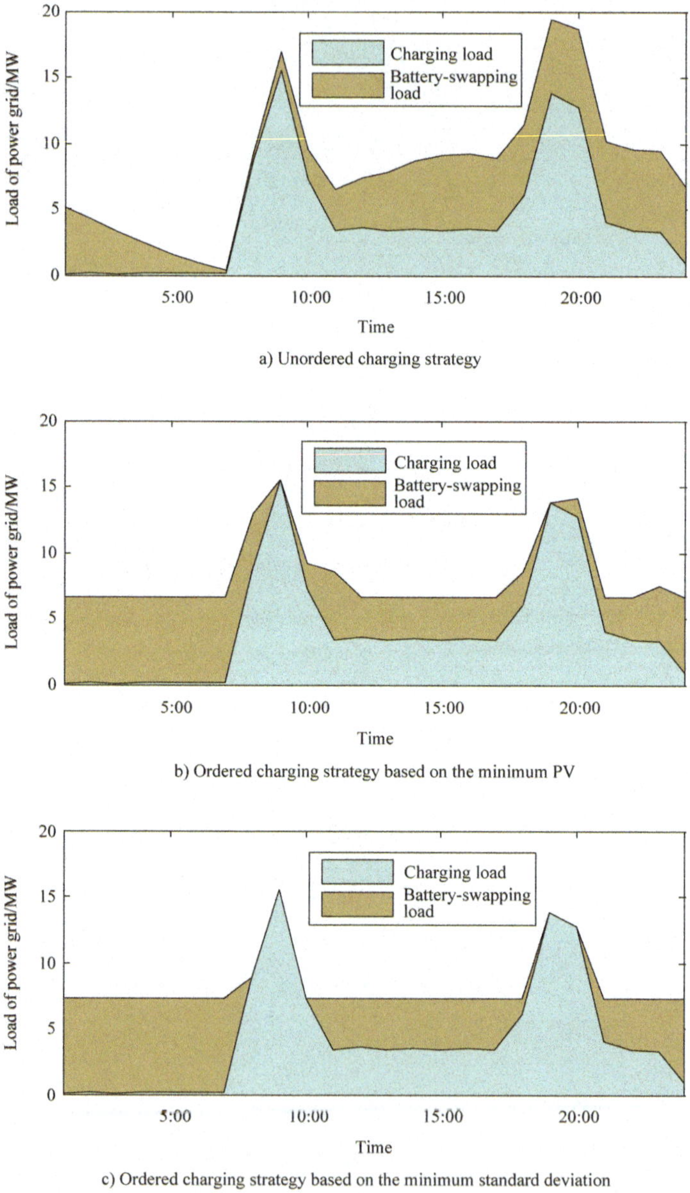

Fig. 8.34 Scheduling result when the proportion of battery-swapping vehicles is 30%

vehicles under different battery charging strategies. The peak load in the power grid is 16.63 MW and the valley load is 0.58 MW under the "unordered charging strategy"; under the "ordered charging strategy" based on the minimum PV, the distribution of load in the power grid is even and the PV of the load is less than 0.01 MW; similarly, under the "ordered charging strategy" based on the minimum standard deviation, the

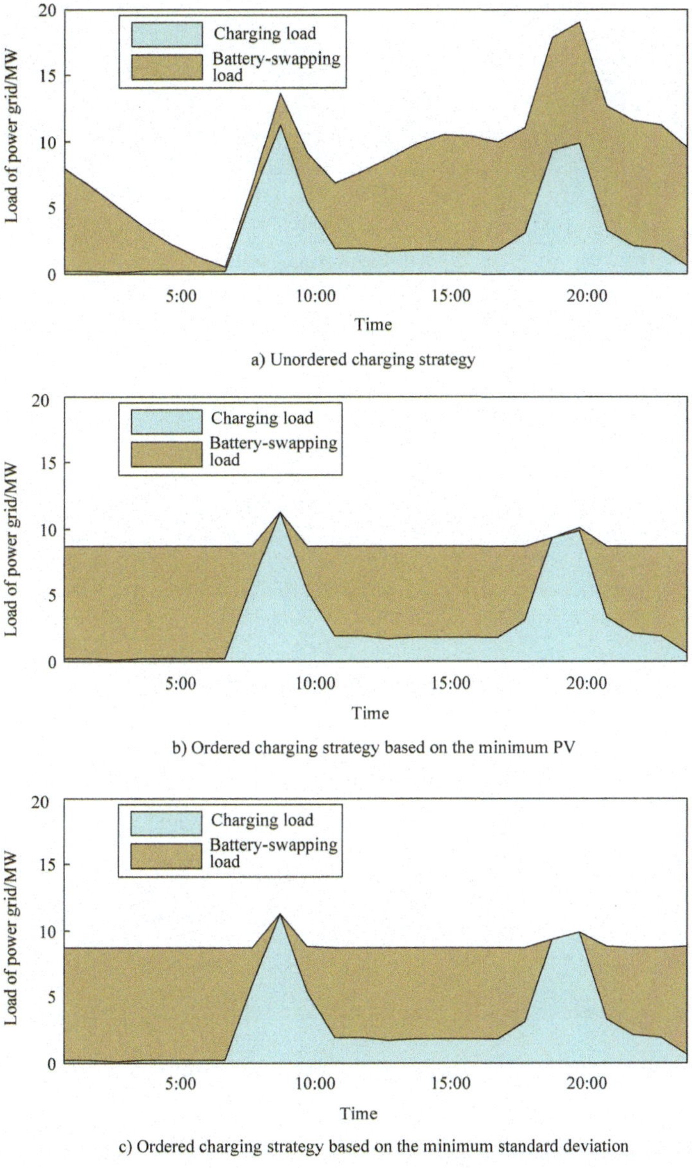

a) Unordered charging strategy

b) Ordered charging strategy based on the minimum PV

c) Ordered charging strategy based on the minimum standard deviation

Fig. 8.35 Scheduling result when the proportion of battery-swapping vehicles is 50%

peak load in the power grid declines by 44.44% and the valley load rises by 14.90 times as compared with the "unordered charging strategy."

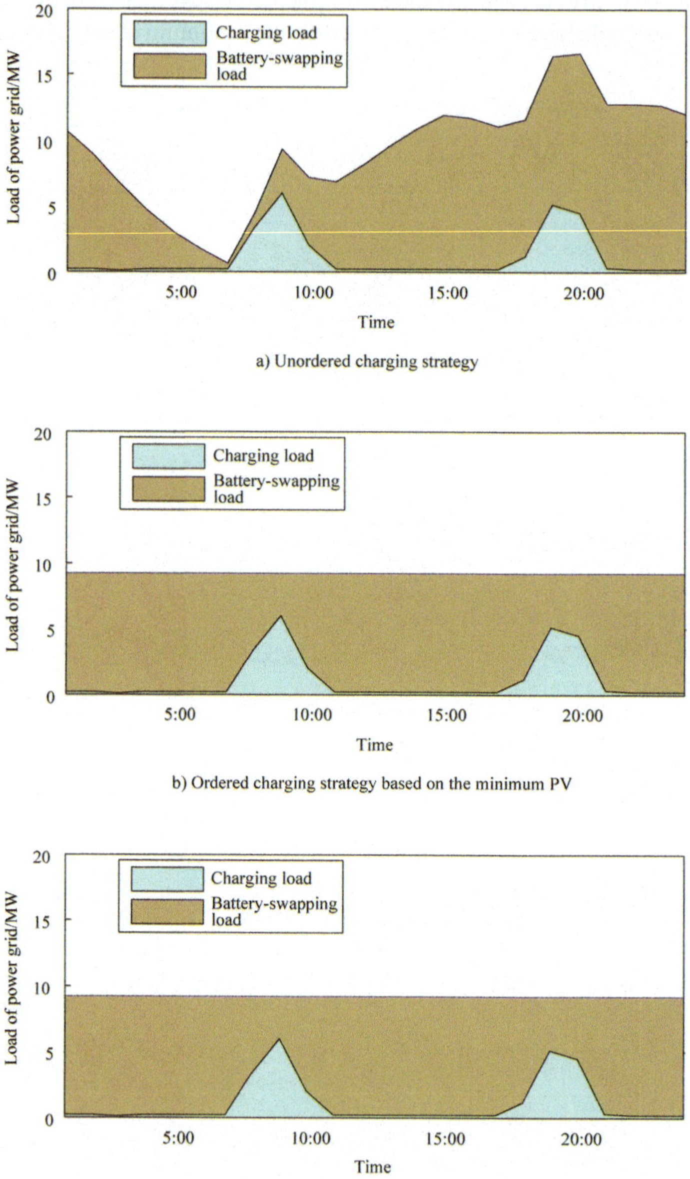

a) Unordered charging strategy

b) Ordered charging strategy based on the minimum PV

c) Ordered charging strategy based on the minimum standard deviation

Fig. 8.36 Scheduling result when the proportion of battery-swapping vehicles is 70%

As shown in Figs. 8.33, 8.34, 8.35 and 8.36, the load of charging stations gradually declines and the load of battery-swapping stations gradually increases, and the peak value of the charging load gradually declines with the increase in the proportion of battery-swapping vehicles. It can be seen from the simulation results that under the "unordered charging strategy", battery-swapping stations immediately charge the batteries after the swapping, leading to the superposition of the load of battery-swapping stations and the peak load of charging stations, which is not beneficial to the operation of the power grid. On the other hand, under the "ordered charging strategies" based on the minimum PV and the minimum standard deviation, the charging time and power of battery-swapping stations are scheduled on the basis of the charging load of charging stations, so as to reduce the peak load growth of the power grid to the greatest extent, make the overall load distribution of the power grid more even, and improve the operation performance of the power grid.

Table 8.24 shows the statistical result of the PV and the standard deviation of the grid load under unordered and ordered charging strategies for the charging load of battery-swapping stations when the proportion of battery-swapping vehicles is 4%, 30%, 50% and 70%. The table indicates that the proposed ordered charging strategies based on the minimum PV the minimum standard deviation can reduce the standard deviation and PV of the grid load to a certain extent, thus optimizing the load curve of the grid.

(c) Analysis of the power grid operation

The voltage offset and the power loss of the power grid are the main indicators of the safety and economic performance of the power grid respectively, so we use them to evaluate the operation of the power grid and count the grid loss and the voltage offset of each power grid in a day, while checking the operation condition of the transmission network.

Since power loss is an economic indicator of the power grid operation, the average power loss at each moment of a day is evaluated; since voltage offset is an indicator of the safety of voltage operation, the maximum voltage offset of each power grid

Table 8.24 Standard deviation and PV of the grid load under different batter-swapping scheduling strategies

Charging strategy for swapped batteries	4%		30%		50%		70%	
	PV	SD	PV	SD	PV	SD	PV	SD
Unordered charging	20.039	6.093	19.063	5.0308	18.489	4.539	16.058	4.215
Ordered charging based on minimum PV	16.667	4.982	8.846	2.814	2.554	0.586	6.71E-05	2.22E-05
Ordered charging based on minimum standard deviation	16.667	4.982	8.214	2.307	2.556	0.564	0.015	0.004

at each moment of a day is evaluated. Figures 8.37 and 8.38 are respectively the statistical graphs of the average power loss rate and the maximum voltage offset rate of each power grid at each moment in a 24-h simulation.

As shown in Fig. 8.37, power grid 8 and power grid 4 have higher average power losses (4.68% and 4.16% respectively) under the "traditional charging scheduling strategy" and the "unordered charging strategy" while under the proposed "comprehensively optimized charging strategy" and the "ordered charging strategy," the

Fig. 8.37 Average power loss of each power distribution grid

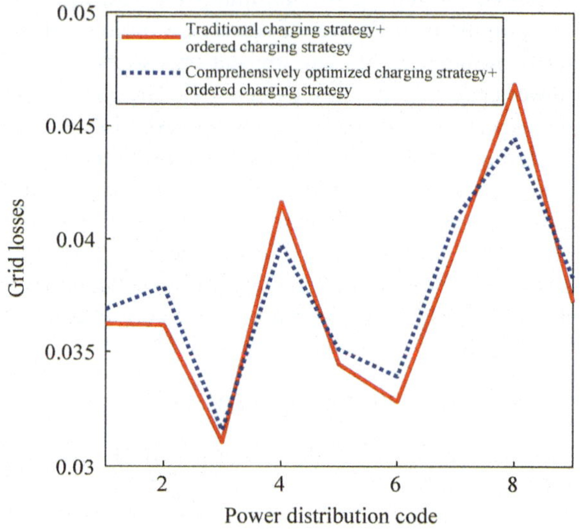

Fig. 8.38 Maximum voltage offset rate of each power distribution grid

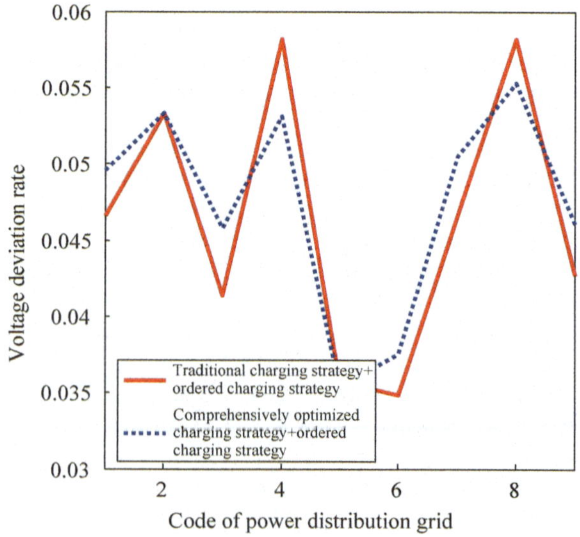

average power losses of power grid 8 and power grid 4 decline to 4.44% by 5.13% and to 4.00% by 3.85% respectively. It also can be seen from the figure that the average power loss rate of all power grids become more even.

As shown in Fig. 8.38, power grid 4 and power grid 8 have higher maximum voltage rates (5.82% and 5.81% respectively) under the "traditional charging scheduling strategy" and the "unordered charging strategy" while under the proposed "comprehensively optimized charging strategy" and the "ordered charging strategy," the higher average power losses of power grid 4 and power grid 8 decline to 5.31% by 8.76% and to 5.53% by 4.82% respectively. The maximum voltage offset is reduced to make the operation of the grid safer.

To evaluate the safety of the transmission network, the maximum offset rate of the power grid is 2.08%, which is within a reasonable range, indicating that the power grid's operation is safe.

(d) Analysis of the simulation result on the charging station side

In this section, the simulation results on the charging station side under different charging scheduling strategies are analyzed. The number of vehicles at charging stations and the waiting time for charging under the "traditional charging strategy" and the "comprehensively optimized charging strategy" are counted and analyzed, and the advantages and disadvantages of the two strategies are compared.

(1) Analysis of the number of vehicles at charging stations

Figures 8.39 and 8.40 show respectively the number of charging vehicles in each electric private car charging station and electric taxi charging station under different charging scheduling strategies.

Figures 8.39a and 8.40b respectively show the number of charging vehicles in each electric private car charging station under the "traditional charging strategy" and the "comprehensively optimized charging strategy." It can be seen from the figure

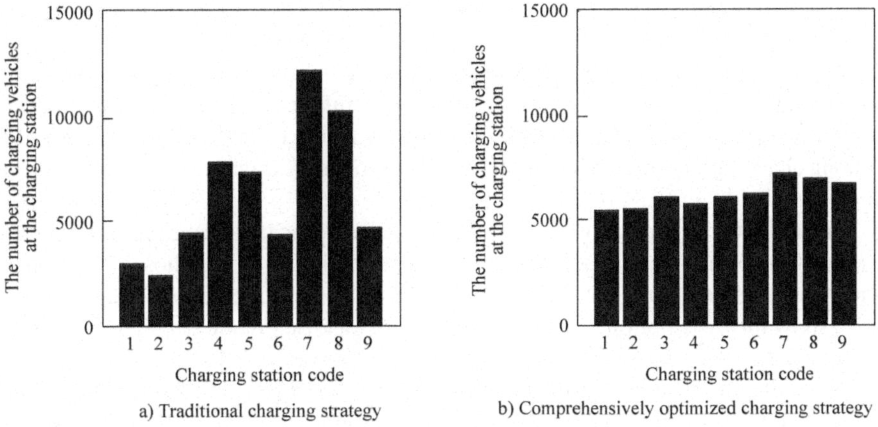

Fig. 8.39 Number of charging vehicles in each electric private car charging station

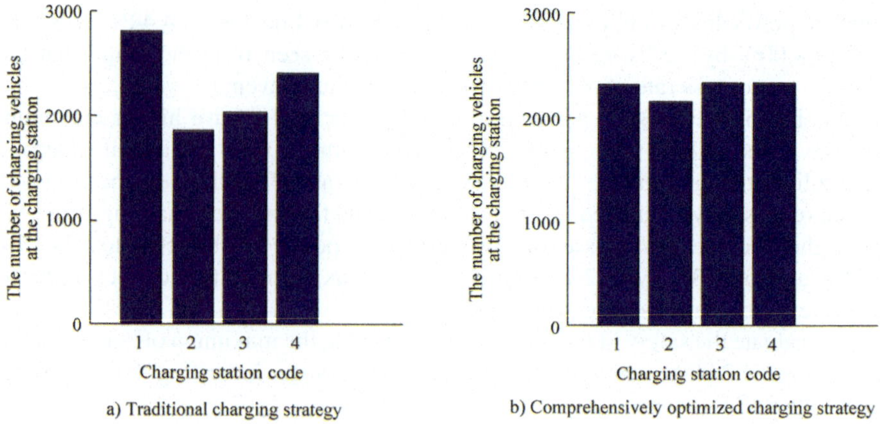

Fig. 8.40 Number of charging vehicles in each electric taxi charging station

that charging station 7 and 8 have larger numbers of vehicles (12,100 and 10,200 respectively) while charging station 2 only has 2400 vehicles; once the "comprehensively optimized charging strategy" is used to schedule the charging of electric private cars, the number of charging vehicles in charging station 7 and 8 declines to 7300 by 39.67% and 7000 by 31.37% respectively, and the same number at charging station 2 rises to 5500 by 1.29 times, resulting in a more even distribution of vehicles among charging stations for electric private cars.

Figure 8.40a and b respectively show the number of charging vehicles in each electric taxi charging station under the "traditional charging strategy" and the "comprehensively optimized charging strategy." It can be seen from the figure that compared with the "traditional charging strategy", using the "comprehensively optimized charging strategy" to schedule the charging of electric taxis makes the peak number of taxis at relevant charging stations decline from 2800 to 2300 by 17.86%, and the valley number rise from 1900 to 2100 by 10.53%, resulting in a more even distribution of vehicles among charging stations for electric taxis.

In summary, the number of charging vehicles in each charging station varies greatly under the"traditional charging strategy" while scheduling with the "comprehensively optimized charging strategy" results in a more even distribution of vehicles among charging stations and an improved operation performance of charging stations the power grid.

Since electric private car charging station 7 and electric taxi charging station 1 have higher vehicle numbers under both charging strategies, they are chosen to compare the distribution of vehicle numbers at different moment; for example, Figs. 8.41 and 8.42 show the distribution of vehicle numbers of electric private car charging station 7 and electric taxi charging station 1 at different moments.

As shown in Fig. 8.41, during the morning and evening peak hours, the maximum vehicle number in the private car charging station is 526 under the "traditional charging strategy" while the same number is only 333 under the "comprehensively

Fig. 8.41 Number of charging vehicles in private car charging station 7

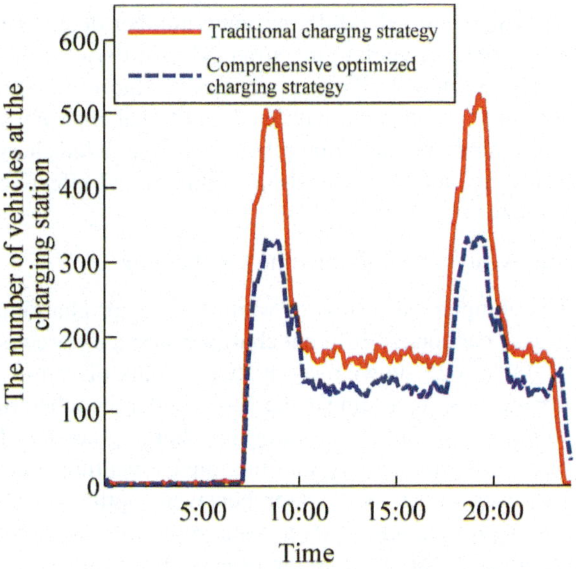

Fig. 8.42 Number of charging vehicles in taxi car charging station 1

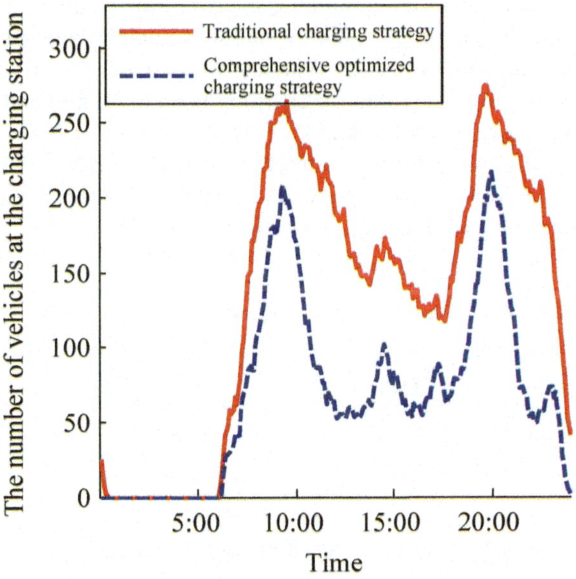

optimized charging strategy"; in off-peak hours, the "comprehensively optimized charging strategy" reduces the vehicle number in private car charging station 7 as compared with the "traditional charging strategy."

It can be seen from Fig. 8.42 that during the morning and evening peak hours, the maximum vehicle number in the taxi charging station is 275 under the "traditional charging strategy" while the same number declines to 217 under the "comprehensively optimized charging strategy"; similarly in off-peak hours, the "comprehensively optimized charging strategy" reduces the vehicle number in taxi charging station 1 as compared with the "traditional charging strategy."

In summary, the "comprehensively optimized charging strategy" can significantly reduce the number of charging vehicles and effectively avoid overload of charging stations.

(e) Analysis of waiting time for charging

The comparison results of waiting time for charging of electric taxis and electric private cars under different charging strategies are respectively counted.

Table 8.25 shows the statistical results of waiting time for charging of electric private cars. According to the Table, under the scheduling of the "traditional charging strategy", private cars waiting for charging account for 4.38% of the total number of electric private cars, and the longest waiting time for charging is up to 45 min. However, under the "comprehensively optimized charging strategy", private cars waiting for charging only account for 0.06% of the total number of private cars, and the longest waiting charging time is only 5 min.

Table 8.25 shows the waiting time for charging of electric taxis. According to the Table, under the "comprehensively optimized charging strategy," the longest waiting time for charging is 40 min and the vehicles waiting for charging account for 0.08% of the total number of vehicles; while under the "traditional charging strategy," the longest waiting time for charging is up to 70 min and 2.2% of vehicles have to wait for charging (Table 8.26).

In summary, the "comprehensively optimized charging strategy" can significantly improve the charging efficiency of electric private cars and electric taxis, and make driving and charging more convenient.

Table 8.25 Wating time for charging of electric private cars

Time (min)	Traditional charging strategy	Comprehensively optimized charging strategy
0–10	3296	156
10–30	6127	0
30–45	1276	0

Table 8.26 Wating time for charging of electric taxis

Time (min)	Traditional charging strategy	Comprehensively optimized charging strategy
0–10	1160	957
10–30	696	915
30–60	1565	93
60–70	203	0

References

1. Ding X, Yonghua S, Zechun H et al (2013) Participation of electric vehicles in the study of the optimal peak-valley electricity price of V2G. Proc Chin Soc Electr Eng 33(31):15–25
2. Ciwei G, Liang Z (2011) Overview of the impacts of electric vehicle charging on the power grid. Power Syst Technol 35(02):127–131
3. Zhiguo W, Jizhen L, Wen T et al (2006) Research on plant-level load distribution of thermal power plant based on rapidity and economy multi-objective optimization. Proc Chin Soc Electr Eng 26(19):86–92
4. Hui W, Guibin W, Junhua Z et al (2013) Planning of charging stations for electric vehicles considering traffic network flow. Autom Electr Power Syst 37(13):63–69, 98
5. Yan C, Chun L, Feifei X et al (2010) Updated scheme of an urban traffic network model and its implementation. J Transp Inf Safety 28(4):39–42, 48
6. Tao Z (2014) Large-scale charging scheduling of electric vehicles with integrated road network information and grid information. Tsinghua University, Beijing
7. Yugong L, Yiyao Y, Tao Z et al (2014) Optimal charging path planning method for intelligent electric vehicles. J Eng Stud Eng Interdisc Perspect 6(1):92–98
8. Yiyao Y, Yugong L, Tao Z et al (2015) Optimal charging path recommendation strategy for electric vehicles based on grid and traffic network information. Proc Chin Soc Electr Eng 35(02):310–318

Chapter 9
Intelligent Travel Planning for Electric Vehicles Based on Multi-network Fusion and CVIS

9.1 Application Background

To solve social problems such as energy security, environmental pollution, traffic congestion, and traffic accidents caused by the widespread use of traditional internal combustion engine vehicles (ICEVs), many countries are vigorously promoting clean and environment-friendly new energy vehicles (NEVs) represented by battery electric vehicles (BEVs). Compared with ICEVs, electric vehicles (EVs) use a wide range of power sources, including wind, solar and nuclear energy, which can significantly reduce carbon dioxide emissions caused by fossil fuel consumption. Also, EVs have high energy efficiency, and braking energy regeneration can further reduce energy consumption during driving. However, some problems hinder the rapid development of EVs, such as limited mileage, limited cycle life of energy storage batteries, long charging time, and insufficient charging stations. Therefore, it's one of the imminent priorities in the development of EVs is to improve the performance of EVs, so that they can fit drivers' travel plans and use habits.

With the rapid development of Intelligent Transportation Systems (ITS) in recent years, including Vehicle to Vehicle (V2V) communication, Vehicle to Infrastructure (V2I) communication, Geographic Information System (GIS), and Global Position System (GPS), etc., a variety of traffic information is available to effectively improve the vehicles' perception of traffic environment ahead and help them to properly arrange the journey in the future. For EVs, a long-timescale travel planning can help them obtain the traffic environment information, road topography information, and energy network information such as the distribution of charging stations in the roads ahead to enable them to optimize travel strategies and travel routes, reduce travel time and energy consumption; in the process of driving and control, by knowing the current position, speed, acceleration, pedal signal and other information of the preceding and surrounding vehicles, EVs can model and predict the movement behaviors of surrounding vehicles, and rationally optimize the power output and driving velocity in combination with the information of road slope changes, to ensure safe driving

K. Li et al., *The Intelligent Environment Friendly Vehicle*,
Key Technologies on New Energy Vehicles,
https://doi.org/10.1007/978-981-19-4851-0_9

while saving energy. Therefore, through the fusion of heterogeneous information of multi-network systems such as the Internet of Vehicles (IOV), road networks, traffic networks, and energy networks, and coordinated control of Cooperative Vehicle Infrastructure System (CVIS), there is a potential for further improvement in the optimization of the travel strategy and the effect of driving process control of EVs. For BEVs, given the inherent limitations of current EVs, how to use the information of ITS to improve the performance of EVs and meet the needs of drivers is an important research topic and frontier research direction.

The research on the travel decision and driving and controlling of EVs based on the multi-network fusion of "road-transport-energy" and the multi-system collaboration of "human-vehicle-infrastructure" includes three key parts:

(1) EV travel planning

There is a strong mutual coupling between the travel constraints, travel objectives, and travel strategies of EVs, and the degree of optimization of travel strategies directly determines whether the travel process of EVs can meet the needs of drivers. However, due to the time-varying stochastic characteristics of the traffic environment, as well as the interaction between travel route and travel speed, charging behavior decision, use of electric accessories, and other behaviors in travel strategy, it is necessary to design a corresponding EV travel optimization method, learn about a variety of information such as traffic information, charging station information, and weather information of the road ahead through V2I communication, and take into consideration multiple travel objectives and travel constraints to optimize the travel strategy and obtain a better travel scheme.

(2) Energy-efficiency driving control of EVs

Energy consumption of EVs during driving is affected by the road gradient and the movement of the preceding vehicle. In actual driving, it is often impossible to know the full movement behavior of the preceding vehicle in advance. Therefore, when energy-efficiency assistant control is performed on EVs, usually only the movement information of the preceding vehicle at the current moment can be used. Based on wireless interactive communication information such as V2V communication and V2I communication, EVs can not only obtain the information of the road ahead and the movement information of multiple vehicles ahead at the current moment, but also the historical driving data and movement model information of the preceding vehicles. Therefore, we need to design a reasonable energy-efficiency control method for EVs, make proper use of gradient information and the current movement information and movement model information of preceding vehicles, to optimize the driving/braking process and speed of EVs, and reduce energy consumption without a compromise on driving safety and comfort.

(3) Energy-efficiency driving control of EV fleet

Using V2V and V2I communication information can not only reduce the energy consumption of individual EVs, but also optimize the driving process of the EV fleet and reduce the energy consumption of the fleet. Since EVs adopt centralized-drive or distributed-drive structures, the optimal control method of EV fleet should be able to take all these two configurations into account. The energy-efficiency control method for EV fleet should be able to reduce the total driving energy consumption of the fleet without compromise on driving safety and comfort, while taking into account various EV configurations, multiple communication architectures, and information input to achieve safe and energy-efficient driving of the EV fleet.

The problem of vehicle travel planning, which is generally simplified into the Shortest Path (SP) problem in the network, is a typical problem in graph theory and network optimization. As early as the 1950s, some scholars conducted in-depth research on the SP problem by using the Bellman [1] principle to calculate the exact SP in a static deterministic network, among which the representative algorithms include the Dijkstra [2] algorithm, the Ford [3] algorithm and the A* [4, 5] algorithm.

Through interaction with the transportation system, connected electric vehicles can obtain information including the road traffic conditions, the location and status of the charging station, and weather changes of the way ahead to make reasonable travel schemes, including the travel path, driving velocity, charging station recommendation, charging mode, charging time, use of electrical accessories, etc., to improve the performance of EVs and meet the travel needs of drivers. The study of travel planning of connected electric vehicles, according to the established forms of optimization, can be divided into two categories: single-objective travel planning and multi-objective travel planning. Among them, the single-objective travel planning only optimizes a main objective in the travel process while ignoring other travel objectives to obtain the optimal travel scheme. In contrast, for the multi-objective travel planning, it is necessary to coordinate multiple travel objectives, formulate the weight coefficient of the optimization objectives, or search for the Pareto solution set in the sense of multi-objective, to ensure that the travel scheme obtained by search optimization has integrated multiple travel objectives.

The energy consumption of EVs in driving is affected by many factors such as driving velocity, road gradient, movement of the preceding vehicle, and traffic lights. With the rapid development of ITS, connected electric vehicles can exchange information with surrounding vehicles, road base stations, traffic light base stations, GPS and GIS, and other information sources, so that EVs can obtain more accurate information about the traffic environment ahead, and further effectively optimize the driving velocity and motor torque of EVs based on such information, and reduce the energy consumption while ensuring driving safety.

Through V2V and V2I communication technologies, vehicles on the road can obtain information such as the speed, position, acceleration, and steering wheel angle of surrounding vehicles, as well as road information such as road curvature, road gradient, road attachment, and traffic light timing, which provide an information basis for the energy-efficiency assistant control of the vehicle, and help to reduce energy consumption. Due to the limited driving distance of EVs, the use of V2V and V2I communication technologies to achieve energy-efficiency assistant control and improve the energy utilization rate is very critical for improving the performance of EVs.

To improve the performance of EVs by using information connectivity technology, some scholars have carried out studies on travel planning of EVs and energy-efficiency assistant driving, but still many problems have not been solved:

(1) Existing travel planning methods for EVs ignore the complex coupling between multiple travel objectives, travel behaviors, travel constraints, as well as the dynamic characteristics of the traffic environment. Existing research on EV travel planning often only considers a single travel objective and ignores the strong mutual coupling between multiple travel objectives, travel behaviors, and travel constraints. The incomplete discussion on the relationship of these three factors makes it difficult for travel planning to meet drivers' travel needs. Also, previous studies related to travel planning of EVs often assume that the traffic environment is static and deterministic, that is, the traffic speed (or road weight) is a definite value which does not change with time. This assumption ignores the dynamic stochastic change characteristics of the average road speed in the traffic environment, resulting in travel optimization results that fail to reflect the real traffic environment.
(2) Existing energy-efficiency assistant control methods for EVs have not yet made full use of V2V communication technology, and only taken into account the use of information of current surrounding vehicles, and cannot predict the movement of preceding vehicles. Existing researches on energy-efficiency driving control of traditional vehicles and new energy vehicles (NEVs) generally assume that there is no preceding vehicle or that the movement of ego vehicle is known before the trip. This assumption can often ensure that the global optimal control strategy is obtained for the optimization objective of the vehicle. However, because this type of method ignores the unknowability of the movement of the preceding vehicle, it is not suitable for working scenarios where the movement of the preceding vehicle is unknown. In existing related studies that take into account the movement behavior of the preceding vehicle, the V2V communication information, the historical travel data of the preceding vehicle and the movement data of several vehicles ahead are not fully utilized in the control process, and the prejudgment of future movement behavior of the preceding vehicle is lacking, which compromises the control effect of the EVs.

9.2 Overall Design of Travel Planning and Energy-Efficiency Control System for Connected Electric Vehicles

The travel planning and intelligent energy-efficiency control system of connected electric vehicles optimize the whole journey of EVs by utilizing various traffic environment information provided by the wireless communication system. Considering the imperfect travel planning system and insufficient utilization of information in the longitudinal control of EVs, a travel planning and intelligent energy-efficiency control system for EVs that makes full use of wireless communication information is designed to comprehensively optimize the travel scheme formulation and driving energy-efficiency effect of EVs.

9.2.1 Systematic Architecture Design

Connecting EVs means using wireless communication information technology to enhance EV's perception of the traffic environment, transfer the long timescale information including road information, charging station information, and weather information, and the short timescale information including current movement information of surrounding vehicles and current road gradient to EVs based on the V2I and V2V communication. Later, the EVs can effectively use this information to reasonably make future travel arrangements and achieve safe and energy-efficient driving. According to the time scale of the obtained information, the key technologies of connected electric vehicles include two parts: (1) travel planning of connected electric vehicles based on traffic environment information. (2) energy-efficiency control of connected electric vehicles based on movement information of surrounding vehicles. The system schematic diagram is shown in Fig. 9.1.

It can be seen from Fig. 9.1 that before a connected electric vehicle travels, it is necessary to clarify its travel needs including destination, estimated time of arrival, and SOC value of the destination, and then obtain information about traffic environment, charging stations and weather based on V2I communication technology to reasonably make a travel scheme, charging scheme, and scheme of using electric accessories. Then, in the travel process of the connected electric vehicles, it is necessary to carry out intelligent energy-efficiency control based on the real-time information of surrounding vehicle movement and road gradient to optimize the

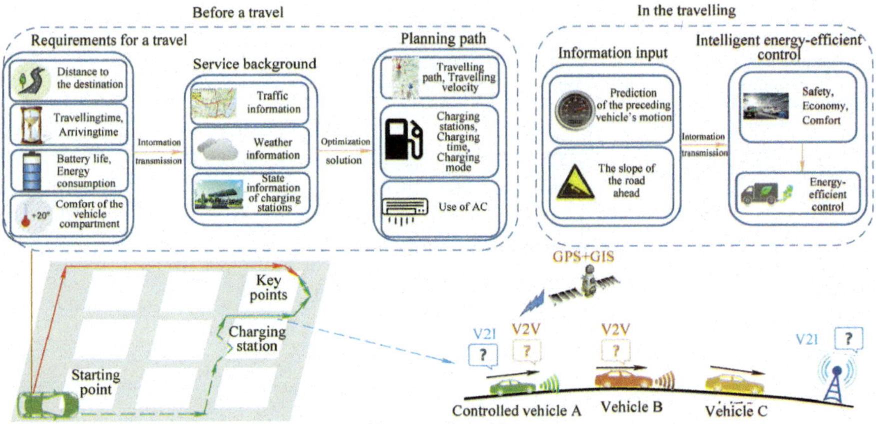

Fig. 9.1 Schematic diagram of travel planning and intelligent energy-efficiency control system

motor torque and driving velocity in real-time, and to reduce energy consumption during driving. It can be seen that throughout the travel task from pre-travel to the process of travel, both V2V and V2I communication technologies can provide EVs with traffic environment information which is a basis for their travel planning and driving control, and enable them to reasonably make travel schemes, optimize energy consumption during driving, and improve the existing performance of EVs.

After the confirmation of the above system functions, the overall structure of the multi-objective travel planning and intelligent energy-efficiency control system for connected electric vehicles is designed, as shown in Fig. 9.2. The information transmission and logical relationship between the sub-parts are clarified, and the overall scheme of multi-objective travel planning and intelligent energy-efficiency control is comprehensively established according to the functions of the system designed in this chapter.

Wireless communication technology is used to obtain road network information $\varsigma = (\Gamma, \gamma, P, T, \phi)$ for connected electric vehicles, including road network node set Γ, node connection relationship γ, charging station information P, time interval set T, average speed of road network ϕ, etc. Then, based on the travel tasks of EVs, appropriate travel planning methods are adopted to coordinately optimize the travel, charging, and electric accessories usage schemes of EVs. The optimization of the travel schemes of the connected electric vehicle provides the basis for its driving control process. Combined with the road length l and gradient information α of the selected path in the optimization scheme of travel planning, intelligent energy-efficiency control can be carried out for the connected electric vehicles and the

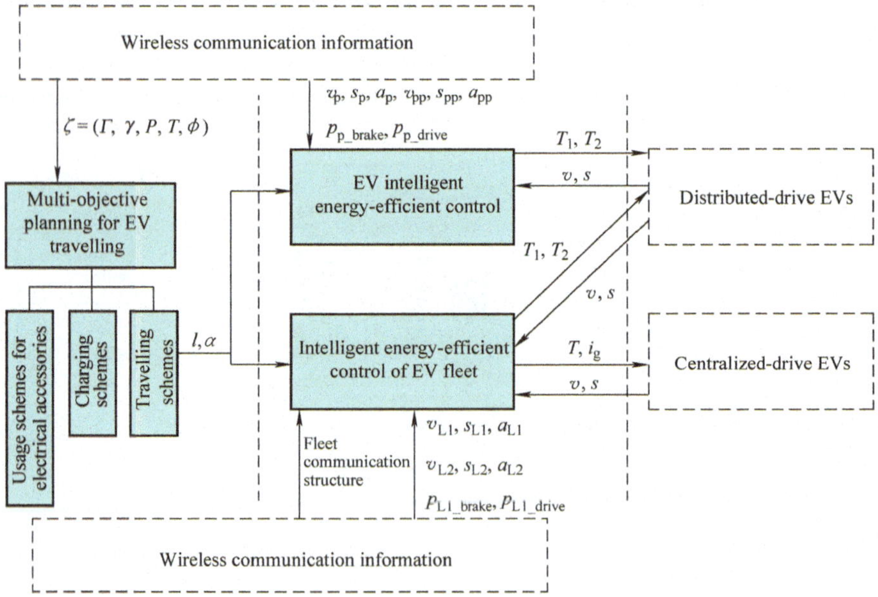

Fig. 9.2 Overall structure of travel planning and intelligent energy-efficiency control system

connected electric vehicle fleet. For intelligent energy-efficiency control of a single EV, in addition to using the existing road information, it is also necessary to use the movement information of the two preceding vehicles, p and pp, including velocity v_{p} and v_{pp}, acceleration a_{p} and a_{pp}, position s_{p} and s_{pp}, pedal position $p_{\mathrm{p\_brake}}$ and $p_{\mathrm{p\_drive}}$ etc., to predict the future movement of the adjacent preceding vehicle p, and then carry out energy-efficiency control of the ego vehicle, and obtain the front and rear wheel drive torque T_1 and T_2 of the distributed-drive EVs after optimization. For intelligent energy-efficiency control of EV fleet, the movement information of preceding vehicles, L_1 and L_2, can also be used to predict the future movement of L_1. At the same time, considering the different communication structures of the fleet, the torque T and gear i_g of the centralized-drive EVs and the torque T_1 and T_2 of distributed-drive EVs in the fleet can be controlled and optimized to reduce the total energy consumption of the EV fleet.

9.2.2 *Design of Travel Planning for EVs*

Considering the unique characteristics of EV travel planning, as well as the complex mutual coupling relationship between the travel objectives, constraints and travel plan, the travel framework of EVs is sorted out, as shown in Fig. 9.3.

It can be seen from Fig. 9.3 that five main objectives need to be considered in the travel process of EVs, which are travel path length, travel time, in-vehicle temperature, battery cycle life, and vehicle energy consumption. For the travel path objective, due to the limited mileage, EVs may need to go to the charging station for charging. Therefore, unlike traditional vehicles, the optimal travel path is no longer the shortest path between the starting point and the destination, but a path that considers both the travel path and the location of charging stations. Time objective is

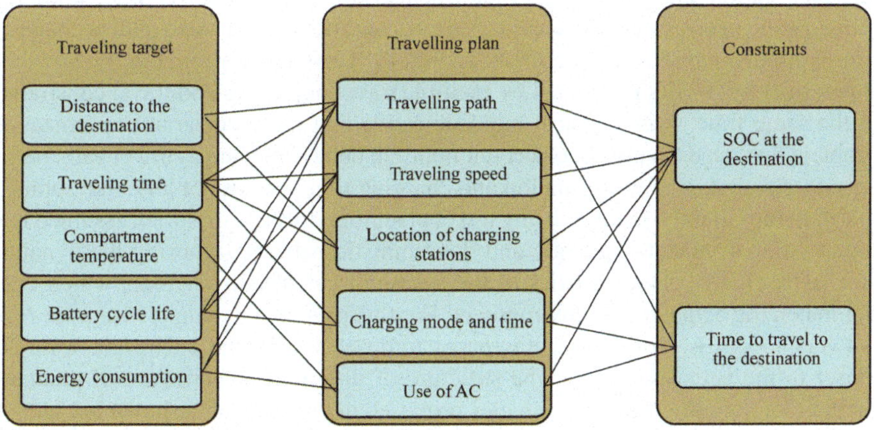

Fig. 9.3 Framework of travel planning for EVs

composed of charging time and travel time, in which the travel time is related to the length of the road and the traffic speed on each road. Average in-vehicle temperature is related to the using of the air conditioning during driving and weather changes, which will affect in-vehicle comfort. Cycle life of the battery is related to the charging and the discharging processes of the battery. The charging process is determined by the charging current (that is, the charging level) and the corresponding charging time, and the discharging process is determined by discharging current value (that is, the travel speed) and the discharging time (travel time). Energy consumption is composed of driving energy consumption and energy consumption of electric accessories. The driving energy consumption is related to the travel path and travel speed in travel strategy, and the air conditioning accounts for a relatively large proportion of the energy consumption of electric accessories, so the use of air conditioning in the process of travel will also affect the total energy consumption of vehicles.

The travel constraints of EVs, considering the driver's needs when arriving at the destination, generally include the SOC value and the time constraints at the destination. To constrain the SOC value upon the arrival at the destination is to ensure that the EVs can complete the potential travel tasks. The constraint is related to the energy consumption process and the charging process in traveling. For example, when the SOC constraint value upon the arrival at the destination is high, the travel scheme will tend to reduce the energy consumption during the travel process, choose the charging station at an appropriate location, higher level of charging mode, and longer charging time, and reduce the use of air conditioning to save battery energy. Therefore, this constraint is related to multiple travel behaviors in the travel strategy. The time constraint upon the arrival at the destination is to ensure that the time for the EV to arrive at the destination is no later than the scheduled arrival time to meet the travel needs of the driver and avoid being late. Arrival time constraint is also associated with multiple travel strategies, in which the main factors influencing the travel time are travel path and speed. At the same time, there may be charge behaviors on the way considering the low battery energy storage, so the location of charging stations and the charging time will also affect the total travel time. Similarly, reducing the use of electrical accessories (air conditioning) can also reduce charging time without compromise of comfort. Therefore, it can be seen from Fig. 11.3 that the travel strategy of EVs is affected by multiple travel objectives and travel constraints at the same time. It is a typical multi-objective and multi-constraint optimization problem that needs to take into account multiple behaviors such as travel path, travel speed, selection of charging station and charging time, and use of air conditioning.

More importantly, in the existing travel planning research, it is often assumed that the road environment is a static and deterministic network, ignoring the dynamic stochastic change characteristics of the average road speed over time, which does not match the actual traffic environment. For this reason, a dynamic stochastic road network model is proposed in this paper to describe a traffic environment model closer to the actual situation. The structure of the multi-objective travel planning method for connected electric vehicles proposed in this chapter is shown in Fig. 9.4.

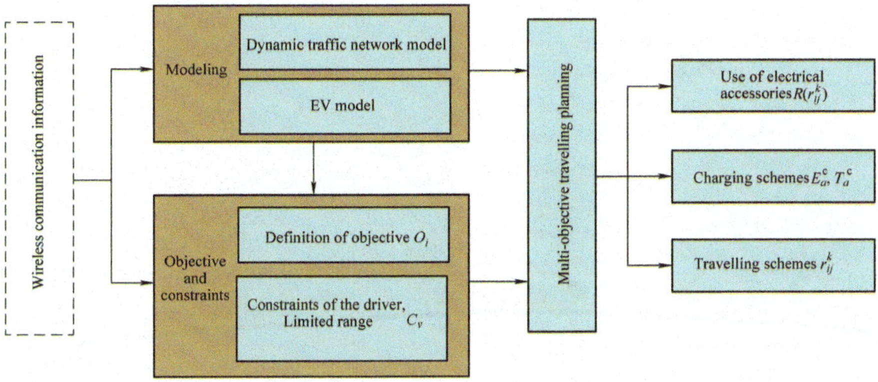

Fig. 9.4 Solution set of travel planning for EVs

It can be seen from Fig. 9.4 that the multi-objective travel planning of connected electric vehicles mainly includes three parts, namely model establishment, optimization objective and constraint establishment, and multi-objective travel scheme optimization solution: (1) To obtain road environment information in combination with wireless communication information. The model establishment includes a traffic network model that can describe dynamic stochastic change characteristics, and an EV model that takes into account vehicle energy consumption and charging characteristics. (2) The optimization objective and constraint establishment define the various objectives O_i involved in travel planning, and define various constraints C_v, such as driver constraints and vehicle mileage constraints. (3) The multi-objective travel planning method is used to obtain the solution set of travel schemes for EVs. When the cloud service platform obtains the model information and the objective and constraint information, the optimization results include the use of air conditioning $R(r_{ij}^k)$ on each path r_{ij}^k, the selection of charging stations, the charging energy E_a^c and time T_a^c, and the selection of travel path r_{ij}^k, etc. Then the cloud service platform transmits the obtained travel scheme back to the EV, and the computing time of the optimization algorithm is guaranteed by the cloud service platform.

9.2.3 *Intelligent Energy-Efficiency Control of EVs*

It helps to reduce the energy consumption of EVs during driving by using V2I and V2V communication to acquire the gradient information of the road ahead and the movement information of surrounding vehicles in real-time, and rationally optimizing the drive torque and velocity of EVs on the premise of ensuring safety. In this study, the application scenario of intelligent energy-efficiency control of EVs is shown in Fig. 9.5.

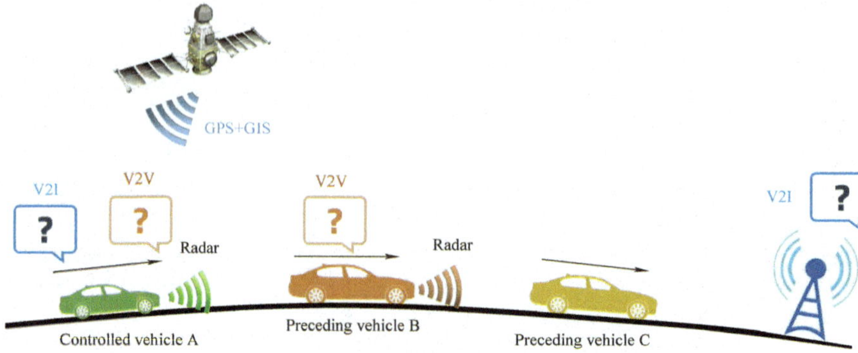

Fig. 9.5 Schematic diagram of intelligent energy-efficiency control condition of EVs

In Fig. 9.5, vehicle A is a controlled EV, and there are two preceding vehicles B and C in front. On the one hand, the driving energy consumption of vehicle A is affected by the gradient of the road ahead. Therefore, vehicle A needs to obtain information on the gradient of the road ahead based on GPS, GIS, and V2I communication. On the other hand, the movement of vehicle A is affected by the movement of the preceding vehicle B, and the movement of the preceding vehicle B is also affected by the movement of the preceding vehicle C. Therefore, it is necessary to obtain the current movement information of vehicle B and vehicle C, as well as vehicle B's historical movement information and movement behavior model information based on the V2V communication technology, and then predict the movement of the preceding vehicle B in each control cycle of vehicle A, optimize and control the movement of ego vehicle A and reduce its driving energy consumption on the basis that a reasonable distance from the preceding vehicle B is guaranteed. Compared with the traditional energy-efficiency assistant driving method, the intelligent energy-efficiency control method makes full use of the V2V communication technology to not only obtain the current movement information of the surrounding vehicles, but also the movement behavior model information of the surrounding vehicles. The future movement behaviors of surrounding vehicles are predicted base on this model, so that the controlled EV can accurately learn the future movement behaviors of surrounding vehicles and improve the control accuracy of the ego vehicle. The structure of the intelligent energy-efficiency control system for EV is shown in Fig. 9.6.

The system acquires the information about the gradient change of the roads ahead through V2I communication, and the current movement information and movement model information of preceding vehicles through V2V communication, and then uses the Bayesian network method to predict the future movement behavior of preceding vehicles; the intelligent energy-efficiency controller takes the longitudinal dynamic model of the EV as the control object to establish a comprehensive performance indicator that coordinates safety, economy, and comfort according to the driving state of the vehicle as well as the gradient information of roads ahead and the prediction of preceding vehicles' movement. The Nonlinear Model Predictive Control (NMPC)

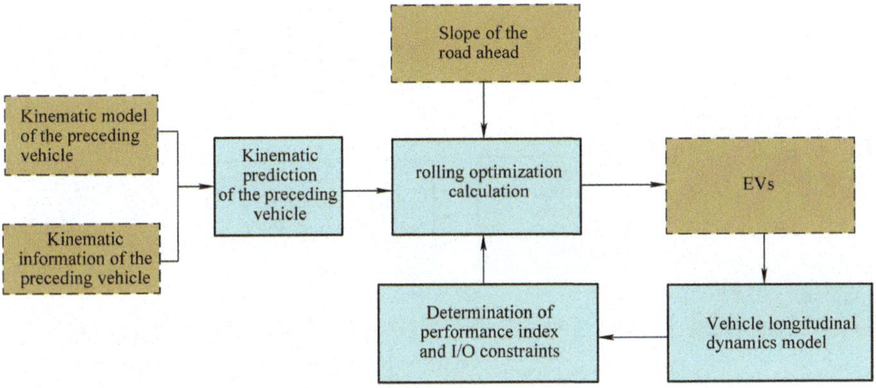

Fig. 9.6 Schematic diagram of the intelligent energy-efficiency controller for EVs

theoretical framework is used as the design basis and the rolling optimization is run to obtain the optimal control quantity, so as to reduce driving energy of the EV under the premise of ensuring proper relative distances between vehicles to extend the driving distance.

9.2.4 *Intelligent Energy-Efficiency Control of EV Fleets*

The intelligent energy-efficiency control system is extended to EV fleets after the research on a single EV is done. The application scenarios of the intelligent energy-efficiency control function of EV fleets proposed in this chapter are shown in Fig. 9.7.

In Fig. 9.7, the EVs coded i, $i-1$, …, 1 together form an EV fleet, and there are two preceding vehicles L1 and L2 in front of the fleet. During the driving process, the energy consumption of each vehicle in the fleet is affected by the gradient of the road ahead, so GPS and GIS systems are needed to transmit the gradient information to the fleet. At the same time, the movements of L1 and L2 are not known in advance, and their movement behaviors also affect the driving energy consumption of the EV fleet. Therefore, it is necessary to use V2V communication to obtain the movement

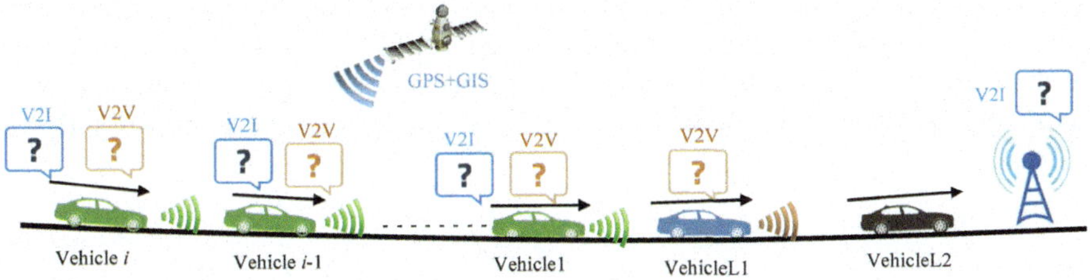

Fig. 9.7 Schematic diagram of the intelligent energy-efficiency control condition of EV fleets

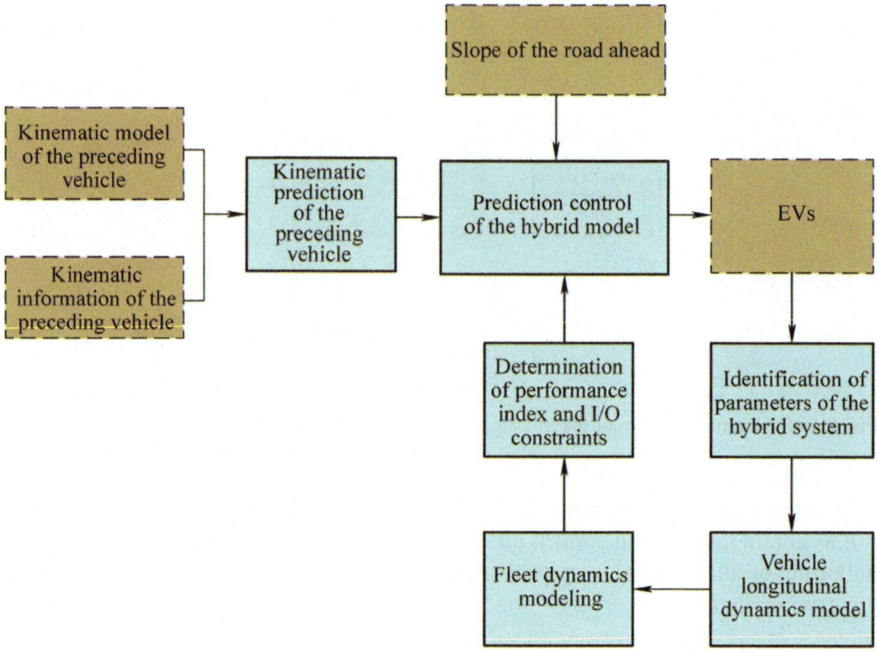

Fig. 9.8 Schematic diagram of the intelligent energy-efficiency controller for EV fleet

information of multiple vehicles ahead, and model the movement of the first vehicle L1 and predict its future movement to ensure more accurate future movement information of L1 to reduce the overall energy consumption of the fleet. The structure of the intelligent energy-efficiency control system for the EV fleet is shown in Fig. 9.8.

The gradient information of the road ahead is obtained through V2I communication, and the movement information of several vehicles in front of the fleet is obtained by V2V communication. The movement of preceding vehicle L1 is modeled by the existing Bayesian network model to predict its future movement. Considering there are EVs with centralized-drive and distributed-drive in the fleet, the longitudinal vehicle dynamic model is transformed into a piecewise affine model with the hybrid system parameter identification to establish a single vehicle dynamic model and expand it to a fleet dynamic model; the optimization objective and constraint function are established based on the economy, safety and comfort performance indicators, and the optimization problem is established by using the hybrid model predictive control method, and the receding horizon optimization is carried out to obtain the optimal control quantity, so as to reduce the total energy consumption of the fleet.

9.2.5 Technical Difficulties and Key Points

The technical difficulties and key points involved in travel planning and intelligent energy-efficiency control system of connected electric vehicles can be divided into three parts: design of multi-objective travel optimization method, intelligent energy-efficiency control of EV, and intelligent energy-efficiency control of EV fleet. Further explanation of the key technologies proposed is shown as follows.

(1) Design of the multi-objective travel optimization method

The purpose of travel planning for EV is to recommend a reasonable travel scheme before the trip, to ensure the completion of travel tasks. If the travel optimization fails to provide a better travel strategy, then the performance of EV will deteriorate, or the travel task cannot even be completed. Therefore, one of the key technologies of connected electric vehicles based on V2V and V2I communication is to recommend a reasonable travel scheme for EV before the trip. Considering the characteristics of EV such as limited mileage, limited battery life, and less battery energy storage, the V2I communication technology should be used to obtain traffic information, location, and state of charging stations, and weather information, etc., to synergistically optimize multiple travel behaviors, so as to meet the travel needs of EV. The travel behaviors that need to be optimized include travel path selection, travel speed, location selection of charging station, charging mode, charging time, use of electrical accessories, etc. Reasonable planning of these travel elements help to reduce the limitations of EV and improve its performance. How to conduct joint optimization of multiple travel behaviors, and obtain travel schemes that meet various travel constraints based on the integration of characteristics of EV and multi-source traffic information is one of the difficulties and focuses of the research.

(2) Intelligent energy-efficiency driving of EV

The main purpose of intelligent energy-efficiency control is to reasonably optimize the drive torque and speed of EV based on the movement information of surrounding vehicles, to reduce energy consumption and expand the reachable scope of EV. The communication information has not been fully utilized in the existing researches on the driving control based on V2V communication, and generally only the current movement information of surrounding vehicles is used. By using V2V communication technology, the movement behavior model of surrounding vehicles can be obtained, and the future movement state of each vehicle can be inferred, which will help to improve the energy-efficiency control effect of the EV controlled and further reduce energy consumption in driving. It's another focus and difficult point of research to establish the movement model of surrounding vehicles based on historical driving information, apply the movement prediction information to the energy-efficiency control optimization algorithm of the controlled EV, and optimize the drive torque and speed to achieve energy-efficiency and safe driving by synergistic consideration of optimization indicators such as economy, safety, and comfort.

(3) Intelligent energy-efficiency driving of EV fleet

The main purpose of the intelligent energy-efficiency driving control of EV fleet is to control vehicles in the fleet by using V2V and V2I communication information to achieve the lowest total energy consumption under the premise of safe driving. Considering that there are centralized-drive and distributed-drive EVs in the fleet, that is, a coexistence of the discrete state (shifting) and the continuous state, so a reasonable modeling method is required for the system to uniformly describe the EV fleet. At the same time, the designed intelligent energy-efficiency controller for EV fleet needs to be able to take into account the predicted movement information input of the vehicles in front of the fleet, different communication structures of the fleet, different communication information, different road gradients, and other different working scenarios, and describe different EV structures to optimize the drive torque of each EV under different working scenarios, ensuring the safe driving of the fleet and reducing driving energy consumption. This is also one of the focuses and difficult points of the research.

9.3 Multi-objective Travel Planning Method for Connected Electric Vehicles

Since BEVs have many limitations in applications, and their performances are unable to meet the driving needs of drivers, the travel behaviors, charging behaviors, and use of electric accessories can be optimized to improve the performance of connected electric vehicles by making full use of multi-source traffic environment information and coordination of multiple travel objectives and constraints.

Current research methods for travel planning are mainly designed for traditional vehicles with little consideration of the unique traveling characteristics of EVs. Given the existing shortcomings, in this section we will study the multi-objective travel planning methods for connected electric vehicles, and propose a travel planning method that can comprehensively consider various travel objectives, constraints, and strategies. Section 9.2 has explained the overall structural design of the travel planning method for connected electric vehicles. Firstly, a dynamic stochastic road network model should be established to describe the dynamic and stochasticity of the traffic environment, and an EV model including energy consumption, battery life, and air conditioning models should be established. Then, multiple travel objectives and constraints are defined considering the driving needs during the trip. Finally, a temperal multi-objective ant colony optimization method is proposed to optimize the travel schemes of EV. The solution set of travel schemes with multiple objectives is obtained through cyclic optimization, which provides the driver with a travel cheme including travel path, speed, charging mode and charging time, use of air conditioning and so on.

9.3.1 Systematic Model of Travel Planning Method

Compared with the existing travel planning methods, this study takes into account the time-varying and stochastic nature of the traffic environment, as well as the unique characteristics of the battery life and charging process during the travel of the EV, and establishes a systematic model of travel planning optimization to ensure a reasonable and effective travel scheme.

9.3.1.1 Building the Dynamic Stochastic Road Network Model

In reality, the road network is often in a dynamic and uncertain traffic environment, subject to various periodic (morning peaks and evening peaks) and non-periodical (weather changes, traffic accidents) influences, thus showing both time-varying and stochastic natures. Therefore, the dynamic stochastic road network model is adopted in this chapter to model the traffic environment. In this network, the average traffic speed of each road is both time-dependent and stochastically variable [6–8], that is, the average traffic speed of the road in this network is a stochastic variable subject to the time-varying probability density function. The dynamic stochastic road network model is defined as $\varsigma = (\Gamma, \gamma, P, T, \phi)$. Γ represents the set of nodes in the road network, $|\Gamma|$ represents the total number of nodes. $\gamma = \{\overrightarrow{pq}\,|p, q \in \Gamma\}$ represents the connection relationship of nodes in the road network, $\overrightarrow{pq}$ means that node p is connected to node q. $l(\overrightarrow{pq})$ and $\alpha(\overrightarrow{pq})$ represent the length and gradient of path $\overrightarrow{pq}$. U_p represents the set of points directly connected to node p, with a total of $|U_p|$ nodes. In the road network model, there are a total of P charging stations, each of which has C_w charging levels and adopts constant current charging. The time interval $T = \{t_0 + h\delta, t_0 + (h+1)\delta\}$ is the set of periods, in which $h = 0, 1, 2, \ldots, H$, and δ is the minimum time interval. In other words, if the moment the vehicle arrives at a certain road section in the road network is within the same time interval δ, the average traffic speed through the road section is subject to the same probability distribution; otherwise, the probability distribution is different. ϕ represents the probability set of the average traffic speed in the road network, including the probability distribution of each road section in various possible states.

$$\phi = \bigcup_{\forall \overrightarrow{pq} \in \gamma, \forall t \in T} \phi_{pq,t} \tag{9.1}$$

In Eq. (9.1), $\phi_{pq,t}$ represents the probability density function distribution that the average traffic speed of the road section $\overrightarrow{pq}$ at time t is subject to.

If nodes i and j are not adjacent, $R_{ij} = \{r_{ij}^k : k = 1, 2, \ldots, |R_{ij}|\}$ represents the set of paths between nodes i and j, where r_{ij}^k represents the path k connecting nodes i and j. The path r_{ij}^k can be expressed as $r_{ij}^k = \{r_{ij}^k(z) : z = 1, 2, \ldots, |r_{ij}^k|\}$, indicating that the path includes $|r_{ij}^k|$ parts, and each part $r_{ij}^k(z)$ has two meanings:

(1) $r_{ij}^k(z)$ represents a section of the path when $r_{ij}^k(z) \in \gamma$; (2) $r_{ij}^k(z)$ represents the charging cycle of an EV when $r_{ij}^k(z) \notin \gamma$. Therefore, the driving behavior and charging behavior during travel for path r_{ij}^k can be expressed in a unified form.

9.3.1.2 Building the EV Model

In the process of travel planning optimization of traditional vehicles, the road length or travel energy consumption is generally used as the road weight in the road network, and the travel path with the shortest travel distance or the minimum energy consumption is selected. When optimizing the travel scheme of EVs, in addition to the path length and energy consumption, the unique characteristics of EVs such as charging needs, battery life, and use of air conditioning should also be taken into account. EV model can be divided into three parts: energy consumption model, battery model, and air conditioning model.

1. Energy consumption model

When the vehicle travels along the path $r_{ij}^k(z)$ (when $r_{ij}^k(z) \in \gamma$), the calculation formulas for driving energy consumption $E_D(r_{ij}^k)$ and time consumption $T_D(r_{ij}^k)$ are shown in Eqs. (9.2) and (9.3):

$$E_D(r_{ij}^k) = \sum_{r_{ij}^k(z)\in\gamma} E_D(r_{ij}^k(z)) \tag{9.2}$$

$$T_D(r_{ij}^k) = \sum_{r_{ij}^k(z)\in\gamma} T_D(r_{ij}^k(z)) \tag{9.3}$$

$E_D(r_{ij}^k(z))$ and $T_D(r_{ij}^k(z))$ are respectively the travel energy consumption and time consumption on the path $r_{ij}^k(z) \in \gamma$. When an EV is decelerating or going downhill, due to the regenerative braking function of EVs, the motor can regenerate energy into the battery and generate braking torque to slow down the vehicle. The energy efficiency function $\eta(x)$ is defined as in Eq. (9.4):

$$\eta(x) = \begin{cases} \frac{x}{\eta_m \eta_d}, \ if\ x \geq 0 \text{ driving condition} \\ \eta_{in} x, \ if\ x \leq 0 \text{ braking condition} \end{cases} \tag{9.4}$$

η_d is the efficiency of the main reducer, η_m the output efficiency, η_{in} the input efficiency of the motor. When $r_{ij}^k(z) \in \gamma$, energy consumption $E_D(r_{ij}^k(z))$ and time consumption $T_D(r_{ij}^k(z))$ in this road section can be defined as in Eqs. (9.5) and (9.6):

$$E_{\mathrm{D}}(r_{ij}^k(z)) = \frac{1}{\eta_{\mathrm{m}}\eta_{\mathrm{d}}}\left(fG\cos\alpha\left(r_{ij}^k(z)\right)l\left(r_{ij}^k(z)\right) + \frac{C_{\mathrm{D}}A\rho}{2}v\left(t\left(r_{ij}^k(z)\right)\right)^2 l\left(r_{ij}^k(z)\right)\right)$$
$$+\eta\left(Gl\left(r_{ij}^k(z)\right)\sin\alpha\left(r_{ij}^k(z)\right)\right)+\frac{1}{2}m\eta\left(v_{\mathrm{T}}\left(t\left(r_{ij}^k(z+1)\right)\right)^2 - v_{\mathrm{T}}\left(t\left(r_{ij}^k(z)\right)\right)^2\right) \tag{9.5}$$

$$T_{\mathrm{D}}(r_{ij}^k(z)) = \frac{l(r_{ij}^k(z))}{v_{\mathrm{T}}(t(r_{ij}^k(z)))} \tag{9.6}$$

where f is the rolling resistance coefficient, m is the mass of the vehicle, G is the gravity of the vehicle, A is the upwind area of the vehicle, C_{D} is the air resistance coefficient, ρ is the air density. Here, $v_{\mathrm{T}}(t(r_{ij}^k(z)))$ represents the recommended driving velocity on the road section $r_{ij}^k(z)$ when the vehicle arrives at moment $t(r_{ij}^k(z))$. $T_{\mathrm{D}}(r_{ij}^k(z))$ represents the time it takes to pass through road section $r_{ij}^k(z)$. If the vehicle does not stop for charging, the moment $t(r_{ij}^k(z+1))$ when it reaches the next node is defined as in Eq. (9.7).

$$\begin{aligned} &t(r_{ij}^k(z+1)) = t(r_{ij}^k(z)) + T_{\mathrm{D}}(r_{ij}^k(z)) \\ &t_0 + l\delta \le t(r_{ij}^k(z)) < t_0 + (l+1)\delta, l \in N, z = 1, 2, \ldots, |r_{ij}^k| \end{aligned} \tag{9.7}$$

2. Air conditioning usage model

The total energy consumption of EV not only includes driving energy consumption, but also the energy consumption of electric accessories, which mainly comes from the use of air conditioning. Air conditioning energy consumption is related to both driving time and external temperature change, so the in-vehicle temperature change model should be first established to reflect the in-vehicle temperature change of EV [9].

$$Q_{\mathrm{T}} = Q_{\mathrm{B}} + Q_{\mathrm{I}} + Q_{\mathrm{P}} + Q_{\mathrm{A}} + \Delta Q_{\mathrm{f}} \tag{9.8}$$

$$Q_{\mathrm{T}} = 1000m_{\mathrm{a}} \cdot \Delta h = 1000\rho V_{\mathrm{v}}c_{\mathrm{p}}\Delta T \tag{9.9}$$

$$Q_{\mathrm{B}} = Q_1 + Q_2 + Q_3 + Q_4 \tag{9.10}$$

$$Q_x = K_x F_x (T_o - T_i), x = 1, 2, 3, 4 \tag{9.11}$$

$$Q_{\mathrm{I}} = 1000(kIF_{\mathrm{I}} + K_1 F_1 \rho_s I) \tag{9.12}$$

$$Q_{\mathrm{p}} = Q_{\mathrm{C}} + 116(n-1)n' \tag{9.13}$$

$$Q_{\mathrm{A}} = 1000G_0\rho c_{\mathrm{p}}(T_o - T_{\mathrm{i}}),\ G_0 = 20n/3600 \tag{9.14}$$

where Q_{T} is the total heat increment, Q_{B} is the heat absorbed by the vehicle, Q_{I} is the heat radiated by the sun, Q_{p} is the heat emitted by the human(s) inside the vehicle, Q_{A} is the heat brought into the vehicle by outdoor air, ΔQ_{f} is the heat generated by air conditioning (ΔQ_{f} is zero when the air conditioning is off). m_{a} is the total air mass in the vehicle, Δh is the enthalpy change of the air in the vehicle, V_{v} is the volume of the vehicle, c_{p} is the specific heat capacity of the air, ΔT is the temperature change in the vehicle, K_1, K_2, K_3 and K_4 are the heat transition coefficients of the top, edge, floor, and window of the vehicle respectively, F_1, F_2, F_3 and F_4 are the heat transition areas of the top, edge, floor, and window of the vehicle respectively. T_{o} is the temperature outside the vehicle, T_{i} is the temperature inside the vehicle. k is the indicator of solar transmittance through the glass, I is the intensity of solar radiation outside the vehicle, F_{I} is the effective area of direct solar radiation, ρ_{s} is the solar radiation absorption indicator of the external structure of the EV, Q_{C} is the radiation energy of the driver, n is the number of people inside the vehicle, n' is the agglomeration coefficient, 116 is the typical thermal radiation value of an adult, G_0 is the volume of fresh air entering the vehicle.

According to the control strategy of the air conditioning system, the comfortable temperature T_{C} should be set first. The temperature inside the vehicle should be detected at the beginning of each path. If the temperature inside the vehicle is higher than the comfortable temperature, the air conditioning is turned on; if the temperature inside the vehicle is lower than the comfortable temperature, the air conditioning is turned off. The air conditioning state variable $R_{\mathrm{A}}(r_{ij}^k(z))$ is defined: when $R_{\mathrm{A}}(r_{ij}^k(z))$ is 1, the air conditioning works normally during period $T_{\mathrm{D}}(r_{ij}^k(z))$ at the rated power P_{A}; when $R_{\mathrm{A}}(r_{ij}^k(z))$ is 0, the air conditioning does not work during period $T_{\mathrm{D}}(r_{ij}^k(z))$. Air conditioning state variable $R_{\mathrm{A}}(r_{ij}^k(z))$ and air conditioning energy consumption $E_{\mathrm{A}}(r_{ij}^k)$ are shown in Eqs. (9.15) and (9.16).

$$R(r_{ij}^k(z)) = \begin{cases} 1, & T_{\mathrm{i}} \geq T_{\mathrm{C}} \\ 0, & T_{\mathrm{i}} < T_{\mathrm{C}} \end{cases} \tag{9.15}$$

$$E_{\mathrm{A}}(r_{ij}^k) = \sum_{r_{ij}^k(z)\in r} E_{\mathrm{A}}(r_{ij}^k(z)) = \sum_{r_{ij}^k(z)\in r} T_{\mathrm{D}}(r_{ij}^k(z))P_{\mathrm{A}}R_{\mathrm{A}}(r_{ij}^k(z)) \tag{9.16}$$

As shown in Eq. 9.16, the sum of the air conditioning energy consumption $E_{\mathrm{A}}(r_{ij}^k(z))$ of all road sections along the path is the energy consumption of air conditioning $E_{\mathrm{A}}(r_{ij}^k)$ on the path r_{ij}^k.

3. Battery model

The travel process of EVs will cause changes in battery SOC and affect battery life. The battery model includes two parts, namely the battery SOC model and the battery life model.

The battery model is described by an equivalent circuit model, which simplifies the battery as an open-loop circuit voltage V_{oc} in series with internal resistance R_{batt}, and both the open-circuit voltage and the internal resistance in series are continuous functions of the SOC value. The power and internal resistance of the battery are defined as P_{batt} and R_{batt} respectively, and Eq. (9.17) can be obtained.

$$\dot{S}_{soc} = -\frac{V_{oc} - \sqrt{V_{oc}^2 - 4P_{batt}R_{batt}}}{2Q_{batt}R_{batt}} \tag{9.17}$$

where S_{soc} is the SOC state of the battery and Q_{batt} is the charge capacity of the battery. The current and output voltage of a battery cell is related to the series and parallel structure of the battery pack. The number of series batteries is defined as n_s, and the number of parallel batteries is defined as n_p. Assuming that the properties of the batteries are the same, the battery pack is equipped with a charge equalization system and the battery temperature is controlled at 25 °C by a constant temperature system, and the current I and voltage V_{cell} of the battery cell can be obtained as shown in Eqs. (9.18) and (9.19).

$$I = \dot{S}_{soc}Q_{batt}/n_p \tag{9.18}$$

$$V_{cell} = (V_{oc} - IR_{batt})/n_s \tag{9.19}$$

The variations of battery open-circuit voltage and battery internal resistance with SOC state are shown in Fig. 9.9 [10].

In regard to battery life, in general, the performance of the battery degrades after recycling, resulting in a decrease in capacity and power. The decline of battery performance is generally believed to be caused by dendritic growth within the battery, changes in electrode structure or chemical composition, decomposition of the electrolyte, and growth of the solid electrolyte interphase (SEI) [11–13], etc. The first cause of battery degradation is dendritic growth inside the battery. In lithium batteries

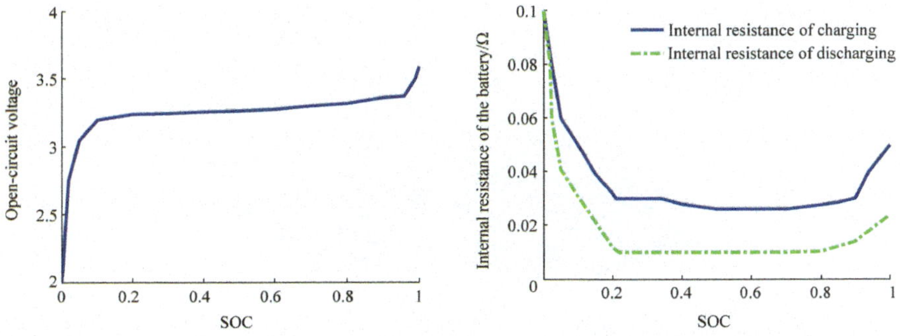

Fig. 9.9 Variation of the open circuit voltage and internal resistance of equivalent battery model with the battery SOC [10]

or lithium-ion batteries, dendritic growth is the uneven growth of lithium metal. These dendritic structures eventually pierce the isolation material and cause a short circuit, resulting in overheat and potential explosion. Dendrite growth is usually caused by overcharging or rapid charging, and the lithium is deposited into the dendrite structures rather than embedded into the negative electrode materials [14]. The second cause is the chemical or structural degradation of the electrode materials at the anode or cathode electrode. For example, the mechanical stresses generated in the embedding and de-embedding of particles will cause the crystal structures to be distorted [15]. Depending on the combination of electrode materials and electrolyte and operating conditions, this degradation may occur at either anode or cathode. The third cause is the decomposition of the electrolyte. The decomposition of the electrolyte material will affect the diffusivity and conductivity of the electrolyte, and it may also produce gas, preceding to increased pressure in the battery, and ultimately may cause an explosion. The fourth cause battery degradation is the growth of SEI film. Relevant studies have proved that the growth of SEI film is an important factor affecting the capacity decline of lithium batteries, which will consume lithium ions and increase internal resistance [16–22]. In this chapter, the battery life model in relevant literature is adopted, and the growth thickness of SEI film is taken as the evaluation criteria of battery life [10]. According to the theoretical derivation in the literature [10], the relationship between the growth rate of the SEI film and the SOC value and the battery current is obtained, as shown in Fig. 9.10.

In Fig. 9.10, the ordinate axis represents the average growth rate of the SEI film $\dot{\delta}_{film}$, and the abscissa axis represents the battery current I and the SOC value. It can be seen from Fig. 9.10 that the growth rate of the SEI film $\dot{\delta}_{film}$ increases as the SOC increases, and when the current shifts from discharging to charging, the growth rate of the SEI film $\dot{\delta}_{film}$ also increases. In EV travel planning, the average growth rate of the

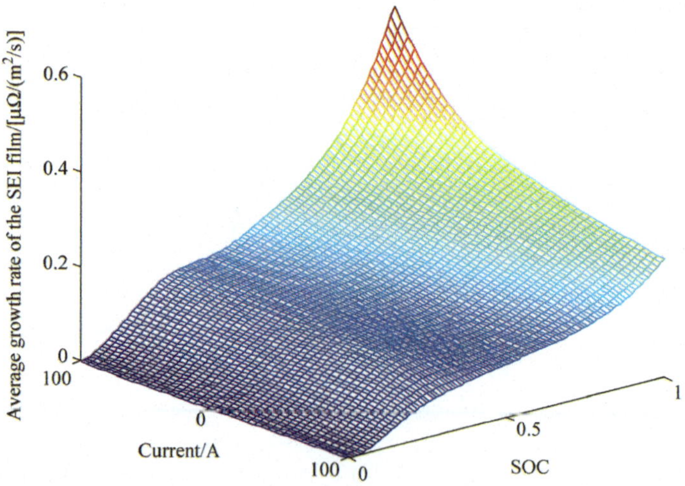

Fig. 9.10 Diagram of the relationship between SEI film growth rate and current and SOC

battery SEI film $\dot{\delta}_{\text{film}}$ can be obtained by checking the battery SOC value table based on the discharge current during driving and the charging current during charging. Then, this value is integrated over time to obtain the SEI film growth thickness of the battery during the whole travel process, which serves as a quantitative indicator for evaluating the degradation of battery cycle life.

For the battery charging process, that is, when $r_{ij}^k(z) \notin \gamma$, it means that the vehicle stops at the charging station for charging, and $E_{\text{D}}(r_{ij}^k(z))$ and $T_{\text{D}}(r_{ij}^k(z))$ are zero. It is constant current charging. When the charging station is a, then the charging time is T_{a}^{c}, and the charging energy is E_{a}^{c}. Therefore, the departure moment when the EV leaves the charging station and is ready to go to the next node is

$$t(r_{ij}^k(z+1)) = t(r_{ij}^k(z)) + T_{\text{a}}^{\text{c}} \tag{9.20}$$

The total energy consumption $E(r_{ij}^k)$ of the EV passing through the path r_{ij}^k includes the energy $E_{\text{D}}(r_{ij}^k)$ consumed to overcome the resistance and the energy consumed by air conditioning $E_{\text{A}}(r_{ij}^k)$ during driving.

$$E(r_{ij}^k) = E_{\text{D}}(r_{ij}^k) + E_{\text{A}}(r_{ij}^k) = \sum_{r_{ij}^k(z)\in\gamma} (E_{\text{D}}(r_{ij}^k(z)) + E_{\text{A}}(r_{ij}^k(z))) \tag{9.21}$$

E_{B} is defined as the energy of the vehicle before departure, and E_{res} is the remaining energy after the path r_{ij}^k is passed. E_{res} is defined as in Eq. (9.22).

$$E_{\text{res}} = E_{\text{B}} + \sum_{a=1}^{P} E_a^{\text{c}} - E(r_{ij}^k) \tag{9.22}$$

where, considering that an EV may stop at multiple charging stations during a trip, $\sum_{a=1}^{P} E_a^{\text{c}}$ represents the total charging energy, and correspondently, $\sum_{a=1}^{P} T_a^{\text{c}}$ represents the total charging time.

9.3.2 *Definition of Travel Objectives and Constraints*

After the EV and traffic environment models are established, it is necessary to set travel objectives and constraints that should be included in the travel scheme according to the travel needs of the EV. The travel constraints include two parts, namely the driver's travel task constraints and reachable scope constraints of the EV.

9.3.2.1 Definition of Travel Objectives

According to the overall framework of EV travel planning, and taking into account the unique characteristics of EV during travel, five travel optimization objectives are selected in this study, as shown in Eq. (9.23).

$$
\begin{aligned}
&\min \quad O_1 = \sum_{z=1}^{\left|r_{ij}^k\right|} l(r_{ij}^k(z)), \quad r_{ij}^k(z) \in \gamma \\
&\min \quad O_2 = T_{\mathrm{D}}(r_{ij}^k) + \sum_{\mathrm{a}=1}^{P} T_{\mathrm{a}}^{\mathrm{c}} \\
&\min \quad O_3 = E(r_{ij}^k) \\
&\min \quad O_4 = \int_0^{O_2} \dot{\delta}_{\mathrm{film}}\, dt \\
&\min \quad O_5 = \left|\overline{T_{\mathrm{i}}} - T_{\mathrm{c}}\right|
\end{aligned} \tag{9.23}
$$

The first objective O_1 has the shortest driving distance to minimize the sum of the length of road sections in the driving process. The second objective O_2 has the shortest total travel time, which includes the total travel time and the total charging time. The third objective O_3 is the minimum total energy consumption, or the sum of the total driving energy consumption and the total air conditioning energy consumption during travel. The fourth objective O_4 is the longest battery cycle life, that is, the increase of SEI film thickness during the trip is taken as the evaluation criteria, and the minimum increase of SEI film thickness is taken as the optimization objective. The fifth objective O_5 is to make the indoor temperature comfortable. This objective uses the absolute value of the difference between the indoor temperature and the average temperature as the evaluation indicator. It can be seen from Eq. (9.23) that there is a mutual coupling relationship between various objectives in the travel process of an EV, so a cooperative optimization of multiple travel objectives is required when a travel scheme is made.

9.3.2.2 Definition of Travel Constraints

When an EV travel scheme is made, it is necessary to take into account the driver's travel task constraints and reachable scope constraints of the EV. Therefore, the driving task constraints of the driver are defined respectively and the mileage of the

EV is estimated to comprehensively reflect the various constraints and restrictions in the travel process of the EV.

(1) Driver's travel task constraints

The driver's travel task constraints include that the arrival time should not be later than the specified time, and the SOC value of the destination should not be lower than the pre-defined SOC limit, as defined in Eq. (9.24).

$$\begin{aligned} C_1 &= \max(t_0 + O_2 - t_j^{\mathrm{d}}, 0) \\ C_2 &= \max(S_{\mathrm{soc},j}^{\mathrm{d}} - S_{\mathrm{soc},j}, 0) \end{aligned} \tag{9.24}$$

where t_j^{d} is the scheduled arrival time of the EV at the node j, $S_{\mathrm{soc},j}^{\mathrm{d}}$ is the scheduled SOC value of the EV at the node j, and $S_{\mathrm{soc},j}$ is the actual SOC value of the EV at the destination node j. The arrival time constraint C_1 is used to ensure that the EV arrives at the destination before the scheduled time t_j^{d} to avoid late arrival. When this constraint is met, C_1 is zero; when this constraint is not met, C_1 is the length of time exceeding the limit. SOC constraint C_2 at the destination is used to ensure that when the EV arrives at the destination node j, the SOC value is higher than the scheduled SOC value $S_{\mathrm{soc},j}^{\mathrm{d}}$, to meet the driver's potential travel tasks in the future. When this constraint is met, C_2 is zero, when this constraint is not met, C_2 is the SOC difference exceeding the value limit of $S_{\mathrm{soc},j}^{\mathrm{d}}$.

In addition, considering the dynamic stochastic change characteristics of the traffic road network, the recommended driving velocity for each road section needs to meet the average driving velocity on the road at the current moment. To set the confidence level ρ_{c}, it is necessary to ensure that the recommended travel speed can reach the average speed of each road under the condition of the confidence level ρ_{c}.

$$g(x) = \begin{cases} 1, \ x > 0 \\ 0, \ x \leq 0 \end{cases} \tag{9.25}$$

$$G_{\mathrm{c}}^n = \sum_{z=1}^{|r_{ij}^k|} g(v_{\mathrm{T}}(t(r_{ij}^k(z)) - v^n(t(r_{ij}^k(z))), \quad n = 1, 2, \ldots, N_{\mathrm{c}}, \quad r_{ij}^k(z) \in \gamma \tag{9.26}$$

$$N_{\mathrm{c}}^x = \sum_{n=1}^{N_{\mathrm{c}}} g(G_{\mathrm{c}}^n) \tag{9.27}$$

$$\rho_{\mathrm{N_c}} = \frac{N_{\mathrm{c}}^x}{N_{\mathrm{c}}} \tag{9.28}$$

where $v^n(t(r_{ij}^k(z))$ is the result of stochastic sampling according to the probability density function $\phi_{r_{ij}^k(z),t(r_{ij}^k(z))}$ the average traffic speed is subject to when the vehicle arrives at the road section $r_{ij}^k(z)$ at moment $t(r_{ij}^k(z)$. When the obtained $\rho_{\mathrm{N_c}}$ is not greater than the preset ρ_{c}, it means that the recommended speed value $v_{\mathrm{T}}(t(r_{ij}^k(z))$ on the road section $r_{ij}^k(z)$ reaches the average traffic speed of the road at the confidence level ρ_c.

(2) Estimation of mileage

In addition to the constraints of the driver, the reachable scope of the EV is also a necessary constraint in travel planning. In the travel planning of EVs, the first thing to do is to estimate the driving distance of EVs. When the EVs cannot reach the destination or the nearest charging station, it means that the EVs are bound to fail to complete the travel task. Only when the EVs can reach one of the two, it is necessary to make a reasonable plan for the travel process. The method for estimating the mileage of EVs includes two parts: (1) Using the genetic algorithm based on stochastic simulation to get the energy consumption to each node of the road network, (2) using the stochastic planning method to analyze the mileage of EVs.

To obtain the energy consumption $E(i, j)$ of the EV from the current node i to each node j in the road network, it is necessary to calculate the energy consumption $E(r_{ij}^k)$ of the vehicle in each path from node i to node j. There are two main problems here: (1) In the stage of estimating the mileage of EV, since the travel path hasn't been planned, there are a lot of paths in the path set R_{ij} from the starting node to any node j in the road network, most of which are detours and turn-back paths, which is inconsistent with real-life driving experience and can barely reflect the energy consumption of the EV by the time it reaches the node j. Therefore, a large number of turn-back paths or detours in the path set R_{ij} connecting node i and node j should be eliminated. (2) In a certain path k ($r_{ij}^k \in R_{ij}$) connecting node i and node j, affected by the dynamic and stochasticity of the traffic network model, even if the EV departs from the starting point and travels along the same path at the same time with the driving velocity conforming to the average speed distribution function of the road, the total energy consumption result $E(r_{ij}^k)$ of the EV is also a stochastic sampling result, not a definite value. Therefore, the genetic algorithm based on stochastic simulation should be adopted to screen the reasonable travel path from the starting point to each node in the road network to avoid detours and turn-back paths. Then, the sampling results of the energy consumption of the EV by the time it reaches each node can be calculated. Figure 9.11 is a flowchart of the genetic algorithm based on stochastic simulation.

For each node j in the road network, a genetic algorithm based on stochastic simulation should be used to calculate the energy consumption of the vehicle from the starting point i to the node j, that is, all nodes in the road network should be looped through. The algorithm includes initialization, fitness calculation, selection, crossover and variation. In the process of estimating the mileage, the charging behavior of EV is not considered, so $r_{ij}^k(z) \in \gamma$ is always valid. Chromosomes are composed of the sequence of path nodes from the starting point to the destination node. Whenever a chromosome sequence is determined, a path connecting the starting point and the node is determined. For this path, since the average traffic speed on each road section is dynamic stochastic, the traffic speed on the same road section during the same period presents stochasticity, which means the time it takes to reach

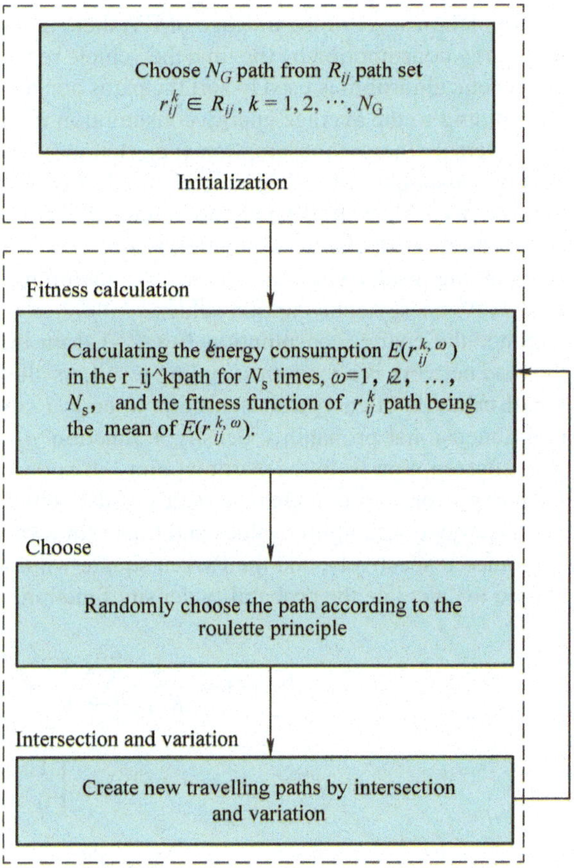

Fig. 9.11 Flowchart of genetic algorithm based on stochastic simulation

the next node is different. Therefore, sampling is performed N_s times during the process from the starting point to the end point along the same path at the same time. Since the average speed of each road section is the result of stochastic sampling, the travel time on each path is different, and the time it takes to reach the next node is different. The energy consumption and time consumption of the N_s sampling results obtained upon the arrival of the destination are different from each other, indicating that the energy consumption and time consumption on a path are stochastically distributed. The main difference between the genetic algorithm based on stochastic simulation and the traditional genetic algorithm is that the expected value of the probability distribution is used as the fitness function value of the chromosome when the fitness function of the chromosome features probability distribution. Therefore, in this study, the expected value of energy consumption through this path is used as the fitness function value of the chromosome. In addition to the stochasticity introduced by the sampling of energy consumption on each path, it is necessary to consider that there are a large number of paths in the road network connecting the starting point to the node. Since the specific driving path to the destination node is not known in the estimation stage of the mileage of EV, there are multiple paths used to represent the energy consumption by the time the vehicle reaches a certain node. In this study, the genetic algorithm is used to sort the paths connecting the starting point and node j according to the average energy consumption indicator, excluding the paths with high average energy consumption (i.e., detours), and reserving K paths with low energy consumption (i.e., directly to the destination). Therefore, for each node in the road network, K travel paths are obtained after optimization by the genetic algorithm based on stochastic simulation where $k = 1, \ldots, K$, and each path contains N_s sampling results where $\omega = 1, \ldots, N_s$; therefore, $E_D(r_{ij}^{k,\omega})$ corresponds to a total of $K \times N_s$ energy consumption values.

Since the energy consumption $E_D(r_{ij}^{k,\omega})$ from the starting point to any node in the road network is a random sampling result, its clustering characteristic can reflect the distribution of energy consumption to node j, so it is necessary to calculate the two-dimensional probability density 1 function $p_{ij}^{E,T}$ of energy consumption and time consumption from the starting point i to node j. To comprehensively consider the energy consumption characteristics of the vehicle by the time it arrives at node j, all energy consumption values and travel time values of the selected K paths are combined respectively, and the Parzen square window kernel function method [23] is used to calculate the probability density function, as shown in Eqs. (9.29–9.32).

$$p_{ij}^{E,T}(x) = \frac{1}{KN_S}\sum_{\varepsilon=1}^{KN_S}\frac{1}{V}\varphi\left(\frac{x - x_{ij}^{\varepsilon}}{h}\right) \tag{9.29}$$

$$\varphi([u_1, u_2, \ldots, u_d]^T) = \begin{cases} 1 & \text{if} |u_i| \le \frac{1}{2} \\ 0 & \text{others} \end{cases} \tag{9.30}$$

$$V = h^d \tag{9.31}$$

$$x_{ij}^{\varepsilon} = (E_D(r_{ij}^{k,\omega}), T_D(r_{ij}^{k,\omega})) \tag{9.32}$$

where

$$\varepsilon = 1, 2, \ldots, K \times N_S$$

The kernel function transformation method is used to transform the discrete stochastic sampling points into a continuous probability distribution function. The probability distribution function comprehensively reflects the stochastic characteristic introduced by the non-unique path to the node j and the time-varying stochastic variation of the driving velocity. Considering the energy consumption $E(i, j)$ between i and j is a stochastic variable subordinate to p_{ij}^E under the dynamic stochastic road network model, it's necessary to compare the relationship between the initial energy value and the energy consumption with stochastic characteristic based on an appropriate criteria when using the battery energy at the starting point to judge the reachable scope of the EV. In this study, the stochastic planning method is used to evaluate the probability density function of energy consumption, and then analyze the reachable scope of EV. The stochastic planning method includes the expected value model, chance constraint model, and correlation chance model [24], each of which represents different analysis strategies and evaluation criteria when making stochastic planning.

In the expected value model, the expected value of the stochastic variable of energy consumption of the vehicle arriving at node j is used as the energy consumption of the EV arriving at the node. To judge whether the EV can reach the node j on the map is to judge whether the current battery energy E_B at the starting point is greater than the expected value of energy consumption $E(i, j)$ by the time the vehicle reaches the node. If it is greater than the expected value, the EV can reach the node; if it is less than the expected value, the EV cannot reach the node. Under this condition, the road network is divided into several areas at equal angles, with the starting point as the center, and the angle of each area is β. $D(i, j, \beta)$ represents the straight-line distance between the starting point i and the node j in the area β on each direction. Find the farthest node that can be reached from the starting point in each area, and then envelop the corresponding farthest node in each area. This range is the reachable scope of EV under the expected value model, as shown in Eq. (9.33).

$$area_1 = \arg\max_{j\in V}\{D(i, j, \beta)|E_B - \mathrm{E}(E(i, j)) > 0,$$
$$\sigma\theta \leq \beta < (\sigma + 1)\theta, \mathrm{mod}(360^\circ, \theta) = 0, \sigma = 0, 1, \ldots, \frac{360^\circ}{\theta} - 1\}. \tag{9.33}$$

In the chance constraint model, to judge whether the EV can reach node j on the map, it is necessary to set the confidence level η in advance, and judge whether the EV can reach the node j with this confidence level η based on the current battery energy value E_B. The level of confidence level η represents the degree of confidence in the estimated results of the reachable scope of the EV. A high confidence level indicates that the driver expects a higher probability for the EV to reach each node within the estimated range of driving distance, as shown in Eq. (9.34).

$$area_2 = \arg\max_{j\in V}\{D(i, j, \beta)| \int_{\min E(i,j)}^{E_B} p_{ij}^E \mathrm{d}E(i, j) > \eta,$$
$$\sigma\theta \leq \beta < (\sigma + 1)\theta, \mathrm{mod}(360^\circ, \theta) = 0, \sigma = 0, 1, \ldots, \frac{360^\circ}{\theta} - 1\}. \tag{9.34}$$

In the correlation chance model, to judge whether the EV can reach node j on the map, it is necessary to compare the current energy E_B of the battery with all the sampling results $E(r_{ij}^{k,\omega})$ of the energy consumption of the EV by the time it reaches this node, and calculate the integral of $p_{i_0 j}^E$ within the tiny area μ near E_B. By this criteria, what is obtained in the area β on each direction is the reachable scope that is most likely to use up the current battery energy. This model aims to find out the envelope range of nodes in the road network that is most likely to consume all the battery energy, that is, the minimum reachable scope of EV, as shown in Eq. (9.35).

$$
\begin{aligned}
area_3 = \ & \underset{j \in V}{\arg\max}\{\mathrm{int}(i, j, \beta)|\mathrm{int}(i, j, \beta) = \int_{E_\mathrm{B}-\mu}^{E_\mathrm{B}} p_{ij}^{E} \mathrm{d}E(i, j), \\
& E_\mathrm{B} \geq E(r_{ij}^{k,\omega}), k = 1, 2, \ldots, K, \omega = 1, 2, \ldots, N_S, \sigma\gamma \leq \beta < (\sigma + 1)\gamma, \\
& \mathrm{mod}\ (360^\circ, \gamma) = 0, \sigma = 0, 1, \ldots, \frac{360^\circ}{\gamma} - 1\}.
\end{aligned} \tag{9.35}
$$

9.3.3 A Temperal Multi-objective Ant Colony Optimization Algorithm is Used to Solve Problems

In this study, a temperal multi-objective ant colony optimization method is proposed to optimize 5 travel objectives, including travel distance, travel time, energy consumption, battery cycle life, and in-vehicle temperature, with S ants simultaneously, and meet the travel constraints. Various travel behaviors in the travel process of EV are optimized, including travel path planning, velocity recommendation on each road, selection of charging station, formulation of charging time and mode, and use of air conditioning, etc. The structure of the temperal multi-objective ant colony optimization algorithm is shown in Fig. 9.12.

It can be seen from Fig. 9.12 that the algorithm's structure mainly includes two parts: initialization and cycling solving. In initialization, each pheromone layer in the road network needs to be initialized; in cycling solving, several steps such as transition probability calculation, ant colony search, weight update value calculation, and pheromone layer update should be carried out in sequence, and the cycling solving should be performed.

The proposed temperal multi-objective ant colony optimization method has the following characteristics:

(1) Given the time characteristics of traffic environment and charging behaviors in the travel process, this algorithm introduces the time factor into the pheromone of ant colony algorithm, that is, the calculation and updating process of pheromone is not only related to the search location of ant, but also related to the time of the search process. This method ensures that time-varying traffic information and charging time information are introduced into the travel planning algorithm.

(2) In each step of searching, the ant needs to decide in turn whether to go to the next node (or stay at this node for charging), the driving velocity in the path (or the charging mode), and whether to turn on the air conditioning in the road section. When the ant search for the destination, it gets a complete travel scheme. In this way, the ant in the charging station node can not only move to the next node,

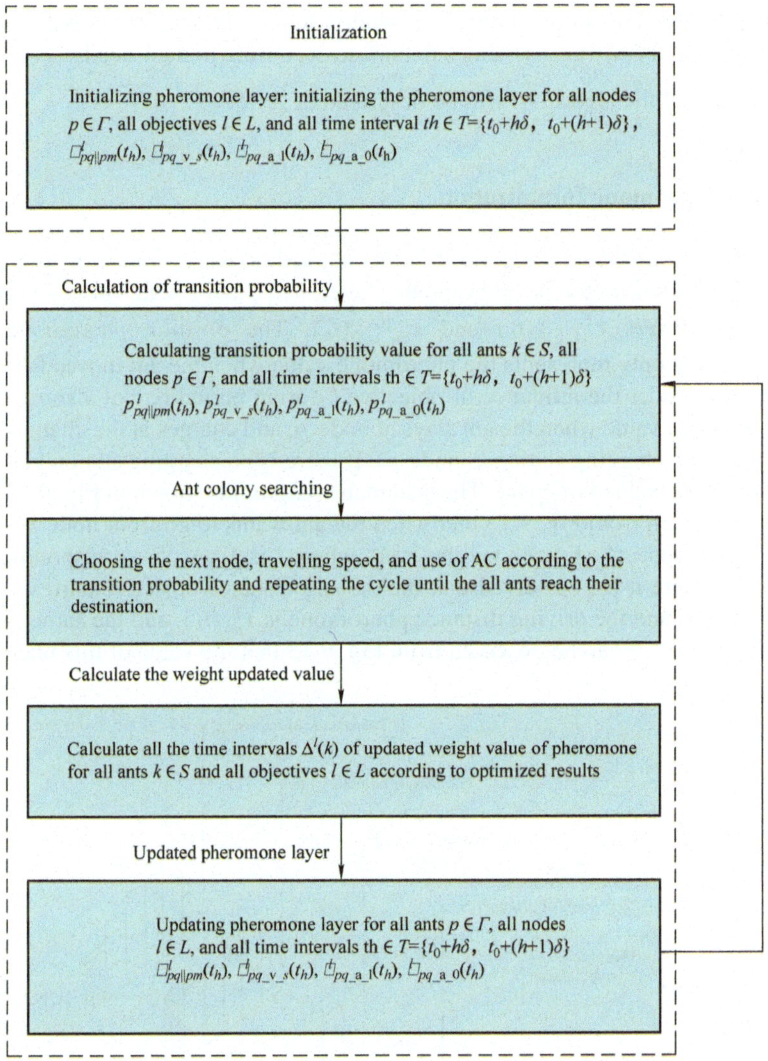

Fig. 9.12 Flowchart of temperal multi-objective ant colony optimization method

but also stay at the current place for charging, which gives a double meaning to the node transition strategy.

(3) To enable the drivers to have a comprehensive understanding of the travel scheme, the optimization results of the multi-objective optimization method form a travel scheme solution set. All the solutions in the set meet the travel constraints and together form the travel scheme solution set in the sense of

multi-objective optimization. The solution set provides the driver with a variety of potential EV travel schemes to choose according to their needs.

Each part of the algorithm will be described as follows.

9.3.3.1 Pheromone Initialization

Three pheromone layers are defined in this algorithm, namely position pheromone layer $\tau^l_{pq||pm}(t_h)$, velocity pheromone layer $\tau^l_{pq\_v\_s}(t_h)$ and air conditioning pheromone layers $\tau^l_{pq\_a\_1}(t_h)$ and $\tau^l_{pq\_a\_0}(t_h)$. The position pheromone layer $\tau^l_{pq||pm}(t_h)$ not only represents the pheromone value when the ant moves from node p to node q under the influence of objective l during period t_h, but also represents the pheromone value when the ant stays at node p, and charges at the charging level m (if there is a charging station at node p). If there is no charging station at node p, the pheromone value is $\tau^l_{pq}(t_h)$. The schematic diagram is shown in Fig. 9.13.

It can be seen from Fig. 9.13 that when the ant wants to go from node p to node q, the pheromone $\tau^l_{pq}(t_h)$ can be observed, where l is $1, \ldots, 5$, corresponding to 5 objectives. Here it is considered that the driving distance objective corresponds to the number 1, and the driving distance pheromone is $\tau^1_{pq}(t_h)$, and the same goes for other objectives. It can be observed from Fig. 9.13 that the value of this pheromone

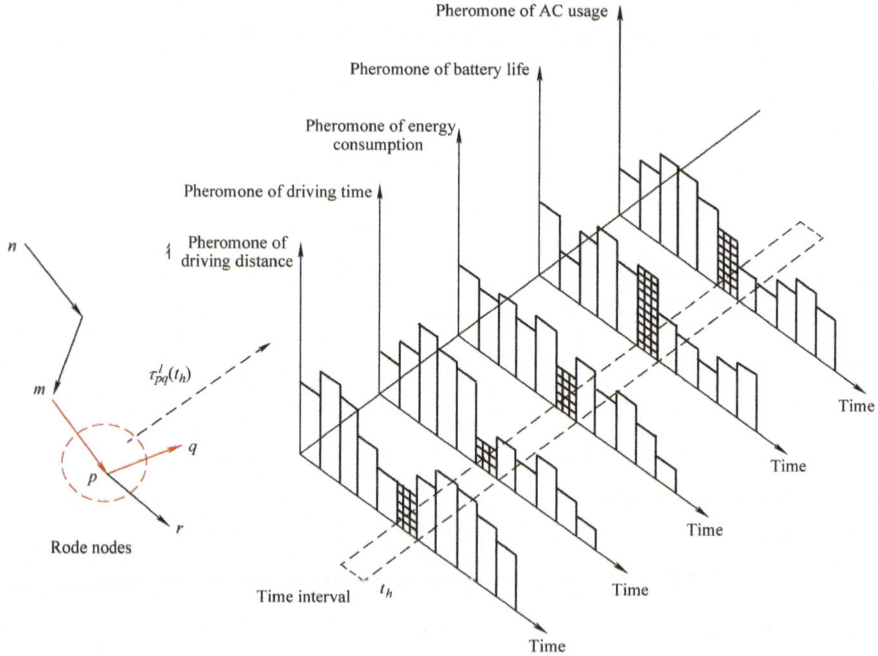

Fig. 9.13 Schematic diagram of position transition pheromone layer $\tau^l_{pq}(t_h)$

varies at different time intervals t_h. The gray part of the driving distance pheromone in Fig. 9.13 represents the pheromone value corresponding to the driving distance objective during the period. In the same way, the gray parts of the pheromone corresponding to four other objectives are all related to the time interval. Through this pheromone structure, the pheromone used to describe the transition from node p to node q is related to the corresponding time interval, that is, the pheromone transferred from node p to node q varies at different time intervals.

When there is a charging station at node p, the pheromone is defined as $\tau^l_{pq||pm}(t_h)$, which means the ant at node p can move to the next adjacent node q or stay at node p and charge at the charging level m. Through this definition method, the ant can freely choose whether to move forward or stay here for charging at node p. The pheromone diagram is shown in Fig. 9.14.

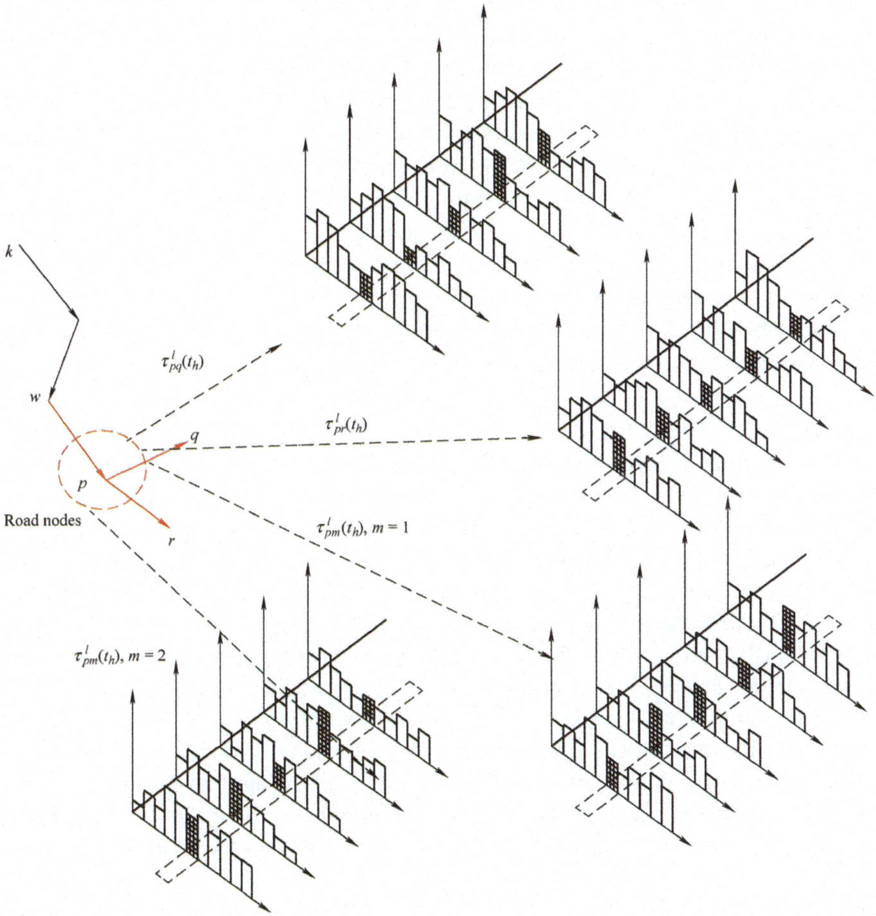

Fig. 9.14 Schematic diagram of the position transition pheromone layer $\tau^l_{pq||pm}(t_h)$ containing a charging station

It can be seen from Fig. 9.14 that when there is a charging station at node p, the pheromone transferred from node p to the next state includes two cases. The first case is to move from node p to the next node q (the corresponding pheromone is $\tau^l_{pq}(t_h)$) or the next node r (the corresponding pheromone is $\tau^l_{pr}(t_h)$); the second case is to stay at node p for charging. Slow charging (m=1) and fast charging (m= 2) are considered here, which respectively correspond to the slow charging pheromone value $\tau^l_{pm}(t_h), m = 1$ and fast charging pheromone value $\tau^l_{pm}(t_h), m = 2$. This pheromone definition method has two advantages: (1) Through the same pheromone structure, it supports both transferring to adjacent nodes and staying at the current node to charge in a certain mode. (2) The pheromone structure considers the time factor, and the pheromone value corresponding to each objective is related to the time rather than a fixed value. This time-related pheromone definition is the "temperal" part of the algorithm. For example, slow charging pheromone $\tau^l_{pm}(t_h), m = 1$ is used to represent staying at node p for charging. If the value of $\tau^l_{pm}(t_h), m = 1$ is lower in the time interval $T_1 < t_h < T_2$, and the value of $\tau^l_{pm}(t_h), m = 1$ is higher in the time interval $T_2 < t_h < T_3$, it indicates that the pheromone carrying out slow charging behavior at node p is higher in the time interval $[T_2, T_3]$ than in the time interval $[T_1, T_2]$. Similarly, when transferring from node p to the next node q, if the value of pheromone $\tau^l_{pq}(t_h)$ is higher in the time interval $T_1 < t_h < T_2$ while the value of pheromone $\tau^l_{pq}(t_h)$ is lower in the time interval $T_2 < t_h < T_3$, it means that the pheromone transferring from node p to the next node q is higher in the time interval $[T_1, T_2]$. Therefore, different periods have different effects on the same behavior with this pheromone structure design.

Similarly, after the position pheromone layer $\tau^l_{pq||pm}(t_h)$ is defined, the algorithm defines the velocity pheromone layer $\tau^l_{pq\_v\_s}(t_h)$, as shown in Fig. 9.15. For the velocity pheromone layer $\tau^l_{pq\_v\_s}(t_h)$, when the ant moves on the road $\overrightarrow{pq}$, the recommended velocity value is determined by level s, where $s = 1, 2, 3 \ldots, V_k$. The velocity pheromone layer varies at different recommended levels. At a certain recommended level s, the value of the velocity pheromone layer is also different during different periods t_h. V_k is set as 4 here, that is, there are 4 recommended velocity levels in each road section, and each velocity level corresponds to a recommended driving velocity value. At each recommended velocity level (e.g. $s = 3$), the pheromone layer is still affected by 5 objectives (5 layers in total). For example, for the energy consumption pheromone layer $\tau^3_{pq\_v\_s}(t_h), s = 3$ of the recommended velocity level $s = 3$, if the pheromone value is lower in the time interval $T_1 < t_h < T_2$, and the pheromone value is higher in the time interval $T_2 < t_h < T_3$, it means that the possibility that the energy consumption pheromone layer has a recommended velocity level of 3 during the period $[T_2, T_3]$ is higher than that during the period $[T_1, T_2]$. This time-related pheromone definition enables the algorithm to take into account the time-varying characteristics of travel speed on the road. For example, on the same road, the travel speed during period $[T_1, T_2]$ is higher, while the travel speed during period $[T_2, T_3]$ is lower. Therefore, a higher speed can be recommended for the period $[T_1, T_2]$, while a lower speed for the period $[T_2, T_3]$. This requires the recommended velocity pheromone value of the same road section to vary in different periods.

Similarly, air conditioning pheromone layers $\tau^{l}_{pq\_a\_1}(t_h)$ and $\tau^{l}_{pq\_a\_0}(t_h)$ are defined to reflect whether the ant needs to turn on the air conditioning on each road section during each period, where 1 means that the air conditioning works normally, and 0 means that the air conditioning is off. Similar to the definition of velocity pheromone, on the road section $\overrightarrow{pq}$, the value of the pheromone is different is

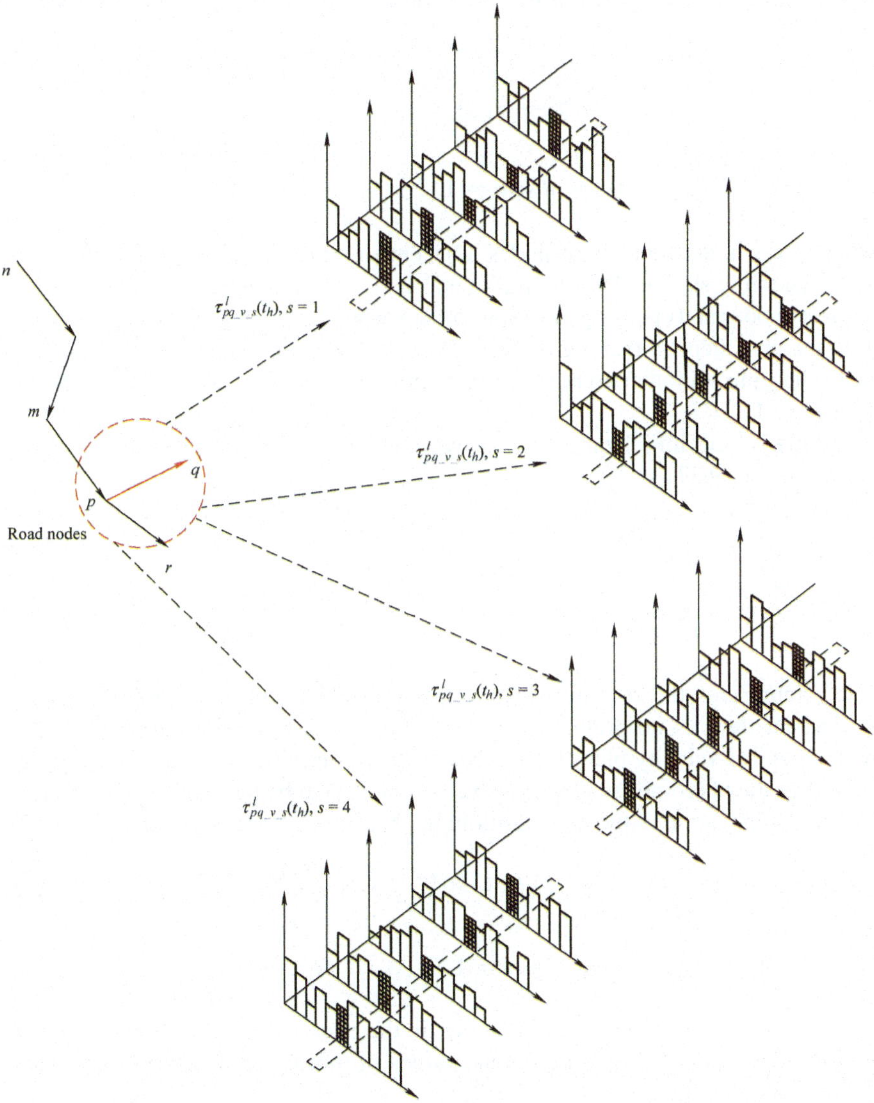

Fig. 9.15 Schematic diagram of the velocity transition pheromone layer

different periods, reflecting that the pheromone is different when the air conditioning is turned on in different periods. For example, if the $\tau^{l}_{pq\_a\_1}(t_h)$ is higher during the period $T_1 < t_h < T_2$ and lower during the period $T_2 < t_h < T_3$, it means that during the period $[T_1, T_2]$, the pheromone about the normal operation of air conditioning is larger, and it is more inclined to turn on the air conditioning. In this way, the behavior of turning on and off the air conditioning on the road section $\overrightarrow{pq}$ is affected by the time.

After the definition, the pheromone needs to be initialized here. The initialization definition of the position transition pheromone layer $\tau^{l}_{pq||pm}(t_h)$ is shown in Eq. (9.36):

$$\tau^{l}_{pq||pm}(t_h) = \frac{1}{|U_p| + C_{\mathrm{w}}} \tag{9.36}$$

$$p, q \in \Gamma, \overrightarrow{pq} \in \gamma, q \in R_p, m = 1, 2, \ldots, C_w, l = 1, 2, \ldots, L, t_h \in T$$

where $|U_p|$ is the number of nodes adjacent to node p, C_w is the charging level of the charging station. This definition makes no distinction between transferring to adjacent nodes and charging in various modes when the the vehicle moves from node p to the next step, so the pheromone value of each action is the same. When there is no charging station at node p, the transition probability at node p is initialized to $\tau^{l}_{pq}(t_h) = 1/|U_p|$.

Similarly, the initialization of the recommended velocity pheromone layer is shown in Eq. (9.37).

$$\tau^{l}_{pq\_v\_s}(t_h) = \frac{1}{V_{\mathrm{K}}} \tag{9.37}$$

$$\overrightarrow{pq} \in \gamma, l = 1, 2, \ldots, L, s = 1, 2, \ldots, V_{\mathrm{K}}, t_h \in T$$

V_{K} represents the recommended velocity level. It can be seen from the definition that there is no preference for velocity levels in the initialization stage of recommended velocity pheromone, that is, in the initialization stage, the algorithm considers the pheromones of various velocity levels to be the same.

The initialization of the air conditioning pheromone layer is shown in Eq. (9.37).

$$\tau^{l}_{pq\_a\_1}(t_h) = \tau^{l}_{pq\_a\_0}(t_h) = 0.5 \tag{9.38}$$

$$\overrightarrow{pq} \in \gamma, l = 1, 2, \ldots, L, t_h \in T$$

It can be seen from the definition that, in the pheromone initialization stage, the initialization results of the pheromone values of turning on and turning off the air conditioning are the same.

9.3.3.2 Calculation of Transition Probability

Similar to the definition of pheromone layer, three transition probability values are defined here, namely the position transition probability $p^k_{pq\|pm}(t_h)$, recommended velocity transition probability $p^k_{pq\_v\_s}(t_h)$, air conditioning usage probability $p^k_{pq\_a\_1}(t_h)$ and $p^k_{pq\_a\_0}(t_h)$, where $p^k_{pq\|pm}(t_h)$ represents the probability that ant k moves from node p to the next adjacent node q or stays at this node p to charge at level m during the period t_h; $p^k_{pq\_v\_s}(t_h)$ means that ant k arriving at road section $\overrightarrow{pq}$ during period t_h and passing through the road section $\overrightarrow{pq}$ at the recommended velocity level s; $p^k_{pq\_a\_1}(t_h)$ and $p^k_{pq\_a\_0}(t_h)$ represent the probability that ant k will turn on or off the air conditioning when it arrives at the road section $\overrightarrow{pq}$ during the period t_h. Therefore, each transition probability value can be calculated by using pheromone layers, as shown in Eqs. (9.39–9.42).

$$p^k_{pq|pm}(t_h) = \frac{\left[\prod_{l=1}^{L} (\tau^l_{pq\|pm)}(t_h))^{\alpha\lambda^l_k}\right] \times (\eta_{pq\|pm)})^{\beta}}{\sum_{|U_p|+C_w} \left[\left[\prod_{l=1}^{L} (\tau^l_{pq\|pm)}(t_h))^{\alpha\lambda^l_k}\right] \times (\eta_{pq\|pm)})^{\beta}\right]} \tag{9.39}$$

$$p^k_{pq\_v\_s}(t_h) = \frac{\left[\prod_{l=1}^{L} (\tau^l_{pq\_v\_s)}(t_h))^{\alpha\lambda^l_k}\right]}{\sum_{V_K} \left[\prod_{l=1}^{L} (\tau^l_{pq\_v\_s}(t_h))^{\alpha\lambda^l_k}\right]} \tag{9.40}$$

$$p^k_{pq\_a\_0}(t_h) = \frac{\prod_{l=1}^{L} (\tau^l_{pq\_a\_0}(t_h))^{\lambda^l_k}}{\prod_{l=1}^{L} (\tau^l_{pq\_a\_0}(t_h))^{\lambda^l_k} + \prod_{l=1}^{L} (\tau^l_{pq\_a\_1}(t_h))^{\lambda^l_k}} \tag{9.41}$$

$$p^k_{pq\_a\_1}(t_h) = \frac{\prod_{l=1}^{L} (\tau^l_{pq\_a\_1}(t_h))^{\lambda^l_k}}{\prod_{l=1}^{L} (\tau^l_{pq\_a\_0}(t_h))^{\lambda^l_k} + \prod_{l=1}^{L} (\tau^l_{pq\_a\_1}(t_h))^{\lambda^l_k}} \tag{9.42}$$

where L is the total number of optimization objectives (a total of 5). $\eta_{pq\|pm}(t_h)$ is the inspiring factor for the ant to move from node p to adjacent node q or charge at level m, and α and β represent the weight value of pheromone and inspiring factor respectively. When the ant moves from the current node to the next node, the inspiring factor is the straight-line distance between the next node and the destination node. When the ant charges at the current node, the inspiring factor is the straight-line distance between the current node and the destination node. Therefore, when the position transition is performed, the inspiring factor can make the probability value of the node closer to the ending point higher than other adjacent nodes, thus improving the convergence speed of the algorithm.

To expand the search range of the ant colony, each ant k has a different search preference λ^l_k, and each ant has a different search weight for each objective. X_{ant} represents the ant colony search indicator value, and λ^l_k is defined as shown in Eq. (9.43).

$$\sum_{l=1}^{L} \lambda_k^l = 1 \tag{9.43}$$

$$\lambda_k^l \in \left\{0, \frac{1}{X_{\text{ant}}}, \frac{2}{X_{\text{ant}}}, \ldots, 1\right\} \tag{9.44}$$

$$(\lambda_k^1, \ldots, \lambda_k^l, \ldots, \lambda_k^L) \neq (\lambda_q^1, \ldots, \lambda_q^l, \ldots, \lambda_q^L), \quad \forall k, q \in S$$

Therefore, the search preference vector of each ant in the ant colony is different from each other. In this way, the search range of the ant colony is expanded.

9.3.3.3 Ant Colony Search

After the transition probability of each behavior in the road network is defined, an ant in the ant colony can depart from the starting point and search for a travel scheme until it reaches the final destination. Then the search of the ant in the current round ends. When all the ants arrive at the destination, it means that the search of this round of ant colony optimization method is over. In the search process, each ant at the current node of the road network need to take roulette to determine the next behavior based on the node transition probability: (1) When it is determined according to the roulette result that the next step is to move to the adjacent node q, it is necessary to use the roulette to determine the recommended velocity level and whether the air conditioning is turned on according to the current period t_h and the recommended velocity probability and air conditioning usage probability of the current road section. Then the ant goes to the next node q. (2) When it is determined according to the roulette result that the next step is to stay at current node p to charge at level m, then the ant stays at the current node for a period to charge. When the period ends, the roulette is used again at the current node to determine whether to move to the next node or continue charging at the current node. The search ends when the ant reaches its final destination. According to the length of each road section, recommended velocity, charging mode and charging time, and air conditioning behavior found in the search process, five objective values (the total length of the travel path, the total travel time, energy consumption, battery cycle life, and the average in-vehicle temperature) can be calculated. When all ants arrive at the destination, each objective value corresponding to the travel scheme of all ants in the search process can be obtained. At the same time, this round of ant colony search ends.

9.3.3.4 Calculating the Weight Update Value

This algorithm uses a multi-objective analysis method to evaluate the quality of each travel scheme in the multi-objective sense based on the Pareto criterion. The optimization objectives and constraint conditions of each solution in the solution set are compared and sorted according to their degree of excellence, and the excellent solutions are used to guide the subsequent search cycle. The objective vector of ant k is defined as $(O_k^1, O_k^2, \ldots, O_k^L)$, and the constraint violation vector is expressed as $(C_k^1, C_k^2, \ldots, C_k^V)$.

The principles for evaluating the Pareto solution set in this algorithm are as follows:

(1) Ant k's travel scheme is a feasible solution, and ant t's travel scheme is an infeasible solution.

In this case, ant k's travel scheme is better than that of ant t.

(2) Ant k's travel scheme and ant t's travel scheme are both feasible solutions.

In this case, it is necessary to compare the respective objective values corresponding to the travel schemes of the two ants.

$$\forall\, O_k^l \leq O_t^l, l = 1, \ldots, L. \;\; \exists O_k^l < O_t^l, l = 1, \ldots, L. \tag{9.45}$$

Since this is a minimization problem, the smaller the objective value, the better the solution. If all the objectives of ant k are not greater than the objectives of ant t, and at least one of the objectives of ant k is absolutely smaller than the objectives of ant t, it is considered that the travel scheme of ant k is better than that of ant t. If all of the objectives of ant k are not smaller than those of ant t, and all of the objectives of ant t are not smaller than those of ant k, it means that ant k and ant t are at the same level, and it is impossible to compare the advantages and disadvantages.

(3) Ant k and ant t are both feasible solutions.

In this case, it is necessary to take into account the relative size of the constraint violation quantitys of the corresponding travel schemes of the two ants, and the solution with smaller constraint violation quantitys has a better effect.

$$\forall\, C_k^v \leq C_t^v, v = 1, \ldots, V. \;\; \exists C_k^v < C_t^v, l = 1, \ldots, V \tag{9.46}$$

If each constraint violation quantity of ant k is not greater than that of ant t, and there is at least one constraint violation quantity of ant k that is smaller than that of ant t, then ant k is considered to be superior to ant t. If the constraint violation quantities of ant k are not all smaller than those of ant t, and the constraint violation quantities of ant t are not all smaller than those of ant k, it means that ant k and ant t are at the same level. In the optimization of solutions, all the infeasible solutions are retained and included in the sorting process. This is because premature elimination of infeasible solutions leads to premature convergence of the whole optimization algorithm to locally feasible solutions. All infeasible solutions will be removed at the end of the solution.

After the completion of the search process, the solution set is sorted according to the results of the current round of search, and the order of the excellent solution is higher than that of the suboptimal solution. Therefore, for the solution in the high

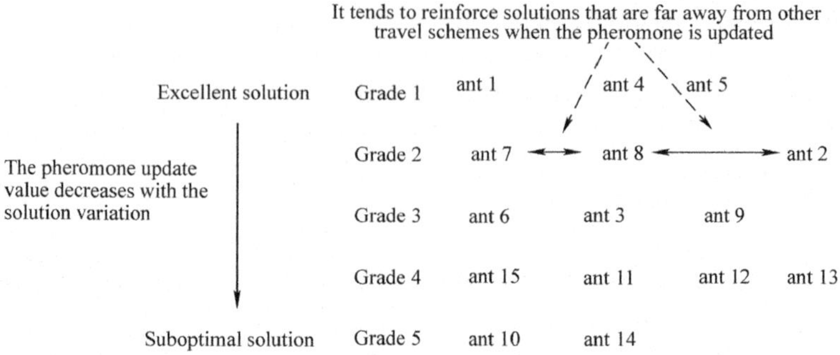

Fig. 9.16 Schematic diagram of the update of pheromone layer

order, the updating effect of pheromone along the way is higher than that in the low order, which can ensure that the algorithm converges towards the better solution while taking into account the search results of the suboptimal solution. The schematic diagram of the update of pheromone is shown in Fig. 9.16.

As shown in Fig. 9.16, the update of pheromone generally includes two parts: (1) Vertically, it is hoped that during the update process, the travel scheme of the excellent solution is updated with greater weight to ensure that the algorithm can converge to the excellent solution. At the same time, the suboptimal travel scheme is considered, so that the entire algorithm does not prematurely converge to the optimal solution of the current generation. (2) Laterally, the algorithm is expected to extend the search scope as wide as possible. Therefore, for solutions of the same level, if a solution is relatively far away from other solutions, the farther solution needs to be strengthened to encourage this search result.

The pheromone update values for path selection, velocity recommendation and air conditioning usage are defined as Ω. Vertically, to strengthen the updating effect of the superior solution and weaken the updating effect of the suboptimal solution, the updating value of each layer solution is defined as $I(i), i = 1, 2, \ldots, e$, where e is the layer number.

$$I(i) = \Omega \times \left(\frac{e - i + 1}{e}\right)^{\mu}, i = 1, 2, \ldots, e, 0 < \mu < 1 \tag{9.47}$$

μ is the coefficient for adjusting the relative size of the update weights between layers. It can be seen from the equation that the update weight of the solution at the high position is greater than the update weight of the solution at the low position.

For solutions in the same layer, to expand the search range of the algorithm, the updated value of several solutions aggregated should be weakened, while the updated value of the solution in a far position should be strengthened. Therefore, the relative distance between each solution in the same layer should be defined here. For layer i, the maximum value $\max(O^l_{rank(i)})$ and minimum value $\min(O^l_{rank(i)})$ of the objective l of each solution in this layer should be calculated. In the same layer, the relative distance d_{kt}^{l} between ant k and t is:

$$d_{kt}^{l} = \left| \frac{O_k^l - O_t^l}{\max(O_{\text{rank}(i)}^l) - \min(O_{\text{rank}(i)}^l)} \right| \tag{9.48}$$

Therefore, for each ant in the same layer, the formula for calculating pheromone update value $\Delta^l(k)$ is as follows:

$$Sh(d_{kt}^{l}) = \begin{cases} 1 - \left(\frac{d_{kt}^l}{\alpha_{\text{share}}}\right)^{\omega}, & if \ \ d_{kt}^l \le \sigma_{\text{share}} \\ 0, & otherwise \end{cases} \tag{9.49}$$

$$nc_k^l = \sum_{t \in rank(i)} Sh(d_{kt}^l) \tag{9.50}$$

$$\Delta^l(k) = \frac{I(i)}{nc_k^l} \tag{9.51}$$

9.3.3.5 Updating the Pheromone Layer

After the updated pheromone value $\Delta^l(k)$ of each travel scheme k on the objective l is obtained, the updated value needs to be reflected in each pheromone layer to guide the next round of cyclic optimization. It should be noted that, for each solution, only the pheromone on the travel scheme corresponding to the solution needs to be updated, so that the solution set found in this round of ant colony search can be used to update each pheromone layer, as shown in Eqs. (9.52–9.55).

$$\tau_{pq||pm}^l(t_h) = \tau_{pq||pm}^l(t_h) + \Delta^l(k) \tag{9.52}$$

$$\tau_{pq\_\text{v}\_s}^l(t_h) = \tau_{pq\_\text{v}\_s}^l(t_h) + \Delta^l(k) \tag{9.53}$$

$$\tau_{pq\_\text{a}\_1}^l(t_h) = \tau_{pq\_\text{a}\_1}^l(t_h) + \Delta^l(k) \tag{9.54}$$

$$\tau_{pq\_\text{a}\_0}^l(t_h) = \tau_{pq\_\text{a}\_0}^l(t_h) + \Delta^l(k) \tag{9.55}$$

$$t_0 + h\delta \le t(\overrightarrow{pq}) \le t_0 + (h+1)\delta, \ \overrightarrow{pq} \in r_{ij}^k, k \in s$$

Here $t(\overrightarrow{pq})$ represents the moment when the ant moves from the current node to the next node, and $t(\overrightarrow{pq})$ is within period t_h. If the vehicle does not move in this step, but stays at node p to charge at level m, then the update of position transition pheromone is defined as $t^l_{pm}(t_h)$, and the velocity recommendation and the air conditioning usage pheromones in this period t_h are not updated.

To avoid premature convergence of each pheromone layer and ensure the search breadth of the algorithm, attenuation should be carried out for each pheromone layer after each cycle is updated, as shown in Eqs. (9.56–9.59).

$$\tau^l_{pq||pm}(t_h) = \rho \times \tau^l_{pq||pm}(t_h) \tag{9.56}$$

$$\tau^l_{pq\_v\_s}(t_h) = \rho \times \tau^l_{pq\_v\_s}(t_h) \tag{9.57}$$

$$\tau^l_{pq\_a\_1}(t_h) = \rho \times \tau^l_{pq\_a\_1}(t_h) \tag{9.58}$$

$$\tau^l_{pq\_a\_0}(t_h) = \rho \times \tau^l_{pq\_a\_0}(t_h) \tag{9.59}$$

$$pq \in V, \quad m = 1, 2, \ldots, C_w, \quad l = 1, 2, \ldots, L, \quad s = 1, 2, \ldots, V_k, \quad 0 < \rho < 1$$

So far, the ant colony has completed an entire search process. The optimization process is iterated, and the excellent solution set is retained as the final output result of the algorithm.

9.3.3.6 Optimal Solution Update

After the ant colony search cycling ends, each ant in the ant colony represents a travel scheme. All the travel schemes are put into the solution set of the optimal travel schemes, and the advantages and disadvantages of the solutions in the set are compared. The comparison strategy is the same as the strategy for calculating the weight update value. Only the optimal solution set at the first layer is retained as the final search result of this generation. With the continuous iteration of the algorithm, the optimal solution set will constantly introduce new optimal solutions and exclude the suboptimal solutions, and always save all the optimal solutions found in the search process of the algorithm, which together constitute the optimal solution set of the optimization algorithm.

9.3.4 Simulation Analysis of Travel Planning Algorithm

The temperal multi-objective ant colony optimization method proposed in this chapter is used to optimize the travel scheme of EV. Combined with the evaluation system of the multi-objective optimization method, the algorithm performance is simulated and analyzed to verify the effectiveness of this method in dealing with the travel problem of EV.

9.3.4.1 Construction of the Travel Planning Simulation Platform

Firstly, to verify the effectiveness of the travel planning method, the calculation results are reasonably evaluated and a simulation analysis platform is established, as shown in Fig. 9.17. A 150 km × 150 km road network simulation model is established, and a charging station is set up, as shown in the diamond shape in Fig. 9.17. The average traffic speed of each road in the road network model follows the characteristics of the dynamic stochastic road network, and the speed data comes from the statistical results of the average traffic speed of Beijing's Fourth Ring Road in the reference [25]. This speed data is based on statistics of the average speed of the Fourth Ring Road; sampled every 5 min, a total of 288 pieces of data are collected every day for two months. Reference [25] analyzed the original speed spectrum and found that it is closer to a time-varying normal distribution, which can update a mean and variance every 5 min. Therefore, in this study, each period δ is set as 5 min. The temperature change outside the vehicle is based on the average temperature change pattern of Beijing in summer in Reference [26]. Each charging station in the road network adopts a constant current charging mode including two charging levels, namely 1C fast charge (full capacity within one hour) and 0.1C slow charge (full capacity within 10 h). Also, the network structure is subject to two basic assumptions: (1) the states of each side of the network are independent of and do not influence each other. (2) EVs can stay for charging at nodes with charging stations, but are not allowed to stop at other nodes in the road network.

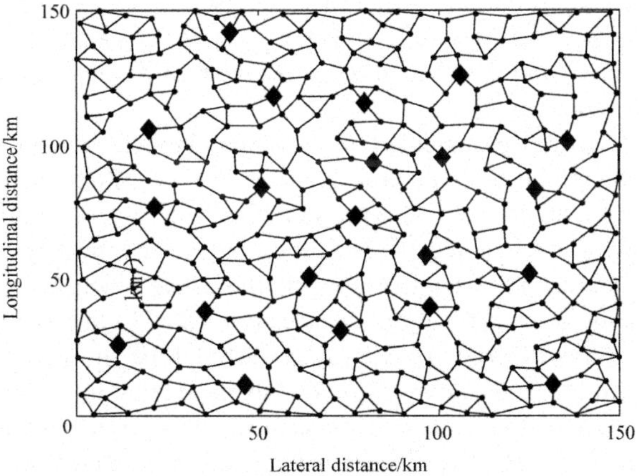

Fig. 9.17 Diagram of road network structure in simulation analysis

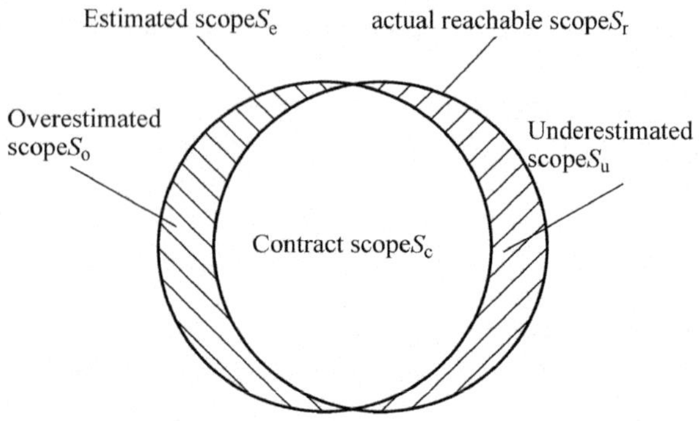

Fig. 9.18 Evaluation method for the estimation results of mileage

9.3.4.2 Simulation Analysis of the Performance of Travel Planning Algorithm

Simulation analysis and evaluation on the calculation effect of the EV travel planning method proposed in this chapter is performed.

1. Simulation analysis of estimation method of mileage

Before an EV departs, it is necessary to first determine whether it can reach the nearest charging station or destination. Therefore, it is necessary to simulate and evaluate the mileage estimation method of EV. As shown in Fig. 9.18, the estimated scope S_e and the actual reachable scope S_r of the vehicle are intersecting. The part where S_e and S_r overlap is called contract scope S_c; the part where S_r exceeds S_e is called underestimated scope S_u; the part where S_e exceeds S_r is called overestimated scope S_o.

It can be seen from Fig. 9.18 that the overestimated scope S_o represents the range where the estimated reachable scope exceeds the actual reachable scope, the underestimated scope S_u represents the range of the estimated reachable scope smaller than the real reachable scope, and the contract scope S_c is the range where the estimated reachable scope and the actual reachable scope overlap. Here, the underestimation ratio λ_u, overestimation ratio λ_o and contract ratio λ_c of the estimation results of driving distance are defined, as shown in Eqs. (9.60–9.63). In the ideal mileage estimation result, it is expected that the estimated reachable scope overlaps the actual reachable scope as much as possible, and the overestimated scope and the underestimated scope are minimized at the same time. Therefore, the comprehensive estimation accuracy rate λ_e is defined here to represent the calculation effect of the algorithm. The higher the value, the better the the algorithm's estimation result of mileage.

$$\lambda_c = S_c / S_r \times 100\% \tag{9.60}$$

$$\lambda_u = S_u / S_r \times 100\% \tag{9.61}$$

$$\lambda_o = S_o/S_r \times 100\% \tag{9.62}$$

$$\lambda_e = \lambda_c - \lambda_o - \lambda_u \tag{9.63}$$

In the process of simulation analysis, the driving distance estimation method proposed in this chapter is firstly used to calculate the probability density function p_{ij}^E of energy consumption of the vehicle arriving at each node in the road network, where $j \in V$. Then, the estimation method based on past driving information and current traffic information is used as the comparison algorithm. For the estimation method based on past driving information, by collecting driving data in a past period and calculating the ratio between driving distance and energy consumption, the future driving energy consumption and the reachable scope can be estimated based on the current remaining battery energy. For the estimation method based on the current traffic information, the energy consumption of the vehicle by the time it arrives at each node is calculated according to the traffic information of the road network at the moment of departure, and then the reachable scope of the vehicle is obtained. The estimation results obtained by these two comparison methods are deterministic, which can clearly tell whether the EV can reach each node in the road network. Here, the initial energy of the EV battery is set as 8 kW · h, and the simulation model is used to generate the actual driving node range of the EV for 100 times at each moment of departure. Compare the estimated results based on past driving information, the estimated results based on current traffic information, and the estimated results based on dynamic stochastic traffic information proposed in this section respectively, with the actual mileage, and then record the means and variances of the contract ratio λ_c, underestimation ratio λ_u, and overestimation ratio λ_o of various mileage estimation methods, and the simulation analysis results are shown in Fig. 9.19.

A good mileage estimation method should be able to ensure a high overlap ratio while minimizing the overestimation and underestimation ratios. It can be seen from the simulation results that, compared with the estimation method based on past travel information and current traffic information, the mileage estimation method proposed in this chapter can effectively control the ratios of overestimation and underestimation on the premise of ensuring a high overlap ratio. For example, at the departure moment of 13:00, the overlap ratios of the estimation methods based on past driving information and current traffic information are lower than those of the expected value model based on the dynamic traffic network. In other cases, such as at the departure moment of 9:00, 11:00, and 15:00, although the overlap ratios of the first two estimation methods are relatively high, their overestimation and underestimation ratios are also very high, leading to a large gap between the estimated result and the actual reachable scope. At the same time, the simulation results obtain the estimated mileages of the three evaluation indicators of the proposed expected value model (E), chance constraint model (C), and relevant chance model (D). The overlap ratios of the expected value model and the chance constraint model are relatively high, while the underestimation ratio and overestimation ratio are suppressed, and the overlap ratio of the expected value model is relatively high in various working scenarios.

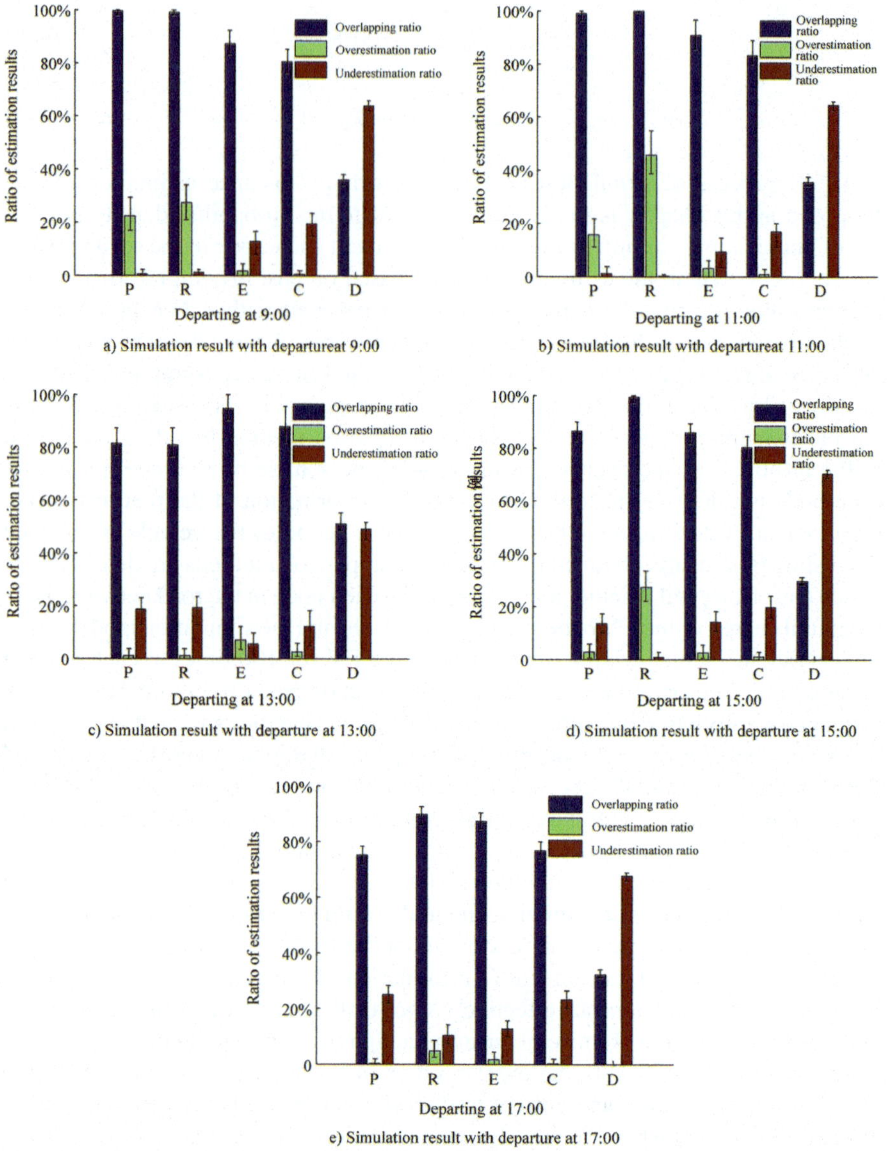

Fig. 9.19 Comparison of results of multiple mileage estimation methods

Compared with other estimation methods, the overestimation ratio of the relevant chance model is the lowest, indicating that the estimated reachable scope of this model is the smallest, which is the most reliable reachable scope for the driver.

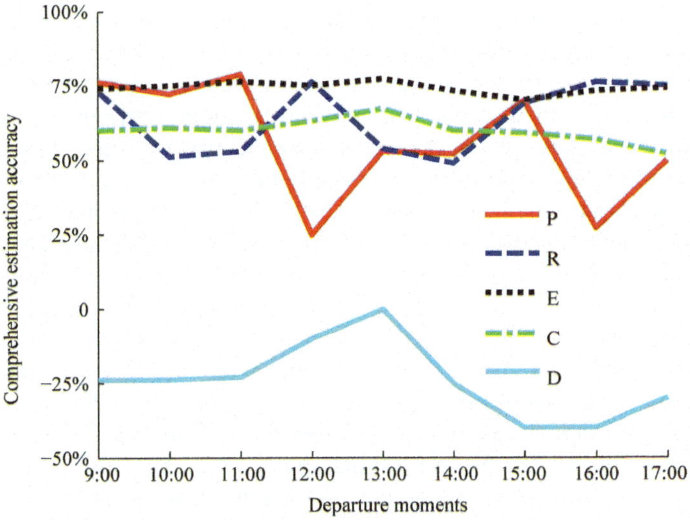

Fig. 9.20 Comparison of consistency of different simulation methods

The consistency of the estimation results is also very important for the mileage estimation method, which means the driver should be able to trust the mileage estimation result no matter when the vehicle departs. By comparing the comprehensive estimation accuracy at different departure moments, the results are shown in Fig. 9.20.

In Fig. 9.20, the comprehensive estimation accuracy of various methods at different departure moments is compared, and it can be seen that the comprehensive estimation accuracy λ_e of the expected value model is relatively high and maintains good stability.

Considering that the air conditioning system may be used in driving, which affects the total energy consumption of the vehicle and the reachable scope, it is necessary to analyze the estimation accuracy of various mileage estimation methods under the condition of using and not using air conditioning. A good estimation method needs to be able to accurately estimate the mileage of EV with or without air conditioning system. To verify the estimation result of each method, the estimation accuracy of various mileage estimation algorithms with and without air conditioning are analyzed respectively. Among them, whether the air conditioning is turned on or not depends on the in-vehicle temperature when the vehicle arrives at each road section along the journey. When the in-vehicle temperature is higher than the outside temperature, the air conditioning is turned on; when the in-vehicle temperature is lower than the outside temperature, the air conditioning is turned off. ξ represents the sum of the absolute error of the overlap ratio, the absolute error of the underestimation ratio, and the absolute error of the overestimation ratio with and without air conditioning.

As shown in Table 9.1, the estimation results show that with or without air conditioning, the estimation errors of simulation results based on past travel information and current traffic information fluctuate greatly, as shown in Columns 1 and 2.

However, the error of the estimation results based on various evaluation criteria in the dynamic stochastic road network model is smaller, as shown in Columns 3, 4 and 5. This is because the temperature outside the vehicle continues to change over time. The traditional method only uses the current traffic environment information and does not take into account the temperature change over time, so it is difficult to consider the on/off state of the air conditioning during travel. Meanwhile, for the same travel path, the change pattern of the outside temperature is different at different departure moments, so the estimation accuracy of traveling along the same path at different departure moments is also different. It can be seen from Table 9.1 that the estimation errors in Columns 1 and 2 fluctuate greatly at different departure moments.

The "inner and outer circle" method is adopted to obtain the final mileage estimation result, as shown in Fig. 9.21. The "inner circle" formed by the estimation results of the relevant chance model is used to describe the reachable scope of EV. By effectively suppressing the overestimation ratio, this estimation method enables the driver to fully trust this conservative estimation result and define the minimum reachable scope of EV. The "outer circle" obtained from the expected value model estimation results provides the driver with a reachable scope that takes into account the expected value of driving energy consumption. Since the expected value model can ensure that the overlap ratio is relatively high in most cases while effectively suppressing the overestimation and underestimation ratios, this "outer circle" estimation result provides the driver with a statistically most accurate reachable scope of EV.

2. Analysis of the performance of travel planning method

The main advantage of multi-objective optimization is that in the optimization process, under the condition of meeting travel constraints, multiple travel objectives are collaboratively optimized to obtain the optimal travel solution set in the multi-objective sense. Compared with the single-objective optimization strategy, since the original multi-objective optimization problem is not simplified in the optimization

Table 9.1 Comparison of mileage estimation results with and without air conditioning

Departure	ξ_P(%)	ξ_R(%)	ξ_E(%)	ξ_C(%)	ξ_D(%)
9:00	23.21	8.45	2.88	2.03	1.83
10:00	101.43	19.37	2.85	6.71	0.68
11:00	58.13	13.91	0.34	4.73	1.36
12:00	58.65	0.80	1.96	4.31	4.89
13:00	41.16	7.31	0.71	4.66	2.03
14:00	44.42	12.43	6.87	5.17	1.39
15:00	36.60	4.40	6.88	2.54	2.44
16:00	37.03	2.43	5.51	3.94	0.67
17:00	44.10	1.40	3.62	6.93	2.01

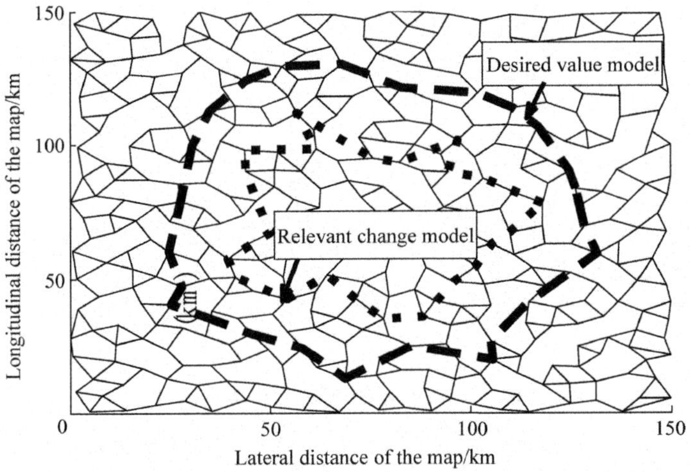

Fig. 9.21 Reachable scope of EV in the dynamic stochastic road network

process, the multi-objective optimization method can take into account multiple travel optimization objectives at the same time, ensuring that the obtained optimization solution set can completely reflect the combination of various travel objectives, thus enabling the driver to choose from multiple travel schemes. To compare the advantages and disadvantages of feasible solutions in single-objective and multi-objective optimization results, Eq. (9.64) is defined to normalize the comparison results of each objective.

$$y_i^k = \frac{O_i^{\mathrm{s}} - O_i^{\mathrm{m},k}}{O_i^{\mathrm{s}}} \tag{9.64}$$

In Eq. (9.64), O_i^{s} represents the No. i objective of single-objective optimization, and $O_i^{\mathrm{m},k}$ represents the No. i objective of the No. k solution in the solution set of multi-objective optimization. y_i^k represents the percentage that the No. i objective of the No. k solution in the multi-objective solution set is better than the No. i objective of the single-objective optimization. Set the initial SOC of the EV to 0.95, and the departure moment to be 8:30 a.m. Randomly select 3 groups of starting-ending point pairs in the road network, and set the SOC value constraint and the time constraintsfor reaching the destination respectively. Use the multi-objective optimization method to obtain the travel scheme, and take the single-objective optimization method with the shortest distance objective and the shortest time objective as the comparison method. The comparison results are shown in Tables 9.2 and 9.3.

Table 9.2 shows the comparison results of single-objective method and multi-objective method in each working scenario when single-objective optimization is performed with the shortest distance as the objective; Table 9.3 shows the comparison results of single-objective method and multi-objective method in each working

Table 9.2 Comparison results of shortest path objective and multi-objective optimization method

Working scenario	Distance (%)	Time (%)	Energy (%)	Film (%)	Temp. (%)
80 km, SOC = 0.2	0	8.02	1.17	16.98	2.10
80 km, SOC = 0.5	0	3.88	− 3.92	14.87	2.87
80 km, SOC = 0.9	0	2.11	2.65	6.37	− 4.46
120 km, no time constraint	3.17	− 1.16	0.74	1.66	− 1.98
120 km, arrive before 13:30	0	2.43	− 2.22	7.82	− 4.20
150 km, no time constraint	0	− 3.48	0.3	2.54	1.13
150 km, arrive before 17:00	0	− 2.91	− 2.4	13.59	1.66

Table 9.3 Comparison results of shortest time objective and multi-objective optimization method

Working scenario	Distance (%)	Time (%)	Energy (%)	Film (%)	Temp. (%)
80 km, SOC = 0.2	3.09	− 4.43	1.74	− 8.84	14.19
80 km, SOC = 0.5	4.23	0.51	0.46	0.24	18.86
80 km, SOC = 0.9	0	− 0.17	− 8.78	4.42	18.50
120 km, no time constraint	0.77	0.88	− 5.81	4.48	0.80
120 km, arrive before 13:30	0.77	1.19	− 7.29	5.90	1.06
150 km, no time constraint	0	0.57	− 5.49	9.19	4.65
150 km, arrive before 17:00	0	0.42	− 5.90	11.63	8.13

scenario when single-objective optimization is performed with the shortest time as the objective; it can be seen from the comparison results in Tables 9.2 and 9.3 that in the travel scheme solution set of multi-objective optimization, there are at least three objectives that are not worse or better than the single-objective optimization result, which shows that the use of multi-objective optimization method can effectively coordinate multiple travel objectives and expand the search range to get a more reasonable solution, reflecting the advantages of the multi-objective optimization method.

3. Comparison of temperal multi-objective ant colony algorithm and other ant colony methods

To verify the effectiveness of the travel method proposed in this chapter compared with the traditional multi-objective ant colony optimization methods, the road network model is extended and two common multi-objective ant colony optimization algorithms are selected to compare the calculation results. Two commonly used multi-objective ant colony optimization methods are adopted in the comparison scheme. Comparison method 1 is the fourth method proposed by Ref. [27] with L pheromone layers, and here ACO1 is used to mark the method. Comparison method 2 is the multi-objective ant colony algorithm in Ref. [28], and here ACO2 is used to mark the method. The longitude and latitude coordinates provided by Baidu Map are used to model the main roads in Beijing, the connection relationship between all nodes are retained in the road network, and 31 charging stations are evenly distributed among them. The road network structure is shown in Fig. 9.22.

a) Schematic diagram of roads in Beijing

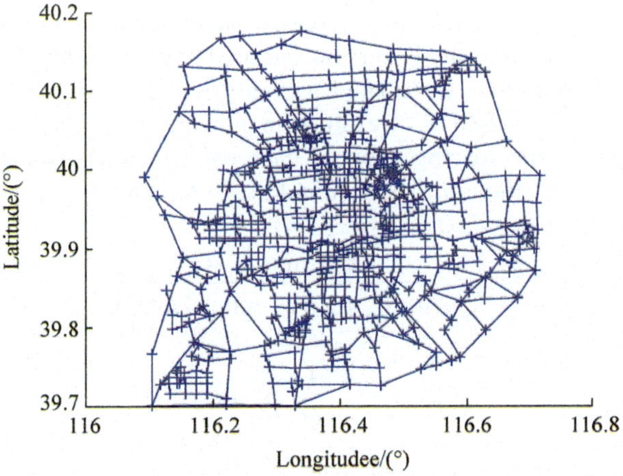

b) Simplified road network map of Beijing

Fig. 9.22 Road network map of Beijing

$Q_{\text{T-MACO}}$ represents the travel scheme solution set searched by the temperal multi-objective ant colony optimization (T-MACO) proposed in this study; Q_{ACO1} and Q_{ACO2} represent the travel scheme solution set obtained by the comparison method ACO1 and the comparison method ACO2 after the search; $|Q_{\text{T-MACO}}|$, $|Q_{\text{ACO1}}|$ and $|Q_{\text{ACO2}}|$ represent the number of solutions in each solution set. For ACO1 and ACO2, when the searched solutions do not meet the constraints, the penalty function is used to punish the violation of the constraints, and it is added to the optimization objective to reflect the penalty for the travel schemes, as shown in Eq. (9.65).

$$O_{k\_f}^{l}=O_{k}^{l}+f_{1}^{l}\times(\max(C_{k}^{1},0)))+f_{2}^{l}\times(\max(C_{k}^{2},0))\cdots+f_{V}^{l}\times(\max(C_{k}^{V},0)) \tag{9.65}$$

In Eq. (9.65), O_k^l represents l objectives corresponding to the travel scheme searched by ant k, and $O_{k\_f}^l$ represents the new optimization objective after considering the penalty of constraints. The penalty factor f_V^l is different for ACO1 and ACO_2.

In view of the three evaluation criteria of multi-objective optimization, three corresponding evaluation indicators are adopted in this research. The first indicator is used to evaluate whether each method is effective in searching for a better solution. Under the same optimization condition, all kinds of multi-objective optimization methods can obtain a Pareto Set. Therefore, all the solutions in these solution sets can be integrated, and a new solution set can be obtained after deleting the suboptimal solutions in the total solution set. Calculate the percentage of the number of solutions of each method in the new solution set. The higher the percentage of an ant colony optimization method, the more excellent solutions can be found by this method compared with other methods, which can represent the quality of the solution set found by each optimization method [29].

The second indicator is used to evaluate whether the solutions in the set are evenly distributed. A good multi-objective optimization algorithm needs to avoid the situation where the solution set is too concentrated in the vicinity of a few solutions, and enable all solutions to be distributed as evenly as possible. Therefore, the spacing between solution sets is used as an evaluation indicator here. The smaller the value, the more consistent the spacing between each solution in the solution set and the more uniform the distribution, as shown in Eqs. (9.66) and (9.67).

$$o_k^l=\frac{O_k^l-\min_{i=1}^{|Q|}(O_i^l)}{\max_{i=1}^{|Q|}(O_i^l)-\min_{i=1}^{|Q|}(O_i^l)} \tag{9.66}$$

$$S_{\text{spacing}}=\sqrt{\frac{1}{|Q|}\sum_{i=1}^{|Q|}(d_i-\bar{d})^2} \tag{9.67}$$

where $d_i=\min_{k\in Q\wedge k\neq i}\sum_{l=1}^{L}|O_k^l-O_i^l|$, $\overline{d}$ represents the average value of the spacing between each other, that is $\overline{d}=\sum_{i=1}^{|Q|}d_i/|Q|$. The standard deviation of

d_i represents the spacing S_{spacing}. The smaller this value, the more uniform the distribution of solutions in the set.

The third indicator is used to evaluate the coverage of each solution set. The larger the range of solution set, the larger the range of Pareto can be obtained by solution set search, which helps describe the solution set of travel scheme more comprehensively. According to the definition of this indicator, the solution set of the optimization method proposed in this paper and the solution set of the comparison method should be compared to obtain the maximum distribution indicator $\overline{D}$. When the value is greater than 1, it indicates that the search range of the proposed method in this study is greater than that of the comparison algorithm. The comparison formula is shown in (Eq 9.68).

$$\begin{aligned} \overline{D}_{\frac{\text{T-MACO}}{\text{ACO1}}} &= \sqrt{\frac{1}{L}\sum_{l=1}^{L}\left(\frac{\max_{i=1}^{|Q_{\text{T-MACO}}|} O_i^l - \min_{i=1}^{|Q_{\text{T-MACO}}|} O_i^l}{\max_{i=1}^{|Q_{\text{ACO1}}|} O_i^l - \min_{i=1}^{|Q_{\text{ACO1}}|} O_i^l}\right)^2} \\ \overline{D}_{\frac{\text{T-MACO}}{\text{ACO2}}} &= \sqrt{\frac{1}{L}\sum_{l=1}^{L}\left(\frac{\max_{i=1}^{|Q_{\text{T-MACO}}|} O_i^l - \min_{i=1}^{|Q_{\text{T-MACO}}|} O_i^l}{\max_{i=1}^{|Q_{\text{ACO2}}|} O_i^l - \min_{i=1}^{|Q_{\text{ACO2}}|} O_i^l}\right)^2} \end{aligned} \tag{9.68}$$

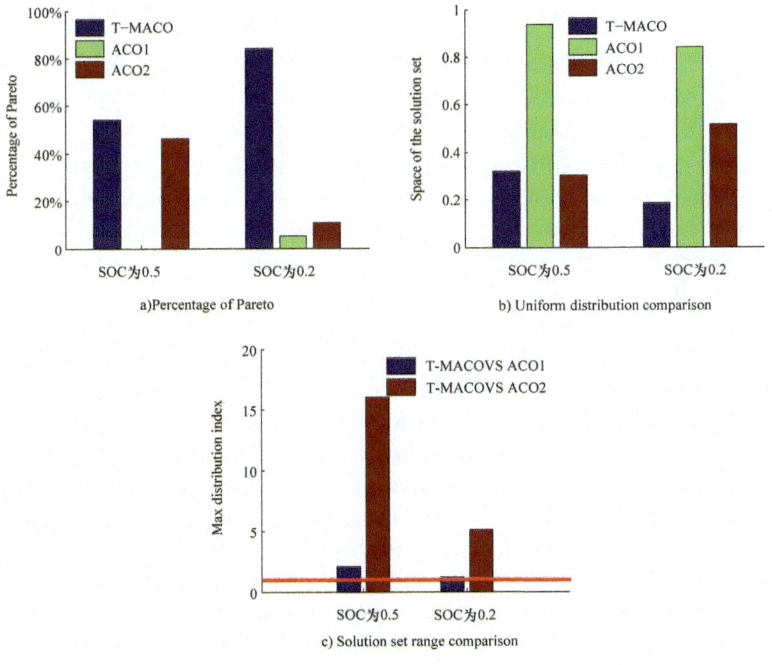

Fig. 9.23 Simulation results with a distance of 30 km between the starting point and the ending point

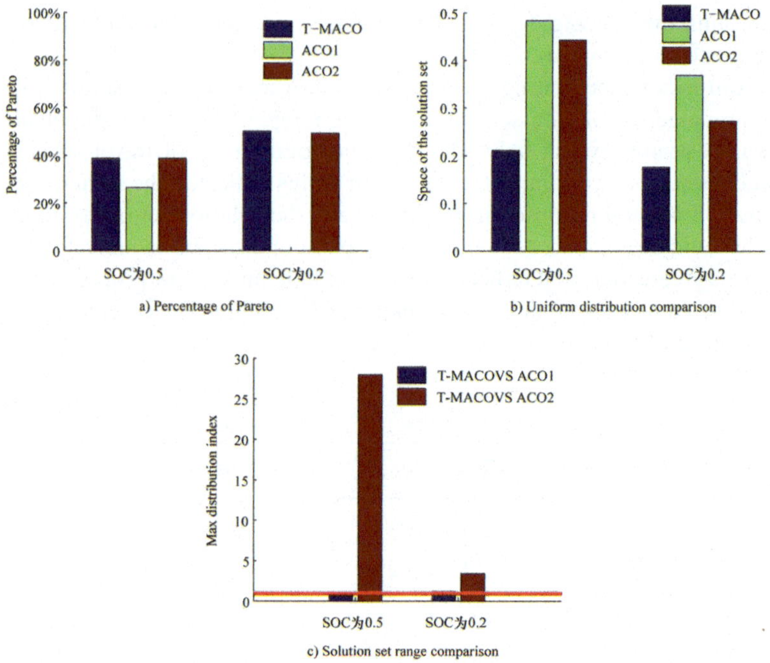

Fig. 9.24 Simulation working scenario with a distance of 45 km between the starting point and the ending point

The comparison scenario is divided into two conditions according to the constraint definition: the scenario of SOC constraint for arrival at the destination node and the scenario of time constraint for arrival at the destination node. Constraint on SOC ensures potential future travel plans, and constraint on arrival time ensures that the driver's arrival time demand is met to avoid being late.

For the SOC constraint for arrival at the destination, two starting-ending point pairs are selected, with straight-line distances of about 30 km and 45 km respectively. The SOC of the battery at the starting point is 0.5, and the SOC constraint at the ending point is 0.5 and 0.2, respectively. The simulation results are shown in Figs. 9.23 and 9.24. As shown from the simulation results, in Figs. 9.23a and 9.24a, the percentage of the Pareto of the temperal multi-objective ant colony optimization method proposed in this paper is higher than that of the other two methods, indicating that the ant colony optimization method proposed in this paper can search more solutions in Pareto Set. As shown in Figs. 9.23b and 9.24b, the solution set spacing S_{spacing} of the optimization method proposed in this paper is smaller than that of the other two methods in most cases, indicating that the internal distribution of the solution set of the proposed T-MACO method is more uniform. In Figs. 9.23c and 9.24c, the value of the horizontal red line is 1. It can be seen from the figure that the maximum distribution indicator $\overline{D}$ of the optimization method proposed in this chapter is greater than 1 in most cases, indicating that the proposed method can obtain a broader Pareto Set.

For the time constraint, to consider the driver's arrival time demand, two starting-ending point pairs at different times are selected. The straight-line distance is about 35 km and 20 km. The time limit is 2.5 h for a starting-ending point pair with a straight-line distance of about 35 km and 1.5 h for a starting-ending point pair with a straight-line distance of about 20 km. The simulation results are shown in Figs. 9.25 and 9.26.

It can be seen from the simulation results that, in Figs. 9.25a and 9.26a, the proposed T-mACO optimization method still maintains a high percentage of Pareto except for the constraint of the time limit in Fig. 9.25a. At the same time, in Figs. 9.25b, c and 9.26b, c, each solution in the solution set of the T-mACO method still maintains a relatively uniform distribution, and the overall distribution of the solution set is wide, indicating that the temperal multi-objective ant colony algorithm can search for better solutions. From the above simulation results, it can be seen that compared with the existing multi-objective ant colony optimization methods, the temperal multi-objective ant colony optimization method proposed in this chapter can effectively search and obtain a better solution set of travel schemes, verifying the effectiveness of the proposed method.

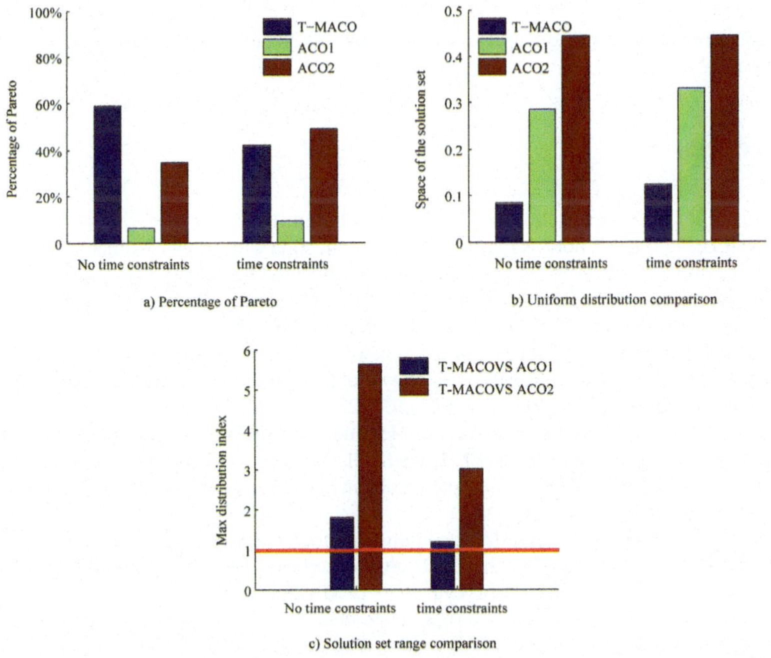

Fig. 9.25 Simulation working scenario with a distance of 35 km between the starting point and the ending point

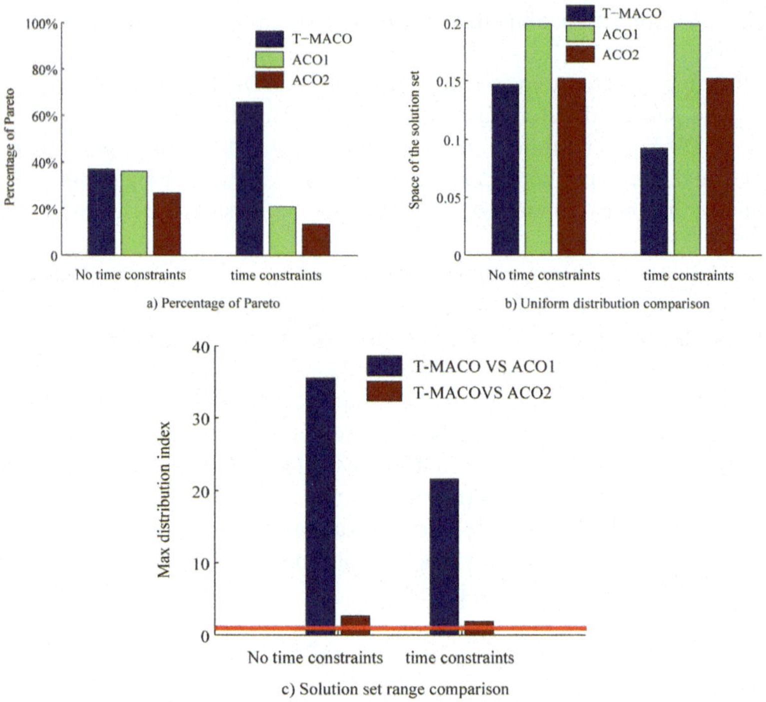

Fig. 9.26 Simulation working scenario with a distance of 20 km between the starting point and the ending point

References

1. Bellman R (1958) On a routing problem. Q Appl Math 16:87–90
2. Dijkstra EW (1959) A note on two problems in connection with graphs. Numer Math 1:269–271
3. Ford L, Fulkerson D (1962) Flows in networks. Princeton University Press, Princeton NJ
4. Hart PE, Nilsson NJ, Raphael B (1968) A formal basis for the heuristic determination of minimum cost routes. IEEE Trans Syst Sci Cybern 4:100–107
5. Nilsson NJ (1999) Artificial intelligence: a new synthesis. China Machine Press, Beijing
6. Gao S, Chabini I (2006) Optimal route policy problems in stochastic time-dependent networks. Transp Res Part B: Methodological 40(2):93–122
7. Gao S (2012) Modeling strategic route choice and real-time information impacts in stochastic and time-dependent networks. IEEE Trans Intell Transp Syst 13(3):1298–1311
8. Feng Y (2009) Study on the shortest path problem of complex dynamic and stochastic network. Zejiang University, Hangzhou
9. Fang G, Li H (2003) Automobile air conditioning technology. China Machine Press, Beijing
10. Moura S (2011) Techniques for battery health conscious power management via electrochemical modeling and optimal control. University of Michigan, Michigan
11. Aurbach D (2000) Review of selected electrode-solution interactions which determine the performance of Li and Li ion batteries. J Power Sources 89(2):206–218

12. Arora P, White RE, Doyle M (1998) Capacity fade mechanisms and side reactions in lithium-ion batteries. J Electrochem Soc 145(10):3647–3667
13. Kanevskii L, Dubasova V (2005) Degradation of lithium-ion batteries and how to fight it: a review. Russ J Electrochem 41(1):1–16
14. Bhattacharyya R, Key B, Chen H et al (2010) In situ NMR observation of the formation of metallic lithium microstructures in lithium batteries. Nat Mater 9(6):504–510
15. Zhang X, Sastry AM, Shyy W (2008) Intercalation-induced stress and heat generation within single lithium-ion battery cathode particles. J Electrochem Soc 155(7):A542–A552
16. Wu HC, Su CY, Shieh DT et al (2006) Enhanced high-temperature cycle life of $LiFePO_4$-based Li-ion batteries by Vinylene carbonate as electrolyte additive. Electrochem Solid-State Lett 9(12):537–541
17. Striebel K, Shim J, Sierra A et al (2005) The development of low cost lifepo4-based high power lithium-ion batteries. J Power Sources 146(1–2):33–38
18. Liu P, Wang J, Hicks-Garner J et al (2010) Aging mechanisms of $LiFePO_4$ batteries deduced by electrochemical and structural analyses. J Electrochem Soc 157(4):A499–A507
19. Dubarry M, Liaw BY (2009) Identify capacity fading mechanism in a commercial $LiFePO_4$ cell. J Power Sources 194(1):541–549
20. Zhang Y, Wang CY, Tang X (2011) Cycling degradation of an automotive $LiFePO_4$ lithium-ion battery. J Power Sources 196(3):1513–1520
21. Amine K, Liu J, Belharouak I (2005) High-temperature storage and cycling of C-$LiFePO_4$/graphite Li-ion cells. Electrochem Commun 7(7):669–673
22. Chang HH, Wu HC, Wu NL (2008) Enhanced high-temperature cycle performance of $LiFePO_4$/carbon batteries by an ion-sieving metal coating on negative electrode. Electrochem Commun 10(2):1823–1826
23. Duda RO, Hart PE, Stork DG (2001) Pattern classification. Wiley, New York
24. Liu B (2001) Uncertain programming: a unifying optimization theory in various uncertain environments. Appl Math Comput 120(1–3):227–234
25. Zhang Z, Yang D, Li T et al (2012) Real-time dynamic traffic information for urban navigation electronic map. Recent Adv Comput Sci Inf Eng 129(1):459–464
26. Xiaodan W (2009) Study on design and parameter matching for air conditioning system of electric bus. Jilin University, Changchun
27. Alaya I, Solnon C, Ghedira K (2007) Ant Colony optimization for multi-objective optimization problems. In: 19th IEEE international conference on tools with artificial intelligence, Paris, France, Oct 2007
28. Doerner K, Gutjahr WJ, Hartl RE et al (2004) Pareto ant colony optimization: a meta heuristic approach to multiobjective portfolio selection. Ann Oper Res 131(1):79–99
29. Siddiqi UF, Shiraishi Y, Sait SM (2012) Multi-objective optimal route selection in electric vehicles. Artif Life Robot 17(1):113–122

The manufacturer's authorised representative in the EU is Springer Nature Customer Service Centre GmbH, Europaplatz 3, 69115 Heidelberg, Germany. If you have any concerns regarding our products, please contact ProductSafety@springernature.com

Printed and bound by CPI Group (UK) Ltd, Croydon, CR0 4YY
01/10/2026
02225868-0018